The
Togaviridae and
Flaviviridae

THE VIRUSES

Series Editors
HEINZ FRAENKEL-CONRAT, *University of California*
Berkeley, California

ROBERT R. WAGNER, *University of Virginia School of Medicine*
Charlottesville, Virginia

THE VIRUSES: Catalogue, Characterization, and Classification
Heinz Fraenkel-Conrat

THE ADENOVIRUSES
Edited by Harold S. Ginsberg

THE HERPESVIRUSES,
Volumes 1–3 • Edited by Bernard Roizman
Volume 4 • Edited by Bernard Roizman and Carlos Lopez

THE PAPOVAVIRIDAE
Volume 1 • Edited by Norman P. Salzman

THE PARVOVIRUSES
Edited by Kenneth I. Berns

THE PLANT VIRUSES
Volume 1 • Edited by R. I. B. Francki
Volume 2 • Edited by M. H. V. Van Regenmortel and Heinz Fraenkel-Conrat

THE REOVIRIDAE
Edited by Wolfgang K. Joklik

THE TOGAVIRIDAE AND FLAVIVIRIDAE
Edited by Sondra Schlesinger and Milton J. Schlesinger

The Togaviridae and Flaviviridae

Edited by
SONDRA SCHLESINGER
and
MILTON J. SCHLESINGER

Washington University School of Medicine
St. Louis, Missouri

PLENUM PRESS • NEW YORK AND LONDON

Library of Congress Cataloging in Publication Data

The Togaviridae and Flaviviridae.

(The Viruses)
Includes bibliographies and index.
1. Togaviruses. 2. Flaviviruses. I. Schlesinger, Sondra. II. Schlesinger, Milton J. III. Series.
QR415.5.T6   1986   616′.0194   86-4914
ISBN 0-306-42176-3

© 1986 Plenum Press, New York
A Division of Plenum Publishing Corporation
233 Spring Street, New York, N.Y. 10013

Printed in the United States of America

Contributors

Margo A. Brinton, The Wistar Institute of Anatomy and Biology, Philadelphia, Pennsylvania 19104

Dennis T. Brown, Cell Research Institute and Department of Microbiology, The University of Texas at Austin, Austin, Texas 78713

Lynn D. Condreay, Cell Research Institute and Department of Microbiology, The University of Texas at Austin, Austin, Texas 78713

Diane E. Griffin, Departments of Medicine and Neurology, The Johns Hopkins University School of Medicine, Baltimore, Maryland 21205

Stephen C. Harrison, Department of Biochemistry and Molecular Biology, Harvard University, Cambridge, Massachusetts 02138

Ari Helenius, Department of Cell Biology, Yale School of Medicine, New Haven, Connecticut 06510

Margaret Kielian, Department of Cell Biology, Yale School of Medicine, New Haven, Connecticut 06510

Thomas P. Monath, Division of Vector-Borne Viral Diseases, Center for Infectious Diseases, Centers for Disease Control, Public Health Service, U.S. Department of Health and Human Services, Fort Collins, Colorado 80522

James S. Porterfield, Sir William Dunn School of Pathology, University of Oxford, Oxford OX1 3RE, England

Charles M. Rice, Division of Biology, California Institute of Technology, Pasadena, California 91125

John T. Roehrig, Division of Vector-Borne Viral Diseases, Center for Infectious Diseases, Centers for Disease Control, Public Health Service, U.S. Department of Health and Human Services, Fort Collins, Colorado 80522

Milton J. Schlesinger, Department of Microbiology and Immunology, Washington University School of Medicine, St. Louis, Missouri 63110

Sondra Schlesinger, Department of Microbiology and Immunology, Washington University School of Medicine, St. Louis, Missouri 63110

Ellen G. Strauss, Division of Biology, California Institute of Technology, Pasadena, California 91125

James H. Strauss, Division of Biology, California Institute of Technology, Pasadena, California 91125

Barbara G. Weiss, Department of Microbiology and Immunology, Washington University School of Medicine, St. Louis, Missouri 63110

Preface

The publication of this volume of The Viruses entitled *The Togaviridae and Flaviviridae* comes at an appropriate time. The structure and replication strategies of these viruses are now known to be sufficiently diverse to warrant the removal of flaviviruses from the Togaviridae family and establish them as an independent family. Flaviviridae have a special place in the history of virology. The prototype virus—yellow fever virus—was the first virus to be identified as the cause of a human disease. Some of the history of this discovery is described in Chapter 1 of this volume; in Chapter 10 the complete sequence of the RNA genome of the virus is presented. This sequence not only defines the primary structure of the viral proteins, it also clarifies the mechanism of translation of the flavivirus genome. Knowledge of the sequence of the structural proteins of these viruses represents an important step in the potential goal of using purified flavivirus glycoproteins as vaccines. Many of the chapters in this volume focus on the structure and replication of the Togaviridae. These viruses have provided valuable models for studies in cell biology, particularly with regard to the cotranslational and posttranslational steps required for the synthesis and localization of membrane glycoproteins. Furthermore, Togaviridae have been pivotal in our growing understanding of how enveloped viruses enter and exit from cells.

The broad outlines of the structure and gene expression of Togaviridae and Flaviviridae are known, but important questions remain. We have relatively little information about the replication of their genomes. Togaviridae synthesize four nonstructural proteins. What are their functions? How is the synthesis of the subgenomic 26S RNA regulated? Major gaps also exist in our understanding of the interactions between these viruses and their hosts. In both families, virulent and avirulent strains of the same virus exist. What actually determines virulence? Is it the ability of a virus to infect a sensitive target cell or are other factors involved? Defective interfering particles can be important in maintaining persistent infections in cultured cells. Do they play a role in modulating an infection in an organism? Genetic traits of the host can be a major factor in the outcome of an infection as evidenced by the ability of a single gene to control the sensitivity or resistance of mice to flaviviruses. What is this gene product and how does it exert its effect? We have compiled this volume as a compendium of what is known about these viruses, but we hope it also serves as an impetus to investigate the unsolved problems.

Sondra Schlesinger
Milton J. Schlesinger

St. Louis, Missouri

Contents

Chapter 3

Structure and Replication of the Alphavirus Genome

Ellen G. Strauss and James H. Strauss

Chapter 4

Entry of Alphaviruses

Margaret Kielian and Ari Helenius

Chapter 5

Formation and Assembly of Alphavirus Glycoproteins

Milton J. Schlesinger and Sondra Schlesinger

Chapter 8

Alphavirus Pathogenesis and Immunity

Diane E. Griffin

Chapter 9

**The Use of Monoclonal Antibodies in Studies of the Structural
Proteins of Togaviruses and Flaviviruses**

John T. Roehrig

Chapter 10

Structure of the Flavivirus Genome

Charles M. Rice, Ellen G. Strauss, and James H. Strauss

Chapter 11

Replication of Flaviviruses

Margo A. Brinton

Chapter 12
Pathology of the Flaviviruses
Thomas P. Monath

CHAPTER 1

Comparative and Historical Aspects of the Togaviridae and Flaviviridae

JAMES S. PORTERFIELD

I. INTRODUCTION

Virology is one of the youngest branches of science, having its roots firmly in the study of infectious diseases (Waterson and Wilkinson, 1978). Yellow fever occupies a unique position in the history of virology, and much of our knowledge of the viruses that are the subject of this volume is derived directly or indirectly from research on yellow fever and its causative virus. Although a viral etiology of diseases of plants and domestic animals antedated similar findings concerning diseases that affect man, yellow fever was the first infectious disease of man to be shown to be due to a filterable agent or virus. Yellow fever was also the first viral infection of man to be shown to be transmitted by a blood-sucking arthropod, making yellow fever (YF) virus the archetypal arthropod-borne virus, or "arbovirus." YF virus was the first arbovirus to be successfully cultivated in the laboratory. Field studies on yellow fever resulted in the isolation of many other arboviruses, some of which in time became known as the Group A and Group B arboviruses. A collective name for these viruses was needed, and the term "togavirus" (from the Latin *toga*, a Roman "mantle" or "cloak," a reference to the possession of a viral envelope) was adopted as a convenient jargon name. The same root later provided the officially approved family name Togaviridae. This family initially contained only two genera, *Alphavirus* (the former Group A ar-

JAMES S. PORTERFIELD • Sir William Dunn School of Pathology, University of Oxford, Oxford OX1 3RE, England.

boviruses) and *Flavivirus* (the former Group B arboviruses), named after the type species, YF virus (from the Latin *flavus*, "yellow"). Ironically, the very recent elevation of the flaviruses to the level of a family has meant their removal from the family Togaviridae, but this change in no way invalidates the dominant role of YF virus in the history of virology in general and of the Togaviridae in the context of this volume. The stages in the foregoing highly condensed and oversimplified history will now be considered in more detail.

II. YELLOW FEVER

A. Early History

Yellow fever almost certainly originated in Africa, where cases continue to occur almost every year (Agadzi *et al.*, 1984). The first clearly recognizable accounts of yellow fever come not from Africa, but from the New World, probably having been introduced there by early sailing ships. Carter (1931) covers this early history and cites the Yucatan epidemic of 1648 as the first recorded outbreak of the disease in the Americas. Throughout the 18th and 19th centuries, epidemics raged in the Caribbean and Central America, extending as far north as Philadelphia, New York, and Baltimore. Europe was also involved at this time, there being deaths from yellow fever in Spain, Portugal, and even England, but the problem was far less severe in Europe than in the Americas or in Africa.

B. Carlos J. Finlay and the Walter Reed Comission

The cause of yellow fever remained highly controversial, and the suggestion put forward in 1881 by Dr. Carlos J. Finlay, a Cuban physician of Scottish-French descent, that the disease was transmitted by mosquitoes, was ridiculed at the time. At the turn of the century, however, Finlay's ideas were vindicated by the dramatic studies carried out in Cuba by the United States Army Yellow Fever Commission (Reed and Carroll, 1902). Using the mosquito species suggested to them by Finlay, which we now know as *Aedes aegypti*, and the production of disease in human volunteers as their only criterion of infectivity, the Reed Commission reached the following conclusions:

1. Yellow fever is due to a filterable agent, i.e., a virus.
2. The virus is present in the blood of infected persons for limited periods only.
3. Mosquitoes that feed on an infected subject are able to transmit the disease to healthy, susceptible individuals only after an interval of about 12 days (this period representing the extrinsic in-

cubation period during which virus replicates in the mosquito and travels to the salivary glands).
4. Infection is not spread by direct contact or by fomites.

These remarkable experiments provided a rational basis for anti-mosquito measures that proved highly successful in disease control in Cuba and later elsewhere. It should be noted that this success was achieved more than a quarter of a century before YF virus was eventually cultivated in the laboratory.

C. Yellow Fever Commissions of the Rockefeller Foundation

Shortly after it was established in 1913 "for the well-being of mankind throughout the world," the Rockefeller Foundation began to take an active interest in yellow fever and its control. Strode (1951) gives a full account of this early period, and Theiler and Downs (1973) and Downs (1982) continue the record. The first Rockefeller Foundation Yellow Fever Commission was set up in 1916; its efforts concentrated on South and Central America. The studies initiated then led much later to the recognition of jungle yellow fever, in which virus was spread by mosquitoes other than *Aedes aegypti*, and control measures directed against urban mosquitoes proved of no avail.

In 1920, the Rockefeller Commission to the West Coast of Africa was established with the object of trying to determine whether the disease seen in Africa was indeed the same as that seen in the New World. That commission recommended that laboratory studies be conducted in Africa, and in 1925 the West Africa Yellow Fever Laboratory was opened in Lagos, Nigeria. Following the successful isolation of YF virus in West Africa, a Yellow Fever Laboratory was set up at the Rockefeller Institute for Medical Research, New York, in 1928, to enable comparative virological studies to be carried out in a region free from the risk of endemic disease.

In 1936, the Rockefeller Foundation, in collaboration with the Colonial Office in London and the Government of the Protectorate of Uganda, opened a Yellow Fever Laboratory at Entebbe, Uganda. Research carried out there did much to unravel the complex epidemiology of yellow fever in Africa (Haddow, 1967–1968).

D. Asibi Virus and the 17D Yellow Fever Vaccine

Although the Reed Commission had provided clear evidence for a viral etiology of yellow fever, no virus had been cultivated in the laboratory, and the claims of the highly respected Japanese scientist Dr. Hideyo Noguchi (1919, 1925) that yellow fever was due to a *Leptospira* profoundly influenced the course of events in West Africa, where Noguchi

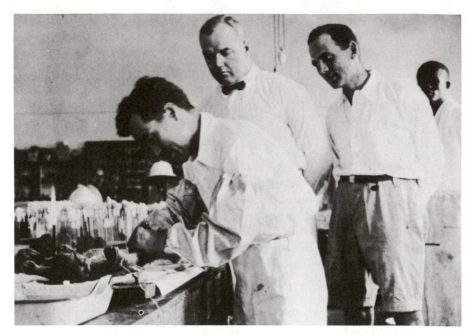

FIGURE 1. Doctor Hideyo Noguchi dissecting a rhesus monkey at the Yellow Fever Laboratory, Yaba, Lagos, Nigeria, in 1927. Watching are Dr. A. F. Mahaffy (with the bow tie), Dr. J. H. Bauer, and an African assistant.

was a member of the Rockefeller Commission. Much effort was devoted to attempts to demonstrate a *Leptospira* in materials from yellow fever cases, and there were repeated failures to transmit the disease to a wide variety of laboratory animals, including African monkeys. Meanwhile, Dr. A. H. Mahaffy, working from the Medical Research Institute at Accra in the Gold Coast (now Ghana), collected blood from a 27-year-old male African named Asibi who had yellow fever and transported it some 100 miles to Accra, where an Indian monkey, *Macacus rhesus*, was inoculated. Four days later, this monkey was moribund, and it was killed on the 5th morning. The Accra laboratory had no facilities for mosquito transmission studies that were needed to strengthen the case that the monkey's death was related to yellow fever, so another rhesus monkey was inoculated in Accra with blood from the moribund animal, and the second monkey was transported to Lagos during the incubation period. In the Lagos laboratory (Fig. 1), mosquitoes that fed on the second monkey successfully transmitted the disease to healthy rhesus monkeys, and the Asibi strain of YF virus was passaged serially in further rhesus monkeys (Stokes *et al.*, 1928). Tragically, this success was achieved only at heavy cost: Dr. Adrian Stokes contracted yellow fever and died of the disease in Lagos, as did Dr. Noguchi the following year.

Rhesus monkeys are expensive laboratory animals, and although their use played a determining role in the initial isolation of YF virus, the demonstration that the virus could be adapted to grow in the brains of mice inoculated intracerebrally greatly accelerated the pace of research and allowed comparisons to be made among different yellow fever isolates. The deaths of laboratory workers in Africa and in South America emphasized the need for a vaccine against yellow fever, and much effort was directed to this end in a number of different centers. Following some early promise with a mouse-grown vaccine used in association with yellow fever immune serum, Dr. Max Theiler, working in the Rockefeller laboratory in New York, achieved a major success when he successfully attenuated the highly virulent Asibi strain of virus by serial passage first in minced mouse embryo cells in culture followed by minced chick embryo cells in culture and later in embryonated hens' eggs (Theiler and Smith, 1937). For this work on what later became known as the 17D yellow fever vaccine, Theiler was awarded the Nobel Prize for Medicine in 1951.

III. EARLY STUDIES ON VIRAL ENCEPHALITIDES

In addition to classic neurological diseases such as rabies and poliomyelitis, early writers recognized a number of other diseases of the central nervous system that were presumed to be of viral origin, such as Japanese and Australian encephalitis, encephalitis lethargica, and postvaccinal encephalitis (Rivers, 1928). During the 1930s, a number of viruses were isolated from cases of encephalitis in different parts of the world. These included louping ill (LI) virus from Scotland; St. Louis encephalitis (SLE) virus, Western equine encephalitis (WEE) virus, and Eastern equine encephalitis (EEE) virus from the United States; Venezuelan equine encephalitis (VEE) virus from South America; Japanese encephalitis (JE) virus from Japan; and Russian spring summer encephalitis (RSSE) virus from the U.S.S.R. Further viruses were added to this list during the next decade. These viruses were all spread by the bites of blood-sucking vectors, either mosquitoes or ticks, and the set became known as the arthropod-borne encephalitis viruses. Later, when further viruses were recognized that had a similar mode of transmission but produced only generalized febrile reactions without involvement of the central nervous system, the more all-embracing term "arthropod-borne viruses" came into use (Hammon and Reeves, 1945). Still later, this was abbreviated to arborvirus, and finally to arbovirus (Casals, 1957, 1966).

IV. ARTHROPOD-BORNE VIRUSES

Arthropods can transmit viruses in two fundamentally different ways. In mechanical transmission, the vector simply functions as a "flying pin," transferring virus from an infected host to a healthy, susceptible one. In biological transmission, on the other hand, the virus undergoes a cycle of replication in the cells of the arthropod vector, which is thus a true host to the virus. The period of replication of virus in the vector, the eclipse phase, was seen in the early studies of the Reed Commission on yellow fever and occurs in all arbovirus infections. It is this biological cycle in an arthropod that distinguishes arthropod-borne animal viruses, or arboviruses, from all other animal viruses. Arboviruses differ from insect viruses in their ability to infect both invertebrate and vertebrate hosts, whereas insect viruses, or invertebrate viruses, are restricted to replication in invertebrates only. In nature, arboviruses usually undergo alternate periods of replication in vertebrate and invertebrate hosts, being transmitted between these hosts when a blood-sucking female arthropod feeds on a vertebrate. Vertical transmission, either transovarial or trans-stadial, occurs with some arboviruses in certain vectors; a few arboviruses also pass from one generation to the next in vertebrate hosts.

The somewhat cumbersome definition of an arbovirus given in the W.H.O. Study Group Report (W.H.O., 1967) is as follows:

> Arboviruses are viruses which are maintained in nature principally, or to an important extent, through biological transmission between susceptible vertebrate hosts by haematophagous arthropods; they multiply and produce viraemia in the vertebrates, multiply in the tissues of arthropods, and are passed on to new vertebrates by the bites of arthropods after a period of extrinsic incubation.

It should be noted that the criteria for accepting a virus as an arbovirus are purely biological and are quite independent of structural, biochemical, and taxonomic considerations.

A. Antigenic Relationships among Arboviruses

Each of the dozen or so arthropod-borne viruses that were known during the 1930s was regarded as a distinct agent, but even during this period, some reports of serological cross-reactivity appeared. Thus, Webster (1938) found that convalescent sera from two cases of Japanese encephalitis neutralized SLE virus in addition to JE virus. A little later, West Nile (WN) virus, which had been isolated in Uganda from a woman with a febrile illness (Smithburn et al., 1940), was shown to bear a complex relationship to both the JE and the SLE virus (Smithburn, 1942). EEE and WEE virus were quite distinct by neutralization tests, but Havens et al., (1943) demonstrated some cross-reactivity by complement-fixation (CF) tests, and Casals (1944) showed that CF tests revealed relationships among JE, SLE, and WN viruses.

As additional arboviruses were isolated, the complexity of the cross-reactions increased, transcending geographic considerations. Thus, Ilheus virus from South America and Murray Valley encephalitis virus from Australia were added to the complex of viruses containing SLE, JE, and

FIGURE 2. Jordi Casals in 1982 (Yale Arbovirus Research Unit).

WN viruses. Sabin (1950) reported that cross-reactions were detectable by CF tests among dengue virus types 1 and 2, YF virus, JE virus, and WN viruses, although potent reagents were required. Other limited cross-reactions were also noted, but no systematic analysis of the problem was made until a new serological test, the hemagglutination-inhibition (HI) test, was introduced. Casals (Fig. 2) and his associates in the Rockefeller Foundation Virus Laboratory in New York, following leads provided by other workers (Hallauer, 1946; Sabin and Buescher, 1950), achieved a major advance when they showed that arboviruses could be divided into two separate serological groups on the basis of their reactions in hemagglutination (HA) and HI tests (Casals and Brown, 1953, 1954).

B. Serological Groups A and B

Having successfully prepared HA antigens from 15 different arboviruses or arbovirus strains, Casals and Brown (1954) proceeded to test each of these antigens against 22 different antisera, using the HI technique. They found that EEE, WEE, and VEE viruses showed extensive cross-reactivity; these viruses they designated Group A arboviruses. Another 10 viruses, namely-dengue types 1 and 2, Ilheus, JE, Ntaya, RSSE, SLE, Uganda S, WN, and YF viruses, fell into a second cross-reacting set, which they designated Group B arboviruses. Antisera against viruses that would now be classified as Bunyaviridae (Anopheles A and B, Bwamba, Bunyamwera, California encephalitis, and Wyeomyia viruses), Picornaviridae (GDVII, Mengo, Coxsackie A and B, polioviruses types 1 and 2), Rhabdoviridae (rabies virus), or Herpesviridae (herpes simplex virus) failed to inhibit HA produced by antigens of viruses in Groups A or B. At that time, no active HA antigens were available against Semliki Forest virus (SFV), LI virus, or RSSE virus. Nevertheless, SFV was placed in Group A and the other two viruses in Group B because antisera against these viruses showed reactivity with Group A or Group B antigens, respectively. This initial subdivision of arboviruses into serogroups A and B was followed by the recognition of further sub-sets of antigenically related arboviruses (Casals, 1971). Casals (1962) defined an antigenic group as "an aggregate of viruses which cross-react in serological tests and therefore possess antigenic constituents in common." Porterfield (1962) reviewed the nature of serological relationships in the light of knowledge available at that time.

V. VIRAL CLASSIFICATION AND NOMENCLATURE

The first formal approach to viral classification and nomenclature occurred during the Fifth International Congress for Microbiology held in Rio de Janeiro in 1950, when the International Committee for Bacteriological Nomenclature set up a Subcommittee on Virus Nomenclature with Dr. C. H. Andrewes as its Chairman (Andrewes, 1953). In 1965, this subcommittee was replaced by the Provisional Committee on the Nomenclature of Viruses (PCNV) under the same chairman, by then Sir Christopher Andrewes. Between these two events, there had been a lively discussion among virologists about possible schemes for classifying viruses, in which Dr. A. Lwoff and his associates played an important role. The Lwoff–Horne–Tournier (LHT) system for the classification of animal, plant, and bacterial viruses was first proposed in 1962 (Lwoff et al., 1962) and was later modified (Lwoff and Tournier, 1966; Tournier and Lwoff, 1966). It was based on the fundamental properties of the virion, namely, the nature of the viral nucleic acid, RNA or DNA, the presence or absence of a viral envelope, and the symmetry of the nucleocapsid, whether helical or cubical. This hierarchical scheme started with the phylum Vira, divided into the subphyla Ribovira and Deoxyvira, the former being further divided into the classes Ribocubica and Ribohelica. The Ribocubica were divided into the orders Gymnovirales, naked, and

Togavirales, enveloped. At the Ninth International Congress for Microbiology held in Moscow in 1966, the PCNV was disbanded and the International Committee for the Nomenclature of Viruses (ICNV) was set up in its place. While the ICNV did not accept the details of the LHT scheme, it nevertheless accepted the principle that viral classification and nomenclature should be based on the fundamental properties of the virion. The ICNV established five subcommittees, one to consider each of the sets of viruses of vertebrates, invertebrates, plants, and bacteria and the fifth to consider the possible uses of a cryptogram in describing viruses. In due course, the Vertebrate Virus Subcommittee set up a number of study groups to examine the problems posed by different sets of viruses; one of these was the Arbovirus Study Group, which was established in 1967 with the writer as its Chairman and Dr. J. Casals, Professor M.P. Chumakov, Dr. C. Hannoun, and Dr. M. Mussgay as its original members. The study group was charged with the task of proposing names for the principal sets of arboviruses that would be in accordance with the rules for viral nomenclature then being formulated and that would also be acceptable to the international community of virologists who worked with arboviruses. It is worth noting that all the other study groups set up then or subsequently established have been oriented toward particular sets of viruses defined according to structural and biochemical criteria, whereas the Arbovirus Study Group was asked to examine a large and heterogeneous set of viruses.

The study groups reported their recommendations to the Vertebrate Virus Subcommittee, which in turn reported to the Executive Committee (ECICNV) and to the ICNV. The ICNV met during international congresses; the ECICNV also met during congresses, but usually between congresses as well. Reports on the decisions of the international committee were published after each congress, the first report by Wildy (1971), the second by Fenner (1976), the third and fourth by Matthews (1979, 1982), and the fifth by Brown (1986). Matthews (1983) has reviewed the evolution of the field of viral taxonomy.

VI. ARBOVIRUS STUDY GROUP: TOGAVIRUSES AND TOGAVIRIDAE

In 1968, the Arbovirus Study Group submitted its first report to the Vertebrate Virus Subcommittee. The need for a name to designate those RNA viruses that are enveloped and possess nucleocapsids of nonhelical, presumed cubical, symmetry was apparent, and the study group favored recognition at the level of a family. However, the ICNV policy at that time was to limit consideration of names to those at the level of genera.

The proposals of the Arbovirus Study Group, and those of other study groups, were received by the ECICNV, and an outline of these proposals was contained in a brief report from the Vertebrate Virus Subcommittee

(Andrewes, 1970). This report noted that the name *Togavirus* was proposed to cover "what is likely to be the great majority of arboviruses having taxonomic characters like those of the A and B groups" (Casals, 1966). When the ICNV met in Mexico City in 1970, it approved the substance of the proposals put forward by the Arbovirus Study Group, but it changed the generic name for the group A arboviruses from *Sindbisvirus* to *Alphavirus*, while approving the retention of Sindbis virus as the type species of the genus. The ICNV also changed the spelling of the generic name for group B from *Flavivirus* to *Flavovirus*. These findings were included in the published first report of the ICNV (Wildy, 1971). Later, the originally proposed spelling *Flavivirus* was approved by the ICNV (Fenner, 1976).

In 1967, Holmes and Warburton (1967) noted the similarity between the fundamental properties of rubella virus and those of the group A and B arboviruses and raised the question, "Is Rubella virus an arbovirus?" Since there is absolutely no evidence to suggest that rubella virus is arthropod-borne, it clearly cannot belong to the biologically determined set of arboviruses. However, if rubella virus is an enveloped RNA virus with nucleocapsids of cubical symmetry, as appeared likely, then it could qualify for consideration as a possible *Togavirus*, as could the non-arthropod-borne viruses that cause hog cholera and bovine virus diarrhea. To accommodate these possibilities, the Arbovirus Study Group was strengthened by the addition of Dr. I. H. Holmes and Dr. M. C. Horzinek, and others later joined for the special expertise that they could contribute, such as Professor N. Oker-Blom for his knowledge of the set of arboviruses with apparent helical symmetry that were being studied in Finland.

During the Ninth International Congress for Tropical Medicine and Malaria held in Athens in 1973, those members of the Arbovirus Study Group who were present met and agreed to recommend the addition of a third genus, *Rubivirus* (from the Latin *ruber*, "red"), containing a single member, rubella virus. Shortly after this, when those members of the study group who had not been present had been consulted, it was agreed to recommend the addition of a fourth genus, *Pestivirus* (from the Latin *pestis*, "plague"), containing the mucosal group of viruses, hog cholera and bovine virus diarrhea virus.

In 1973, the name of the main committee was changed to International Committee on Taxonomy of Viruses (ICTV), rather than International Committee for the Nomenclature of Viruses. The following year, a postal vote of members of the ICTV approved certain family and generic names (Fenner *et al.*, 1974). The family name Togaviridae was approved, with the following definition:

> Virions contain single-stranded RNA, 3×10^6 to 4×10^6 daltons, have isometric, probably icosahedral, nucleocapsids surrounded by a lipoprotein envelope containing host cell lipid and virus-specific polypeptides including one or more glycopeptides. Virions yield infectious RNA.

Two genera were approved, *Alphavirus* and *Flavivirus* (note the spelling), but the study group's proposal that rubella virus be admitted as the third genus, *Rubivirus*, and that bovine diarrhea, hog cholera, and equine arteritis viruses be admitted as the fourth genus, *Pestivirus*, was referred back to the study group for further consideration.

In preparation for the International Congress for Virology to be held in Madrid in 1975, the Arbovirus Study Group concentrated its attention on two separate objectives. One was to clarify the status of the non-arthropod-borne viruses that were possible members of the familiy Togaviridae. The other was to clarify the status of those arboviruses that were clearly not Togaviridae, nor were they Reoviridae or Rhabdoviridae (since these two families also contained some members that were arboviruses). Evidence was accumulating that some mosquito-borne and some tick-borne viruses have nucleocapsids with helical symmetry (von Bonsdorff *et al.*, 1969; Saikku *et al.*, 1971). Murphy *et al.* (1973) showed that many viruses within the Bunyamwera supergroup share similar morphological and morphogenetic characteristics that they felt justified the recognition of a new family, Bunyaviridae. The Arbovirus Study Group agreed with this suggestion, which they put forward as a formal proposal to the ICTV. Whereas Murphy *et al.* (1973) had suggested the formation of two genera, the study group proposed only a single genus, *Bunyavirus*, the remaining members of the family being "other possible members" until such time as separation into further genera appeared justified (Porterfield *et al.*, 1973–1974).

At the Madrid Congress, the ICTV approved formal recognition of the family Bunyaviridae, with a single genus, *Bunyavirus*, of at least 85 members and at least 48 other possible members of the family in at least two different genera (Fenner, 1976). The family Togaviridae was redefined with the addition of two new genera, *Rubivurus* and *Pestivirus*, in addition to the earlier genera *Alphavirus* and *Flavivirus*. Also, equine arteritis virus and lactic dehydrogenase virus were listed as possible members of the family Togaviridae, outside existing genera. Shortly thereafter, the Arbovirus Study Group's description of the family Bunyaviridae was published (Porterfield *et al.*, 1975–1976), and that of the family Togaviridae followed (Porterfield *et al.*, 1978).

VII. NON-ARTHROPOD-BORNE TOGAVIRUSES

The description of the family Togaviridae given in Matthews (1982) named four genera, *Alphavirus*, *Flavivirus*, *Rubivirus*, and *Pestivirus*, and five additional possible members, the *Aedes albopictus* cell fusing agent, carrot mottle, equine arteritis, lactic dehydrogenase, and simian hemorrhagic fever viruses. All 25 members of the genus *Alphavirus* are arboviruses, and all are mosquito-transmitted. The genus *Flavivirus* contains 28 mosquito-borne members, 11 tick-borne members, and 18 members for which no vector is known. This last subset of the flaviviruses contains viruses such as Modoc and Rio Bravo viruses, isolated respectively from small rodents and bats; these viruses appear to spread in nature by direct vertebrate-to-vertebrate transmission, and they fail to replicate in arthropods or in arthropod cell cultures. The members of this subset therefore cannot be arboviruses, but they are clearly flaviviruses. The viruses in the genera *Rubivirus* and *Pestivirus* are not arboviruses, and the additional members, equine arteritis, lactic dehydrogenase, and simian hemorrhagic fever viruses, are also not arboviruses. The non-arthropod-borne togaviruses have been described in detail elsewhere (Brinton, 1980; Horzinek, 1981), but since they are not covered elsewhere in this volume, they are described briefly below.

Genus Pestivirus.   The viruses in this genus are collectively known as the mucosal disease viruses, and they include bovine virus diarrhea (BVD) virus, hog cholera virus, and the virus of border disease. They remain comparatively little studied as compared to the alphaviruses and flaviviruses, but they are agents of important veterinary diseases. Virions are 50–60 nm in diameter and have 10 to 12-nm ringlike subunits on the surface. The genomic RNA is about 3.2×10^6 daltons. There are two envelope glycoproteins, E1 ($M_r \approx 55K$) and E2 ($M_r \approx 45K$), and there is a core protein, C ($M_r \approx 35K$). The replication strategy is not known. Recent data (Purchio *et al.*, 1983, 1984a,b) have indicated that BVD virus is closer to the flaviviruses than to the alphaviruses or rubivirus. Cells infected with BVD virus contain a single species of RNA of 8.2 kilobases, which appears to code for three major virus specific proteins of molecular weights 115, 80, and 55K, of which the last is glycosylated. In a reticulocyte cell-free translation system, native RNA was unable to serve as an efficient message unless it was denatured immediately before translation. When polyribosomes were used in a cell-free translation system, the two larger structural proteins were formed, but the 55K glycoprotein was not detected, possibly due to incomplete glycosylation *in vitro*.

Genus Rubivirus.   Rubella virus is the sole member of this genus. Virus particles are about 60 nm in diameter and have surface spikes. The genome is a molecule of single-stranded, capped, and polyadenylated 40 S RNA, and a 24 S sub genomic messenger RNA (mRNA) is present in infected cells. This mRNA codes for a precursor protein, p110, which is proteolytically cleaved to yield the core protein, C (M_r 33K), and two envelope glycoproteins, E1 (M_r 53K) and E2 (M_r 30K). The E_2 protein is subjected to heterogeneous modification to give E2a (M_r 47K) and E2b (M_r 42K), and the E1 is processed and glycosylated to give E1 of M_r 58K. Three genes appear to code for the four structural proteins (Kalkkinen *et al.*, 1984; Oker-Blom *et al.*, 1984).

Genus Arterivirus.   Equine arteritis virus causes disease in equine species throughout the world. Virions are 60 nm in diameter and have

12 to 15-nm ringlike structures on the surface. The genome is a molecule of 48 S single-stranded RNA that is polyadenylated. A genomic-sized RNA and five smaller polyadenylated RNAs are found in infected cells. There are three structural proteins, a glycosylated envelope protein, E1 (M_r 21K); a nonglycosylated envelope protein, E2 (M_r 14K); and a core protein, C (M_r 12K). E1 and E2 are antigenically distinct. The virus causes necrosis in muscle cells of small arteries, hence the name, and also causes abortion in pregnant mares. Both horizontal and vertical transmission occur; there is no evidence suggesting arthropod transmission (Horzinek, 1981; Van der Zeijst *et al.*, 1975; van Berlo *et al.*, 1982).

In 1984, the ICTV approved the recommendation of the Togaviridae Study Group that equine arteritis be recognized as the type species and only member of a newly defined genus, *Arterivirus*, within the newly defined family Togaviridae.

Lactic Dehydrogenase Virus. This virus remains relatively poorly characterized, but its properties are certainly compatible with its remaining a member of the family Togaviridae. Virions are about 55 nm in diameter, and the core is about 35 nm in diameter. Three structural proteins have been reported, an envelope glycoprotein, E1 (M_r 24–44K); a nonglycosylated envelope protein, E2 (M_r 16–18K); and a core protein, C (M_r 13–15K) (Michaelides and Schlesinger, 1973; Riley, 1974; Rowson and Mahy, 1975, 1985; Brinton-Darnell and Plagemann, 1975).

The host range of lactic dehydrogenase virus is limited to mice, and viral replication is restricted to cells of the macrophage system, which are destroyed, liberating large amounts of lactic dehydrogenase enzyme (Steuckemann *et al.*, 1982). The virus produces a number of changes in host immune responses (Isakov *et al.*, 1982).

Simian Hemorrhagic Fever Virus. This virus remains relatively poorly characterized. Virions are about 45–50 nm in diameter, with a core of about 25 nm. A single envelope glycoprotein with a molecular weight of about 50K has been reported, with four additional smaller proteins (Leon *et al.*, 1982). In the recent taxonomic revision, simian hemorrhagic fever virus has been placed as a possible member of the newly defined family Flaviviridae.

Carrot Mottle Virus. The status of this virus remains uncertain.

Cell Fusing Agent. This virus, originally isolated from mosquito cell cultures, remains relatively little studied. It is suggested that it joins simian hemorrhagic fever virus as a possible member of the family Flaviviridae.

VIII. CONTRIBUTION OF THE WORLD HEALTH ORGANIZATION

An informal meeting held in Lisbon in 1958 during the Sixth International Congress for Tropical Medicine and Malaria made a number of recommendations that were acted on by a scientific group that met in Geneva in November 1958. In September 1960, a study group met in Geneva under W.H.O. auspices, and its report was published a little later (W.H.O., 1961). Impelled by this initiative, the W.H.O. designated a number of laboratories as reference centers for arbovirus studies. The Rockefeller Foundation Virus Laboratory, initially in New York, and later in Yale University, became the International Reference Center for Arboviruses.

In 1967, a second W.H.O. report was published, this time on "arboviruses and human diseases" (W.H.O., 1967). In later years, the W.H.O. continued to support a program designed to provide field laboratories with diagnostic and typing reagents; it also sponsored many training schemes in arbovirus studies and acted as a coordinating agent for field and reference centers.

The W.H.O. has sponsored many scientific meetings related to the problems of arboviruses. Technical guides on the diagnosis, prevention, and control of dengue hemorrhagic fever were prepared in 1975 and revised in 1908. Bres (1980) and Self (1982) have given brief accounts of some of the W.H.O. activities in different parts of the world. Through its own publications, such as the *Bulletin of the World Health Organization*, many research reports and reviews of public health problems are brought to the attention of the scientific community.

IX. CONTRIBUTION OF THE AMERICAN COMMITTEE ON ARTHROPOD-BORNE VIRUSES

Taylor (1962) described the origins of the American Committee on Arthropod-borne Viruses (ACAV), which also arose as a result of the 1958 meeting in Lisbon. Among its many achievements, perhaps the most important is that of creating and revising a catalogue of arthropod-borne viruses. This began with submissions from workers active in the field of arbovirus research, and a great deal of valuable information on individual viruses was assembled using a common format. A "catalogue of arthropod-borne viruses of the world" was published (Taylor, 1967), followed by a supplement (ACAV, 1970). In 1975, a revised catalogue was published (Berge, 1975), and a further supplement followed (Karabatsos, 1978). In parallel with this cataloguing activity, the ACAV ran an information exchange that allowed rapid interchange of information among active scientists in many parts of the world. Other services provided by the ACAV have been the standardization of reagents and that of the use of arbovirus names (ACAV, 1969). The arbovirus field extends far beyond the Togaviridae, but the activities of the ACAV have been of very great assistance to all virologists in this general area.

X. IMPACT OF MOLECULAR BIOLOGY ON VIRAL CLASSIFICATION

The LHT system of classification recognized the importance of the viral nucleic acid, but emphasized the structural virion or the phenotype of the virus, rather than its genotype. In 1971, Baltimore (1971) described a classification of viruses based on the expression of the viral genome. Viruses were divided into six classes based on the strategies of genome structure and transcription into mRNA. In the Baltimore classification, Togaviridae are Class IV viruses, as are Picornaviridae. The initiation unit for replication in viruses in this class is positive-stranded RNA that also functions as +mRNA.

When viral families were first being defined, little was known about the different strategies of viral replication, but it was assumed that viruses that were structurally similar also had similar replication patterns. As new data became available, some inconsistencies appeared. The caliciviruses closely resemble the picornaviruses in morphology and were initially placed within the family Picornaviridae. When it was realized that the caliciviruses had only a single structural protein, whereas the picornaviruses had four proteins, it was decided to remove the caliciviruses from the Picornaviridae and to establish Caliciviridae as a separate family.

The genotype of a virus is determined by the nucleotide sequences of its genome, which have now been determined for a number of viruses, as discussed elsewhere in this volume. A fuller knowledge of virion structure at the molecular level should contribute to our understanding of the basis of antigenic interrelationships among viruses.

XI. FLAVIVIRIDAE AS A SEPARATE FAMILY

The hierarchical scheme of viral classification suggested a quarter of a century ago implied that the set of RNA viruses with envelopes and nucleocapsids of cubical symmetry were homogeneous, and the term "togavirus," and later the family name Togaviridae, were introduced to embrace all such viruses. From the very beginning, differences between the alphaviruses and the flaviviruses were known. Thus, the flaviviruses are a little smaller than the alphaviruses, and the two genera were known to have different patterns of intracellular development. Probably because the alphaviruses were somewhat simpler to study, they were the first to be characterized (Kaariainen and Soderlund, 1978). There was little doubt that alphaviruses have nucleocapsids of cubical symmetry (Horzinek and

Mussgay, 1969), but the symmetry of the flaviviruses was less certain. Whereas alphaviruses have at least two envelope proteins, only a single envelope protein could be found in flaviviruses. Westaway (1980) stressed the differences between the replication strategies of the flaviviruses as compared with the alphaviruses and argued that this was sufficient to justify their separation, possibly by the creation of a new family. The alphaviruses specify a subgenomic 26 S mRNA corresponding to the 3' end of the genome that codes for the structural proteins; these are translated as a polyprotein that is subsequently cleaved and processed (Kaariainen and Soderlund, 1978). No such subgenomic mRNA is found in flavivirus-infected cells (Westaway, 1973; Westaway and Shew, 1977). In cell-free systems, flavivirus genomic mRNA yields only structural proteins (Wengler *et al.*, 1979; Svitkin *et al.*, 1981; Monkton and Westaway, 1982), whereas genomic mRNA of alphaviruses yields only nonstructural proteins (Kaariainen and Soderlund, 1978).

In 1984, the Togaviridae Study Group proposed that the genus *Flavivirus* be removed from the family Togaviridae and be recognized as the basis for a new family, Flaviviridae. These proposals were considered by the ICTV in 1984 and were approved (Westaway *et al.*, 1985).

REFERENCES

Agadzi, V. K., Boatin, B. A., Appawu, M. A., Mingle, J. A. A., and Addy, P. A., 1984, Yellow fever in Ghana, 1977–1980, *Bull. W.H.O.* **62:**577–583.

American Committee on Arthropod-borne Viruses, 1969, Arbovirus names, *Am. J. Trop. Med. Hyg.* **18:**731–734.

American Committee on Arthropod-borne Viruses, 1970, Catalogue of arthropod-borne viruses of the world, *Am. J. Trop. Med. Hyg.* (Suppl.) **19:**1079–1160.

Andrewes, C. H. 1953, The Rio Congress decisions with regard to study of selected groups of viruses, *Ann. N. Y. Acad. Sci.* **56:**428–432.

Andrewes, C. H., 1970, Generic names of viruses of vertebrates, *Virology* **40:**1070–1071.

Baltimore, D., 1971, Expression of animal virus genomes, *Bacteriol. Rev.* **35:**235–241.

Berge, T. D., 1975, *International Catalogue of Arboviruses*, U. S. Department of Health, Education, and Welfare, DHEW Publ. No. (CDC) 75-8301.

Bres, P., 1980, Development of the WHO programme in the field of arboviruses and allied viruses, in: *Arboviruses in the Mediterranean Countries* (J. Vesenjak-Hirjan, J. S. Porterfield, and E. Arslanagic, eds.), pp. 1–6, *Zentralbl. Bakteriol. Suppl. 9,* Gustav Fischer Verlag, Stuttgart.

Brinton, M., 1980, Non-arbo togaviruses, in: *The Togaviruses: Biology, Structure, Replication* (R. W. Schlesinger, ed.), pp. 623–666, Academic Press, New York.

Brinton-Darnell, M., and Plagemann, P. G. W., 1975, Structural and chemical–physical characteristics of lactate-dehydrogenase-elevating virus and its RNA, *J. Viol.* **16:**420–433.

Brown, F., 1986, Classification and nomenclature of viruses: 5th Report of ICTV, *Intervirology* (in press).

Carter, H. R., 1931, *Yellow Fever: An Epidemiological and Historical Study of Its Place of Origin*, Williams and Wilkins, Baltimore.

Casals, J., 1944, Immunological relationships among central nervous system viruses, *J. Exp. Med.* **79:**341–359.

Casals, J., 1957, The arthropod-borne viruses of vertebrates, *Trans. N. Y. Acad. Sci.* **19:**219–235.

Casals, J. 1962. Antigenic relationships among arthropod-borne viruses: Effect on diagnosis and cross immunity, in: *Biology of the Tick-borne Encephalitis Complex* (H. Libikova, ed.), pp. 56–66, Czechoslovak Academy of Sciences, Prague.

Casals, J., 1966, Special points about classification of arboviruses, in: IX International Congress of Microbiology, Moscow, pp. 441–452, Ivanovski Institute of Virology, Moscow.

Casals, J., 1971, Arboviruses: Incorporation in a general system of virus classification, in: *Comparative Virology* (K. Maramarosch and E. Kurstak, eds.), pp. 307–333, Academic Press, New York.

Casals, J, and Brown, L. V., 1953, Hemagglutination with certain arthropod-borne viruses, *Proc. Soc. Exp. Biol. Med.* **83**:170–173.

Casals, J., and Brown, J. V., 1954, Hemagglutination with arthropod-borne viruses, *J. Exp. Med.* **99**:429–449.

Downs, W. G., 1982, The Rockefeller Foundation virus program 1951–1971, with an update to 1981, *Annu. Rev. Med.* **33**:1–29.

Fenner, F., 1976, Classification and nomenclature of viruses: 2nd Report of ICTV, *Intervirology* **7**:1–115.

Fenner, F., Pereira, H. G,. Porterfield, J. S., Joklik, W. K., and Downie, A. W., 1974, Family and generic names for viruses approved by the International Committee on Taxonomy of Viruses, June 1974, *Intervirology* **3**:193–198.

Haddow, A. J., 1967–1968, The natural history of yellow fever in Africa, *Proc. R. Soc. Edinburgh Sect. B.* **70**:191–227.

Hallauer, C., 1946, Über den Virusnachweis mit dem Hirst-Test, *Schweiz. Z. Pathol. Bakteriol.* **9**:553–554.

Hammon, W. McD., and Reeves, W. C., 1945, Recent advances in the epidemiology of the arthropod-borne virus encephalitides, including certain exotic types, *Am. J. Public Health* **35**:994–1004.

Havens, W. P., Watson, D. W., Green, R. H., Lavin, G. I., and Smadel.,J. E., 1943, Complement fixation with neurotropic viruses, *J. Exp. Med.* **77**:139–153.

Holmes, I. H., and Warburton, M. F. 1967, Is rubella an arbovirus?, *Lancet* **2**:1233–1236.

Horzinek, M. C., 1981, *Non-Arthropod-borne Togaviruses*, Academic Press, New York.

Horzinek, M. C., and Mussgay, M., 1969, Studies on the nucleocapsid structure of a group A arbovirus, *J. Virol.* **4**:514–520.

Isakov, N., Feldman, S., and Segal, S., 1982, The mechanism of modulation of humoral immune responses after infection of mice with lactic dehydrogenase virus, *J. Immunol.* **128**:969–975.

Kaariainen, L., and Soderlund, H., 1978, Structure and replication of alphaviruses, *Curr. Top. Microbiol. Immunol.* **82**:15–69.

Kalkkinen, N., Oker-Blom, C., and Pettersson, R. F., 1984, Three genes code for rubella virus structural proteins, E1, E2a, E2b and C, *J. Gen. Virol.* **65**:1549–1557.

Karabatsos, N., 1978, International catalogue of arboviruses including certain other viruses of vertebrates, *Am. J. Trop. Med. Hyg.* (Suppl.) **27**(2):372–440.

Leon, M., Gravell, O., Gutenson, R., Hamilton, R., and London, W., 1982, Classification of simian hemorrhagic fever virus, in Fifth International Congress of Virology (L. Hirth and F. A. Murphy, eds.), Institute de Biologie Moleculaire et Cellulaire, Strasbourg, Abstracts, p. 385.

Lwoff, A., and Tournier, P., 1966, The classification of viruses, *Annu. Rev. Microbiol.* **20**:45–73.

Lwoff, A., Horne, R., and Tournier, P., 1962, A system of viruses, *Cold Spring Harbor Symp. Quant. Biol.* **27**:51–55.

Matthews, R. E. F., 1979, Classification and nomenclature of viruses: 3rd Report of ICTV, *Intervirology* **7**:132–296.

Matthews, R. E. F., 1982, Classification and nomenclature of viruses: 4th Report of ICTV, *Intervirology* **17**:1–199.

Matthews, R. E. F., 1983, *A Critical Appraisal of Viral Taxonomy*, CRC Press, Cleveland, Ohio.

Michaelides, M. C., and Schlesinger, S., 1973, Structural proteins of lactic dehydrogenase virus, *Virology* **55**:211–217.

Monkton, R. P., and Westaway, E. G., 1982, Restricted translation of the genome of the flavivirus Kunjin *in vitro, J. Gen. Virol.* **63**:227–232.

Murphy, F. A., Harrison, A. K., and Whitfield, S. G., 1973, Bunyaviridae: Morphologic and morphogenetic similarities of Bunyamwera serologic supergroup viruses and several other arthropod-borne viruses, *Intervirology.* **1**:297–316.

Noguchi, H., 1919, Etiology of yellow fever: Transmission experiments on yellow fever, *J. Exp. Med.* **29**:565–584.

Noguchi, H., 1925, Yellow fever research, 1918–1924: Summary, *L. icteroides, J. Trop. Med.* **28**:185–193.

Oker-Blom, C., Ulmanen, I Kaariainen, L. and Pettersson, R.F., 1984, Rubella virus 40S genome RNA specifies a 24S subgenomic mRNA that codes for a precursor to structural proteins, *J. Virol.* **49**:403–408.

Porterfield, J. S., 1962, The nature of serological relationships among arthropod-borne viruses, *Adv. Virus Res.* **9**:127–156.

Porterfield, J. S., Casals, J., Chumakov, M. P., Gaidamovich, S. Ya., Hannoun, C., Holmes, I. H., Horzinek, M. C., Mussgay, M., and Russell, P. K., 1973–1974, Bunyaviruses and Bunyaviridae, *Intervirology* **2**:270–272.

Porterfield, J. S., Casals, J. Chumakov, M. P., Gaidamovich, S. Ya., Hannoun, C., Holmes, I. H., Horzinek, M. C., Mussgay, M., Oker-Blom, N., and Russell, P. K., 1975–1976, Bunyaviruses and Bunyaviridae, *Intervirology* **6**:13–24.

Porterfield, J. S., Casals, J., Chumakov, M. P., Gaidamovich, S. Ya., Hannoun, C., Holmes, I. H., Horzinek, M. C., Mussgay, M., Oker-Blom, N., Russell, P. K., and Trent, D. W., 1978, Togaviridae, *Intervirology* **9**:129–148.

Purchio, A. F., Larson, R., and Collett, M. S., 1983, Characterization of virus-specific RNA synthesized in bovine cells infected with bovine viral diarrhea virus, *J. Virol.* **48**:320–324.

Purchio, A. F., Larson, R., and Collett, M. S., 1984a, Characterization of bovine viral diarrhea virus proteins, *J. Virol.* **50**:666–669.

Purchio, A. F., Larson, R., Torborg, L. L., and Collett, M. S., 1984b, Cell free translation of bovine viral diarrhea virus RNA, *J. Virol.* **52**:973–975.

Reed, W., and Carroll, J., 1902, Etiology of yellow fever: Supplemental note, *Am. Med.* **3**:301–315.

Riley, V., 1974, Persistence and other characteristics of lactate-dehydrogenase-elevating virus (LDH-virus), *Prog. Med. Virol.* **18**:198–213.

Rivers, T. M., 1928, *Filterable Viruses*, Williams and Wilkins, Baltimore.

Rowson, K. E. K., and Mahy, B. W. J., 1975, *Lactic Dehydrogenase Virus*, Springer-Verlag, New York.

Rowson, K. E. K., and Mahy, B. W. J., 1985, Lactate dehydrogenase-elevating virus, *J. Gen. Virol.* **66**:2297–2312.

Sabin, A. B., 1950, The dengue group of viruses and its family relationships, *Bacteriol. Rev.* **14**:225–232.

Sabin, A. B., and Buescher, E. L., 1950, Unique physico-chemical properties of Japanese B encephalitis virus hemagglutinin, *Proc. Soc. Exp. Biol. Med.* **74**:222–230.

Saikku, P., von Bonsdorff, C.-H., Brummer-Korvenkontio, M., and Vaheri, A., 1971, Isolation of non-cubical ribonucleoprotein from Inkoo virus, a Bunyamwera supergroup arbovirus, *J. Gen. Virol.* **13**:335–337.

Self, L. S., 1982, WHO activities on controlling vectors of arboviral diseases in the Western Pacific region, in: *Viral Diseases in South-East Asia and the Western Pacific* (J. S. Mackenzie, ed.), pp. 494–497, Academic Press, Sydney.

Smithburn, K. C., 1942, Differentiation of the West Nile virus from the viruses of St. Louis and Japanese encephalitis, *J. Immunol.* **44**:25–31.

Smithburn, K. C., Hughes, T. B., Burke, A. W., and Paul, J. H., 1940, A neurotropic virus isolated from the blood of a native of Uganda, *Am. J. Trop. Med.* **20**:471–492.

Stokes, A., Bauer, J. H., and Hudson, N. P., 1928, Experimental transmission of yellow fever to laboratory animals, *Am. J. Trop. Med.* **8**:103–164.

Strode, G. K., 1951, *Yellow Fever*, McGraw-Hill, New York, 710 pp.

Stueckemann, J. A., Holth, M., Swart, W. J., Kowalchyk, K., Smith, M. S., Wolstenholme, A. J., Cafruny, W. A., and Plagemann, P. G. W., 1982, Replication of lactate dehydrogenase-elevating virus in macrophages. 2. Mechanism of persistent infection in mice and cell culture, *J. Gen. Virol.* **59**:263–272.

Svitkin, Y. V., Ugarova, T. Y., Chernovskaya, T. V., Lyapustin, V. N., Lashkevich, V. A., and Agol, V. I., 1981, Translation of tick-borne encephalitis virus (flavivirus) genome *in vitro*: Synthesis of two structural polypeptides, *Virology* **110**:26–34.

Taylor, R. M., 1962, Purpose and progress in cataloguing and exchanging information on arthropod-borne viruses, *Am. J. Trop. Med. Hyg.* **11**:169–174.

Taylor, R. M., 1967, Catalogue of arthropod-borne viruses of the world: A collection of data on registered arthropod-borne animal viruses, USPHS Publ. No. 1760, U.S. Government Printing Office, Washington, D.C.

Theiler, M., and Downs, W. G., 1973, *The Arthropod-borne Viruses of Vertebrates*, Yale University Press, New Haven, 578 pp.

Theiler, M., and Smith, H. H., 1937, Use of yellow fever virus modified by *in vitro* cultivation for human immunization, *J. Exp. Med.* **65**:787–800.

Tournier, P., and Lwoff, A., 1966, Systematics and nomenclature of viruses: The PCNV proposals in: IX International Congress of Microbiology, Moscow, pp. 417–422. Ivanovski Institute of Virology, Moscow.

Van Berlo, M. F., Horzinek, M. C., and van der Zeijst, B. A. M., 1982, Equine arteritis virus-infected cells contain six polyadenylated virus-specific RNA's, *Virology* **118**:345–352.

Van der Zeijst, B. A. M., Horzinek, M. C., and Moennig, V., 1975, The genome of equine arteritis virus, *Virology* **68**:418–425.

Von Bonsdorff, C.-H., Saikku, P., and Oker-Blom, N., 1969, The inner structure of Uukuniemi virus and two Bunyamwera supergroup arboviruses, *Virology* **39**:342–344.

Waterson, A. P., and Wilkinson, L., 1978, *An Introduction to the History of Virology*, Cambridge University Press, Cambridge, 237 pp.

Webster, L. T., 1938, Japanese B encephalitis virus: Its differentiation from St. Louis encephalitis virus and relationship to louping ill virus, *J. Exp. Med.* **67**:609–618.

Wengler, G., Beato, M., and Wengler, G., 1979, *In vitro* translation of 42S virus specific RNA from cells infected with the flavivirus West Nile virus, *Virology* **96**:516–529.

Westaway, E. G., 1973, Proteins specified by group B togaviruses in mammalian cells during proliferative infections, *Virology* **51**:454–465.

Westaway, E. G., 1980, Replication of flaviviruses, in: *The Togaviruses: Biology, Structure, Replication* (R. W. Schlesinger, ed.), pp. 531–581, Academic Press, New York.

Westaway, E. G., and Shew, M., 1977, Proteins and glycoproteins specified by the flavivirus Kunjin, *Virology* **80**:309–319.

Westaway, E. G., Brinton, M. A., Gaidamovich, S. Ya., Horzinek, M. C., Igarashi, A., Kaariainen, L., Lvov, D. K., Porterfield, J. S., Russell, P. K., and Trent, D. W., 1985, Flaviviridae, *Intervirology* **24**:183–192.

W.H.O., 1961, Arthropod-borne viruses, W.H.O. Technical Report Series, No. 219.

W.H.O., 1967, Arboviruses and human disease, W.H.O. Technical Report Series, No. 369.

Wildy, P., 1971, Classification and nomenclature of viruses: 1st Report of ICNV, *Monogr. Virol.* Vol. 5, S. Karger, Basel.

CHAPTER 2

Alphavirus Structure

Stephen C. Harrison

I. INTRODUCTION

Alphavirus particles are roughly spherical, with three principal substructures—an outer glycoprotein shell, a lipid bilayer, and an RNA-containing core (nucleocapsid). The radial organization of these components is summarized in Fig. 1. The most extensive structural studies have been carried out on Sindbis virus and Semliki Forest virus (SFV). It is on these very similar viruses that the description below will focus.

II. GENERAL PROPERTIES

The following characteristics of the substructures are well established: (1) The glycoproteins are arranged in an icosahedral surface lattice with 240 structure units per particle—a so-called $T = 4$ lattice (von Bonsdorff, and Harrison, 1975, 1978; Adrian et al., 1984). The lipid bilayer, derived from the cell membrane during budding, contains a representative sample of cellular plasma membrane lipids (Laine et al., 1972; Hirschberg and Robbins, 1974). Hydrophobic sequences near the C termini of the E1 and E2 polypeptide chains form structures that cross the bilayer (Garoff and Simons, 1974; Garoff and Söderlund, 1978; Rice et al., 1982). (3) The core is an isometric particle, with a radius of about 200 Å. It probably has icosahedral symmetry, although the details of its surface organization are not certain. There is, however, reason to believe that like the glycoprotein shell, it has a $T = 4$ lattice and 240 subunits.

Microcrystals of SFV have been observed in the electron microscope, and the packing of virions in these lattices has been analyzed by Wiley

STEPHEN C. HARRISON • Department of Biochemistry and Molecular Biology, Harvard University, Cambridge, Massachusetts 02138.

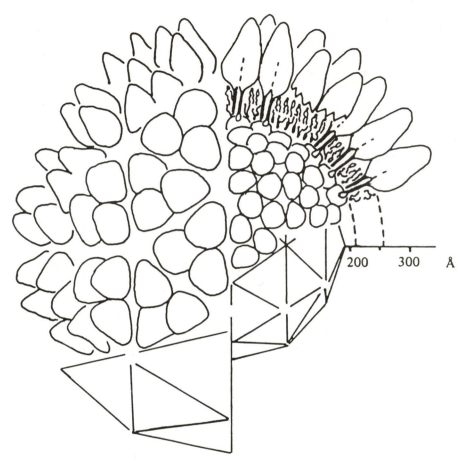

FIGURE 1. Organization of protein and lipid in Sindbis virus particles. Semliki Forest virus is essentially identical, but about 10 Å smaller in diameter. The pear-shaped units represent (E1 + E2) heterodimers. Their clustering in trimers gives the grooved patterns seen in electron micrographs. The hydrophobic C-terminal roots of E1 and E2 penetrate the lipid bilayer. Those of E2 have a small internal domain, shown here making contact with a site on a nucleocapsid subunit. The bilayer is symbolized in the diagram by an array of lipid molecules between radii of 210 and 255 Å. The icosahedral surface lattices that characterize the outer glycoprotein layer and the inner core are shown in the lower part of the drawing.

and von Bonsdorff (1978). Larger crystals, giving weak low-resolution X-ray diffraction, have also been obtained (F. K. Winkler, personal communication). Crystallization shows that the virus particle is a precise and regular structure and that all particles are identical.

A. Radial Organization

Small X-ray scattering from Sindbis virus and SFV (Fig. 2) shows the position of the lipid bilayer and the radial dimensions of the core and of

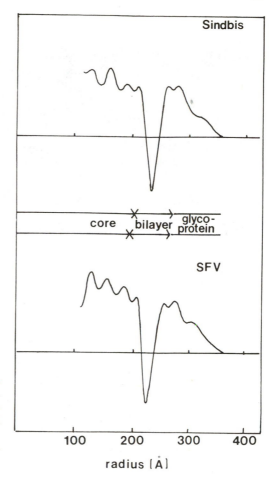

FIGURE 2. Electron densities, as functions of radius, for Sindbis virus (Harrison *et al.*, 1971) and SFV (Harrison and Kääriänen, unpublished data), as determined by small-angle X-ray scattering. The deep trough shows the position of the lipid bilayer.

the glycoprotein coat. The SFV particle appears to be slightly smaller than that of Sindbis virus, but there is otherwise no fundamental difference. The lipid content is sufficient to make essentially continuous bilayers with head groups occupying shells at the indicated radii (Harrison *et al.*, 1971; Laine *et al.*, 1972). The area of the outer leaflet is about 40% larger than that of the inner leaflet, imparting a substantial asymmetry.

B. Glycoprotein Shell

Figures 3–5 show electron micrographs of Sindbis virus. The principal contrast comes from the arrangement of glycoprotein in the outer shell. When negatively stained with phosphotungstate (Fig. 3a–e), the virus appears to be significantly flattened, and most of the image comes from one side of the particle. Stain penetrates along grooves in the surface and at nodes where five or six grooves meet. The grooves delineate a

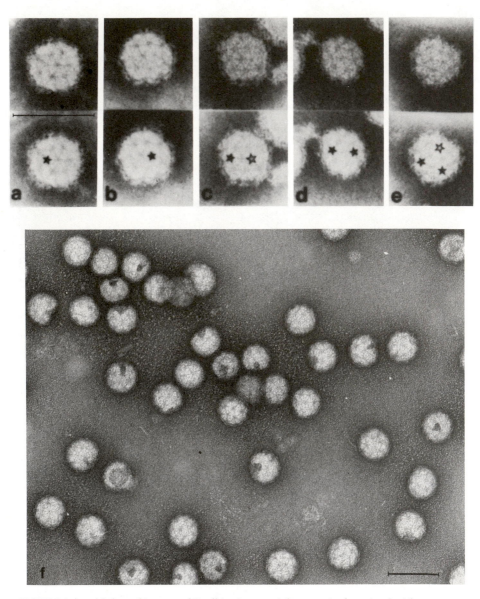

FIGURE 3. (a–e) Selected images of Sindbis virus particles negatively stained with potassium phosphotungstate. Each particle is shown twice, with fivefold nodes of the surface lattice marked by stars on the lower image. (c–e) These images show two or more fivefold positions, in a relationship corresponding to a T = 4 lattice. (f) Field of Sindbis particles negatively stained with uranyl acetate. The particles are less deformed than in potassium phosphotungstate. Scale bars: 1000 Å. From von Bonsdorff and Harrison (1975).

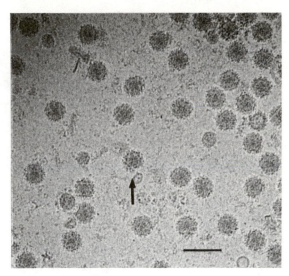

FIGURE 4. Sindbis virus particles in vitreous ice. The contrast
is uniform throughout the particle, making the surface grooves
less prominent. Nonetheless,
clear lines can be seen on many
particles. The arrow shows a particle viewed down a twofold axis.
Note the clear imaging of trimeric clusters. Scale bar: 1000 Å.
Micrograph courtesy of R. Mil-
ligan, Stanford University.

$T = 4$ icosahedral surface lattice. The characteristic pattern for such a lattice is a sequence 5-6-6 for the coordination of nodes along a lattice line. Several images displaying this criterion are included in Fig. 3. Neg-ative staining in uranyl acetate preserves the particle somewhat better (Fig. 3f), but two-sided contrast often makes the images less clearly interpretable.

Recent advances have yielded images of unstained SFV (Adrian *et al.*, 1984) and of Sindbis virus (Fig. 4) using direct transmission micros-copy of specimens in amorphous ice at liquid N_2 temperatures. The virus particles are much better preserved than in negative stain, and the $T = 4$ icosahedral surface lattice is clearly revealed in particles viewed along appropriate symmetry axes. The complete superposition of contrast through the entire particle makes other views appear confusing on sub-jective inspection, although the preservation of detail is in fact equally good.

Shadowing of frozen specimens, uncovered by deep-etching after cleavage, gives a dramatic view of the surface topography of Sindbis virus (Fig. 5); (von Bonsdorff and Harrison, 1978). The pattern of fivefold and sixfold nodes, characteristic of a $T = 4$ lattice, is again present.

The glycoproteins of Sindbis virus and SFV are thus associated into roughly triangular clusters in a $T = 4$ icosahedral lattice on the surface of the alphavirus particle. One such cluster consists of three (E1 + E2) or (E1 + E2 + E3) units, grouped about a local threefold symmetry axis. Further evidence for the organization of these trimer groups comes from analysis of hexagonal glycoprotein arrays, which are described in Section II.E. The local association of E1 and E2 as heterodimers in virions and in Triton-solubilized glycoprotein has been demonstrated by chemical cross-linking, both in SFV (Ziemiecki and Garoff, 1978) and in Sindbis

FIGURE 5. Freeze–etch images of Sindbis virus particles and of a hexagonal glycoprotein array. Scale bar: 1000 Å. From von Bonsdorff and Harrison (1978).

virus (Rice and Strauss, 1982). In the case of SFV, it is plausible to suppose that one copy of E3 is associated with each E1–E2 heterodimer, although no cross-linking is actually observed (Ziemiecki and Garoff, 1978). The cross-linking of intact Sindbis virus not only shows that E1 and E2 are closely associated, but also demonstrates very clearly that three such heterodimers form a cross-linkable cluster in the viral surface (Rice and Strauss, 1982). This clustering does not survive Triton X-100 treatment, since species larger than the heterodimer are not observed when cross-linking detergent-solubilized glycoprotein. The chemical results are thus in complete agreement with the interpretation of micrographs given above. Assuming that all sites in the lattice are occupied (vacant sites are not seen in electron micrographs), the virion must contain 240 copies of each glycoprotein chain.

Gahmberg et al. (1972) and Utermann and Simons (1974) first used proteolytic cleavage of SFV to show that small portions of E1 and E2 lie buried in the lipid bilayer. Subsequent cross-linking experiments indicated that E2 does indeed interact directly with the core (Garoff and Simons, 1974). Analysis of the amino acid sequences of SFV and Sindbis virus glycoproteins, derived from sequences of complementary DNA, shows probable transmembrane segments near the C termini of both E1 and E2 (Garoff et al., 1980a; Rice and Strauss, 1981). It is likely that these hydrophobic sequences form α-helices, since they are of suitable length

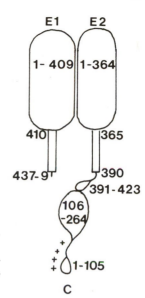

FIGURE 6. Diagram showing the approximate modular division of Sindbis virus structural proteins. Numbers refer to positions in the amino acid sequence.

to span a bilayer as a single helix. The only internal part of E1 is the dipeptide Arg-Arg. A larger segment of E2 (33 residues in Sindbis virus, 31 residues in SFV) faces the core, presumably interacting with a binding site on its surface (Rice *et al.*, 1982). The sequence of this segment is reasonably well conserved between Sindbis virus and SFV, as is the non-basic, C-terminal domain of the core polypeptide. The glycoprotein domain organization is summarized in Fig. 6.

Details of the glycoprotein structure at high resolution are not yet known. Some inferences may be made by analogy with the influenza virus glycoproteins, the structures of which have been determined by X-ray crystallography (Wilson *et al.*, 1981; Varghese *et al.*, 1983). As in the influenza virus proteins, E1 and E2 extend more than 100 Å outward from the membrane surface. Oligosaccharides (two each on E1 and E2 of Sindbis virus; one on E1 and two on E2 of SFV) are significantly "buried" in the virus, as measured by the ratio of glycosidase sensitivity of intact virus to sensitivity of detergent-solubilized glycoprotein (McCarthy and Harrison, 1977). This result suggests that as in the influenza virus proteins, some of the glycosylation sites are on lateral surfaces of the projecting protein, rather than at the outer tip. A low-pH-induced fusion activity in SFV (Helenius *et al.*, 1980) implies that a conformational change in one or both of the major glycoproteins may reveal a sequence or structure that facilities fusion of viral membranes with cell membranes to which they are bound (see Chapter 4). There is evidence that in influenza virus hemagglutinin, a dramatic conformational change occurs at low pH (Skehel *et al.*, 1982), exposing a "fusion peptide" that is normally buried in the folded structure.

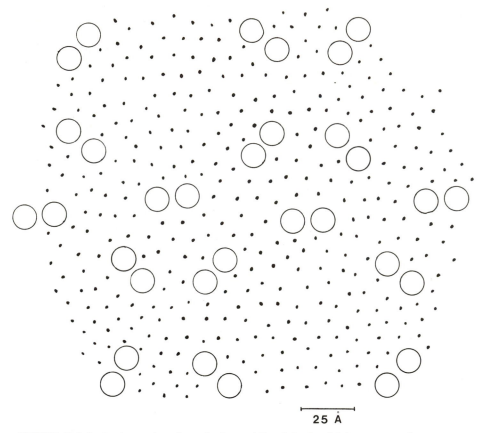

25 Å

FIGURE 7. Schematic section through the middle of the lipid bilayer in Sindbis virus or SFV. The transmembrane segments of E1 and E2 are represented as large circles, with the van der Waals radius of an α-helix in cross sections. Each small solid circle represents one phospholipid and one cholesterol head group. The molar ratio of phospholipid and cholesterol in SFV is about 1:1 (Laine *et al.*, 1973), and the area of a phospholipid–cholesterol pair is taken as 90 Å^2 (Rand and Luzzati, 1968).

C. Lipid Bilayer

The lipid composition of alphaviruses represents rather closely a random sample of plasma membrane lipids in the cells in which they are grown (Laine *et al.*, 1973). As in most enveloped viruses, there is little evidence for strong preferential exclusion or inclusion of any particular class of lipids, and by growing virus in different host cells, appropriate variation of viral lipid composition is obtained (Luukkonen *et al.*, 1976). The difference in areas of inner and outer leaflets may generate some overall compositional differences, due to the asymmetry of the cell membrane itself. An asymmetry in composition of the SFV bilayer has been shown by van Meer *et al.* (1981). Figure 7 shows a schematic cross section through the bilayer, indicating the relative areas occupied by lipid molecules [taken as equimolar phospholipid and cholesterol (Laine *et al.*, 1973)] and by transmembrane segments of the glycoprotein "roots" (assumed to be α-helical). The figure was drawn assuming a surface area of about 90 Å^2 for a phospholipid–cholesterol pair (Rand and Luzzati, 1968). We do not know whether or not the transmembrane segments of E1 and E2 are closely paired (forming for example, an α-helical coiled coil), but they are drawn here assuming some interaction.

D. Core (Nucleocapsid)

The cores of Sindbis virus and SFV may be isolated by gentle detergent treatment of virus particles. When negatively stained with uranyl acetate, cores appear in the electron microscope as spherical particles, about 400 Å in diameter. The lack of striking surface features has prevented definitive analysis of the surface lattice. The reported chemical composition of both viruses and cores is consistent with the presence of either 240 core subunits (a T = 4 lattice) or 180 subunits (T = 3). An icosahedral arrangement congruent with the glycoprotein surface lattice is a plausible structure, since one-to-one interaction of E2 "roots" with core subunits accounts in a simple way for specificity in budding (see below). Better micrographs, taken of unstained cores embedded in vitreous ice, will probably resolve the issue.

The amino acid sequences of Sindbis virus (Rice and Strauss, 1981) and SFV (Garoff *et al.*, 1980b) core proteins suggest a modular organization, similar to that of a number of plant-virus coat proteins (Harrison, 1983, 1984a,b). The N-terminal regions—about 115 residues in Sindbis virus and SFV—have sufficient positive charge to neutralize over half the approximately 55 RNA phosphates per subunit. These parts show moderate sequence similarity between the two viruses, and they appear to correspond to the flexibly tethered, inward-projecting "R domains" of icosahedral plant viruses such as tomato bushy stunt virus (Hopper *et al.*, 1984; Harrison, 1984b). There is a striking frequency of proline, as well as of lysine and arginine, in this part of the chain. The remainder of the core protein sequences, about 150 residues, do not have strong positive charges, but they do show very marked conservation between Sindbis virus and SFV. They probably correspond structurally to the parts of the subunits that form the rigid shell of the core and that interact with the internal peptide of E2.

The RNA of spherical plant viruses has been shown to be tightly condensed within a cavity defined by the inner surfaces of coat subunits. The RNA retains most or all of its secondary structure, and the helical stems may be locally well packed against each other. There is no fixed geometry of interaction with the rigid part of the coat subunit, and RNA segments bind principally to flexibly linked, N-terminal R domains. It is

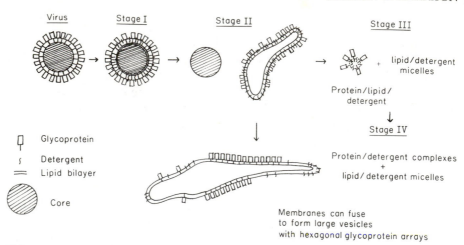

FIGURE 8. Stages of disruption of SFV with Triton X-100. Redrawn from Helenius and Söderlund (1973).

likely that this description also applies to the RNA in Sindbis virus and SFV cores.

The cores of SFV shrink about 20% in diameter when exposed to low pH (von Bonsdorff, 1973). The mechanism of this transition is not clear. Comparable treatment of Sindbis nucleocapsids (pH 6 or lower) appears to lead to dissociation (Harrison, unpublished observation).

E. Hexagonal Glycoprotein Arrays

Treatment of SFV and Sindbis virus with increasing quantities of Triton X-100 produces the different stages of virion disruption that are diagrammed in Fig. 8 (Helenius and Söderlund, 1973). Detergent adds to the bilayer, causing viral membranes to expand and ultimately to fuse. Cores appear to be expelled in the process. In the case of Sindbis virus, glycoproteins in the fused membrane vesicles can be induced to crystallize under suitable conditions, forming two-dimensional hexagonal arrays (see Fig. 5). The arrangement of glycoprotein in these arrays is similar to that found in the intact particles. The (E1 + E2) units are clustered in trimers, with six trimers grouped around each sixfold axis of the hexagonal unit cell. Identical trimers are found in the surface of the virus particle, grouped in sixes and in fives, and in that case, the fivefold groupings generate the curvature of the icosahedral surface lattice.

Formation of extensive arrays, such as shown in Fig. 8, on the surface of an infected cell might be expected to inhibit budding, since formation of fivefold vertices would require removal of a large wedge of protein. It is therefore unlikely that such arrays represent an intermediate stage in viral assembly. Their spontaneous formation does indicate, however, precise and reasonably strong glycoprotein–glycoprotein contacts.

III. VIRION ASSEMBLY

Assembly of alphaviruses is compartmentalized. Nucleocapsids form in the cytoplasm from 49 S RNA and core subunits, and the completed nucleocapsids acquire lipid and glycoprotein by budding out through the cell membrane. Host membrane proteins are efficiently excluded, presumably by the close packing of glycoproteins in the regular surface lattice.

What drives the process of viral budding? Budding involves formation of contacts between E2 cytoplasmic domains and presumptive binding sites on the nucleocapsid (see Fig. 1), as well as acquisition of glycoprotein interactions in the surface lattice. The glycoproteins tend by themselves to form hexagonal arrays, as just described, so the icosahedral geometry of the surface lattice appears to be determined by the symmetry and curvature of the nucleocapsid, rather than by glycoprotein interactions alone. There is no evidence, either morphological or biochemical, that host-cell factors or metabolic energy are required in the process (see Simons and Warren, 1984). The following self-assembly picture therefore seems to be the simplest description of likely events at the plasma membrane: A cluster of glycoproteins interacts, via a set of internal domains, with a nucleocapsid. The strongest lateral interactions are between E1 and E2 in heterodimers, and there are also strong contacts within trimes of these (E1 + E2) units. It is therefore possible that a trimer of heterodimers, presenting three internal domains to the nucleocapsid, can form a contact with the core sufficiently stable to initiate assembly. Recruitment of further glycoprotein units causes the membrane to wrap around the nucleocapsid. Addition of these units is stabilized by contacts between the E2 internal domain and the core. If the core is a 240 subunit, T = 3 structure, these contacts are identical to the initial one. If the core is a T = 3 particle, however, then not all the E2 internal domains can bind to identical sites. Some flexibility of the connection between external domains and transmembrane segments—or of the internal domains themselves—could allow a significant number of the E2 "roots" to contact sites on the core, with the remaining glycoprotein units held in the particle by lateral interactions such as those found in hexagonal arrays. The final pinching off may not require any additional mechanism, since the strain on the membrane could be sufficiently great to cause the bilayer to close around the nucleocapsid as the last glycoproteins add to the growing shell. There is, however, no definite evidence on this point.

Nucleocapsid assembly is probably analogous to assembly of simple RNA viruses, such as tomato bushy stunt and turnip crinkle viruses (Harrison, 1983; Sorger et al., 1986). In these cases, an initiation complex, composed of a defined cluster of subunits and a particular RNA "packaging sequence," appears to determine specific incorporation of viral RNA. Assembly proceeds by incorporation of coat subunit dimers, and interactions of N-terminal arms determine which of two conformations a dimer will adopt when it adds to the growing shell (Harrison, 1984a,b; Sorger *et al.*, 1986). Reassembly of Sindbis virus cores *in vitro* (Wengler *et al.*, 1982) shows that in this case also, core subunits and viral RNA are sufficient to determine correct assembly and that no further components are required. A specific model for T = 4 self-assembly has been outlined (Harrison, 1983). It involves addition of subunit *trimers* to an appropriate nucleus. It postulates two conformations for the trimer, with 20 trimers in one state and 60 in the other, and it involves, as in T = 3 self-assembly, an interaction (e.g., of N-terminal arms) that correctly determines which conformations a trimer will adopt when incorporated into the structure.

To demonstrate that the mechanisms suggested herein are correct, it is clearly important to establish conclusively the structure of the core and to examine directly the postulated interaction between the E2 internal domain and the nucleocapsid. Indeed, only by obtaining budding *in vitro* of purified components will it be possible to rule out participation of other factors in the assembly of the viral membrane.

ACKNOWLEDGMENT.   The author acknowledges support from NIH Grant CA-13202.

REFERENCES

Adrian, M., Dubochet, J., Lepault, J., and McDowell, A. W., 1984, Cryo-electron microscopy of viruses, *Nature (London)* **308:**32–36.

Gahmberg, C. G., Utermann, G., and Simons, K., 1972, The membrane proteins of Semliki Forest virus have a hydrophobic part attached to the viral membrane, *FEBS Lett.* **28:**179–182.

Garoff, H., and Simons, K., 1974, Location of the spike glycoproteins in the Semliki Forest virus membrane, *Proc. Natl. Acad. Sci. U.S.A.* **71:**3988–3992.

Garoff, H., and Söderlund, H., 1978, The amphiphobic membrane glycoproteins of Semliki Forest virus are attached to the lipid bilayer by their COOH-terminal ends, *J. Mol. Biol.* **124:**535–549.

Garoff, H., Frischauf, A.-M., Simons, K., Lehrach, H., and Delius, H., 1980a, Nucleotide sequence of cDNA coding for Semliki Forest virus membrane glycoproteins, *Nature (London)* **28:**236–241

Garoff, H., Frischauf, A.-M., Simons, K., Lehrach, H., and Delius, H., 1980b, The capsid protein of Semliki Forest virus has clusters of basic amino acids and prolines in its amino-terminal region, *Proc. Natl. Acad. Sci. U.S.A.* **77:**6376–6380.

Harrison, S. C., 1983, Virus structure: High resolution perspectives, in: *Advances in Virus Research*, Vol. 28 (M. Lauffer and K. Maromorosch, eds.), pp. 175–240, Academic Press, New York.

Harrison, S. C., 1984a, Structure of viruses, in: *The Microbe 1984*, Part 1, *Viruses* (B. W. J. Mahy and J. R. Pattison, eds.), pp. 29–73.

Harrison, S. C., 1984b, Multiple modes of subunit association in the structures of simple spherical viruses, *TIBS* **9**(1):345–351.

Harrison, S. C., David, A., Jumblatt, J., and Darnell, J. E., 1971, Lipid and protein organization in Sindbis virus, *J. Mol. Biol.* **60:**532.

Helenius A., and Söderlund, H., 1973, Stepwise dissociation of the Semliki Forest virus membrane with Triton X-100, *Biochim. Biophys. Acta* **307**:287–300.

Helenius, A., Kartenbeck, J., Simons, K., and Fries, E., 1980, On the entry of Semliki Forest virus into BHK-21 cells, *J. Cell Biol.* **84**:404–420.

Hirschberg, C. B., and Robbins, P. W., 1974, The glycolipids and phospholipids of Sindbis virus and their relation to the lipids of the host-cell plasma membrane, *Virology* **61**:602–608.

Hopper, P., Harrison, S. C., and Sauer, R., 1984, Amino acid sequence of the coat protein of tomato bushy stunt virus: Chemical determination and structural implications, *J. Mol. Biol.* **177**:701–713.

Laine, R., Hettunen, M.-L., Gahmberg, C. G., Kaariainen, L., and Renkonen, O., 1972, *J. Virol.* **10**:433–438.

Laine, R., Soderlung, H., and Renkonen, O., 1973, Chemical composition of Semliki forest virus, *Intervirology* **1**:110–118.

Luukkonen, A., Kaariainen, L., and Renkonen, O., 1976, Phospholipids of Semliki forest virus grown in cultured mosquito cells, *Biochim. Biophys. Acta* **450**:109–120.

McCarthy, M., and Harrison, S. C., 1977, Glycosidase susceptibility: A probe for the distribution of glycoprotein oligosaccharides in Sindbis virus, *J. Virol.* **23**:61.

Rand, R. P., and Luzzati, V., 1968, X-ray diffraction study in water of lipid extracted from human erythrocytes, *Biophys. J.* **8**:125–137.

Rice, C. M., and Strauss, J. H., 1981, Nucleotide sequence of the 26S mRNA of Sindbis virus and deduced sequence of the encoded virus structural proteins, *Proc. Natl. Acad. Sci. U.S.A.* **78**:2062–2066.

Rice, C. M., and Strauss, J. H., 1982, Association of Sindbis virus glycoproteins and their precursors, *J. Mol. Biol.* **154**:325–348.

Rice, E. M., Bell, J. R., Hunkapiller, M. W., Strauss, E. G., Strauss, J. H., 1982, Isolation and characterization of the hydrophobic COOH-terminal domains of Sindbis virion glycoproteins, *J. Mol. Biol.* **154**:355–378.

Simons, K., and Warren, G., 1984, Semliki Forest virus: A probe for membrane traffic in the animal cell, *Adv. Protein Chem.* **36**:79–132.

Skehel, J. J., Bayley, P. M., Brown, E. B., Martin, S. R., Waterfield, M. D., White, J. M. Wilson, I. A., and Wiley, D. C., 1982, Changes in the conformation of influenza virus haemagglutinin at the pH optimum of virus mediated membrane fusion, *Proc. Natl. Acad. Sci. U.S.A.* **79**:968.

Sorger, P. K., Stockley, P. G., and Harrison, S. C., 1986, Structure and assembly of turnip crinkle virus. II. Mechanisms of *in vitro* reassembly, *J. Mol. Biol.* (in press).

Utermann, G., and Simons, K., 1974, Studies on the amphipathic nature of the membrane proteins in Semliki Forest virus, *J. Mol. Biol.* **85**:569–587.

Van Meer, G., Simons, K., Op den Kamp, J., and van Deenen, L. L. M., 1981, Phospholipid assymmetry in Semliki Forest virus grown on BHK-21 cells, *Biochemistry* **20**:1974–1981.

Varghese, J. N., Laver, W. G., and Coleman, P. M., 1983, Structure of the influenza virus glycoprotein antigen neuraminidase at 2.9 Å resolution, *Nature (London)* **303**:35–40.

Von Bonsdorff, C.-H., 1973, The structure of Semliki forest virus, *Commentat. Biol. Soc. Sci. Fenn.* **74**:1–53.

Von Bonsdorff, C.-H., and Harrison, S. C., 1975, Sindbis virus glycoprotiens form a regular icosahedral surface lattice, *J. Virol.* **16**:141.

Von Bonsdorff, C.-H., and Harrrison, S. C., 1978, Crystalline arrays of Sindbis virus glycoprotein, *J. Virol.* **28**:578.

Wengler, G., Boege, U., Wengler, G., Bischoff, H., and Wahn, K., 1982, The core protein of the alphavirus Sindbis virus assembles into core-like nucleoproteins with the viral genome RNA and with other single-stranded nucleic acids *in vitro*, *Virology* **118**:401–410.

Wiley, D. C., and von Bonsdorff, C.-H., 1978, Three-dimensional crystals of the lipid enveloped Semliki Forest virus, *J. Mol. Biol.* **120**:375–379.

Wilson, I. A., Wiley, D. C., and Skehel, J. J., 1981, Structural identification of the antibody-binding sites of Hong Kong influenza haemagglutinin and their involvement in antigenic variation, *Nature* (*London*) **289:**373–378.

Ziemiecki, A., and Garoff, H., 1978, Subunit composition of the membrane glycoprotein complex of Semliki Forest virus, *J. Mol. Biol.* **122:**259–269.

CHAPTER 3

Structure and Replication of the Alphavirus Genome

ELLEN G. STRAUSS AND JAMES H. STRAUSS

I. INTRODUCTION

In the last few years, our knowledge of the molecular biology of viruses has been greatly expanded by the technology of nucleic acid sequencing. Determination of the complete sequence of virus genomes coupled with mapping of the virus-encoded proteins on those genomes has resulted in a wealth of information about the structure of the genome, the nature of the encoded proteins, the translation strategy used by the virus, and the nature of proteolytic processing or other modification events involved in maturation of viral proteins. Comparison of the nucleic acid and deduced amino acid sequences of related viruses can reveal conserved domains, suggesting that these regions play key roles in either virus replication or morphology. Recombinant DNA technology makes it possible to design experiments to test the function of such domains directly; in particular, manipulation of viral genomes may lead to a more directed approach to vaccine production than the empirical strategies used heretofore. In a number of cases, the single base changes (and resulting amino acid substitutions) responsible for the temperature-sensitive phenotype of certain mutants have been determined. Nucleic acid sequencing is also being used to locate immunological epitopes as well as protein domains involved in virulence and specific tissue tropisms.

The expanding data base of virus genomic sequences is beginning to allow us to explore the higher-order relationships among RNA viruses, not only within families but also among families, and to develop a better understanding of the evolution of these viruses. Genome organization,

ELLEN G. STRAUSS AND JAMES H. STRAUSS • Division of Biology, California Institute of Technology, Pasadena, California 91125.

replication strategy, and protein homologies may soon form the basis of a more rational approach to taxonomy than the morphological and serological tools in use today. Our current knowledge of alphavirus and flavivirus genomes supports the recent decision to place the flaviviruses in their own family (Flaviviridae) removed from the family Togaviridae.

There are limits of course, to what sequence information can tell us, and there is a need for more direct experiments designed to probe the biological properties and functions of virus proteins and nucleic acids—in particular the interaction of togaviruses and flaviviruses with a single cell in culture or with an animal host and its immune system and the effects of the alternate challenges of efficient replication and transmission in vertebrates and invertebrates. In addition, we need more information on the plasticity of the viral genome and the interplay of selective pressures that result in virus evolution or stasis. In this chapter, we have presented much of the recent sequence data obtained for alphaviruses and tried to interpret its significance for alphavirus structure, replication, and evolution. We have attempted to interpret much of the older literature on alphavirus replication in light of the new sequence information available, and we hope that the result will stimulate new approaches and new ideas for the study of alphavirus biology.

II. STRUCTURE OF THE ALPHAVIRUS GENOME

A. Physical Structure

The RNA genome of alphaviruses consists of one continuous chain of single-stranded RNA (ssRNA) of plus polarity. This RNA is capped on the 5' end and polyadenylated on the 3' end. The naked RNA is infectious, and thus the RNA genome alone is sufficient to initiate the complete replication cycle. All alphavirus genomic RNAs examined to date appear to be of roughly the same size. The sedimentation coefficient of the genomic RNA has been variously reported as 42–49 S; we will use 49 S in this review. The complete genome of Sindbis virus has now been sequenced (E. G. Strauss *et al.*, 1984). It is 11,703 nucleotides in length exclusive of the cap nucleotide and the polyadenylate [poly(A)] tract and has a molecular weight of 4.06 million [calculated for the Na^+ form of the RNA and including the poly(A) tract]. The base composition is 28.3% A, 20.8% U, 24.8% G, and 26.1% C.

The cap structure of the alphavirus RNA resembles that of other eukaryotic messenger RNAs (mRNAs), but differs in some details. RNA isolated from the virion has m^7G as the cap nucleotide linked through the 5'-5' triphosphate to an A residue; this A lacks the ribose methylation commonly found at this position (and the cap of alphaviruses is termed a type 0 cap) (Hefti *et al.*, 1976). RNA isolated from infected cells, on the other hand, has in addition to the predominant m^7G cap small amounts of $m^{2,7}G$ and $m^{2,2,7}G$ as cap nucleotides (HsuChen and Dubin, 1976; Dubin *et al.*, 1977). The significance of the extra methylation events is not known; di- and trimethylated caps occur in eukaryotic cells in certain small nuclear RNAs, but are not known to occur in other mRNAs. These di- and trimethylated caps are more common in the 26 S subgenomic RNA (Dubin *et al.*, 1977), the only function of which is as mRNA (see below). The alphavirus RNAs also lack the adenine methylation often seen in eukaryotic mRNAs. Because alphavirus RNA replication occurs in the cytoplasm with no known nuclear involvement, it is probable that capping is a function of the alphavirus replicase encoded by the virus and that the nature of the cap structure is a characteristic specified by this enzyme system (see also Section V.B).

The poly(A) tract found at the 3' end of alphavirus RNA (Eaton and Faulkner, 1972; Johnston and Bose, 1972) is shorter on average than that found on eukaryotic mRNAs and other virus mRNAs, averaging about 70 nucleotides. This poly(A) tract is believed to be synthesized by the alphavirus replicase from a polyuridylic acid [poly(U)] template at the 5' end of the minus strand (D. L. Sawicki and Gomatos, 1976; Frey and Strauss, 1978). Virus RNAs with shorter poly(A) tracts are infectious [although RNAs totally lacking the poly(A) are apparently not infectious] (Frey and Strauss, 1978). The average length of 70 nucleotides for the poly(A) tract appears to represent an equilibrium in which the poly(A) length can expand or contract, presumably due to inexact copying of the poly(U) tract by the replicase.

As noted, the naked alphavirus genomic RNA is infectious (Wecker, 1959). Specific infectivities up to 5×10^{-7} plaque-forming unit (PFU)/RNA molecule have been achieved (Frey and Strauss, 1978), although more efficient delivery systems for getting the RNA inside the cell would almost certainly result in higher specific infectivities. Specific infectivities for the virion up to 0.5 PFU/virion have been achieved (E. G. Strauss *et al.*, 1977).

B. Genome Organization

The genome of the alphaviruses is organized into two distinct regions. The 5' two thirds of the genome encodes the nonstructural proteins (E. G. Strauss *et al.*, 1984), which are translated as polyproteins from a messenger that is indistinguishable from the virion RNA. These precursors are then cleaved posttranslationally to produce, ultimately, four products, nonstructural protein 1 (nsP1), nsP2, nsP3, and nsP4 (Lachmi and Kääriäinen, 1976; Bracha *et al.*, 1976; Collins *et al.*, 1982; Keränen and Ruohonen, 1983; E. G. Strauss *et al.*, 1984; Lopez *et al.*, 1985), which function as the replicase/transcriptase for the virus (Clewley and Kennedy, 1976; Ranki and Kääriäinen, 1979; Keränen and Kääriäinen, 1979; Gomatos *et al.*, 1980). The 3' one third of the genome encodes the struc-

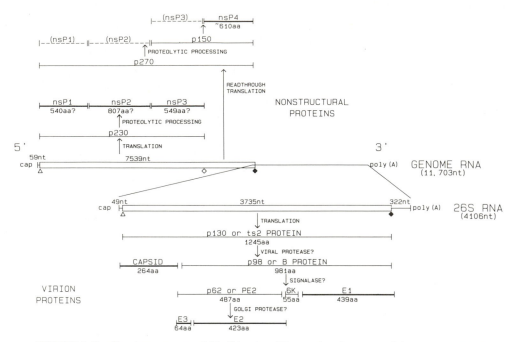

FIGURE 1. Replication strategy of Sindbis virus. Untranslated regions of the genomic RNA are shown as single lines and the translated region as a narrow open rectangle. The subgenomic RNA region is expanded below, using the same convention. All translation products are indicated, and the final protein products, both virion and nonstructural, are shown as bold lines. (△) Initiation codons; (◆) termination codons; (◇) UGA codon readthrough to produce nsP4. Reproduced from E. G. Strauss *et al.* (1984) with the permission of Academic Press.

tural proteins, i.e., the polypeptide components of the mature virion (Garoff *et al.*, 1980a,b; Rice and Strauss, 1981a). These are translated from a subgenomic mRNA, 26 S RNA (Simmons and Strauss, 1974; Cancedda *et al.*, 1974; Wengler and Wengler, 1974; Lachmi *et al.*, 1975; Clegg and Kennedy, 1975; Garoff *et al.*, 1978; Bonatti *et al.*, 1979), as a polyprotein that is also cleaved posttranslationally to form five polypeptides. Of these, the capsid protein C and the envelope glycoproteins E1 and E2 are found in the mature virion (J. H. Strauss *et al*, 1968; Schlesinger *et al.*, 1972; Kääriäinen *et al.*, 1969; Garoff *et al.*, 1974; Pedersen *et al.*, 1974), whereas the small glycoprotein E3 has been found virion-associated only in the case of Semliki Forest virus (Garoff *et al.*, 1974) and the small 6K polypeptide (Welch and Sefton, 1979) is not known to be present in any alphavirion. The genome organization of alphaviruses and the translation strategy used are illustrated in Fig. 1 for the type alphavirus, Sindbis virus.

The region between the nonstructural genes and the structural genes has been called the junction region (Ou *et al.*, 1982a; Riedel *et al.*, 1982). It probably contains signals for the initiation of transcription of the subgenomic 26 S RNA and contains about 50 untranslated nucleotides

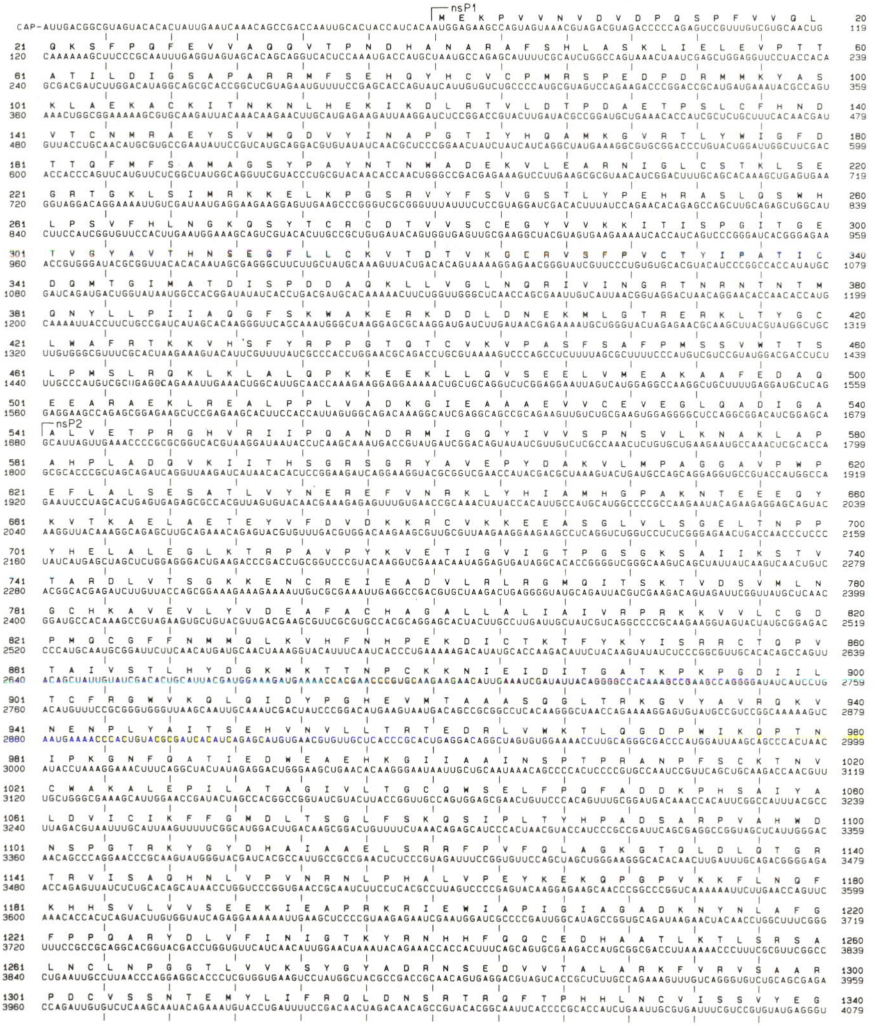

FIGURE 2. Complete nucleotide sequence of Sindbis virus. The complete translated sequence is shown (E. G. Strauss *et al.*, 1984). In the nonstructural region, amino acids are numbered from the Met encoded by nucleotides 60–62 to the end of the polyprotein at amino acid 2513. In the structural region, amino acids are numbered from the beginning of the polyprotein precursor. The start of each of the protein products is shown; for the structural-protein region, N-terminal amino acid sequencing has been used to align the proteins (Bell *et al.*, 1978, 1982; Bell and Strauss, 1981; Boege *et al.*, 1981; Bonatti and Blobel, 1979; Welch *et al.*, 1981; Mayne *et al.*, 1984); for the nonstructural proteins, the start points are provisional (E. G. Strauss *et al.*, 1984). Relevant termination codons are boxed. Single-letter amino acid code (here and in Figs. 13 and 17): (A) Ala; (C) Cys; (D) Asp; (E) Glu; (F) Phe; (G) Gly; (H) His; (I) Ile; (K) Lys; (L) Leu; (M) Met; (N) Asn; (P) Pro; (Q) Gln; (R) Arg; (S) Ser; (T) Thr; (V) Val; (W) Trp; (X) Termination; (Y) Tyr.

```
                         ┌─nsP3
1341  T R D G V G A A P S Y R T K R E N I A D C Q E E A V V N A A N P L G R P G E G V  1380
4080  ACAAGAGAUGGAGUUGGAGCCGCGCCGUCAUACCGCACCAAAAGGGAGAAUAUUGCUGACUGUCAAGAGGAAGCAGUUGUCAACGCAGCCAAUCCGCUGGGUAGACCAGGCGAAGGAGUC  4199

1381  C R A I Y K R W P T S F T D S A T E T G T A R M T V C L G K K V I H A V G P D F   1420
4200  UGCCGUGCCAUCUAUAAACGUUGGCCGACCAGUUUUACCGAUUCAGCCACGGAGACAGGCACCGCAAGAAUGACUGUGUGCCUAGGAAAGAAAGUGAUCCACGCGGCGGCCCUGAUUUC   4319

1421  R K H P E A E A L K L L Q N A Y H A V A D L V N E H N I K S V A I P L L S T G I   1460
4320  CGGAAGCACCCAGAAGCAGAAGCCUUGAAAUUGCUACAAAACGCCUACCAUGCAGUGGCAGACUUAGUAAAUGAACAUAACAUCAAGUCUGUCGCCAUUCCACUGCUAUCAACAGGCAUU   4439

1461  Y A A G K D R L E V S L N C L T T A L D R T D A D V T I Y C L D K K W K E R I D   1500
4440  UACGCAGCCGGAAAAGACCGGCUUGAAGUAUCACUUAACUGCUUGACAACCGCGCUAGACAGAACGGACGCAGAUGUCACAAUCAUAUGUCUGGAUAAGAAGUGGAAGGAAAGAAUCGAC   4559

1501  A A L Q L K E S V T E L K D E D M E I D D E L V W I H P D S C L K G R K G F S T   1540
4560  GCGGCACUCCAACUUAAGGAGUCUGUAACGGAGUUGAAGGAUGAAGAUAUGGAGAUCGAUGAUGAGUUAGUGUGGAUCCAUCCAGACAGUUGCUUGAAGGGAAGAAAGGGAUUCAGUACU   4679

1541  T K G K L Y S Y F E G T K F H Q A A K D M A E I K V L F P N D Q E S N E Q L C A   1580
4680  ACAAAAGGAAAAUUGUAUUCGUAUUUCGAAGGCACCAAAUUCCAUCAAGCAGCAAAAGACAUGGCGGAGAUAAAGGUCCUGUUCCCUAAUGAUCAGGAAAGUAAUGAACAGUUGUGUGCC   4799

1581  Y I L G E T M E A I R E K C P V D H N P S S S P P K T L P C L C M Y A M T P E R   1620
4800  UACAUAUUGGGUGAGACCAUGGAAGCAAUCCGCGAAAAGUGCCCGGUCGACCAUAACCCGUCGUCUAGCCCGCCCAAAACGUUGCCGUGCCUUUGCAUGUAUGCCAUGACGCCAGAAAGG   4919

1621  V H R L R S N N V K E V T V C S S T P L P K H K I K N V Q K V Q C T K V V L F N   1660
4920  GUCCACAGACUUAGAAGCAAUAACGUCAAAGAAGUAACAGUGUGCUCCUCCACCCCCCUUCCUAAGCACAAAAUUAAGAAUGUUCAGAAGGUUCAGUGCACGAAAGUAGUCCUGUUUAAU   5039

1661  P H T P A F V P A R K Y I E V P E Q P T A P P A Q A E E A P E V V A T P S P S T   1700
5040  CCGCACACUCCCGCAUUCGUUCCCGCCCGUAAGUACAUAGAAGUGCCAGAACAGCCUACCGCUCCUCCUGCACAGGCCGAGGAGGCCCCGGAAGUUGUAGCAACCCCAUCACCAUCUACA   5159

1701  A D N T S L D V T D I S L D M D D S S E G S L F S S F S G S D N S I T S M D S W   1740
5160  GCUGAUAAACACCUGCCUUGAUGUCACAGACAUCAUCUCAUUGGAUAUGGAUGACUCGUCCGAAGGCUCAUUGUUUUCGUCAUUCUCCGGAUCAGACAAUUCUAUCACUAGUAUGGAUUCA   5279

1741  S S G P S S L E I V D R R Q V V V A D V H A V Q E P A P I P P P R L K K M A R L   1780
5280  UCGUCAGGACCAAGUUCACUAGAGAUAGUAGACCGAAGGCAGGUGGUGGUCGCUGAUGUCCAUGCUGUCCAAGAGCCUGCCCCAAUCCCGCCUCCACGGCUCAAGAAGAUGGCCCGCCUG   5399

1781  A A A R K E P T P P A S N S S E S L H L S F G G V S M S L G S I F D G E T A R Q   1820
5400  GCAGCGGCAAGAAAAGAGCCUACACCGGCAAGCAAUAGCUCUGAGAGCUUGCACCUCUCCUUUGGCGGGGUAAGCAUGUCUCUGGGAUCAAUCUUUGACGGAGAGACUGCCCGCCAG     5519

1821  A A V Q P L A T G P T D V P M S F G S F D Q G E I D E L S R A V T E S E P V L F   1860
5520  GCAGCGGUAACAACCCCUGGCACAGGCCCCACAGAUGUGCCUAUGUCUUUCGGAUCAUUUGAUCAGGGUGUUUCCGACGAGCUAAGCCGCGCAGUCACAGAGUCAGAGCCAGUCCUGUUU   5639

1861  G S F E P G E V N S I I S S R S A V S F P L R K Q R R R R R S R R T E Y X L T G   1900
5640  GGAUCAUUUGAACCUGGGGAAGUGAACUCAAUCAUAUCGUCCCGAUCAGCCGUAUCUUUCCCACUACGCAAGCAAAGACGAAGACGAAGAGCAGGAGCAGGAUCUGAAUACAUGAGCGGG   5759
                                                      ┌──nsP4
1901  V G G Y I F S T D T G P G H L Q K K S V L Q N Q L T E P T L E R N V L E R I H A   1940
5760  GUAGGUGGGUACAUAUUCAGCACUGACACUGGGCCAGGGCAUCUGCAAAAGAAAUCGGUUCUGCAGAACCAGCUUACAGAACCGACCUUGGAGCGCAAUGUCCUGGAAAGAAUUCAUGCC   5879

1941  P V L D T S K E E Q L K L R Y Q M M P T E A N K S R Y Q S R K V E N Q K A I T T   1980
5880  CCGGUGCUCGACACGUCGAAAGAGGAACAACUCAAACUCAGGUACCAGAUGAUGCCCACCGAAGCCAAUAAGAGUCGCUACCAGAGUCGGAAAGUGGAAAAUCAAAAAGCCAUAACCACU   5999

1981  E R L L S G L R L Y N S A T D Q P E C Y K I T Y P K L Y S S S V P A N Y S D P   2020
6000  GAGCGACUACUGUCAGGACUGCGACUGUAUAACUCUGCCACAGAUCAGCCAGAAUGCUAUAAGAUCACCUAUCCGAAACUCAUGUACUCCAGUAGCGUACCGGCGAACUACUCCGAUCCA   6119

2021  Q F A V A V C N N Y L H E N Y P T V A S Y Q I T D E Y D A Y L D M V D G T V A C   2060
6120  CAGUUCGCUGUAGCUGUCUGUAACAAUUACUUGCAUGAAAAUUAUCCUACUGUAGCUAGCUAUCAGAUCACUGAUGAGUAUGAUGCAUAUUUGGAUAUGGUAGACGGGACAGUCGCCUGC   6239

2061  L D T A T F C P A K L R S Y P K K H E Y R A P N I R S A V P S A M Q N T L Q N V   2100
6240  CUGGAUACGGCAACCUUCUGCCCAGCUAAGUUAAGAUCCUAUCCGAAGAAACAUGAGUAUAGAGCCCCGAAUAUCCGCAGUGCGGUUCCAUCAGCGAUGCAGAACACGCUACAAAAUGUG   6359

2101  L I A A T K R N C N V T Q M R E L P T L D S A T F N V E C F R K Y A C N D E Y W   2140
6360  CUCAUUGCCGCAACUAAAAGAAAUUGCAUCGUCACGCAGAUGCGUGAACUGCCAACACUGGACUCAGCUACAUUCAAUGUCGAAUGCUUUCGAAAAUACGCAUGUAAUGACGAGUAUUGG   6479

2141  E E F A R K P I R I T T E F V T A Y V A R L K G P K A A A L F A K T Y N L V P L   2180
6480  GAGGAGUUCGCUCGGAAGCCAAUUAGGAUUACCACAGAGUUUGUCACGGCAUAUGUAGCUAGACUGAAAGGCCCUAAAGCCGCGCUACAUUUGCAAAGACGUAAAUUUGGCCUCAUUG   6599

2181  Q E V P M D R F V M D M K R D V K V T P G T K H T E E R P K V Q V I Q A A E P L   2220
6600  CAAGAAGUGCCUAUGGAUAGGUUCGUCAUGGACAUGAAAAGAGACGUGAAGGUAACACCAGGCACAAAGCACACAGAAGAAAGACCGAAAGUACAAGUGAUACAAGCCGCAGAACCCCUG   6719

2221  A T A Y L C G I H R E L V R R L T A V L L P N H H D F D M S A E D F D A I I A   2260
6720  GCGACAGCAUACUUAUGUGGGAUUCACCGGGAAUUAGUGCGUAGGCUAACAGCUGUUCUGCUCCCAAAUCAUCACGACUUUGAUAUGUCGGCAGAAGACUUUGAUGCGAUUAUAGCA     6839

2261  E H F K Q G D P V L E T D I A S F D K S Q D A M A L T G L M I L E D L G V D Q   2300
6840  GAACACUUCAAGCAAGGCGAUCCGGUACUGGAGACAGAUAUCGCGAGCAUCGAUAAAUCGCAAGAUGCGAUGGCUUAUGGCCUUUAUGGGGACCUUGGUGUUGAUCAA           6959

2301  P L L D L I E C A F G E I S S T H L P T G T R F K F G A M M K S G M F L T L F V   2340
6960  CCACUACUGGACUUGAUCGAAUGUGCGUUUGGCGAGAUUAGCUCCACCCAUCUACCUACGGGACUCAGGUUCAAAUUUGGGGCGAUGAUGAAAUCGGGAAUGUUUCUCACUCUUUUUGUC   7079

2341  N T V L N V V I A S R V L E E R L K T S R C A A F I G D D N I I H G V V S D K E   2380
7080  AACACAGUUUUGAAUGUCGUUAUCGCCAGCAGAGUACUAGAAGAGCGGCUUAAAACGUCCAGAUGUGCAGCGUUCAUUGGCGAUGACAACAUCAUCCAUGGAGUAGUAUCUGACAAAGAA   7199

2381  M A E R C A T W L N M E V K I I D A V I G E R P P Y F C G G F I L Q D S V T S T   2420
7200  AUGGCUGAGAGGUGCGCCACCUGGCUCAACAUGGAGGUUAAGAUCAUCGACGCAGUCAUCGGUGAACGACCCCCGUAUUUCUGCGGCGGAUUCAUACUACAAGACUCGGUUACUUCCACA   7319

2421  A C R V A D P L K R L F K L G K P L P A D D E Q D E D R R A L L D E T K A W F   2460
7320  GCGUGCCGCGUGGCGGAUCCCCUGAAAAGGCUGUUUAAAUUGGGUAAACCCCUCCCAGCCGACGAGCAAGAUGAAGAUCGUCGCGCAUUACUUGAUGAGACUAAAGCGUGGUUC     7439

2461  R V G I T G T L A V A V T T R Y E V D N I T P V L L A L R T F A Q S K R A F Q A   2500
7440  AGAGUAGGUAUAACAGGCACGCUAGCUGUGGCCGUGACCACACGUUAUGAGGUAGACAAUAUUACACCUGUUUUACUGGCACUCAGGACUUUUGCCCAGUCCAAGCGAGCCUUCCAAGCC   7559

2501  I R G E I K H L Y G G P K ┌──26S RNA BEGINS─────────────────M N R G F F N M L G R   11
7560  AUCAGAGGGGAAAUCAAACAUCUCUACGGUGGUCCUAAAUAGUCAGCAUAGUACAAUUUCAACUGUCAUUUGAUAGUACCACCAUGAAUAGAGGAUUCUUUAACAUGCUCGGCCGC     7679
                                              C
12  R P F P A P T A M W R P R R R R Q A A P M P A R N G L A S Q I G Q L T T A V S A   51
7680  CGCCCCUUCCCGGCCCCAACUGCCAUGUGGAGGCCGCGGAGAAGGCAGGCGGCCCCGAACGGGCUGGCCUCACAAAUCGGGCAAUUAACUACUGCAGUUUCAGCU           7799

52  L V I G Q A T R P Q P P R P P P R P Q K K Q A P K Q P P K P K K P K T Q E K K   91
7800  CUAGUCAUUGGACAAGCAACUAGACCUCAGCCUCCUCGGCCGCCACGCCCGCCUCAGAAGAAGCAGGCCCCGAAGCAGCCGCCAAAGCCGAAGAAGCCCAAAACUCAGGAGAAGAAG     7919

92  K K Q P A K P K P G K R Q R M A L K L E A D R L F D V K N E D G D V I G H A L A   131
7920  AAGAAGCAACCUGCAAAACCCAAACCCGGAAAGAGACAGCGCAUGGCACUCAAGUUGGAGGCCGACAGAUUGUUCGACGUCAAGAACGAGGACGGAGAUGUCAUCGGGCACGCACUGGCC   8039

132  M E G K V M K P L H V K G T I D H P V L S K L K F T K S S A Y D M E F A Q L P V   171
8040  AUGGAAGGAAAGGUAAUGAAACCUCUGCACGUGAAAGGAACCAUCGAUCACCCUGUGCUAUCAAAGCUCAAAUUCACCAAGUCGUCAGCAUACGACAUGGAGUUCGCACAGCUGCCAGUC   8159
```

FIGURE 2. (*continued*)

that form the 5' end of 26 S RNA. The compactness of the genome is illustrated by the fact that the stop codon for the nonstructural read-through polyprotein (see below) actually lies within the region transcribed into 26 S RNA. Depending on the virus, in addition to 40–50 nucleotides untranslated in the junction region (Ou *et al.*, 1982a), there are 60–80 nucleotides untranslated at the 5' end of the genome (Ou *et al.*, 1983; Lehtovaara *et al.*, 1982) and 121–524 nucleotides at the 3' end (Garoff *et al.*, 1980b; Ou *et al.*, 1982b; Rice and Strauss, 1981a; Dalgarno *et al.*, 1983; R. M. Kinney, B. J. Johnson, and D. W. Trent, personal communication).

The complete translated sequence of the genomic RNA of Sindbis

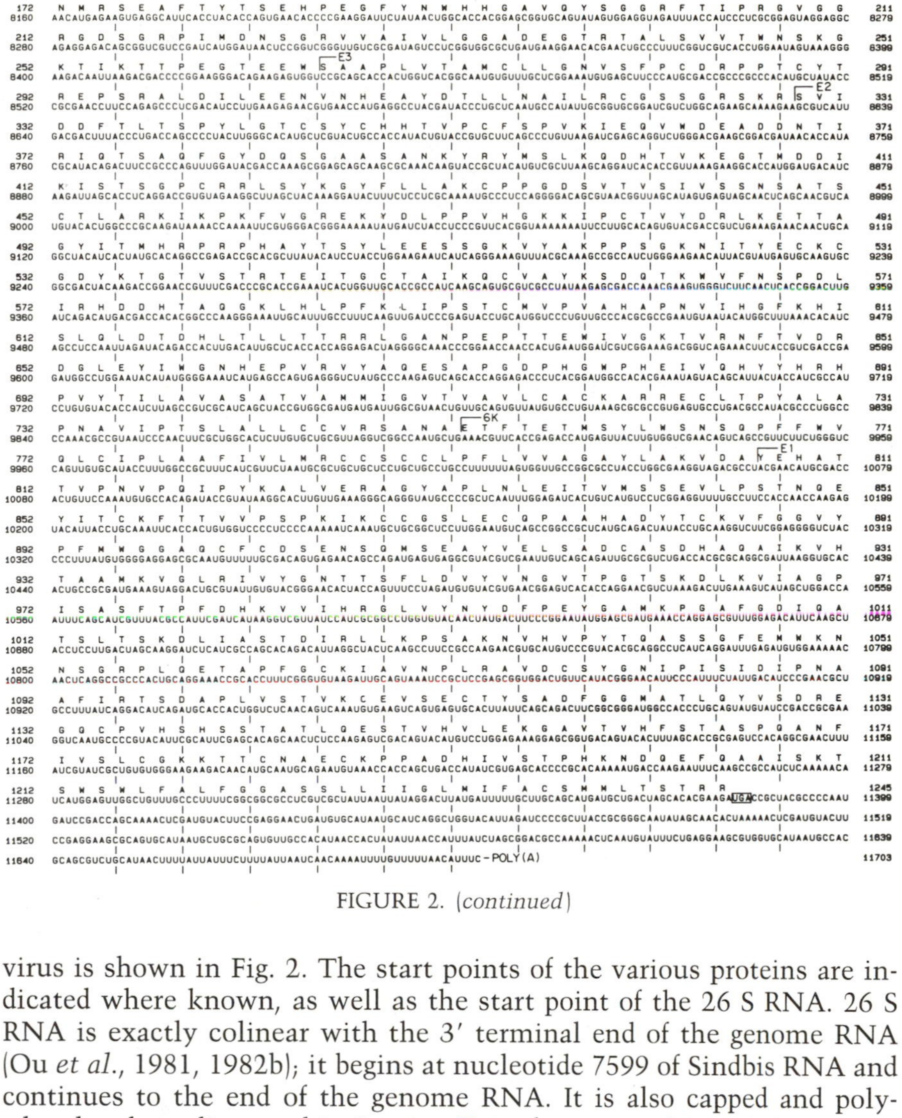

FIGURE 2. (continued)

virus is shown in Fig. 2. The start points of the various proteins are indicated where known, as well as the start point of the 26 S RNA. 26 S RNA is exactly colinear with the 3' terminal end of the genome RNA (Ou *et al.*, 1981, 1982b); it begins at nucleotide 7599 of Sindbis RNA and continues to the end of the genome RNA. It is also capped and polyadenylated; as discussed in Section II.A, the cap on 26 S RNA consists of a mixture of mono-, di-, and trimethylated G. The poly(A) tract has the same length distribution as that of genome RNA. The transcription of 26 S RNA from a full-length minus strand, and its capping and polyadenylation, are believed to be carried out by the viral replicase/transcriptase formed by the nonstructural proteins (see Section V, A–C).

Another way of visualizing the overall genome structure is the open reading frame analysis presented in Fig. 3. The entire nonstructural region is open in reading frame 1, except for the opal stop codon at nucleotides 5748–5750 (discussed in more detail in Section III.B). The first methionine (*) in this open reading frame is encoded by nucleotides 60–62. Note also

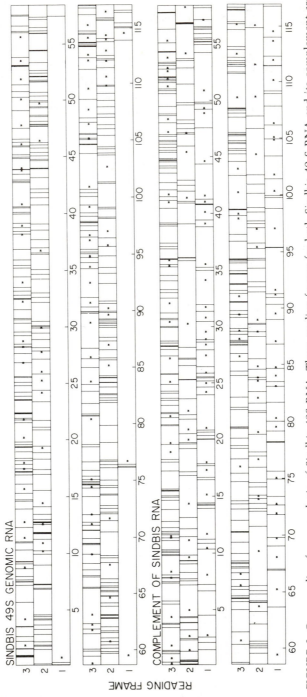

FIGURE 3. Open reading frame analysis of Sindbis 49S RNA. Three reading frames for both Sindbis 49 S RNA and its complement are shown. (*) First methionine encoded in each open frame. Vertical lines are termination codons. The horizontal scale indicates the number of nucleotides ($\times 10^{-2}$).

that there are three closely spaced termination codons in the junction region before the open reading frame begins again at nucleotide 7647 for the structural polyprotein. In the case of Sindbis, the polyprotein translated from 49 S RNA and that from the 26 S region are in phase; this, however, is the exception, since these two open reading frames are out of phase in three other alphaviruses examined (Ou *et al.*, 1982a). There are no other long open reading frames in the Sindbis genome, in either the plus strand or the minus strand, and there is no reason to believe that any protein products are translated from alphavirus RNAs other than those encoded in these two very long open reading frames. Furthermore, within the 26 S region, open reading frame analysis for four alphavirus RNAs has shown that the short open reading frames that could potentially encode minor virus proteins are not conserved, and thus these are almost certainly not translated.

In contemplating the structure of the genome and the gene order, it is obvious that the order of the nonstructural vs. the structural genes in alphaviruses is dictated by the fact that the genomic RNA must be translated to produce the viral replicase and thus that these genes must be 5' terminal. The structural proteins can be translated only from the subgenomic 26 S mRNA, which requires prior translation of the transcriptase necessary for 26 S RNA transcription.

III. NONSTRUCTURAL PROTEINS

A. Translation and Processing

The four nonstructural proteins, or some subset of these proteins, form a replicase/transcriptase that is responsible for transcribing a minus-strand copy of the genomic RNA and for transcribing genomic RNA and 26 S subgenomic RNA from this minus-strand copy. In addition, one or more host-cell proteins may be integral components of this enzyme system (see Section V.D). Because the naked RNA is infectious, the viral components of this enzyme system must be translated from the infecting 49 S RNA genome.

Four final nonstructural polypeptides have been described in cells infected with Semliki Forest virus (Lachmi and Kääriäinen, 1976; Keränen and Ruohonen, 1983), and the complete sequence of Sindbis virus RNA (E. G. Strauss *et al.*, 1984), together with results on proteins found in infected cells (Brzeski and Kennedy, 1977; Schlesinger and Kääriäinen, 1980; Lopez *et al.*, 1985) or translated *in vitro* (Collins *et al.*, 1982), also suggests that there are four final nonstructural proteins in Sindbis-infected cells. We have referred to these four polypeptides as nsP1, nsP2, nsP3, and nsP4 in the order in which they are encoded in the genome from 5' to 3' (E. G. Strauss *et al.*, 1984).

Sequence data for Sindbis virus and Middelburg virus (E. G. Strauss *et al.*, 1983, 1984), as well as from precursor polyproteins described in infected cells (Clegg *et al.*, 1976; Kääriäinen *et al.*, 1978; Schlesinger and Kääriäinen, 1980; Lopez *et al.*, 1985) or during *in vitro* translation (Collins *et al.*, 1982), lead to the conclusion that the nonstructural proteins of Sindbis and Middelburg viruses, and by inference of most alphaviruses, are translated as two polyproteins (See Fig. 1). The first polyprotein contains the sequences of nsP1, nsP2, and nsP3 and terminates at an opal (UGA) termination codon. In the case of Sindbis virus, this precursor polyprotein is 1896 amino acids in length. Cleavage of this polyprotein to produce nsP1, nsP2, and nsP3 is believed to be catalyzed by a virus-encoded protease probably located within one of these three proteins. In the case of Sindbis virus, the two cleavages required both probably occur within the sequence Gly-Ala ↓ Ala at the point denoted with the arrow (E. G. Strauss *et al.*, 1984). The second polyprotein is produced by read-through of the opal termination codon (Collins *et al.*, 1982; Lopez *et al.*, 1985) to produce a polyprotein 2513 amino acids long containing the sequences of nsP1, nsP2, nsP3, and nsP4. On cleavage, possibly by the same virus-encoded protease, nsP4 is produced (presumably nsP1, nsP2, and nsP3 are also produced, but these are not expected to contribute appreciably to the pool of these products). Thus, nsP4, which requires readthrough of an opal termination codon, is produced in smaller amounts than the other nonstructural proteins. In Sindbis-infected cells, this protein cannot be seen except by specific immunoprecipitation (Lopez *et al.*, 1985); in Semliki-Forest-virus-infected cells, on the other hand, nsP4 has been seen without immunoprecipitation and appears to be produced in relatively larger amounts (Keränen and Ruohonen, 1983). The recent finding that there is no opal codon in the Semliki forest virus genome between nsP3 and nsP4 (K. Takkinen, personal communication) is thus of great interest. Why Semliki Forest virus differs in what would appear to be a fundamental landmark of the alphavirus genome (Strauss *et al.*, 1983) is unclear, especially given that Middelburg virus is closely related to Semliki Forest virus (Ou *et al.*, 1982a; Bell *et al.*, 1984). It would be of considerable interest to determine if the presence or absence of the opal codon to regulate production of nsP4 affects the host range or virulence of the virus. Because nsP4 is produced in lesser relative amounts and is highly conserved between Sindbis and Middelburg virus, it has been postulated to have a regulatory role in virus RNA replication (E. G. Strauss *et al.*, 1983) (see also Section VI.D).

B. Opal Codon

The mechanism by which readthrough of the opal codon occurs is unknown at present. Naturally occurring opal suppressor transfer RNAs (tRNAs) have been isolated from rabbit reticulocytes (Geller and Rich, 1980), bovine liver (Diamond *et al.*, 1981), and chicken liver (Hatfield *et al.*, 1983) that insert either tryptophan (rabbit tRNA) or serine (bovine or chicken tRNA) in response to the opal codon. These suppressor tRNAs are present in small concentrations and lead to limited suppression of opal codons (the chicken gene has been shown to be single-copy). Furthermore, the serine tRNA can be phosphorylated, and phosphoserine may be inserted. Geller and Rich (1980) have postulated that these naturally occurring suppressor tRNAs function to produce small amounts of readthrough products required for normal cell functions or for developmental pathways. It is possible that such opal suppressors are widespread in animal cells and that the alphaviruses take advantage of this cellular pathway to regulate their growth cycle by producing nsP4 in small amounts by readthrough. If this model is correct, then the concentrations of the opal suppressor tRNA could influence virus growth and affect the host range or tissue tropisms of a particular alphavirus. It is also conceivable that virus infection could induce the synthesis of opal suppressor tRNAs. As discussed in Section IIIA, Semliki Forest virus appears to be exceptional in that it lacks the modulating opal codon and produces relatively larger amounts of nsP4 than does Sindbis virus or Middelburg virus or, by inference, most alphaviruses. How this affects the growth cycle or epidemiology of the virus is unknown.

Although the use of naturally occurring suppressor tRNAs to suppress the opal termination codon in alphavirus RNAs is an appealing hypothesis, it is also possible that suppression could be caused by misreading of the codon, especially by tRNA^{Trp}. Such misreading is known to occur in bacteria, leading to insertion of tryptophan at UGA codons about 3% of the time (Hirsh and Gold, 1971), and in fact is required for the production of minor products necessary for replication of phage Qβ (Weiner and Weber, 1971) and of other phages (Engelberg-Kulka *et al.*, 1979). It is not known whether such misreading also occurs in eukaryotes.

The use of readthrough of a stop codon to regulate the production of replicase proteins is not unique to the alphaviruses. A number of plant viruses, including tobacco mosaic virus (TMV) and turnip crinkle virus, use readthrough of an amber termination codon to produce small amounts of a replicase product [although in the plant virus, no posttranslational cleavage events take place to separate the various domains of the replicase (reviewed in E. G. Strauss and J. H. Strauss, 1983)]. Intriguingly, the readthrough product of TMV shares amino acid homology with nsP4 (Haseloff *et al.*, 1984) suggesting that these two viruses are descendants of a common ancestor that have retained this mechanism for regulating replicase production and that plants and animals may share the property of partial suppression of amber or opal codons.

An amber codon also punctuates the retrovirus genome between the *gag* and *pol* genes (Shinnick *et al.*, 1981). Readthrough may be used to produce small amounts of the *pol* gene product, although a splicing event to produce a minor mRNA for *pol* that has deleted the stop codon has not been completely ruled out.

C. Functions of Nonstructural Proteins

Four activities can be distinguished in the alphavirus replicase/transcriptase: (1) initiation of minus-strand synthesis, (2) initiation of genome-length plus-strand synthesis, (3) initiation of subgenomic 26 S RNA synthesis, and (4) elongation and completion of initiated chains. It is possible that these activities are found in different nonstructural polypeptides, although there is no evidence for this at present. Catalogues of temperature-sensitive (ts) mutants have been isolated for Sindbis, Semliki Forest, and Western equine encephalitis viruses (reviewed in E. G. Strauss and J. H. Strauss, 1980). These have been divided into RNA$^+$ mutants, which are capable of synthesizing RNA at the nonpermissive temperature and hence are presumed to have defects in structural proteins, and RNA$^-$ mutants, which synthesize reduced amounts of RNA at the nonpermissive temperature and are presumed to have defects in the nonstructural proteins that form the replicase. The RNA$^-$ mutants of Sindbis virus have been grouped by complementation assay into four groups. The presence of four compolementation groups and four nonstructural polypeptides is intellectually satisfying and suggests that each complementation group defines mutations in one of the nonstructural proteins, but this has yet to be shown and may not be correct. The replicase activities of the nonstructural proteins will be discussed in more detail in Section V, A–C.

As noted earlier, the nonstructural proteins are produced as polyproteins that are cleaved posttranslationally. There is no compelling evidence to date of host proteases that function in the cytosol to process virus polyproteins or other protein precursors (although host proteases localized in subcellular organelles have been described, some of which are probably active in processing viral precursor proteins). On the other hand, there are numerous examples of virus-encoded proteases (Palmenberg et al., 1979; von der Helm, 1977; Bhatti and Weber, 1979; Hahn et al., 1985). It seems intuitively unlikely that so many different viral proteases would have evolved if cellular proteases were readily available to perform such functions. Hence, it appears likely that the cleavages of the nonstructural polyproteins will be catalyzed by a virus-encoded protease, and we have postulated that all cleavages of viral precursor proteins that occur in the cytosol are catalyzed by virus-encoded enzymes (Rice and Strauss, 1981a). This viral protease is presumably contained within nsP1, nsP2, or nsP3. An appealing hypothesis is that the protease resides in nsP3, because this protein, unlike nsP1, nsP2, and nsP4, shares no amino acid sequence homology with the nonstructural proteins of several plant viruses (discussed in Section VI.D) that appear to lack proteolytic activity, but its exact location has yet to be determined. Expression of the nonstructural region of the genome, with or without modification, from cloned copies of the genes inserted into either eukaryotic or prokaryotic expression vectors and/or mapping of *ts* mutants that lead to accumulation of uncleaved precursors could be used to map the (hypothetical) virus nonstructural protease.

IV. STRUCTURAL PROTEINS

A. Translation and Processing

The structural proteins are translated from a subgenomic 26 S mRNA as shown schematically in Fig. 1. This RNA has been sequenced in the case of four alphaviruses: Sindbis virus (Rice and Strauss, 1981a,b), Semliki Forest virus (Garoff *et al.*, 1980a,b), Ross River virus (Dalgarno *et al.*, 1983) and Venezuelan equine encephalitis virus (R. M. Kinney, B. J. Johnson, and D. W. Trent, personal communication), and in each case is exactly 3' coterminal with the genomic RNA (Ou *et al.*, 1981, 1982a; Riedel *et al.*, 1982). The length of the 26 S RNA, exclusive of the 5' cap and the 3' poly(A), is 4106 nucleotides in Sindbis, 4074 in Semliki forest, 4334 in Ross River virus, and 3913 in Venezuelan equine encephalitis virus [molecular weights including the poly(A) are $1.43 - 10^6$, $1.43 - 10^6$, $1.52 - 10^6$ and $1.35 - 10^6$, respectively, expressed as the Na^+ form]. The differences in length reflect primarily the variability in the length of the 3' untranslated region which is 322 nucleotides long in Sindbis, 264 in Semliki Forest, 524 in Ross River, and 121 in Venezuelan equine encephalitis virus. [The 3' untranslated region of an additional alphavirus, Western equine encephalitis virus, is known to be 300 nucleotides (Ou *et al.*, 1982b).]

Use of a subgenomic mRNA to produce the structural proteins of the alphaviruses allows for the production of these proteins in great molar excess over the nonstructural (replicase) proteins required to replicate virus RNA. Examination of mRNAs found on polysomes has shown that more than 90% of virus-specific messenger is 26 S, the rest being genome-length RNA (reviewed in J. H. Strauss and E. G. Strauss, 1977; Schlesinger and Kääriäinen, 1980). The preponderance of 26 S subgenomic mRNA in the polyribosome population is due to the fact that 26 S RNA is produced in a 3-fold molar excess and to the fact that genomic 49 S RNA is quickly sequestered into nucleocapsids inside the infected cell. In addition, 26 S RNA appears to be more efficiently translated, so that the end result is that the structural proteins are quite prominent in infected cells, whereas the nonstructural proteins are produced in much lower amounts. Because several hundred copies of each structural protein are required for each RNA molecule encapsidated into a virion, such amplification of structural proteins leads to a more efficient use by the virus of the available resources of an infected cell. Replication of alphaviruses in vertebrate cells is indeed quite efficient, with up to 10^5 virus particles released per cell during a normal infection cycle.

Translation of 26 S RNA, as of 49 S RNA, produces a polyprotein that is cleaved to produce, ultimately, five polypeptides. Cleavage begins while the polyprotein is nascent, and the complete precursor polyprotein is never seen except in the case of certain *ts* mutants that are defective in an early cleavage step (reviewed in J. H. Strauss and E. G. Strauss, 1977; Schlesinger and Kääriäinen, 1980). The first cleavage event removes the capsid protein from the N terminus of the nascent polyprotein. This event is believed to be autocatalytic, carried out by the capsid protein itself (Simmons and Strauss, 1974; Scupham *et al.*, 1977; Aliperti and Schlesinger, 1978). In agreement with this hypothesis, mapping of *ts* mutants that fail to cleave the capsid protein from the polyprotein precursor, and that therefore accumulate a structural polyprotein representing the complete translated product of 26 S mRNA, has shown that the *ts* lesions lie in the C-terminal half of the capsid protein (Hahn *et al.*, 1985) (see Section IV.C). It has been postulated that the alphavirus capsid protein precursor is a serine protease the catalytic triad of which is formed by His-141, Asp-147, and Ser-215 in Sindbis virus (Hahn *et al.*, 1985). It is possible that during folding of the protein during synthesis, the substrate Trp at the C terminus of the mature capsid protein is brought into the active site and that after the autocatalytic cleavage occurs, the protease activity is diminished or lost when the protein-folding changes. The requirements for autocatalysis can now be tested by recombinant DNA techniques using constructs in which only parts of the polyprotein can be translated or in which site-specific mutagenesis has been used to alter certain residues and examining the protein produced for autoproteolytic activity. Mapping of additional *ts* mutants would also be useful to further explore the nature of the proteolytic activity.

Removal of the capsid protein from the N terminus of the polyprotein allows an N-terminal signal sequence to function and leads to insertion of the nascent chain (and thus PE2) into the endoplasmic reticulum (Wirth *et al.*, 1977; Garoff *et al*, 1978; Bonatti *et al.*, 1979). The 6K polypeptide (Welch and Sefton, 1979) located between PE2 and E1 (Garoff *et al.*, 1980b; Rice and Strauss, 1981a) evidently functions as an internal signal sequence that leads to the insertion of E1 into the endoplasmic reticulum (Hashimoto *et al.*, 1981). These two proteins are core-glycosylated as they are inserted (Sefton, 1977) and are then transported through the Golgi apparatus to the cell surface (Chapter 5). Both glycoproteins are oriented with the N terminus and the bulk of the protein outside the membrane bilayer and a transmembrane anchor located at the C terminus (E1) or near it (PE2) (Garoff and Söderlund, 1978; Rice *et al.*, 1982; Wirth *et al.*, 1977; Blobel, 1980).

The enzyme responsible for cleaving after PE2 and after the 6K polypeptide to separate PE2 from E1 is not known with certainty at present. Both cleavages occur after alanine (see Fig. 2) in the case of all four alphaviruses the 26 S RNAs of which have been sequenced to date. It is probable that signalase, the enzyme responsible for cleaving signal sequences from nascent polypeptides in the lumen of the endoplasmic reticulum, is responsible for one or both cleavages (Garoff *et al*,. 1980b; Rice and Strauss, 1981a). Alanine is known to be one of the preferred substrates for this enzyme (reviewed in Blobel *et al.*, 1979), and the cleavage after the 6K polypeptide appears analogous to the removal of a signal peptide from an integral membrane glycoprotein. The cleavage after PE2 is more problematical. If signalase is responsible for this cleavage, it implies that the C terminus of PE2 is external to the membrane, at least transiently (Rice and Strauss, 1981a). A second possibility is that the virus nonstructural protease, which appears to cleave after Gly-Ala (E. G. Strauss *et al.*, 1984), is responsible, although this seems unlikely in view of expression experiments in which the structural proteins are synthesized and cleaved in the absence of nonstructural proteins (Kondor-Koch *et al.*, 1982; Huth *et al.*, 1984; Rice *et al.*, 1985). Finally, it is possible that an unidentified cellular or viral protease performs this cleavage.

B. Functions of Structural Proteins

The nucleocapsid protein binds within 5–7 min after synthesis (Söderlund, 1973) to genome-length RNA and assembles it into a nucleocapsid that apparently possesses icosahedral symmetry. Newly synthesized capsid protein has also been found in association with the large ribosomal subunit (Ulmanen *et al.*, 1976), which might serve as a form of storage or as a precursor to nucleocapsid-associated capsid protein, and has been reported to interfere with host protein synthesis (Van Steeg *et al.*, 1984). The capsid protein possesses two domains: (1) The N-terminal part of the molecule is highly basic, containing strings of Arg or Lys residues, and appears in general to lack a defined tertiary structure, since there are in this region many prolines, which break α-helices, and most of this region shows little conservation among alphaviruses (Garoff *et al.*, 1980a; Rice and Strauss, 1981a; Dalgarno *et al.*, 1983; R. M. Kinney, B. J. Johnson, and D. W. Trent, personal communication) (see also Section VI.A). It probably interacts with the RNA in a nonspecific fashion through electrostatic bonds and may penetrate into the interior of the nucleocapsid. There is one domain within the N-terminal region that does show reasonable conservation, however, corresponding to amino acids 39–52 of Sindbis virus, and that could have a defined function such as a specific interaction with the genomic RNA to initiate encapsidation (as opposed to nonspecifc RNA–protein interactions that might occur during capsid assembly after a specific interaction event). (2) The C-terminal half of the molecule, in contrast, is highly conserved among alphaviruses and probably possesses a highly defined structure. Functions ascribed to this region include subunit–subunit interactions involved in assembling the nucleocapsid, capsid–glycoprotein interactions involved in virus budding, an autoproteolytic activity, and possibly a specific interaction with the genomic RNA to initiate encapsidation.

The N-terminal domain of E3 serves as an uncleaved signal sequence (Garoff *et al.*, 1978; Bonatti *et al.*, 1979). The function or functions of the C-terminal region of E3 are unclear. Presumably, the production of the precursor PE2 and its subsequent cleavage to produce E2 serves an important function in virus assembly. The cleavage event occurs after the consensus sequence Arg-X-(Arg/Lys)-Arg (Dalgarno *et al.*, 1983), where X can vary, and is similar to the cleavages undergone by glycoproteins of several other viruses both in timing and in the recognition sequence (reviewed in E. G. Strauss and J. H. Strauss, 1983, 1985). It has been postulated to be mediated by a protease located in the Golgi apparatus (Garoff *et al.*, 1980b; Rice and Strauss, 1981a).

Glycoproteins E1 and E2 form the outer antigens of the virus. They are believed to form a functional dimer (Bracha and Schlesinger, 1976; Jones *et al.*, 1977), and E1 and E2 can be specifically cross-linked to one another in the virion or in the infected cell to form a heterodimer (Ziemiecki *et al.*, 1980; Rice and Strauss, 1982). After transport to the plasma membrane, they are acquired when the nucleocapsid buds through the cell surface and is enveloped in a lipoprotein envelope (see Chapter 2). The interaction between capsid protein and glycoproteins appears to be quite specific: Proteins other than alphavirus glycoproteins are excluded from the envelope (E. G. Strauss, 1978), and the ratio of proteins in the virion is equimolar. It has been suggested that the cytoplasmic domain of glycoprotein E2 binds specifically to the capsid protein, the result being incorporation of one E1–E2 heterodimer for each capsid protein subunit. Analysis of the N-terminal regions of the glycoproteins of eight alphaviruses has shown that although they are variable in amino acid sequence, residues important in secondary structure formation, such as cysteine, proline, glycine, and aromatic amino acids, tend to be conserved, suggesting that the three-dimensional structures of all the alphavirus E1s and E2s are similar, while considerable variation in primary amino acid sequence is present (Bell *et al.*, 1984).

Although the E1–E2 heterodimer appears to be the functional unit in the alphavirion, attempts have been made to assign certain functions to domains of either E1 or E2. Thus, glycoprotein E2 appears to carry the major neutralization epitope of alphaviruses (Dalrymple *et al.*, 1976), although a neutralization epitope is also found on E1 (Schmaljohn *et al.*, 1982; Chanas *et al.*, 1982). In fact, E2 is less highly conserved throughout its length than E1 and may be involved in generating new strains of alphaviruses (reviewed in J. H. Strauss and E. G. Strauss, 1985). On the other hand, glycoprotein E1 is postulated to carry the fusion activity of the virus in a conserved, hydrophobic domain (Garoff *et al.*, 1980b; Rice and Strauss, 1981a). It may also carry the site involved in cell attachment, since for three alphaviruses, Sindbis (Dalrymple *et al.*, 1976), Semliki Forest (Helenius *et al.*, 1976), and chikungunya (Simizu *et al.*, 1984), isolated E1 can hemadsorb. However, various antibodies to either E1 or E2 of different alphaviruses will block hemagglutination (Dalrymple *et al.*, 1976; Chanas *et al.*, 1982; France *et al.*, 1979), and specific attachment to host cells may therefore be a property of the E1–E2 dimer rather than of glycoprotein E1 as postulated above.

C. Mapping of Temperature-Sensitive Mutants

Temperature-sensitive mutants that make normal amounts of RNA at the nonpermissive temperature have been isolated for a number of alphaviruses, and these mutants have therefore been assigned to the structural protein region of the genome. In the case of Sindbis virus, these RNA$^+$ mutants have been divided by complementation into three groups (Burge and Pfefferkorn, 1966; reviewed in E. G. Strauss and J. H. Strauss, 1980). Complementation Group C fails to cleave the precursor polyprotein at the nonpermissive temperature (J. H. Strauss *et al.*, 1969; Schlesinger and Schlesinger, 1973; Simmons and Strauss, 1974) and was assigned to the capsid protein. Cells infected with complementation Group D mutants fail to hemadsorb at the nonpermissive temperature (Burge and Pfefferkorn, 1968) because the glycoproteins are not transported to the cell surface (Saraste *et al.*, 1980), and this group was assigned to glycoprotein E1. Finally, Group E mutants were assigned to glycoprotein E2 in part by elimination and in part because PE2 is not cleaved at the nonpermissive temperature; in the only known Group E mutant, the glycoproteins are transported to the cell surface, but PE2 is not cleaved and no virions bud. Nucleotide sequencing has now shown that these initial assignments were correct. Arias *et al.* (1983) sequenced two Group D mutants, *ts*10 and *ts*23, of Sindbis virus, Lindqvist *et al.* (submitted) sequenced the single Group E mutant, *ts*20; and Hahn *et al.* (1985) sequenced three Group C mutants, *ts*2, *ts*5, and *ts*13. The results are summarized in Fig. 4. In the case of the Group C mutants, the mutations mapped are three residues removed from the serine or the histidine postulated to form part of the catalytic triad of a serine protease (Hahn *et al.*, 1985). In the case of Group D, widely spaced mutations were found that render the glycoproteins unable to be transported, suggesting that many different changes can alter the overall conformation that is important for proper transport. In the case of the Group E mutant, the change must identify a domain of PE2(E2) crucial for function. It is of interest that all revertants isolated from these mutants showed same-site reversion to the original amino acid (or, in one case, same-site reversion to a functionally equivalent amino acid, arginine in place of lysine) and that in every case the *ts* mutation is a change in an amino acid that is invariant among Sindbis, Semliki Forest, and Ross River viruses (in the case of Venezuelan equine encephalitis virus, these amino acids are also conserved, with the exception of that involved in the *ts*20 mutation). Conditional lethal mutations,

A. SINDBIS VIRUS GLYCOPROTEIN E1

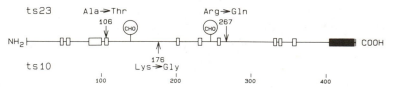

B. SINDBIS VIRUS GLYCOPROTEIN E2

C. SINDBIS VIRUS CAPSID PROTEIN

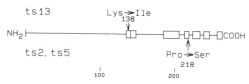

FIGURE 4. Summary of locations of *ts* mutants of complementation groups C, D, and E in the structural proteins of Sindbis virus. In the sequences for all three proteins, the open boxes denote regions in which five or more consecutive amino acids are conserved among Sindbis, Semliki Forest, and Ross River viruses. The solid boxes denote the hydrophobic membrane-spanning anchors. (CHO) Sites of carbohydrate attachment. (A) Location of the amino acid changes in Group D mutants *ts*10 and *ts*23 in glycoprotein E1 (data from Arias *et al.* (1983). (B) Location of the amino acid change in *ts*20, the only member of Group E, in glycoprotein E2 (data from Lindqvist *et al.*, submitted). (C) Location of the amino acid changes in Group C mutants *ts*2, *ts*5, and *ts*13 in the capsid protein (data from Hahn *et al.*, 1985).

such as *ts* mutations, must by definition affect essential virus functions; virus is not produced under nonpermissive conditions. By locating such mutations, it is possible to define amino acids of particular importance in viral replication. That the particular amino acid at these sites is conserved among even the most distantly related alphaviruses (Bell *et al.*, 1984) lends even greater importance to these residues.

V. REPLICATION AND TRANSCRIPTION OF VIRAL RNA

A. *In Vivo* Studies of RNA Replication

Alphavirus infection is normally a catastrophic event for a vertebrate cell. Host-cell macromolecular synthesis is inhibited beginning about 3

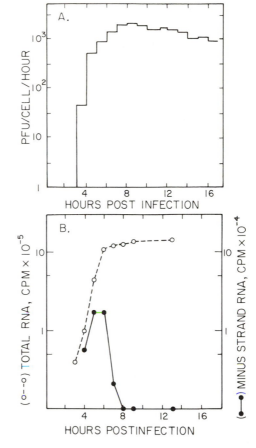

FIGURE 5. Growth curve of Sindbis virus in chicken embryo fibroblast cells at 30°C. (A) Release of progeny virus into the extracellular fluid. Adapted from E. G. Strauss *et al.* (1977) with the permission of Academic Press. (B) Cells infected as in (A) and pulsed for 1 hr at the times specified with radioactive uridine. (O) Total radioactivity incorporated into acid-insoluble form; (●) incorporation into minus-strand RNA (note that the scale differs by 10-fold from that for total RNA). Adapted from S. G. Sawicki *et al.* (1981) with the permission of Academic Press and the authors.

hr after infection, and the synthetic machinery of the infected cell becomes a factory for virus multiplication. By the time the cell eventually dies and disintegrates at some 10–20 hr post infection, up to 10^5 progeny virus particles have already budded from the cell surface. Differential growth curves of Sindbis virus replication in vertebrate cells are shown in Fig. 5. Mature viruses are first detected at about 3 hr postinfection (Fig. 5A), at which time they can also be seen budding from the plasmalemma by electron microscopy (Birdwell *et al.*, 1973). The release of virus becomes maximal by 5–7 hr postinfection and continues at a high rate until the cell disintegrates.

The first event in RNA replication is the synthesis of a full-length minus strand on the incoming genomic RNA template. This occurs during the 1st hour after infection, but it is not seen in labeling studies such as those shown in Fig. 5B. At 3 hr postinfection, RNA synthesis is easily detectable, and both plus and minus strands are being made. Note that even during the period of maximal minus-strand synthesis, minus strands constitute less than 10% of total RNA synthesis. At 5–6 hr postinfection, minus-strand synthesis is shut off, while plus-strand synthesis continues for several hours. During the entire infection cycle, the plus-stranded RNA synthesized consists of both 49 S genomic RNA and 26 S subgenomic RNA. Early in infection, however, relatively more of the subgenomic 26 S message for the structural proteins is made, while late in infection, 49 S genomic RNA predominates (Simmons and Strauss, 1972). The synthesis of both 49 and 26 S RNAs was found to occur by the transcription of a full-length minus-strand template. For 26 S RNA transcription, an internal initiation site in the minus strand is used to produce the subgenomic RNA as such, without splicing or other processing events besides capping. The templates that give rise to 49 S genome-length RNA and 26 S subgenomic RNA were found to be distinguishable by RNase treatment. RNase treatment of a template producing 49 S RNA gave rise to a full-length double-stranded RNA (dsRNA) molecule (called RFI), whereas the template producing 26 S RNA was cut by RNase so that one-third-length (RFIII) and two-thirds-length (RFII) dsRNAs were produced. The two types of templates were interconvertible during infection (Simmons and Strauss, 1972). In addition, work with *ts* mutants has suggested that the enzyme complexes that synthesize 49 and 26 S RNAs are different (see Section V.C).

Although the polypeptides that make up the functional polymerase have not been unambiguously determined (see Section V.B), physiological differences and different patterns of RNA synthesis have been described for various classes of RNA$^-$ mutants (reviewed in E. G. Strauss and J. H. Strauss, 1980). The plus-strand replicase is apparently synthesized early in infection and, once made, is stable, since plus-strand RNA synthesis will continue after the addition of protein-synthesis inhibitors. In contrast, minus-strand synthesis requires continuing translation (D. L. Sawicki and S. G. Sawicki, 1980).

Replication of alphavirus RNA in vertebrate cells has been found to be associated with cytoplasmic membranes in structures called "cytopathic vacuoles" (Grimley *et al.*, 1968, 1972; Friedman *et al.*, 1972). The nature of this association is unclear. In mosquito cells, virus replication and assembly also appear to be associated with vacuoles that appear after infection (Raghow *et al.*, 1973; Gliedman *et al.*, 1975). In mosquito cells, a persistent infection is established, in contrast to the cytocidal infection of vertebrate cells, implying a differential host response to virus infection (Stollar, 1980).

B. *In Vitro* Studies of RNA Replication

To date, *in vitro* studies of alphavirus RNA replication have met with only partial success; *in vitro* systems have been developed, but no laboratory has succeeded in purifying an active replicase that is free of endogenous template and capable of specific initiation of RNA synthesis.

Indeed, it has been difficult to determine which of the nonstructural polypeptides are components of the active replicase, although there seems to be little doubt that nsP1 is a major constituent in most purified preparations (Clewley and Kennedy, 1976; Ranki and Kääriäinen, 1979; Gomatos *et al.*, 1980). Both nsP2 and nsP4 are found as minor components, and it has been suggested that nsP2 is responsible for regulation of 26 S RNA transcription (Brzeski and Kennedy, 1978; D. L. Sawicki *et al.*, 1978). *In vitro* studies have been hampered by the fact that the active complex appears to be particulate and closely associated with membranes, in agreement with the *in vivo* studies, and is found in the mitochondrial pellet during cell fractionation. Solubilization of the particulate complex leads to partial or complete inactivation of the replicase.

In early attempts to isolate an alphavirus replicase from infected cells, a particulate preparation capable of incorporating label into dsRNA was obtained (Martin and Sonnabend, 1967; Sreevalsan and Yin, 1969). Michel and Gomatos (1973) isolated from Semliki-Forest-virus-infected cells a particulate complex that was capable of synthesizing both 49 and 26 S ssRNAs as well as dsRNAs. The ssRNAs appeared to be released from the template as ribonucleoproteins of unknown structure or composition.

Clewley and Kennedy (1976) further purified a replicative complex from Semliki-Forest-virus-infected cells by differential centrifugation, isopycnic centrifugation, and affinity chromatography using immobilized virion RNA. They found that whereas the crude postnuclear supernatant led to the synthesis of both 49 and 26 S RNA, further purification resulted in the loss of the capability to synthesize ssRNAs and led to the synthesis of only dsRNA. Their purified soluble complex, which bound to genomic RNA presumably because it contained minus-strand RNA (although this has not been shown), contained nsP1 and nsP2 (then called ns63 and ns90). Ranki and Kääriäinen (1979) also isolated from Semliki-Forest-virus-infected cells a solubilized complex containing replicative intermediate RNAs that were labeled *in vitro* under the conditions used for RNA synthesis; this complex sedimented at 40 S and contained nsP1, nsP2, and nsP4.

The role of nsP4 in alphavirus replication has not been resolved. It is synthesized only by readthrough and is present in very small quantities in infected cytoplasm (Lopez *et al.*, 1985). It has been suggested that nsP4 undergoes posttranslational cleavage and has a significantly shorter half-life than the other nonstructural proteins (Keränen and Ruohonen, 1983). Recently, it was found that domains within nsP4 share amino acid homology with polymerase proteins of a number of plant and animal viruses (Kamer and Argos, 1984; Ahlquist *et al.*, 1985), as discussed in greater detail in Section VI.D. Thus, although the particular function of nsP4 is unknown, its extreme conservation among alphaviruses (see Section VI.A) and its lesser but distinct conservation among widely divergent RNA virus groups make it clear that this polypeptide must be fundamentally important for RNA synthesis and replication. Which of the non-structural polypeptides possesses the major catalytic function of the RNA polymerase is still unclear. The conservation of nsP4 suggests it for such a function, but it would appear unlikely that the major polymerase sub-unit would be encoded in a region that can be translated only by read-through of a termination codon and that, in the case of Sindbis virus, is found in minuscule quantities in infected cells (Lopez *et al.*, 1985). That all the replicase complexes isolated have contained some nsP1 suggests that nsP1 might contain the elongation activity, and domains of nsP1 are also conserved between alphaviruses and certain plant viruses (Ahlquist *et al.*, 1985) (see also Section VI.D). The activity or components required to initiate polynucleotide chains are apparently lost during fractionation, perhaps accounting for the absence of some of the other nonstructural polypeptides. It has also been proposed on the basis of its large size and large positive charge that nsP2 is the elongation enzyme (E.G. Strauss *et al.*, 1984). Thus, at various times, three of the four nonstructural poly-peptides, nsP1, nsP2, and nsP4, have been proposed as the major elon-gation enzyme, and these three have all been found as components of *in vitro* RNA-synthesizing complexes, although nsP1 has been the most consistently reported in various preparations. The only point that is clear at present is that the replicase functions reside in several proteins prob-ably associated in a complex that is necessarily bound to a cellular mem-brane. It is possible that the functional integrity of this complex is de-stroyed when its template is removed, making it difficult or impossible to purify a system that replicates RNA *in vitro* and that is dependent on exogenously supplied RNA molecules.

The remaining nonstructural polypeptide of alphaviruses, nsP3, has not been directly implicated in the RNA-replication complex and appar-ently contains a function or functions not shared with the plant viruses (Section VI.D) (Ahlquist *et al.*, 1985). It behaves anomalously on poly-acrylamide-gel electrophoresis, and the C-terminal regions of the nsP3s of Sindbis and Middelburg viruses have no homology (Section VI.D) (E. G. Strauss *et al.*, 1983). One possible function for nsP3 is to serve as the virus-encoded protease to process the nonstructural polyprotein.

Two other enzymatic activities have been ascribed to the alphavirus nonstructural proteins, but again have not been localized in a particular polypeptide. One is a guanine-7-methyl transferase that has been isolated from Semliki-Forest-virus-infected cells (Cross, 1983) and the other is a nucleoside triphosphate phosphohydrolase (NTPase) that has been iso-lated from chicken cells, mosquito cells, and mammalian cells infected with Western equine encephalitis virus (Koizumi *et al.*, 1979). The 7-methyl transferase is associated with the membrane-bound replication complex found in the mitochondrial pellet of alphavirus-infected cells and could consist of a complex of nonstructural polypeptides. It is pre-sumably involved in capping the viral RNAs. The NTPase from Western-equine-encephalitis-infected cytoplasm copurifies with a nonstructural polypeptide of 82K daltons (Ishida *et al.*, 1981) that is presumably nsP3. Its function in RNA replication is unknown.

C. Genetic Studies of Replicase Functions

As discussed in previous sections, efforts to isolate functional RNA replicases from alphavirus-infected cells have met with only limited success. The isolated replicase preparations are not template-dependent and have the ability to complete preinitiated chains, but have not been shown unambiguously to initiate new polynucleotide chains. In part because of the lack of an *in vitro* replication system, three other approaches have been developed to study the mechanisms of alphavirus RNA replication: (1) use of *ts* mutants to define the steps involved in RNA replication, (2) study of conserved nucleotide sequences to define areas of the genome that might interact specifically with virus proteins, and (3) study of defective interfering (DI) RNAs of alphaviruses to define segments of the genome that are crucial for RNA replication and encapsidation. The results of genetic studies will be discussed in this section, conserved sequences in Section V.E, and DI RNAs in Section V.G and Chapter 6.

Expressed in their simplest form, the phenotypes of the four RNA$^-$ complementation groups of Sindbis virus mutants are as follows: Group F mutants cease all RNA synthesis (both plus- and minus-strand, 49 and 26 S) on shifting from the permissive temperature to the nonpermissive temperature; on shifting, Group B mutants cease production of minus-strand RNA only; Groups A and G mutants synthesize reduced levels of 26 S RNA on shifting (Keränen and Kääriäinen, 1979; D. L. Sawicki *et al.*, 1981; S. G. Sawicki *et al.*, 1981). These phenotypes can be interpreted in light of the events that occur during RNA replication by postulating that Group F mutants have a defect in the elongation activity of the replicase; that Group B mutants are defective in the initiation factor for minus-strand synthesis, and that Group A or G or both are required to initiate 26 S RNA synthesis. Unfortunately, however, only a few mutants have been examined for some groups, and in particular, *ts*11 is the only known member of Group B (E. G. Strauss and J. H. Strauss, 1980). For this reason, it is unclear whether the phenotypes found to date represent the major biological activities of the four nonstructural polypeptides or whether they represent peculiarities of individual mutants.

In the one case, Group A mutants, in which a large number of representatives of a single complementation group have been examined, the situation appears to be quite complex (D. L. Sawicki and S. G. Sawicki, 1985). Members of Group A can be split into two subclasses. One set includes *ts*24, *ts*15, *ts*17, *ts*21, and *ts*133. These accumulate precursors to the nonstructural proteins (with the exception of *ts*21) at the nonpermissive temperature and are deficient in 26 S mRNA synthesis. Temperature-insensitive revertants no longer accumulate precursors, and the

26 S/49 S RNA ratio is normal. Members of the second subclass, ts4, ts14, ts16, ts19, and ts138, are also unable to replicate RNA at the nonpermissive temperature and belong to Group A by complementation analysis, but process the nonstructural proteins and have a normal 26 S/49 S RNA ratio at the nonpermissive temperature. In addition, several representatives of Group A, ts24, ts17, and ts133, have another variant phenotype in that they are temperature-sensitive for the regulation of minus-strand synthesis; i.e., they fail to shut off minus-strand synthesis late in infection at the nonpermissive temperature (S. G. Sawicki et al., 1981). [A similar mutant (that fails to shut off minus-strand synthesis) has been described for Semliki Forets virus, ts4 (D. L. Sawicki et al., 1978).] However, this particular aberrant behavior does not appear to be responsible for either their ts phenotype or their assignment to complementation Group A because ts+ revertants of these mutants still fail to shut off minus-strand synthesis at 40°C, although RNA synthesis and virus replication are otherwise normal (D. L. Sawicki and S. G. Sawicki, 1985).

Experiments of Fuller and Marcus (1980) in which the complementation groups were ordered by the relative rates of UV inactivation of their ability to complement established the gene order of the complementation groups as NH_2-G-A-B/F-COOH. Mutants of B and F could not be precisely localized, but both occurred downstream of Groups G and A. This would suggest that Group G mutants have mutations in nsP1 and Group A mutants in nsP2. As noted above, Group A mutants lead to reduced levels of 26 S RNA synthesis, and the conclusion that the Group A mutation lies in nsP2 correlates with the previous suggestion of Brzeski and Kennedy (1978) and D. S. Sawicki et al. (1978) that this protein might be responsible for regulation of 26 S RNA transcription (assuming we have correctly assigned nsP2 as the protein referred to by these authors). The UV mapping data also support the assignment of nsP4 as the polymerase (Group F). The validity of UV mapping in assigning gene order is open to some question, however, for systems in which the primary translation product is a polyprotein that contains its own protease; e.g., UV mapping apparently gives an erroneous description of the flavivirus genome (Chapter 10). If a functional protease must be translated before the polyprotein can be processed to produce gene products identifiable by the assay used, in this case to produce proteins that can complement ts lesions (which would be the case for a cis-acting autoprotease but presumably not for a diffusible trans-acting protease), then the UV mapping results may be influenced by the position of the protease. More precise mapping data, like those obtained for structural protein mutants (see Section IV.C), will be required to map the RNA⁻ mutants to particular nonstructural polypeptide chains and to deduce the relationship between the phenotypes of particular mutants and the normal function of each polypeptide chain. Moreover, one or more of the replicase components may well be multifunctional, and the specific interactions of these polypeptides with one another, with host elements (see Section V.D), or with both may be required for the proper temporal regulation of plus- and minus-strand synthesis and of 26 S message vs. 49 S genomic RNA synthesis.

D. Host-Cell Proteins as Components of the Viral Replicase

The results of several different studies have suggested that one or more host proteins are components of the viral replicase. Kowal and Stollar (1981) isolated two mutants of Sindbis virus that were temperature-sensitive in chick cells, growing at 34.5 but not at 40°C, and that were grouped by complementation analysis into Group F, the putative elongation enzyme. These mutants were restricted in their host range, failing to grow in mosquito cells at 34.5°C, although this was a permissive condition in chick cells. One interpretation of these results is that a host protein forms part of the viral replicase and that the altered virus F protein cannot bind the mosquito component at 34.5°C, but will bind the chick protein at 34.5 but not at 40°C.

A second set of studies implicating host components in the viral replicase are those that show that the time of appearance of alphavirus DI RNAs during repeated high-multiplicity passages depends on the host cell (Stark and Kennedy, 1978; Holland et al., 1980). A host component of the replicase could affect the frequency of replicase switching events that are presumed to be required for DI RNA generation.

Finally, Baric et al. (1983) have shown that pretreatment of cells with actinomycin D or α-amanitin renders the cells incapable of supporting the replication of Sindbis virus. Such pretreated cells are competent for the replication of vesicular stomatitis virus, and actinomycin D or α-amanitin added at the time of infection has no effect on alphavirus replication. Furthermore, these authors have isolated Sindbis mutants able to replicate in such pretreated cells. One interpretation of these results is again that a host component forms part of the replicase, and pretreatment with inhibitors reduces the concentration to such a low level that virus replication cannot occur; in this model, the mutants would possess a higher affinity for the cell component. Mapping studies of the mutants would be very valuable for the interpretation of these data.

E. Conserved Sequences in Alphavirus RNAs

It seems intuitively obvious that segments of the alphavirus genome that are bound in a specific fashion to virus proteins during RNA replication or encapsidation will tend to be conserved during alphavirus evolution. Domains involved in specific interactions should change more slowly than other regions of the genome because changes in the nucleo-

FIGURE 6. Location of conserved sequences in alphavirus RNAs. The four boxes show the location of the four regions of conserved sequences. These are (right to left) the 19-nucleotide (19nt) sequence adjacent to the 3' poly(A) tract [present in both 26 and 49 S RNA (Fig. 7A)], the 21nt sequence in the junction region (Fig. 7C), the 51nt domain near the 5' end (Fig. 7D), and the 5' terminus (Fig. 7B).

tide sequence or structure would require compensating changes in the structure of the protein that binds to that domain. Furthermore, as discussed earlier, host proteins, which evolve much more slowly than viral components, may form a component of the replicase. In such cases, the interacting viral RNA sequence would be under constraints not to change. Thus, conserved nucleotide sequences or structures in alphavirus RNAs may serve as regulatory elements for initiation of RNA transcription or for initiation of encapsidation of the genomic RNA.

To date, only one alphavirus genome (Sindbis virus) has been sequenced in its entirety. However, comparative studies have been undertaken in which selected regions of several alphavirus genomes have been sequenced, and four regions of conserved nucleotide sequence or structure or both have been identified (Ou *et al.*, 1982a,b, 1983). The locations of these four regions in the alphavirus genome are indicated schematically in Fig. 6; the conserved sequences or structures or both are shown in Figs. 7 and 8.

There is a conserved sequence of 21 nucleotides found in the junction region that includes the 19 nucleotides preceding the start point of the subgenomic 26 S RNA and the first two nucleotides of the sequence transcribed into 26 S RNA (Ou *et al.*, 1982a). This sequence is virtually invariant in four viruses examined (Fig. 7C), with only a single nucleotide in this sequence showing any variability whatsoever within this 21-nucleotide stretch. It is possible that two or more additional nucleotides at either end of the 21-nucleotide element are also important, since these are also highly but not absolutely conserved. We have postulated that the complement of this 21-nucleotide sequence in the minus strand acts as a promoter for initiation of transcription of 26 S RNA. According to this model, one of the virus nonstructural polypeptides, possibly acting as a component of a multifunctional replicase/transcriptase enzyme complex, specifically recognizes this promoter (in the minus strand) and binds to it, leading to initiation of 26 S RNA transcription at a defined nucleotide, corresponding to nucleotide 7599 of the genomic 49 S RNA. This 26 S RNA transcript is presumably capped by the same enzyme complex dur-

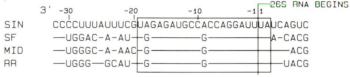

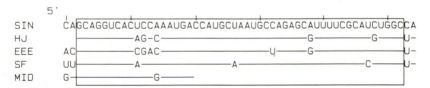

A. 3'TERMINAL SEQUENCES IN GENOME RNAs

```
                3'            10          20          30
     SIN   poly(A) CUUUACAAUUUUUGUUUUAAAACAACUAAUUAUUUUCUUU
     WEE          ————A——————————————UUUUUGU-UUC-U——AA-A-
     HJ           ————A——————————————UUUUCUU-UUA——CA-UAGA
     EEE          ————U——————————————UUUUUGUAUU—C———UAA-
     VEE          ————U——————————————GGCU—GCCU—UC——UC—
     SF       C———U————————G ——CGUUUUAUU————————UAGG
     MID      —C——U————————G——ACCGCUG-U-GUUAGCAGCG—A-
     RR           AU———U————UG——————UCUGCGG-UGC-GGGG-CAC
```

B. SEQUENCES COMPLEMENTARY TO THE 5'TERMINI

```
                3'            10          20          30
     SIN       UAACCGCCGCAUCA UGUGUGAUAACUUAGU
     HJ        —U—CPYA-CAU- -C-CACU-PCU
     EEE       —U—CAUA-CA— -C-CC-UCGGUGGGC-
     VEE       —U————————P-PC -C-C-U
     SF        — ————UACA—C——A—UGCUGCGGU-
     MID       ———————A—AAUG— ——CACGGUGG-G-G
```

C. SEQUENCES COMPLEMENTARY TO THE JUNCTION

```
                                                  ┌26S RNA BEGINS
                3'  -30      -20      -10     -1│1
     SIN       CCCCUUUAUUUCGUAGAGAUGCCACCAGGAUUUAUCAGUC
     SF        —UGGAC-A-AU┤G————————G————————┤A-CACG
     MID       —UGGGC-A-AAC┤G————————G————————┤—ACG
     RR        —UGGG—GCAU┤G————————G————————┤—UACG
```

D. 51 NUCLEOTIDE CONSERVED SEQUENCE

```
             5'
     SIN   CAGCAGGUCACUCCAAAUGACCAUGCUAAUGCCAGAGCAUUUUCGCAUCUGGCCA
     HJ    ————————AG-C——————————————G————————G——U-
     EEE   AC————————CGAC——————————————U————G————————U-
     SF    UU————A————————————A————————————C————U-
     MID   G————————G———————
```

FIGURE 7. Conserved sequences in alphavirus RNAs. (A) 3'-Terminal sequences of genomic RNAs. Sequences read from 3' [poly(A)] to 5' and are shown as the plus strand (genomic RNA). (B) Sequences complementary to the 5' termini of genomic RNAs. The sequences shown read from 3' to 5' on the minus strand. (C) Sequences complementary to the junction region. The sequences shown are in the minus strand in the region containing the start of the 26 S subgenomic RNA and read 3' to 5'. (D) Conserved sequences in the 51-nucleotide double hairpin structure, reading 5' to 3' in the genomic RNA (plus strand). In all cases, boxes enclose regions of highest conservation and horizontal lines indicate that the nucleotides at the corresponding positions are identical to those shown in the complete sequence at the head of the group. Gaps have been introduced for alignment. Data are from Ou *et al.* (1982a,b, 1983) and Dalgarno *et al.* (1983). Here and in Figs. 9, 13, 14, and 17: (SIN) Sindbis virus; (WEE) Western equine encephalitis virus; (HJ) Highlands J virus; (EEE) Eastern equine encephalitis virus; (VEE) Venezuelan equine encephalitis virus; (SF) Semliki Forest virus; (MID) Middelburg virus; (RR) Ross River virus.

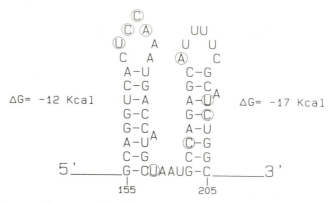

FIGURE 8. Conserved structure in alphavirus RNAs. A possible structure is shown for the 51-nucleotide conserved sequence of Figs. 6 and 7D. The sequence shown is that of Sindbis virus, HR strain, and the nucleotides are numbered from the 5' end of this RNA. Nucleotides that vary in one or more of the sequences are circled; others are invariant for four viruses (Fig. 7D). Reproduced from E. G. Strauss and J. H. Strauss (1983) with the permission of Springer-Verlag. Here and in Fig. 9, free energies were calculated according to the method of Tinoco *et al.* (1973).

ing or shortly after initiation of 26 S RNA transcription. *In vitro* methylation studies of Cross and Gomatos (1981) suggest that capping occurs after initiation and that the capping activity can be separated from the transcription activity. Transcription continues to the end of the minus strand, where the poly(A) tract is added by copying the poly(U) tract at the 5' end of the minus strand. It has been shown that 26 S RNA is exactly coterminal with 49 S RNA (Ou *et al.*, 1981).

It should be noted here that the conservation of nucleotide sequences found cannot be explained simply by the necessity to conserve a stretch of amino acid sequence in nsP4. As will be discussed in more detail in Section VI.B, the alphaviruses have diverged so extensively that even in regions in which amino acid sequence is conserved, the nucleotide sequence encoding this amino acid sequence is not conserved (Ou *et al.*, 1982a, 1983; E. G. Strauss *et al.*, 1983).

The transcription of the minus strand from the genome-length plus strand is postulated to involve two sequence elements. The first of these is a conserved sequence of 19 nucleotides found at the 3' end of the genomic RNA immediately adjacent to the poly(A) tract (Fig. 7A). In five of the eight viruses examined, only one nucleotide (6th from the end) shows any variability at all. In the three remaining viruses, which as discussed in more detail below form a separate branch of the alphavirus evolutionary tree, there is more divergence but nonetheless strong overall conservation of this sequence. As with the conserved sequence in the junction region, the promoter model postulated that one of the alphavirus nonstructural polypeptides, again probably as a component of a multifunctional enzyme, binds to this sequence element to begin transcription of a minus strand from a plus-strand template. In this case, transcription begins upstream of the conserved sequence, in the poly(A) tract, to produce a poly(U) tract at the 5' end of the minus strand, and it is possible that some or all of the poly(A) tract forms part of the promoter initiation element. The length of the poly(A) tract, and presumably that of the poly(U) tract as well, is not fixed. Rather, the length seems to fluctuate around a mean, and RNA molecules with short tracts replicate to produce molecules with a normal-length tract (Frey and Strauss, 1978; Spector and Baltimore, 1974). One possibility is that the replicase binds to the 19-nucleotide conserved sequence, and on initiation of transcription with a poly(U) tract, the length of the tract produced is related to the length of poly(A) spanned by the enzyme. Chattering or slippage could increase the length of the tract, whereas an overly long tract could be brought back to the mean either by deletion during replication or by initiation within the complementary tract (rather than at the 3' end of the RNA). Other mechanisms are also possible. Wengler *et al.* (1982) have found that the 5' end of the minus-strand RNA found in double-stranded molecules isolated from infected cells is not capped, so no capping occurs during or after initiation of minus strand. After initiation, elongation presumably follows until the end of the plus-strand template is reached. The 3' end of the minus-strand product has been shown to be exactly complementary to the 5' end of genome-length RNA except for the presence of an unpaired, 3'-terminal G in the minus strand (Wengler *et al.*, 1979, 1982). How this extra G is added to the 3' end of the minus strand is unknown; it has been found in the case of two different alphaviruses, Sindbis and Semliki Forest viruses, and presumably occurs in all alphavirus minus-stranded RNAs. Reverse transcriptase also adds an extra nucleotide to the 3' end of transcripts (Gupta and Kingsbury, 1984), and it has been suggested that this results from inefficient copying of the cap nucleotide.

There would appear to be some additional recognition element involved in minus-strand synthesis other than the 3' end of the genome-length plus strand. As noted, 26 S subgenomic RNA is exactly coterminal with 49 S RNA and thus possesses any recognition elements found at or near the 3' end. Yet all evidence indicates that 26 S RNA is not used as a template for minus-strand production, and no 26 S-size minus-strand copy of the subgenomic RNA has been found in infected cells. Although it is conceivable that compartmentalization of RNAs could be used to achieve this result, we have postulated that some additional sequence element in the genomic RNA, found in the non-26 S region of this RNA, is also required for minus-strand production. Hsu *et al.* (1973) found some years ago that the RNA isolated from Sindbis virions was circular, but could be linearized by mild denaturation conditions. Frey *et al.* (1979) studied the thermodynamics of cyclization in some detail, and the cyclization reaction is sufficiently rapid that linear molecules would cyclize at a significant rate in an infected cell (see Section V.F). Since alphavirus

RNAs can cyclize, the 5' and 3' ends of the RNA must be relatively close together much of the time, and the obvious place to search for a second sequence element is thus at or near the 5' end of the RNA.

Such a conserved sequence element is shown in Figs. 7D and 8. This element is 51 nucleotides long, begins about 150 nucleotides from the 5' end of 49 S RNA, and is highly conserved in five viruses examined. In addition to being a conserved sequence, this element is able to form a double hairpin structure (Fig. 8) that is also conserved in the five viruses. Thus, we have postulated that either the conserved sequence (or some segment of it) or the conserved structure (or some segment of it) is recognized by the virus replicase simultaneously with the conserved sequence at the 3' end of the genome RNA to initiate transcription of minus strand (see, however, Section V.G and Chapter 5). The hypothesis of requiring the viral replicase to recognize sequence elements at the two ends of the RNA molecule in order that replication may proceed is attractive, because in this way replication is restricted to full-length genomic RNA.

The 51-nucleotide element is so much longer than the 19- and 21-nucleotide elements discussed earlier that it seems likely that it could have more than one function. An appealing hypothesis is that it also serves as a nucleation site for encapsidation (but see Section V.G and Chapter 5). In the well-studied case of tobacco mosaic virus (TMV), it is known that coat protein subunits recognize and bind in a specific fashion to a sequence in the RNA genome (Zimmern, 1977). After nucleation, encapsidation proceeds by extension of the helix in both directions until the ends of the RNA are reached (Otsuki *et al.*, 1977; Fukuda *et al.*, 1978). The coat protein is capable of encapsidating RNAs that lack this sequence, but at a much slower rate. Similarly, a nucleation sequence in the alphavirus RNA genome could lead to the rapid, efficient, and specific encapsidation observed (Brown, 1980). After nucleation, encapsidation could proceed by protein–protein interactions to assemble the icosahedral shell, in the process enveloping the RNA. Because 26 S RNA is not encapsidated, this encapsidation sequence (or at least some elements of it) would be postulated to reside in the non-26-S region. The recent finding that alphavirus capsid protein can encapsidate *in vitro* RNAs that lack alphavirus sequences could be analogous to the case for TMV (Wengler *et al.*, 1982).

If the 51-nucleotide element does serve as an encapsidation element as well as a replication element, binding of capsid protein to this element would prevent the RNA from being transcribed into minus strand and commit the RNA to encapsidation.

Finally, there is the matter of transcribing genome-length plus strands from a minus-strand template. Sequences at the 3' terminus of the minus strand (deduced for the most part from the sequences at the 5' end of the plus strand) are shown in Fig. 7B. Sequence homologies are apparent, but not as striking as those of the other sequence elements discussed so far. There is, however, a conserved stem-and-loop structure

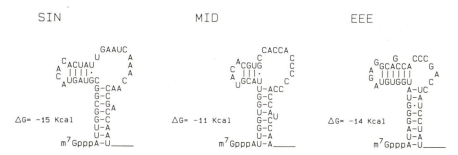

FIGURE 9. Conserved stem-and-loop structures at the 5' termini of alphavirus RNAs. Moderately stable secondary structures can be formed by the sequences at the 5' ends of alphavirus genomic RNAs. See the Fig. 7 caption for the virus abbreviations. Free energies were calculated as noted in the Fig. 8 caption. Reproduced from E. G. Strauss and J. H. Strauss (1983) with the permission of Springer-Verlag.

found at the 3' end of each minus strand (Fig. 9) that could be recognized by the replicase as a binding site for initiation of plus-strand transcription. Whether cyclization of the RNA is required for initiation of plus strands is unknown.

It is clear that the model presented above constitutes simply a hypothesis for alphavirus RNA replication, but a hypothesis that is capable of being tested by the use of recombinant DNA technology or other methods. It also seems clear that even if the function of one or more of the conserved sequence elements has not been properly identified, the conservation *per se* suggests that these sequence elements must have some important role in the virus life cycle. Moreover, there may be additional sequence elements, as yet unidentified, that have essential functions in RNA replication and encapsidation.

F. Cyclization of Alphavirus RNAs

It has been known from some time that the RNAs of Sindbis virus and of Semliki Forest virus, and presumably of all alphaviruses, are circular when extracted from virions (Hsu *et al.*, 1973; Kennedy, 1976; Frey *et al,*. 1979), and it has been supposed that cyclization is important for virus RNA replication or encapsidation or both. Possible stages in RNA replication in which cyclization could be important have been discussed in the last section.

Even though the RNA of Sindbis virus has been completely sequenced and both the 5' and 3' ends of a number of other alphaviruses have been sequenced, the sequences in the RNA responsible for panhandle formation have not been identified. A representative electron micrograph of circular RNA (Fig. 10) illustrates that these cyclization sequences must be found near the ends of the molecule. No branching is seen, and

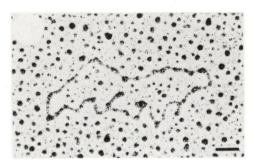

FIGURE 10. Electron micrograph of Sindbis virus 49 S RNA. The RNA was treated with 0.5 M glyoxal in 0.01 M sodium phosphate buffer (pH 7.0) for 30 min at 35°C before spreading and microscopy according to the methods of Davis *et al.* (1971). Scale bar: 100 nm. Reproduced from Frey *et al.* (1979) with the permission of Academic Press.

only short hairpins remain that could be panhandles derived from hybridization of complementary sequences found at the ends of the molecule (more rigid denaturation conditions to remove residual RNA secondary structure results in linearization of the molecules). Hsu *et al.* (1973), in their study of Sindbis RNA circles, concluded that the panhandles they observed had a length of 250 ± 50 base pairs. The thermodynamic studies of Frey *et al.* (1979) indicated that the cyclization sequences are much shorter than this, probably only 10–20 nucleotides long. Taken together, these studies suggest that short stretches of complementary sequence located 200–300 nucleotides from the ends of the molecules are responsible for cyclization.

Ou *et al.* (1983) examined the RNA of Sindbis virus and several other alphaviruses for possible cyclization sequences. No perfectly matched complementary sequences were found, but several mismatched sequences were located that had a calculated free energy for cyclization similar to the measured value and that were found 150–300 nucleotides from the ends as predicted from the electron-microscopic studies. More work will be required to determine whether these are in fact the cyclization sequences or whether other sequences not yet identified are involved.

Thermal-denaturation profiles in which the fraction of circular molecules is assayed in samples heated to various temperatures are illustrated in Fig. 11. In unheated samples, all the molecules are circular (these RNA preparations were derived from purified virions). With increasing temperature, a larger proportion of the molecules become linear, and the denaturation temperature depends on the ionic strength. In 0.1 M NaCl, the mean temperature for linearization is 53.5°C, whereas in 0.023 M NaCl, it is 39.5°C. From these and other data, it can be calculated that the length of the cyclization sequences is probably on the order of 10–20 nucleotides and that the free energy of cyclization is −13 kcal/mole at 25°C in 0.1 M NaCl.

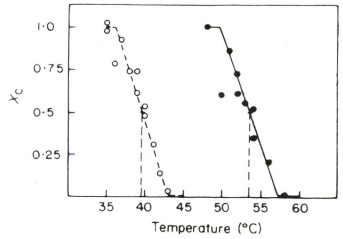

FIGURE 11. Melting curve of the circular form of Sindbis 49 S RNA. The mole fraction of circles (X_c) present in sucrose gradients of 49 S RNA that had been heated for 5 min at various temperatures in either 0.023 M NaCl, 0.01 M Tris (pH 7.4), 0.001 M EDTA (○) or in 0.1 M NaCl, 0.01 M Tris (pH 7.4), 0.001 M EDTA (●) was quantitated (Frey *et al.*, 1979). The vertical dashed lines indicate the mean melting temperature under the two conditions. Reproduced from Frey *et al.* (1979) with the permission of Academic Press.

Linear alphavirus RNA molecules will cyclize readily in solution. The kinetics of cyclization in 0.1 M NaCl at two temperatures are illustrated in Fig. 12. Cyclization follows first-order kinetics, and the half-time for cyclization under these conditions is 1 hr at 30°C and 6 min at 40°C. Thus, the RNAs will cyclize under physiological conditions within a period of time that is short with respect to the virus life cycle in an infected cell, although interactions with virus or host proteins could influence the rapidity of this reaction.

G. Alphavirus Defective Interfering RNAs

Alphavirus DI RNAs have been studied in a number of laboratories in an attempt to define sequences within the RNA required for RNA replication and encapsidation (Kennedy, 1976; Lehtovaara *et al.*, 1981, 1982; Monroe *et al.*, 1982; Monroe and Schlesinger, 1983; Tsiang *et al.*, 1985). These studies are presented in some detail in Chapter 5, but here we would like to review very briefly those aspects of DI genome organization that are directly related to the possible function of conserved RNA sequences in the replication of alphavirus RNAs.

First the model (Section V.E) predicts that DI RNAs would contain intact the sequence at the 3' terminus for initiation of minus strands, and indeed all DIs sequenced to date have a 3' end identical to that of the parental infectious viral RNA for at least 50 nucleotides after the poly(A) tract. Thus, the 3'-terminal 19-nucleotide sequence element is

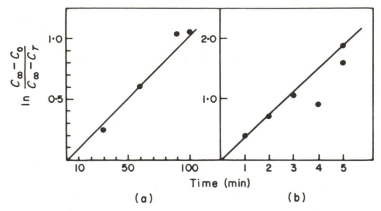

FIGURE 12. Kinetics of cyclization of Sindbis virus RNA as a function of temperature. Denatured (linear) Sindbis RNA was allowed to cyclize in 0.1 M NaCl, 0.01 M Tris (pH 7.4), 0.001 M EDTA for various periods of time at either 30°C (a) or 45°C (b), and the amounts of circular and linear RNA were assayed on sucrose gradients. The natural logarithm of the function $(C_\infty - C_0)/(C_\infty - C_T)$, where C is the fraction of circles at 0 time, infinite time, or time T, was plotted as a function of the time of renaturation. The straight lines obtained show that the reaction follows first-order kinetics, and from the slope of this line, a rate constant for cyclization can be calculated. Reproduced from Frey *et al.* (1979) with the permission of Academic Press.

present adjacent to the poly(A) tract in the same configuration as in the parental virus. Second, for efficient genome-length plus-strand replication as well as possible encapsidation functions, the 51-nucleotide sequence element should be present in a position where a replicase molecule can interact simultaneously with it and the 3′ end. All naturally arising DI RNAs sequenced to date retain this sequence element, and in some DIs, this sequence element is amplified and rearranged. However, recent studies in which DIs are modified by *in vitro* gene-splicing technology suggest that the sequence element can be deleted and the DI remain functional (S. Schlesinger, personal communication) (reviewed in Chapter 5), and this element may therefore be dispensable for DI RNA replication and packaging. Third, since DI RNAs are not translated or transcribed to produce a subgenomic RNA, the 21-nucleotide sequence element in the junction region is superfluous. All DI RNAs sequenced to date have deleted the junction region as well as most of the 26 S region. Since no templates are being used for transcription, they are all available for replication, increasing the relative replication efficiency of DI RNAs. Finally, comparative sequencing of 5′ ends suggested that a stem-and-loop structure might be important in RNA replication, but that sequence homology was not required (see Fig. 9). DI RNAs sequenced to the 5′ end display a variety of 5′-terminal sequences. Some retain the parental 5′ sequence, some have a 5′ terminus derived from the 5′ terminus of 26 S RNA, and some have as their 5′ terminus nucleotides 10–75 of a cellular tRNA for aspartic

FIGURE 13. Amino-terminal amino acid sequences of glycoprotein E1 for eight different alphaviruses. See the Fig. 7 caption for the virus abbreviations. (BF) Barmah Forest virus. See the Fig. 2 caption for the amino acid code. The top line shows a consensus sequence in which boxed residues are probably those of the ancestral protoalphavirus. A dot means the amino acid is the same as in the consensus sequence; if the amino acid at any position differs from this consensus sequence, the changed amino acid is shown. Gaps have been introduced as necessary to maintain the alignment. (?) No assignment was made. Under-lining indicates some uncertainty in the assignment. Certain highly conserved residues or areas are shaded. For SIN, SF, RR, VEE, and MID, amino acid sequence has been confirmed and extended by nucleotide sequence data. This figure is adapted from Bell *et al.* (1984) and includes RNA sequence data for VEE from R. M. Kinney, B. J. Johnson, and D. W. Trent (unpublished data).

acid (Monroe and Schlesinger, 1983; Tsiang *et al.*, 1985). Thus, several different 5'-end sequences can lead to effective DI RNA replication.

VI. EVOLUTION OF THE ALPHAVIRUS GENOME

A. Evolution within the Genus *Alphavirus*

It has been known for some time that all alphaviruses are closely related on the basis of serological cross-reactions, similarities in virus size and structure, and similarities in the details of the molecular biology of replication. Nucleotide sequence studies and amino acid sequence an-alysis have begun to put these relationships on a more defined footing and to trace the evolutionary relationships among these viruses.

About 25 alphaviruses are currently recognized, and these have been grouped into six serological subgroups (Calisher *et al.*, 1980). Comparison of the amino-terminal amino acid sequences of glycoproteins E1 and E2 of eight alphaviruses representing five different subgroups has shown that these N-terminal domains of the glycoproteins are 26–82% conserved between any two viruses and has led to an evolutionary tree for these viruses (Bell *et al.*, 1984). The data for E1 are shown in Fig. 13, in which the most highly conserved regions have been shaded. It is noteworthy that all the cysteines in this region are conserved [as they are in general throughout E1 and E2 (Rice and Strauss, 1981a; Dalgarno *et al.*, 1983)] and that at many other positions, amino acids important in secondary structure formation have been conserved, suggesting that whereas the three-dimensional structure of the glycoprotein is the same or very sim-ilar for all alphaviruses, the primary sequence has evolved to the point

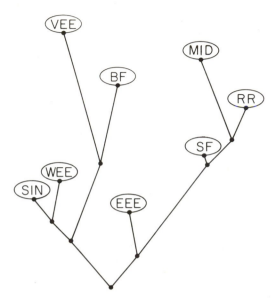

FIGURE 14. An evolutionary tree for eight alphaviruses deduced from the N-terminal amino acid sequences shown in Fig. 13. The distance along any line is proportional to the number of amino acid substitutions. See the Fig. 7 caption for the virus abbreviations. (BF) Barmah Forest virus. Data from Bell *et al.* (1984).

that there is considerable variability. An evolutionary tree derived from these data is shown in Fig. 14. The tree is in reasonable agreement with relationships deduced from serological cross-reaction studies; note, however, that Ross River, Semliki Forest, and Middelburg viruses are closely related to one another and form a separate branch of the tree.

The evolutionary divergences of the structural proteins are illustrated in a different way in Fig. 15, in which the percentage homology between the structural proteins of Sindbis and Ross River viruses is plotted as a moving average over a string of 15 amino acids (Fig. 15A). It can be seen in Fig. 14 that these are distantly related alphaviruses. It is clear from Fig. 15 that homology is not uniform; there are domains that are highly conserved, domains that are quite divergent, and domains that show an intermediate degree of conservation. The greatest degree of conservation is exhibited by the C-terminal half of the capsid protein and by several domains in the N-terminal third of E1. The least conservation is found in regions near the N terminus of the capsid protein and in the domains that cross the lipid bilayer to anchor the glycoproteins in the bilayer. It is of interest that E2 is less highly conserved overall (42%) than is E1 (51%). That E2 possesses the major neutralization epitope of the virus suggests that E2 evolves more rapidly than E1 and the function of E2 in alphavirus evolution might be to generate strain diversity. It is also of interest that the Ross River and Sindbis glycoproteins average 47% homology, a degree of homology similar to that exhibited by the glycoproteins of two flaviviruses belonging to different subgroups, Murray Valley encephalitis and yellow fever, the glycoproteins of which share 45% sequence homology (see Chapter 10). In flaviviruses, however, the func-

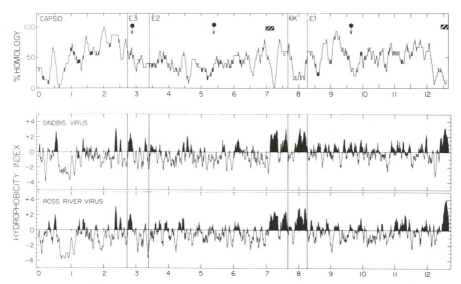

FIGURE 15. Graphic representation of amino acid homology between the structural proteins of Sindbis virus and Ross River virus (A) and hydrophobicity plots for each virus (B). (A) The percentage homology is plotted as a moving average using a string length of 15. Amino acids in the polyprotein are numbered in hundreds. Conserved glycosylation sites are shown as arrows; membrane-spanning domains are indicated by cross-hatched bars; vertical lines separate the final protein products. (B) Computer-generated hydrophobicity plots using the program of Kyte and Doolittle (1982) with a string length of 7. Hydrophobic regions are plotted above the center line. Data from Rice and Strauss (1981a,b) and Dalgarno *et al.* (1983).

tions of both E1 and E2 in the mature alphavirion are for the most part expressed by the single E protein.

Figure 15B shows hydrophobicity profiles for Sindbis and Ross River virus structural proteins. Even in regions of low or moderate homology, the hydrophobicity profiles are quite similar, suggesting that characteristics of particular domains are preserved even when the linear amino acid sequences have diverged. Indeed, if we could examine the native three-dimensional conformation of these polypeptides, they would probably appear even more similar.

A similar comparison of nonstructural proteins nsP3 and nsP4 for Sindbis and Middelburg viruses is shown in Fig. 16. It can be seen in Fig. 14 that Sindbis and Middelburg are distantly related, and Middelburg and Ross River are more closely related, so that Figs. 15 and 16 should be directly comparable. As was the case for the structural-protein region, the homology is nonuniform, but overall averages much higher in nsP3 and nsP4 than in the structural region. Nonstructural protein 3 has regions of low homology near the N terminus and almost no homology in the C-terminal third, but consistently high homology elsewhere. The region of zero homology between amino acids 455 and 530 is due to the presence of 88 residues in the Sindbis protein that are absent in the Mid-

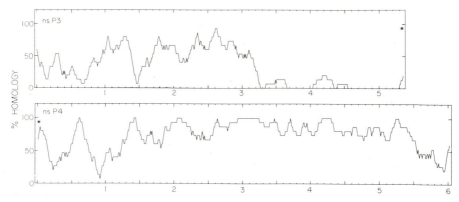

FIGURE 16. Graphic representation of amino acid homology between Sindbis virus and Middelburg virus for two nonstructural proteins. A string length of 15 was used. Amino acids are numbered in hundreds from the beginning of nsP3 and nsP4, respectively. (*) Position of the opal codon. Data from E. G. Strauss *et al.* (1983) and E. G. Strauss (unpublished).

delburg sequence. The nsP4s are very highly conserved. There is a conserved core within nsP4 in which there are only five amino acid substitutions within a 99-amino-acid stretch, and overall the proteins exhibit 73% homology (E. G. Strauss *et al.*, 1983) (Fig. 16). Thus, the nonstructural proteins are changing much more slowly than the structural proteins.

The very high error frequency during viral RNA replication, on the order of 10^{-4}, that results from a lack of proofreading functions by RNA replicases, has some interesting effects on virus evolution. For one, it probably sets an upper limit to the size of RNA genomes, a subject reviewed recently by Reanney (1982). For another, it signifies that a virus RNA population is in fact a population of molecules that differ slightly in sequence from one another but that has a defined average RNA sequence that is selected for (i.e., noncoding changes, as well as coding changes, are selected against). This situation has been explored in a paper by Domingo *et al.* (1978), who studied the population dynamics of phage Qβ RNA. These authors found that populations of Qβ RNA had an average sequence that was selected for and maintained during repeated passage, even though, on average, each RNA molecule differed at 1.7 nucleotides from this average sequence. Many of these changes from the average sequence must have been silent mutations, but these were nonetheless selected against, and thus RNA secondary structure must contribute to replication efficiency and be maintained. For alphavirus evolution, this means that there exists a gradient of selective pressures against changes in sequence, from quite mild pressures against silent changes, to varying degrees of stronger pressures against coding changes that may affect protein function to varying degrees, to absolute selection against changes that are lethal. In nature, there is also the very compli-

TABLE I. Sindbis Virus 49 S RNA Dinucleotide
Frequencies

5' Member of doublet	Second member of doublet			
	A	G	T	C
A	922	867	664	855
G	870	633	605	800
T	552	726	577	582
C	963	682	592	811

cated interplay of changing selective pressures as the virus cycles through alternate arthropod and vertebrate hosts.

B. Codon Usage in Alphaviruses

Vertebrate DNA has been found to be deficient in the dinucleotide CG, and a number of animal viruses have also been found to have RNA or DNA that mimics their vertebrate hosts and is similarly deficient (Russell et al., 1976; Grantham et al., 1981). Alphaviruses, however, have a CG doublet frequency very close to that predicted from their base composition. The dinucleotide frequencies in Sindbis virus RNA are given in Table I. The dinucleotide CG occurs 682 times in the RNA, compared to the predicted occurrence of 757. Thus, alphaviruses have a dinucleotide frequency more similar to that of their mosquito hosts than to that of their vertebrate hosts.

The codon usage in the genome of Sindbis virus is given in Table II. Codon usage is nonrandom; note that the leucine codon CUG is used five times as frequently as UUA and twice as frequently as CUU, CUC, or CUA. Similarly, the alanine codon GCC is used twice as often as GCU, the cysteine codon UGC is used for 78% of the cysteines, and so forth. These patterns of codon preferences have been found to be a general feature of alphaviruses (Rice and Strauss, 1981b; Dalgarno et al., 1983), so that the nonrandom usage reflects some adaptation of the viruses to growth in their hosts, rather than to the peculiarities of Sindbis virus RNA (see also below). It is worth noting, however, that codons containing the CG doublet are not infrequently used, as might be expected from the CG frequency in Sindbis RNA; that the serine codon UCG is one of the preferred codons, as is the proline codon CCG; and that more than half the arginine is encoded by the CGN codons. At the current time, the reasons for codon selection are not well understood, but it may be that preference for certain codons by alphaviruses reflects their adaptation for maximal replication efficiency in both their arthropod and their vertebrate hosts.

Although all alphaviruses are probably descendants of a single an-

TABLE II. Codon Usage in the Genome of Sindbis Virus[a]

Phe	UUU	61	Ser	UCU	26	Tyr	UAU	47	Cys	UGU	22
	UUC	70		UCC	37		UAC	78		UGC	79
Leu	UUA	18		UCA	49	Och	UAA	0	Opal	UGA	2
	UUG	61		UCG	53	Amb	UAG	1	Trp	UGG	37
Leu	CUU	43	Pro	CCU	48	His	CAU	49	Arg	CGU	25
	CUC	40		CCC	46		CAC	52		CGC	46
	CUA	41		CCA	80	Gln	CAA	54		CGA	14
	CUG	87		CCG	73		CAG	78		CGG	17
Ile	AUU	61	Thr	ACU	52	Asn	AAU	47	Ser	AGU	36
	AUC	78		ACC	96		AAC	90		AGC	54
	AUA	44		ACA	78	Lys	AAA	113	Arg	AGA	62
Met	AUG	86		ACG	47		AAG	130		AGG	36
Val	GUU	48	Ala	GCU	50	Asp	GAU	64	Cly	GGU	32
	GUC	86		GCC	117		GAC	115		GGC	51
	GUA	73		GCA	96	Glu	GAA	119		GGA	89
	GUG	75		GCG	65		GAG	98		GGG	38

[a] Codon usage in both long open reading frames of Sindbis virus. There is a total of 3760 codons.

cestral protovirus, such extensive nucleotide sequence divergence has occurred during evolution that even in domains in which amino acid sequence is highly conserved, the codon usage for any given amino acid has been virtually randomized (Ou *et al.*, 1982b, 1983; E. G. Strauss *et al.*, 1983). Outside regions of conserved nucleotide sequence discussed in Section V.E, conservation is at the level of amino acid sequence (and thus protein function), and no evolutionary remnants are left in the nucleotide sequence *per se* to reveal the evolutionary relationships of these viruses. This extensive divergence is presumably due in part to the rapid evolution of RNA viruses (Holland *et al.*, 1982) caused by the high error frequencies during RNA replication discussed above. It could also be due in part to a long evolutionary history, for we have no way at present to guess the time of origin of the ancestral alphavirus. The end result is that any two alphavirus RNAs demonstrate little or no cross-hybridization (Wengler *et al.*, 1977), even though the proteins encoded may share extensive sequence homology.

The randomization of codon usage that has arisen among alphaviruses is illustrated in two ways below. Figure 17 shows a short stretch of nucleotide sequence for four alphaviruses within a highly conserved region of protein nsP4. Note that in this stretch of 24 nucleotides, the 8 amino acids encoded are identical in all four viruses. However, the three leucines in this sequence are encoded by three, three, and four different codons in the four virus RNAs; the proline and glycine are encoded by two codons (of the four possible in each case); and both possible codons are used for the two lysines and the phenylalanine.

A second approach to studying codon randomization is illustrated in

FIGURE 17. Randomization of codons used for conserved amino acids
among alphaviruses. A short stretch
of conserved protein sequence found
in nsP4 about 80 amino acids from
the C terminus, alongwith the
nucleotide sequence encoding it, is
shown for four alphaviruses (the first

```
          L   F   K   L   G   K   P   L
SIN     CUG UUU AAG UUG GGU AAA CCG CUC
          L   F   K   L   G   K   P   L
MID     CUC UUU AAG CUC GGA AAA CCG CUG
          L   F   K   L   G   K   P   L
RR      UUA UUU AAA CUA GGU AAA CCU UUA
          L   F   K   L   G   K   P   L
SF      CUG UUC AAG UUG GGU AAG CCG CUA
CODONS    3   2   2   3   2   2   2   4
```

leucine is amino acid 2431 in the Sindbis nonstructural polyprotein). The bottom line gives
the number of different codons used at each position. See the Fig. 2 caption for the amino
acid code. See the Fig. 7 caption for the virus abbreviations. Data from Ou *et al.* (1982).

Table III. In this table, codons used for conserved amino acids were compared for two proteins. One comparison was of the nsP4 of Sindbis and
Middelburg viruses. As noted, these two viruses are distantly related alphaviruses, but the two proteins are highly conserved (73% amino acid
sequence homology). Furthermore, the two nsP4s exhibit perfect alignment, so that no gaps or deletions are necessary to maintain alignment.
The second comparison was for the E1 glycoproteins of Semliki Forest
and Ross River viruses. In this case, the two viruses are more closely
related (and presumably have diverged more recently), and the two glycoproteins are 79% homologous and again exhibit perfect alignment. It
is clear from Table III that randomization of codons used is almost complete in both cases. In each example, there is only a slight bias toward
having greater codon matching for conserved amino acids than expected
by random chance, which may reflect the evolutionary divergence of the
alphaviruses from a common ancestor, or it could be due to chance.

C. Evolution within the Family Togaviridae

Currently, there are three genera (*Alphavirus*, *Rubivirus*, and *Pestivirus*) within the family Togaviridae, which also contains several un-

TABLE III. Codon Randomization[a]

Protein	Amino acids		Codons used[b]	
	Total	Conserved	Same[c]	Expected[d]
SIN/MID nsP4	626	448	181	171
RR/SF E1	438	345	143	134

[a] Comparison of Sindbis (SIN) and Middelburg (MID) protein nsP4 and of
Ross River (RR) and Semliki Forest (SF) glycoprotein E1.
[b] Codons used for conserved amino acids.
[c] Number of conserved amino acids encoded by the same codon.
[d] Expected number of conserved amino acids encoded by the same codon
based on random chance, calculated using the actual composition of conserved amino acids and the average codon usage distribution exhibited
by the alphaviruses Sindbis, Semliki Forest, and Ross River.

classified viruses. The genus *Flavivirus*, formerly classified as belonging to this family, is now classified as the sole genus of the family Flaviviridae.

Only very limited sequence data exist for rubella virus (genus *Rubivirus*). However, the genome organization, including transcription of a subgenomic RNA to produce the structural proteins, is very similar to that of alphaviruses (Oker-Blom *et al.*, 1983, 1984; Oker-Blom, 1984). The size of the genome and the size and function of the structural proteins are quite similar to the corresponding alphavirus entities. This suggests that rubiviruses and alphaviruses are in fact evolutionarily related and belong to one virus family. Nucleic acid and amino acid sequence data of rubella virus, when it is obtained, will be of great interest in defining in greater detail the relationships of these viruses.

Insufficient data exist for other viruses currently classified as togaviruses to judge their evolutionary relationship to the alphaviruses.

D. Relationship of Alphaviruses to Other RNA Viruses

With the increasing number of RNA virus sequences being obtained, it has become possible to explore the evolutionary relationships among different groups of RNA viruses. From the previous discussion of codon usage randomization among alphaviruses, it is obvious that sequence relationships should be sought at the amino acid level, rather than at the nucleotide level. Examination of amino acid sequence homologies by computer search programs has shown that the replicase of Sindbis virus shares three domains of sequence homology with the replicases of three plant viruses, tobacco mosaic virus (TMV), bromegrass mosaic virus (BMV), and alfalfa mosaic virus (AMV) (Haseloff *et al.*, 1984; Ahlquist *et al.*, 1985). Figure 18 illustrates in a schematic fashion the genome organization of these four groups of viruses and the domains in which amino acid sequence homology has been found. These four groups of viruses all share similarities in their replication strategies. The similarities between alphaviruses and TMV and are particularly striking (Fig. 18). Both produce two large proteins by translation of the infectious genomic RNA. The first protein terminates at an amber codon in TMV and at an opal codon in alphaviruses and contains two of the conserved domains (Ahlquist *et al.*, 1985). In alphaviruses, but not in TMV, posttranslational cleavage occurs to separate the two conserved domains into nsP1 and nsP2 and to produce a third nonstructural protein, nsP3, that shares no detectable sequence homology with the plant virus proteins. In both viruses, a second nonstructural protein is translated by readthrough of the termination codon (E. G. Strauss *et al.*, 1983; Pelham, 1978), and the readthrough portions share amino acid sequence homology (Haseloff *et al.*, 1984). Proteolytic cleavage separates the readthrough region as nsP4 in the alphaviruses, but not in TMV. In both viruses, subgenomic RNAs are tran-

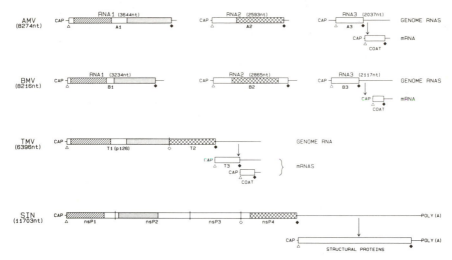

FIGURE 18. Genome organization and amino acid homologies of Sindbis virus (SIN) and three plant viruses: alfalfa mosaic virus (AMV), bromegrass mosiac virus (BMV), and tobacco mosaic virus (TMV). The conventions are the same as those used in Fig. 1. Within the translated regions (open boxes), there are three areas of homology indicated with different types of shading (hatching, stippling, and cross-hatching). All genomes are shown to scale. Reproduced from Ahlquist *et al.* (1985) with the permission of the American Society for Microbiology.

scribed that are translated into the virion structural proteins. In TMV, there is also a second subgenomic RNA produced that is translated into a protein of unknown function. BMV and AMV differ from this scheme only in having the genome divided into three segments (Fig. 18). Thus, to produce the domain corresponding to nsP4 (A2 or B2) requires translation of a separate genomic segment, rather than readthrough; the amount produced could be regulated by a differential efficiency of translation of RNA2. A subgenomic RNA is also transcribed that is translated into the structural protein.

These four groups of viruses have quite different structures, however. Alphaviruses, as detailed in Chapter 2, have an icosahedral nucleocapsid containing the RNA that is surrounded by a lipid bilayer containing two virus-specific glycoproteins. TMV is a rod-shaped virus built on helical symmetry principles, and only a single species of coat protein is used to construct the virion. BMV is a simple icosahedron that again is constructed with a single species of coat protein. Finally, AMV is bacilliform. Perhaps for this reason, at least in part, no sequence homologies can be detected among the virion structural proteins. As noted earlier, the structural proteins of the alphaviruses are more divergent than the nonstructural proteins, and if these four groups of viruses are descended from a common ancestor, it would appear that the structural proteins have diverged so extensively that new forms of virus structure have evolved and no residual sequence homology exists. As we have discussed in a recent

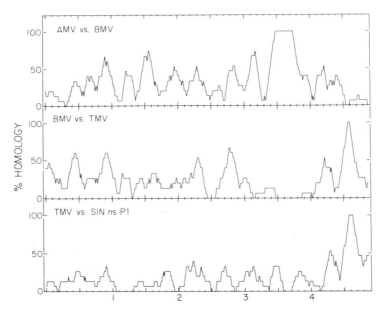

FIGURE 19. Graphic representation of amino acid homologies between nsP1 of Sindbis virus (SIN) and comparable polypeptides encoded by three plant viruses, in the regions shown by hatching in Fig. 18. The sequences reported for BMV (Ahlquist *et al.*, 1984a), AMV (Cornelissen *et al.*, 1983a,b), TMV (Goelet *et al.*, 1982), and SIN (E. G. Strauss *et al.*, 1984) were aligned as reported in Ahlquist *et al.* (1985). The AMV and BMV sequences begin at amino acids 55 and 36, respectively, in the open reading frame of RNA 1; the TMV sequence begins at amino acid 43 of the 126K protein; and the SIN sequence begins with the initiating Met of nsP1. *Ordinate:* percentage homology as a moving average with a search string of 15; *abscissa:* amino acid residues in hundreds. Gaps have been introduced for alignment and treated as nonmatches.

review (E. G. Strauss and J. H. Strauss, 1983), similarities in replication strategy and genome organization may be a more accurate predictor of evolutionary relationships than virus structure.

The amino acid homologies among the conserved domains of the replicases for BMV, AMV, TMV, and Sindbis virus were compare by plotting in pairwise combinations the percentage homology as a moving average in the region corresponding to nsP1 (Fig. 19), to nsP2 (Fig. 20), and to nsP4 (Fig. 21). In each figure, the top panel illustrates AMV and BMV, the two tripartite viruses, the second panel is BMV vs. TMV, and the bottom panel is TMV vs. Sindbis virus. In each case, the homology is by no means uniform along the region, but rather consists of a number of conserved domains interspersed with regions of little or no homology.

As expected, overall the homology is highest between the two tripartite plant viruses, followed by TMV and BMV; Sindbis virus has fewer regions in common with plant viruses. However, there are a few surprises in this analysis. One example is the region of nsP1 between amino acids 340 and 380 as plotted (Fig. 19), in which AMV and BMV are identical

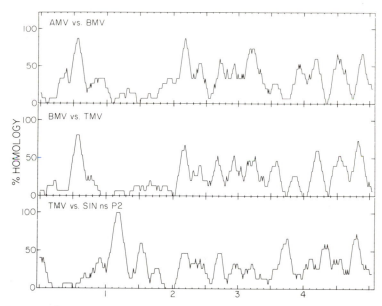

FIGURE 20. Graphic representation of amino acid homologies between nsP2 of Sindbis virus (SIN) and portions of the plant virus replicases, in the region shown by stippling in Fig. 18. Conventions and sequence sources are the same as in Figs. 18 and 19. The AMV and BMV sequences begin at amino acids 641 and 510, respectively, of the open reading frame of RNA 1; the TMV sequence begins at amino acid 735 of the 126K protein; the SIN sequence begins with amino acid 30 of nsP2.

but there is no homology with TMV or Sindbis virus. This region presumably encodes some function that is required only for a tripartite genome. Similarily, in the nsP2 region (Fig. 20), the highest homology between Sindbis and TMV occurs in a region of nonhomology among plant viruses (amino acids 110–130 as plotted) and may encode some function related to readthrough as a translation strategy or to having a nonsegmented genome. The homology profiles in the region corresponding to nsP4 are, on the contrary, similar for all three pairwise comparisons. In each case, the homologies extend over hundreds of amino acids, so the matchups cannot be due to chance. It seems clear that the replicases of these four groups of viruses share a common origin and that in all probability these four groups of viruses have diverged from a common ancestor. (A further discussion of the homologies to flaviviruses will be found in Chapter 10.)

The similarities in replication and translation strategy suggest that certain domains of the viral replicases have been conserved because they serve equivalent functions. Even though evolution of RNA viruses is rapid, the internal milieu of eukaryotic cells changes only slowly with time, and the replicases, once evolved to replicate virus RNA efficiently and rapidly in the cell cytoplasm, might evolve much more slowly than

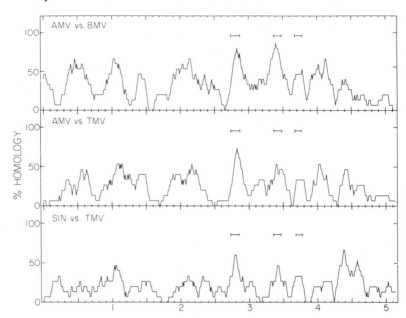

FIGURE 21. Graphic representation of amino acid homologies between nsP4 of Sindbis virus (SIN) and those portions of the plant virus replicases indicated by cross-hatching in Fig. 18. Conventions are the same as in Figs. 18 and 19. The AMV and BMV sequences begin with amino acids 265 and 202, respectively, of the open reading frame of RNA 2. The TMV sequence begins at the amber codon punctuating the open reading frame. The SIN sequence starts with amino acid 101 of nsP4. Bars show the location of the conserved domains illustrated in Fig. 15 of Chapter 10.

structural proteins, which interact with the external environment, including vertebrate immune systems.

Although we believe that the alphaviruses and these three groups of plant viruses diverged from a common ancestral protovirus, and moreover that all RNA viruses probably diverged from a common ancestor, it is possible that RNA viruses emerged more than once and in so doing captured the same host polymerase or related host polymerases that had in turn evolved from a common ancestral protein. If we take the viewpoint that RNA viruses arose but once (or arose only a very limited number of times), the known rapidity of RNA evolution (Holland *et al.*, 1982) suggests that the alphaviruses and the plant viruses TMV, BMV, and AMV diverged in the not too distant past and, indeed, that perhaps all the current eukaryotic RNA viruses are a relatively recent invention. Two other groups of RNA plant viruses and RNA animal viruses have also been shown to be related: the picornaviruses to the cowpea mosaic virus group (Franssen *et al.*, 1984) and the retroviruses to the cauliflower mosaic virus group (Toh *et al.*, 1983). It also appears that minus-stranded RNA viruses are all related in that influenza and vesicular stomatitis virus matrix proteins have been shown to be homologous (Rose *et al.*, 1982) and they all possess similar replication strategies (reviewed in E. G. Strauss and J. H. Strauss, 1983). Furthermore, Kamer and Argos (1984) have recently found amino acid sequence homologies among a large group of RNA viruses (see Chapter 10 for a more complete discussion of this topic). Thus, it appears possible that the eukaryotic RNA viruses currently extant diverged from one or a few protoviruses. It is attractive to speculate that these protoviruses may have arisen in insects. An argument for an insect virus ancestor has been made for the alphaviruses based on the symptomless infection of insects by these viruses contrasted with the often severe symptoms induced in vertebrates. The hypothesis is also attractive from a radiation standpoint. Many viruses other than the alphaviruses are known that are able to replicate in insects and in vertebrates. Similarly, many plant viruses are known that can replicate in both plants and insects. Thus, an insect protovirus could radiate in principle to both plants and animals. Once radiation occurred, the viruses may in some cases retain the ability to replicate in insects as well as in their new hosts, whereas in other cases, the ability to replicate in insects may have been lost. Although lack of a fossil record for viruses may prevent us from ever answering such questions on the origin of viruses, further analysis of RNA sequence data may lead to some surprising conclusions.

VII. CONCLUDING REMARKS

The recent advent of rapid techniques for nucleotide and amino acid sequence analysis that require only small amounts of material has led to the gathering of a great deal of data on the structure of the alphavirus genome and on the evolutionary relationships of these viruses to one another and to other RNA viruses. It is to be expected that as more data are obtained, we will understand in greater detail the replication events that occur in alphavirus infection and be able to use more direct approaches for developing alphavirus vaccines. It is obvious that techniques are lacking in two areas if we are to understand alphavirus replication completely. One is the need to develop an *in vitro* RNA replication system that is template-dependent and capable of initiation; ultimately, the most useful system would be one that would replicate alphavirus RNA to produce an infectious molecule. This has proven to be a most difficult area of endeavor, but an *in vitro* system will eventually be required if we are to understand the functions of the replicase proteins. It is to be hoped that some laboratory succeeds in doing this in the not too distant future. The second area is the need to express cloned copies of alphavirus complementary DNA. The expression of DI genomes described in Chapter 6 is a good start in this direction, as is the expression of the structural protein region of Sindbis virus in a vaccinia vector to produce proteins that can be incorporated into infectious virions (Rice *et al.*, 1985). Ultimately, however, it would be very useful to be able to obtain an infectious

RNA molecule from a cloned DNA copy, as has been done for poliovirus (Racaniello and Baltimore, 1981) and for bromegrass mosaic virus (Ahlquist *et al.*, 1984b). Such systems would allow testing of the models that implicate certain RNA sequences in RNA replication and encapsidation. For example, control sequences could be put next to unrelated genes, and if correctly identified, the synthetic molecule should replicate and be encapsidated when coinfected with a helper virus to furnish the replicase and the structural proteins. Such a system would also allow methods such as site-specific mutagenesis to be used to change key nucleotides and assay the effect on replication efficiency, as well as numerous other experiments that will be obvious to the reader.

ACKNOWLEDGMENTS.   The work of the authors is supported by NIH grants AI 10793 and AI 20612 and by NSF grant DMB-8316856. We are grateful to Drs. D. Sawicki, R. Kinney, B. Johnson, and D. Trent for making data available to us prior to their publication.

REFERENCES

Ahlquist, P., Dasgupta, R., and Kaesberg, P., 1984a, Nucleotide sequence of the brome mosaic virus genome and its implications for viral replication, *J. Mol. Biol.* **172:**369–383.

Ahlquist, P., French, R., Janda, M., and Loesch-Fries, L. S., 1984b, Multicomponent RNA plant virus infection derived from cloned viral cDNA, *Proc. Natl. Acad. Sci. U.S.A.* **81:**7066–7070.

Ahlquist, P., Strauss, E. G., Rice, C. M., Strauss, J. H., Haseloff, J., and Zimmern, D., 1985, Sindbis virus proteins nsP1 and nsP2 contain homology to nonstructural proteins from several RNA plant viruses, *J. Virol.* **53:**536–542.

Aliperti, G., and Schlesinger, M. J., 1978, Evidence for an autoprotease activity of Sindbis virus capsid protein, *Virology* **90:**366–369.

Arias, C., Bell, J. R., Lenches, E. M., Strauss, E. G., and Strauss, J. H., 1983, Sequence analysis of two mutants of Sindbis virus defective in the intracellular transport of their glycoproteins, *J. Mol. Biol.* **168:**87–102.

Baric, R. S., Carlin, L. J., and Johnston, R. E., 1983, Requirement for host transcription in the replication of Sindbis virus, *J. Virol.* **45:**200–205.

Bell, J. R., and Strauss, J. H., 1981, *In vivo* NH_2-terminal acetylation of Sindbis virus proteins, *J. Biol. Chem.* **256:**8006–8011.

Bell, J. R., Hunkapiller, M. W., Hood, L. E., and Strauss, J. H., 1978, Amino-terminal sequence analysis of the structural proteins of Sindbis virus, *Proc. Natl. Acad. Sci. U.S.A.* **75:**2722–2726.

Bell, J. R., Rice, C. M., Hunkapiller, M. W., and Strauss, J. H., 1982, The N-terminus of PE2 in Sindbis virus-infected cells, *Virology* **119:**255–267.

Bell, J. R., Kinney, R. M., Trent, D. W., Strauss, E. G., and Strauss, J. H., 1984, An evolutionary tree relating eight alphaviruses based on amino-terminal sequences of their glycoproteins, *Proc. Natl. Acad. Sci. U.S.A.* **81:**4702–4706.

Bhatti, A. R., and Weber, J., 1979, Protease of adenovirus type 2: Partial characterization, *Virology* **96:**478–485.

Birdwell, C. R., Strauss, E. G., and Strauss, J. H., 1973, Replication of Sindbis virus. III. An electron microscopic study of virus maturation using the surface replica technique, *Virology* **56:**429–438.

Blobel, G., 1980, Intracellular protein topogenesis, *Proc. Natl. Acad. Sci. U.S.A.* **77:**1496–1500.

Blobel, G., Walter, P., Chang, C.-N., Goldman, B. M., Erickson, A. H., and Lingappa, V. R., 1979, Translocation of proteins across membranes: The signal hypothesis and beyond, in: *Secretory Mechanisms*, pp. 9-36, Cambridge University Press, Cambridge.

Boege, U., Wengler, G., Wengler, G., and Wittmann-Liebold, B., 1981, Primary structures of the core proteins of the alphaviruses Semliki Forest virus and Sindbis virus, *Virology* **113:**293–303.

Bonatti, S., and Blobel, G., 1979, Absence of a cleavable signal sequence in Sindbis virus glycoprotein PE$_2$, *J. Biol. Chem.* **254:**12261–12264.

Bonatti, S., Cancedda, R., and Blobel, G., 1979, Membrane biogenesis: *In vitro* cleavage, core glycosylation, and integration into microsomal membranes of Sindbis virus glycoproteins, *J. Cell Biol.* **80:**219–224.

Bracha, M., and Schlesinger, M. J., 1976, Defects in RNA$^+$ temperature-sensitive mutants of Sindbis virus and evidence for a complex of PE2–E1 viral glycoproteins, *Virology* **74:**441–449.

Bracha, M., Leone, A., and Schlesinger, M. J., 1976, Formation of a Sindbis virus nonstructural protein and its relation to 42 S mRNA function, *J. Virol.* **20:**612–620.

Brown, D. T., 1980, The assembly of alphaviruses, in *The Togaviruses* (R. W. Schlesinger, ed.), pp. 473–501, Academic Press, New York.

Brzeski, H., and Kennedy, S. I. T., 1977, Synthesis of Sindbis virus nonstructural polypeptides in chicken embryo fibroblasts, *J. Virol* **22:**420–429.

Brzeski, H., and Kennedy, S. I. T., 1978, Synthesis of alphavirus-specified RNA, *J. Virol.* **25:**630–640.

Burge, B. W., and Pfefferkorn, E. R., 1966, Complementation between temperature-sensitive mutants of Sindbis virus, *Virology* **30:**214–223.

Burge, B. W., and Pfefferkorn, E. R., 1968, Functional defects of temperature-sensitive mutants of Sindbis virus, *J. Mol. Biol.* **35:**193–205.

Calisher, C. H., Shope, R. E., Brandt, W., Casals, J., Karabatsos, N., Murphy, F. A., Tesh, R. B., and Wiebe, M. E., 1980, Proposed antigenic classification of registered arboviruses. I. Togaviridae, alphavirus, *Intervirology* **14:**229–232.

Cancedda, R., Swanson, R., and Schlesinger, M. J., 1974, Effects of different RNAs and components of the cell-free system on *in vitro* synthesis of Sindbis viral proteins, *J. Virol.* **14:**652–663.

Chanas, A. C., Gould, E. A., Clegg, J. C. S., and Varma, M. G. R., 1982, Monoclonal antibodies to Sindbis virus glycoprotein E1 can neurtralize, enhance infectivity and independently inhibit haemagglutination of haemolysis, *J. Gen. Virol.* **58:**37–46.

Clegg, J. C. S., and Kennedy, S. I. T., 1975, Translation of Semliki-Forest-virus intracellular 26-S RNA, *Eur. J. Biochem.* **53:**175–183.

Clegg, J. C. S., Brzeski, H., and Kennedy, S. I. T., 1976, RNA polymerase components in Semliki Forest virus-infected cells: Synthesis from large precursors, *J. Gen. Virol.* **32:**413–430.

Clewley, J. P., and Kennedy, S. I. T., 1976, Purification and polypeptide composition of Semliki Forest virus RNA polymerase, *J. Gen. Virol.* **32:**395–411.

Collins, P. L., Fuller, F. J., Marcus, P. I., Hightower, L. E., and Ball, L. A., 1982, Synthesis and processing of Sindbis virus nonstructural proteins *in vitro*, *Virology* **118:**363–379.

Cornelissen, B. J. C., Brederode, F. T., Moormann, R. J. M., and Bol, J. F., 1983a, Complete nucleotide sequence of alfalfa mosaic virus RNA1, *Nucleic Acids Res.* **11:**1253–1265.

Cornelissen, B. J. C., Brederode, F. T., Veeneman, G. H., van Boom, J. H., and Bol, J. F., 1983b, Complete nucleotide sequence of alfalfa mosiac virus RNA 2, *Nucleic Acids Res.* **11:**3019–3025.

Cross, R. K., 1983, Identification of a unique guanine-7-methyltransferase in Semliki Forest virus (SFV) infected cell extracts, *Virology* **130:**452–463.

Cross, R. K., and Gomatos, P. J., 1981, Concomitant methylation and synthesis *in vitro* of

Semliki Forest virus (SFV) ss RNAs by a fraction from infected cells, *Virology* **114**:542–554.

Dalgarno, L., Rice, C. M., and Strauss, J. H., 1983, Ross River virus 26S RNA: Complete nucleotide sequence and deduced sequence of the encoded structural proteins, *Virology* **129**:170–187.

Dalrymple, J. M., Schlesinger, S., and Russell, P. K., 1976, Antigenic characterization of two Sindbis envelope glycoproteins separated by isoelectric focusing, *Virology* **69**:93–103.

Davis, R. W., Simon, M., and Davidson, N., 1971, Electron microscope heteroduplex methods for mapping regions of base sequence homology in nucleic acids, in: *Methods of Enzymology*, Vol. XX (L. Grossman and K. Moldave, eds.), pp. 413–428, Academic Press, New York.

Diamond, A., Dudock, B., and Hatfield, D., 1981, Structure and properties of a bovine liver UGA suppressor serine tRNA with a tryptophan anticodon, *Cell* **25**:497–506.

Domingo, E., Sabo, D., Taniguchi, T., and Weissmann, C., 1978, Nucleotide sequence heterogeneity of an RNA phage population, *Cell* **13**:735–744.

Dubin, D. T., Stollar, V., HsuChen, C.-C., Timko, K., and Guild, G. M., 1977, Sindbis virus messenger RNA: The 5′-termini and methylated residues of 26 and 42S RNA, *Virology* **77**:457–470.

Eaton, B. T., and Faulkner, P., 1972, Heterogeneity in the poly(A) content of the genome of Sindbis virus, *Virology* **50**:865–873.

Engelberg-Kulka, H., Dekel, L., Israeli-Reches, M., and Belfort, M., 1979, The requirement of nonsense suppression for the development of several phages, *Mol. Gen. Genet.* **170**:155–159.

France, J. K., Wyrick, B. C., and Trent, D. W., 1979, Biochemical and antigenic comparisons of the envelope glycoproteins of Venezuelan equine encephalomyelitis virus strains, *J. Gen. Virol.* **44**:725–740.

Franssen, H., Leunissen, J., Goldbach, R., Lomonossoff, G., and Zimmern, D., 1984, Homologous sequences in non-structural proteins from cowpea mosaic virus and picornaviruses, *Eur. Mol. Biol. Org. J.* **3**:855–861.

Frey, T. K., and Strauss, J. H., 1978, Replication of Sindbis virus. VI. Poly(A) and poly(U) in virus-specific RNA species, *Virology* **86**:494–506.

Frey, T. K., Gard, D. L., and Strauss, J. H., 1979, Biophysical studies on circle formation by Sindbis virus 49S RNA, *J. Mol. Biol.* **132**:1–18.

Friedman, R. M., Levin, J. G., Grimley, P. M., and Berezesky, I. K., 1972, Membrane-associated replication complex in arbovirus infection, *J. Virol.* **10**:504–515.

Fukuda, M., Ohno, T., Okada, Y., Otsuki, Y., and Takebe, I., 1978, Kinetics of biphasic reconstitution of tobacco mosaic virus *in vitro*, *Proc. Natl. Acad. Sci. U.S.A.* **75**:1727–1730.

Fuller, F. J., and Marcus, P. I., 1980, Sindbis virus. I. Gene order of translation *in vivo*, *Virology* **107**:441–451.

Garoff, H., and Söderlund, H., 1978, The amphiphilic membrane glycoproteins of Semliki Forest virus are attached to the lipid bilayer by their COOH-terminal ends, *J. Mol. Biol.* **124**:535–549.

Garoff, H., Simons, K., and Renkonen, O., 1974, Isolation and characterization of the membrane proteins of Semliki Forest virus, *Virology* **61**:493–504.

Garoff, H., Simons, K., and Dobberstein, B., 1978, Assembly of the Semliki Forest virus membrane glycoproteins in the membrane of the endoplasmic reticulum *in vitro*, *J. Mol. Biol.* **124**:587–600.

Garoff, H., Frischauf, A.-M., Simons, K., Lehrach, H., and Delius, H., 1980a, The capsid protein of Semliki Forest virus has clusters of basic amino acids and prolines in its amino-terminal region, *Proc. Natl. Acad. Sci. U.S.A.* **77**:6376–6380.

Garoff, H., Frischauf, A.-M., Simons, K., Lehrach, H., and Delius, H., 1980b, Nucleotide sequence of cDNA coding for Semliki Forest virus membrane glycoproteins, *Nature* (*London*) **288**:236–241.

Geller, A. I., and Rich, A., 1980, A UGA termination suppression tRNA^Trp active in rabbit reticulocytes, *Nature (London)* **283**:41–46.

Gliedman, J. B., Smith, J. F., and Brown, D. T., 1975, Morphogenesis of Sindbis virus in cultured *Aedes albopictus* cells, *J. Virol.* **16**:913–926.

Goelet, P., Lomonossoff, G. P., Butler, P. J. G., Akam, M. E., Gait, M. J., and Karn, J., 1982, Nucleotide sequence of tobacco mosaic virus RNA, *Proc. Natl. Acad. Sci. U.S.A.* **79**:5818–5822.

Gomatos, P. J., Kääriäinen, L., Keränen, S., Ranki, M., and Sawicki, D. L., 1980, Semliki Forest virus replication complex capable of synthesizing 42S and 26S nascent RNA chains, *J. Gen. Virol.* **49**:61–69.

Grantham, R., Gautier, C., Gouy, M., Jacobzone, M., and Mercier, R., 1981, Codon catalog usage is a genome strategy modulated for gene expressivity, *Nucleic Acids Res.* **9**:r43–r74.

Grimley, P. M., Berezesky, I. K., and Friedman, R. M., 1968, Cytoplasmic structures associated with an arbovirus infection: Loci of viral ribonucleic acid synthesis, *J. Virol.* **2**:1326–1338.

Grimley, P. M., Levin, J. G., Berezesky I. K., and Friedman, R. M., 1972, Specific membranous structures associated with the replication of Group A arboviruses, *J. Virol.* **10**:492–503.

Gupta, K. C., and Kingsbury, D. W., 1984, Complete sequences of the intergenic and mRNA start signals in the Sendai virus genome: Homologies with the genome of vesicular stomatitis virus, *Nucleic Acids Res.* **12**:3829–3841.

Hahn, C. S., Strauss, E. G., and Strauss, J. H., 1985, Sequence analysis of three Sindbis virus mutants temperature-sensitive in the capsid protein autoprotease, *Proc. Natl. Acad. Sci. U.S.A.* **82**:4648–4652.

Haseloff, J., Goelet, P., Zimmern, D., Ahlquist, P., Dasgupta, R., and Kaesberg, P., 1984, Striking similarities in amino acid sequence among nonstructural proteins encoded by RNA viruses that have dissimilar genomic organization, *Proc. Natl. Acad. Sci. U.S.A.* **81**:4358–4362.

Hashimoto, K., Erdei, S., Keränen, S., Saraste, J., and Kääriäinen, L., 1981, Evidence for a separate signal sequence for the carboxy-terminal envelope glycoprotein E1 of Semliki Forest virus, *J. Virol.* **38**:34–40.

Hatfield, D. L., Dudock, B. S., and Eden, F. C., 1983, Characterization and nucleotide sequence of a chicken gene encoding an opal suppressor tRNA and its flanking DNA segments, *Proc. Natl. Acad. Sci. U.S.A.* **80**:4940–4944.

Hefti, E., Bishop, D. H. L., Dubin, D. T., and Stollar, V., 1976, 5′Nucleotide sequence of Sindbis viral RNA, *J. Virol.* **17**:149–159.

Helenius, A., Fries, E., Garoff, H., and Simons, K., 1976, Solubilization of the Semliki Forest virus membrane with sodium deoxycholate, *Biochim. Biophys. Acta* **436**:319–334.

Hirsh, D., and Gold, L., 1971, Translation of the UGA triplet *in vitro* by tryptophan transfer RNA's, *J. Mol. Biol.* **58**:459–468.

Holland, J. J., Kennedy, S. I. T., Semler, B. L., Jones, C. L., Roux, L., and Grabau, E. A., 1980, Defective interfering RNA viruses and the host-cell response, in: *Comprehensive Virology*, Vol. 16 (H. Fraenkel-Conrat and R. R. Wagner, eds.), pp. 137–192, Plenum Press, New York.

Holland, J., Spindler, K., Horodyski, F., Grabau, E., Nichol, S., and VandePol, S., 1982, Rapid evolution of RNA genomes, *Science* **215**:1577–1585.

Hsu, M. T., Kung, H. J., and Davidson, N., 1973, An electron microscope study of Sindbis virus RNA, *Cold Spring Harbor Symp. Quant. Biol.* **38**:943–950.

HsuChen, C.-C., and Dubin, D. T., 1976, Di- and trimethylated cogeners of 7-methylguanine in Sindbis virus mRNA, *Nature (London)* **264**:190–191.

Huth, A., Rapoport, T. A., and Kääriäinen, L., 1984, Envelope proteins of Semliki Forest virus synthesized in *Xenopus* oocytes are transported to the cell surface, *Eur. Mol. Biol. Org. J.* **3**:767–771.

Ishida, I., Simizu, B., Koizumi, S., Oya, A., and Yamada, M., 1981, Nucleoside triphosphate phosphohydrolase produced in BHK cells infected with Western equine encephalitis virus is probably associated with the 82 K dalton nonstructural polypeptide, *Virology* **108:**13–20.

Johnston, R. E., and Bose, H. R., 1972, An adenylate-rich segment in the virion RNA of Sindbis virus, *Biochem. Biophys. Res. Commun.* **46:**712–718.

Jones, K. J., Scupham, R. K., Pfeil, J. A., Wan, K., Sagik, B. P., and Bose, H. R., 1977, Interaction of Sindbis virus glycoproteins during morphogenesis, *J. Virol.* **21:**778–787.

Kääriäinen, L., Simons, K., and von Bonsdorff, C.-H., 1969, Studies in subviral components of Semliki Forest virus, *Ann. Med. Exp. Fenn.* **17:**25–248.

Kääriäinen, L., Sawicki, D., and Gomatos, P. J., 1978, Cleavage defect in the nonstructural polyprotein of Semliki Forest virus has two separate effects on viral RNA synthesis, *J. Gen. Virol.* **39:**463–473.

Kamer, G., and Argos, P., 1984, Primary structural comparison of RNA-dependent polymerases from plant, animal and bacterial viruses, *Nucleic Acids Res.* **12:**7269–7282.

Kennedy, S. I. T., 1976, Sequence relationships between the genomic and the intracellular RNA species of standard and defective-interfering Semliki Forest virus, *J. Mol. Biol.* **108:**491–511.

Keränen, S., and Kääriäinen, L., 1979, Functional defects of RNA-negative temperature-sensitive mutants of Sindbis and Semliki Forest virus, *J. Virol.* **32:**19–29.

Keränen, S., and Ruohonen, L., 1983, Nonstructural proteins of Semliki Forest virus: Synthesis, processing and stability in infected cells, *J. Virol.* **47:**505–515.

Koizumi, S., Simizu, B., Ishida, I., Oya, A., and Yamada, M., 1979, Inhibition of DNA synthesis in BHK cells infected with Western equine encephalitis virus. 2. Properties of the inhibitory factor of DNA polymerase induced in infected cells, *Virology* **98:**439–447.

Kondor-Koch, C., Riedel, H., Söderberg, K., and Garoff, H., 1982, Expression of the structural proteins of Semliki Forest virus from cloned cDNA microinjected into the nucleus of the baby hamster kidney cells, *Proc. Natl. Acad. Sci. U.S.A.* **79:**4525–4529.

Kowal, K. J., and Stollar, V., 1981, Temperature-sensitive host-dependent mutants of Sindbis virus, *Virology* **114:**140–148.

Kyte, J., and Doolittle, R. F., 1982, A simple method for displaying the hydropathic character of a protein, *J. Mol. Biol.* **157:**105–132.

Lachmi, B., and Kääriäinen, L., 1976, Sequential translation of nonstructural proteins in cells infected with a Semliki Forest virus mutant, *Proc. Natl. Acad. Sci. U.S.A.* **73:**1936–1940.

Lachmi, B., Glanville, N., Keränen, S., and Kääriäinen, L., 1975, Tryptic peptide analysis of nonstructural and structural precursor proteins from Semliki Forest virus mutant-infected cells. *J. Virol.* **16:**1615–1629.

Lehtovaara, P., Söderlund, H., Keränen, S., Pettersson, R. F., and Kääriänien, L., 1981, 18S defective interfering RNA of Semliki Forest virus contains a triplicated linear repeat, *Proc. Natl. Acad. Sci. U.S.A.* **78:**5353–5357.

Lehtovaara, P., Söderlund, H., Keränen, S., Pettersson, R. F., and Kääriäinen, L., 1982, Extreme ends of the genome are conserved and rearranged in the defective interfering RNAs of Semliki Forest virus, *J. Mol. Biol.* **156:**731–748.

Lopez, S., Bell, J. R., Strauss, E. G., and Strauss, J. H., 1985, The nonstructural proteins of Sindbis virus as studied with an antibody specific for the C terminus of the nonstructural readthrough polyprotein, *Virology* **141:**235–247.

Martin, E. M., and Sonnabend, J. A., 1967, Ribonucleic acid polymerase catalyzing synthesis of double-stranded arbovirus ribonucleic acid, *J. Virol.* **1:**97–109.

Mayne, J. T., Rice, C. M., Strauss, E. G., Hunkapiller, M. W., and Strauss, J. H., 1984, Biochemical studies of the maturation of the small Sindbis virus glycoprotein E3, *Virology* **134:**338–357.

Michel, M. R., and Gomatos, P. J., 1973, Semliki Forest virus-specific RNAs synthesized *in vitro* by enzyme from infected BHK cells, *J. Virol.* **11:**900–914.

Monroe, S. S., and Schlesinger, S., 1983, RNAs from two independently isolated defective interfering particles of Sindbis virus contain a cellular tRNA sequence at their 5' ends, *Proc. Natl. Acad. Sci. U.S.A.* **80:**3279–3283.

Monroe, S. S., Ou, J.-H., Rice, C. M., Schlesinger, S., Strauss, E. G., and Strauss, J. H., 1982, Sequence analysis of cDNA's derived from the RNA of Sindbis virions and of defective interfering particles, *J. Virol.* **41:**153–162.

Oker-Blom, C., 1984, The gene order for Rubella virus structural proteins is NH$_2$-C-E2-E1-COOH, *J. Virol.* **51:**354–358.

Oker-Blom, C., Kalkkinen, N., Kääriäinen, L., and Pettersson, R. F., 1983, Rubella virus contains one capsid protein and three envelope glycoproteins, E1, E2a, and E2b, *J. Virol.* **46:**964–973.

Oker-Blom, C., Ulmanen, I., Kääriäinen, and Pettersson, R. F., 1984, Rubella virus 40S genome RNA specifies a 24S subgenomic mRNA that codes for a precursor to structural proteins, *J. Virol.* **49:**403–408.

Otsuki, Y., Takabe, I., Ohno, T., Fukuda, M., and Okada, Y., 1977, Reconstitution of tobacco mosaic virus rods occurs bidirectionally from an internal initiation region: Demonstration by electron microscopic serology, *Proc. Natl. Acad. Sci. U.S.A.* **74:**1913–1917.

Ou, J.-H., Strauss, E. G., and Strauss, J. H., 1981, Comparative studies of the 3' terminal sequences of several alphavirus RNAs, *Virology* **109:**281–289.

Ou, J.-H., Rice, C. M., Dalgarno, L., Strauss, E. G., and Strauss, J. H., 1982a, Sequence studies of several alphavirus genomic RNAs in the region containing the start of the subgenomic RNA, *Proc. Natl. Acad. Sci. U.S.A.* **79:**5235–5239.

Ou, J.-H., Trent, D. W., and Strauss, J. H., 1982b, The 3'-non-coding regions of alphavirus RNAs contain repeating sequences, *J. Mol. Biol.* **156:**719–730.

Ou, J.-H., Strauss, E. G., and Strauss, J. H., 1983, The 5'-terminal sequences of the genomic RNAs of several alphaviruses, *J. Mol. Biol.* **168:**1–15.

Palmenberg, A. C., Pallansch, M. A., and Rueckert, R. R., 1979, Protease required for processing picornaviral coat protein resides in the viral replicase gene, *J. Virol.* **32:**770–778.

Pedersen, C. E., Jr., Marker, S. C., and Eddy, G. A., 1974, Comparative electrophoretic studies on the structural proteins of selected Group A arboviruses, *Virology* **60:**312–314.

Pelham, H. R. B., 1978, Leaky UAG termination codon in tobacco mosaic virus RNA, *Nature (London)* **272:**469–471.

Racaniello, V. R., and Baltimore, D., 1981, Cloned poliovirus complementary DNA is infectious in mammalian cells, *Science* **214:**916–919.

Raghow, R. S., Grace, T. D. C., Filshie, B. K., Bartley, W., and Dalgarno, L., 1973, Ross River virus replication in cultured mosquito and mammalian cells: Virus growth and correlated ultrastructural changes, *J. Gen. Virol.* **21:**109–122.

Ranki, M,. and Kääriäinen, L., 1979, Solubilized RNA replication complex from Semliki Forest virus-infected cells, *Virology* **98:**298–307.

Reanney, D. C., 1982, The evolution of RNA viruses, *Annu. Rev. Microbiol.* **36:**47–73.

Rice, C. M., and Strauss, J. H., 1981a, Nucleotide sequence of the 26S mRNA of Sindbis virus and deduced sequence of the encoded virus structural proteins, *Proc. Natl. Acad. Sci. U.S.A.* **78:**2062–2066.

Rice, C. M., and Strauss, J. H., 1981b, Synthesis, cleavage, and sequence analysis of DNA complementary to the 26S messenger RNA of Sindbis virus, *J. Mol. Biol.* **150:**315–340.

Rice, C.M., and Strauss, J. H., 1982, Association of Sindbis virion glycoproteins and their precursors, *J. Mol. Biol.* **154:**325–348.

Rice, C. M., Bell, J. R,. Hunkapiller, M. W., Strauss, E. G., and Strauss, J. H., 1982, Isolation and characterization of the hydrophobic COOH-terminal domains of the Sindbis virion glycoproteins, *J. Mol. Biol.* **154:**355–378.

Rice, C. M., Franke, C. A., Strauss, J. H., and Hruby, D. E., 1985, Expression of Sindbis virus structural proteins via recombinant vaccinia virus: Synthesis, processing, and incorporation into mature Sindbis virions *J. Virol.* **56:**227–239.

Riedel, H., Lehrach, H., and Garoff, H., 1982, Nucleotide sequence at the junction between the nonstructural and the structural genes of the Semliki Forest virus genome, *J. Virol.* **42**:725–729.

Rose, J. K., Doolittle, R. F., Anilionis, A., Curtis, P.J., and Wunner, W. H., 1982, Homology between the glycoproteins of vesicular stomatitis virus and rabies virus, *J. Virol.* **43**:361–364.

Russell, G. J., Walker, P. M. B., Elton, R. A., and Subak-Sharpe, J. H., 1976, Doublet frequency analysis of fractionated vertebrate nuclear DNA, *J. Mol. Biol.* **108**:1–23.

Saraste, J., von Bonsdorff, C.-H., Hashimoto, K., Kääriäinen, L., and Keränen, S., 1980, Semliki Forest virus mutants with temperature-sensitive transport defect of envelope proteins, *Virology* **100**:229–245.

Sawicki, D. L., and Gomatos, P. J., 1976, Replication of Semliki Forest virus: Polyadenylate in plus-strand RNA and polyuridylate in minus-strand RNA, *J. Virol.* **20**:446–464.

Sawicki, D. L., and Sawicki, S. G., 1980, Short-lived minus-strand polymerase for Semliki Forest virus, *J. Virol.* **34**:108–118.

Sawicki, D. L., and Sawicki, S. G., 1985, Functional analysis of the A complementation group mutants of Sindbis HR virus, *Virology* **144**:20–34.

Sawicki, D. L., Kääriäinen, L., Lambek, C., and Gomatos, P. J., 1978, Mechanism for control of synthesis of Semliki Forest virus 26S and 42S RNA, *J. Virol.* **25**:19–27.

Sawicki, D. L., Sawicki, S. G., Keränen, S., and Kääriäinen, L., 1981, Specific Sindbis virus-coded function for minus-strand RNA synthesis, *J. Virol.* **39**:348–358.

Sawicki, S. G., Sawicki, D. L., Kääriäinen, L., and Keränen, S., 1981, A Sindbis virus mutant temperature-sensitive in the regulation of minus-strand RNA synthesis, *Virology* **115**:161–172.

Schlesinger, M. J., and Schlesinger, S., 1973, Large-molecular-weight precursors of Sindbis virus proteins, *J. Virol.* **11**:1013–1016.

Schlesinger, M. J., and Kääriäinen, L., 1980, Translation and processing of alphavirus proteins, in: *The Togaviruses* (R. W. Schlesinger, ed.), pp. 371–392, Academic Press, New York.

Schlesinger, M. J., Schlesinger, S., and Burge, B W., 1972, Identification of a second glycoprotein in Sindbis virus, *Virology* **47**:539–541.

Schmaljohn, A. L., Johnson, E. D., Dalrymple, J. M., and Cole, G. A., 1982, Non-neutralizing monoclonal antibodies can prevent lethal alphavirus encephalitis, *Nature (London)* **297**:70–72.

Schupham, R. K., Jones, K. J., Sagik, B. P., and Bose, H. R., Jr., 1977, Virus-directed post-translational cleavage in Sindbis virus-infected cells, *J. Virol.* **22**:568–571.

Sefton, B. M., 1977, Immediate glycosylation of Sindbis virus membrane proteins, *Cell* **10**:659–668.

Shinnick, T. M., Lerner, R. A., and Sutcliffe, J. G., 1981, Nucleotide sequence of Moloney murine leukaemia virus, *Nature (London)* **293**:543–548.

Simizu, B., Yamamoto, K., Hashimoto, K., and Ogata, T., 1984, Structural proteins of chikungunya virus, *J. Virol.* **51**:254–258.

Simmons, D. T., and Strauss, J. H., 1972, Replication of Sindbis virus. II. Multiple forms of double-stranded RNA isolated from infected cells, *J. Mol. Biol.* **71**:615–631.

Simmons, D. T., and Strauss, J. H., 1974, Translation of Sindbis virus 26S RNA and 49S RNA in lysates of rabbit reticulocytes, *J. Mol. Biol.* **86**:397–409.

Söderlund, H., 1973, Kinetics of formation of Semliki Forest virus nucleocapsid, *Intervirology* **1**:354–361.

Spector, D. H., and Baltimore, D., 1974, Requirement of 3'-terminal poly(adenylic acid) for the infectivity of poliovirus RNA, *Proc. Natl. Acad. Sci. U.S.A.* **71**:2983–2987.

Sreevalsan, T., and Yin, F. H., 1969, Sindbis virus induced viral ribonucleic acid polymerase, *J. Virol.* **3**:599–604.

Stark, C., and Kennedy, S. I. T., 1978, The generation and propagation of defective-interfering particles of Semliki Forest virus in different cell types, *Virology* **89**:285–299.

Stollar, V., 1980, Togaviruses in cultured arthropod cells, in: *The Togaviruses* (R. W. Schlesinger, ed.), pp. 583–621, Academic Press, New York.

Strauss, E. G., 1978, Mutants of Sindbis virus. III. Host polypeptides present in purified HR and ts103 virus particles, *J. Virol.* **28**:466–474.

Strauss, E. G., and Strauss, J. H., 1980, Mutants of alphaviruses: Genetics and physiology, in: *The Togaviruses* (R. W. Schlesinger, ed.), pp. 393–426, Academic Press, New York.

Strauss, E. G., and Strauss, J. H., 1983, Replication strategies of the single stranded RNA viruses of eukaryotes, *Curr. Top. Microbiol. Immunol.* **105**:1–98.

Strauss, E. G., and Strauss, J. H., 1985, Assembly of enveloped animal viruses, in: *Virus Structure and Assembly* (S. Casjens, ed.), pp. 205–234, Jones and Bartlett, Portola Valley, California.

Strauss, E. G., Birdwell, C. R., Lenches, E. M., Staples, S. E., and Strauss, J. H., 1977, Mutants of Sindbis virus. II. Characterization of a maturation-defective mutant, ts103, *Virology* **82**:122–149.

Strauss, E. G., Rice, C. M., and Strauss, J. H., 1983, Sequence coding for the alphavirus nonstructural proteins is interrupted by an opal termination codon, *Proc. Natl. Acad. Sci. U.S.A.* **80**:5271–5275.

Strauss, E. G., Rice, C. M., and Strauss, J. H., 1984, Complete nucleotide sequence of the genomic RNA of Sindbis virus, *Virology* **133**:92–110.

Strauss, J. H., and Strauss, E. G., 1977, Togaviruses, in: *The Molecular Biology of Animal Viruses*, Vol. I (D. P. Nayak, ed.), pp. 111–166, Marcel Dekker, New York.

Strauss, J. H., and Strauss, E. G., 1985, Antigenic structure of Togaviruses, in: *Immunochemistry of Viruses—The Basis for Serodiagnosis and Vaccines* (M. H. V. van Regenmortel and A. R. Neurath, eds.), pp. 407–424, Chapter 22, Elsevier, Amsterdam.

Strauss, J. H., Burge, B. W., Pfefferkorn, E. R., and Darnell, J. E., 1968, Identification of the membrane protein and "core" protein of Sindbis virus, *Proc. Natl. Acad. Sci. U.S.A.* **59**:533–537.

Strauss, J. H., Burge, B. W., and Darnell, J. E., 1969, Sindbis virus infection of chick and hamster cells: Synthesis of virus-specific proteins, *Virology* **37**:367–376.

Tinoco, I., Borer, P. N., Dengler, B., Levine, M. D., Uhlenbeck, O. C., Crothers, D. M., and Gralla, J., 1973, Improved estimation of secondary structure in ribonucleic acids, *Nature (London) New Biol.* **246**:40–41.

Toh, H., Hayashida, H., and Miyata, T., 1983, Sequence homology between retroviral reverse transcriptase and putative polymerases of hepatitis B virus and cauliflower mosaic virus, *Nature (London)* **305**:827–829.

Tsiang, M., Monroe, S. S., and Schlesinger, S., 1985, Studies of defective interfering RNAs of Sindbis virus with and without rRNA[Asp] sequences at their 5' termini, *J. Virol.* **54**:38–44.

Ulmanen, I., Söderlund, H., and Kääriäinen, L., 1976, Semliki Forest virus capsid protein associates with the 60S ribosomal subunit in infected cells, *J. Virol.* **20**:203–210.

Van Steeg, H., Kasperaitis, M., Voorma, H. O., and Benne, R., 1984, Infection of neuroblastoma cells by Semliki Forest virus: The interference of viral capsid protein with the binding of host messenger RNAs into initiation complexes is the cause of the shutoff of host protein synthesis, *Eur. J. Biochem.* **138**:473–478.

Von der Helm, K., 1977, Cleavage of Rous sarcoma viral polypeptide precursor into internal structural proteins *in vitro* involves viral protein p15, *Proc. Natl. Acad. Sci. U.S.A..* **74**:911–915.

Wecker, E., 1959, The extraction of infectious virus nucleic acid with hot phenol, *Virology* **7**:241–243.

Weiner, A. M., and Weber, K., 1971, Natural readthrough at the UGA termination signal of Qβ coat protein cistron, *Nature (London) New Biol.* **234**:206–209.

Welch, W. J., and Sefton, B. M., 1979, Two small virus-specific polypeptides are produced during infection with Sindbis virus, *J. Virol.* **29**:1186–1195.

Welch, W. J., Sefton, B. M., and Esch, F. S., 1981, Amino-terminal sequence analysis of alphavirus polypeptides, *J. Virol.* **38**:968–972.

Wengler, G., and Wengler, G., 1974, Studies on the polyribosome-associated RNA in BHK21 cells infected with Semliki Forest virus, *Virology* **59**:21–35.

Wengler, G., Wengler, G., and Filipe, A. R., 1977, A study of nucleotide sequence homology between the nucleic acids of different alphaviruses, *Virology* **78**:124–134.

Wengler, G., Wengler, G., and Gross, H. J., 1979, Replicative form of Semliki Forest virus RNA contains an unpaired guanosine, *Nature (London)* **282**:754–756.

Wengler, G., Boege, U., Wengler, G., Bischoff, H., and Wahn, K., 1982, The core protein of the alphavirus Sindbis virus assembles into core-like nucleoproteins with the viral genome RNA and with other single-stranded nucleic acids *in vitro*, *Virology* **118**:401–410.

Wirth, D. F., Katz, F., Small, B., and Lodish, H. F., 1977, How a single Sindbis virus mRNA directs the synthesis of one soluble protein and two integral membrane glycoproteins, *Cell* **10**:253–263.

Ziemiecki, A., Garoff, H., and Simons, K., 1980, Formation of the Semliki Forest virus membrane glycoprotein complexes in the infected cell, *J. Gen. Virol.* **50**:111–123.

Zimmern, D., 1977, The nucleotide sequence at the origin for assembly on tobacco mosaic virus RNA, *Cell* **11**:463–482.

CHAPTER 4

Entry of Alphaviruses

Margaret Kielian and Ari Helenius

I. INTRODUCTION

To replicate, viruses must deliver their genomes into the cytoplasm of a host cell, entailing the transport of large macromolecular assemblies through one or more membrane barriers. The problem is not a trivial one, in view of the large size and polar nature of the viral components to be delivered and the fact that both the cell and viral components must remain intact. It is not yet known how most viruses have solved this dilemma, but in the case of enveloped animal viruses, the overall pathway is becoming increasingly clear. The membranes of enveloped viruses serve as transport vesicles between infected cells and new host cells, and the process depends on well-regulated membrane fission and membrane fusion events. The membrane fission reaction occurs when the virus buds from a membrane of the infected host and the membrane fusion reaction when the virus interacts with a membrane of the recipient cell (Fig. 1). During the voyage between the two cells, the viral envelope serves to protect the nucleocapsid. As shown in Fig. 1, the fusion reaction responsible for releasing the genome into the host cell can occur either at the plasma membrane or in the organelles of the endocytotic pathway. The main advantage of this general mechanism seems to be that the genome and accessory proteins do not at any stage need to undergo a direct transfer through a bilayer membrane.

The entry of alphaviruses is exceptionally well understood. Semliki Forest virus (SFV) has served as the prototype for most recent virus-entry studies, and the pathway that has emerged is used not only by alphaviruses, but also by other virus families including orthomyxo- rhabdo-, and retroviruses (Maeda and Ohnishi, 1980; Matlin *et al.*, 1981, 1982; White

MARGARET KIELIAN AND ARI HELENIUS • Department of Cell Biology, Yale School of Medicine, New Haven, Connecticut 06510.

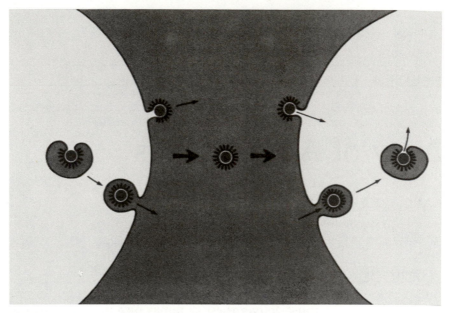

FIGURE 1. Role of membrane fusion and fission in the transmission of virus infection. Enveloped viruses are typically formed by budding and *membrane fission* at a cellular membrane. The membrane can be the plasma membrane or some other membrane in the secretory pathway. Penetration into the cells takes place by *membrane fusion*. It can occur either at the plasma membrane (for viruses with fusion activity at neutral pH) or after endocytosis in intracellular vacuoles. Endocytosis is apparently a requirement for acid-activated viruses including the alphaviruses.

et al., 1981; Redmond *et al.*, 1984). For recent reviews on the topic of virus entry, see Helenius *et al.*, (1980a,b,c), Dimmock (1982), Lenard and Miller (1982), White *et al.* (1983), and Marsh (1984). This chapter will focus on the cellular and molecular aspects of the early interactions of alphaviruses with their host cells in tissue culture. The emphasis will be on the cellular mechanisms exploited by the virus during entry and the mechanism of virus membrane fusion.

II. ATTACHMENT

The first step in any virus-entry process is binding to the plasma membrane of the host cell. The nature of this obligatory step depends on the properties of viral and cellular surface components, and its characteristics are quite diverse among viruses (see Lonberg-Holm and Philipson, 1974; Gallaher and Howe, 1976; Meager and Hughes, 1977). While some viruses display extreme specificity of binding, others utilize common surface groups as receptors. The presence or absence of appropriate binding sites obviously constitutes one of the factors that determine whether a cell can be infected by a virus.

In the case of alphaviruses, the nature of the binding reaction is only partially understood. In keeping with their wide host range, they bind to cells from a variety of tissues and species. Some cultured cell lines can bind up to 10^6 virus particles per cell, almost fully covering the surface of each cell with particles (Birdwell and Strauss, 1974; Fries and Helenius, 1979; Smith and Tignor, 1980), whereas other cell lines show no detectable binding. Where binding does occur, its magnitude does not necessarily correlate with susceptibility to infection; many cells that are refractory to infection bind viruses, and others that are efficiently infected show low binding (Hilfenhaus, 1976; Fries and Helenius, 1979; A. Helenius, unpublished observation).

Studies on virus attachment have usually been performed at low temperatures to block endocytosis and penetration. Aldehyde-fixed cells have also been used. The most important results obtained from such binding studies can be summarized as follows:

1. *Attachment occurs relatively slowly.* Depending on the volume of the medium and the amount of virus added, it takes, for example, 20–40 min to reach half-maximal binding to baby hamster kidney (BHK)-21 cells (Hahon and Cooke, 1967; Fries and Helenius, 1979). The binding step is probably the rate-limiting step in infection at low multiplicity.

2. *Attachment is virtually irreversible.* Only a small fraction of the bound viruses can be eluted off the cell surface by washing. Our estimate for the apparent binding constant for SFV binding to BHK-21 cells is in the 10^{11}–10^{10}/mole range (Fries and Helenius, 1979). Since isolated spike glycoproteins bind to the cells with a more modest affinity ($K_a = 10^7$ M^{-1}), we have concluded that the high-affinity binding of intact viruses is, at least in part, a result of a high valency of attachment (Helenius *et al.*, 1978; Fries and Helenius, 1979). It is clear that numerous spikes participate in tethering the particle to the cell surface (Helenius *et al.*, 1980a).

3. *The viruses are not evenly distributed.* The surface distribution of bound viruses depends on the cell type and the architecture of the cell surface. For most cells, the viruses seem to concentrate on, or near, microvilli (Fig. 2) (Fries and Helenius, 1979; Helenius *et al.*, 1980a,b). Lateral movement of bound Sindbis viruses to form paracrystalline arrays has been described (Birdwell and Strauss, 1974). Individual cells in a culture also vary in their capacity to bind. Despite its high affinity, the attachment does not seem to cause any changes in the shape of the virus particles or the curvature of the plasma membrane. The distance between the unit membranes of the cell and the bound viruses, as shown in thin sections, is $9.0^{\pm 2}$ nm, which is about 2 nm more than the length of the viral spike glycoprotein (Fries and Helenius, 1979). Occasionally, electron-dense material connecting the virus surface with the cell membrane is observed (Helenius *et al.*, 1980a).

4. *Attachment depends on the composition of the medium.* In general, it is found that the extent of binding increases with decreasing pH

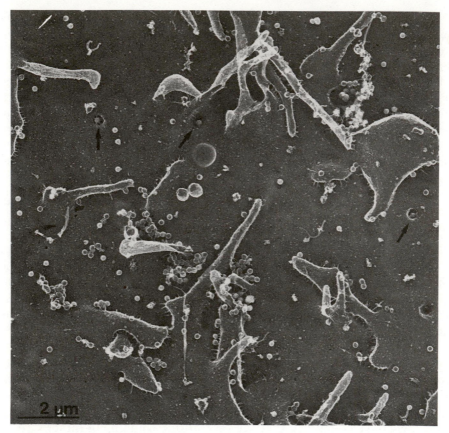

FIGURE 2. SFV attachment and endocytosis in BHK-21 cells. The virus particles can be seen attached to the cell surface with preferential location on or close to the microvilli. Some of the viruses are located within coated pits (see arrows). Transmission electron microscopy after metal shadowing was performed by John Heuser, Washington University, St. Louis.

(Clarke and Casals, 1958; Fries and Helenius, 1979; Marker *et al.*, 1977). Binding at pH 6.2 and below is frequently quite high, but interpretation of this finding (as well as hemagglutination data at acid pH) is complicated by the fact that low pH activates viral membrane fusion. Increasing ionic strength tends to decrease binding, particularly at pH values above 7 (Marker *et al.*, 1977) (but see also Johnston and Faulkner, 1978). The ionic composition can also be important; sodium depletion will elevate attachment by about 20% (Helenius *et al.*, 1985). It is frequently observed, moreover, that the presence of serum is inhibitory (Hahon and Cooke, 1967).

5. *The binding sites are proteinaceous.* Not only is binding to most mammalian cells inhibited by prior digestion of the cells with proteases, but also bound viruses can be efficiently detached from cells by digestion

with the appropriate proteolytic enzymes (Smith and Tignor, 1980; Marsh and Helenius, 1980; Helenius, unpublished results). In contrast, neuraminidase treatment of cells only marginally affects Sindbis binding, indicating that the binding is not dependent on the presence of sialic acid (Smith and Tignor, 1980). Attempts at identifying receptor proteins for SFV and Sindbis virus on lymphoblastoid cells have given some suggestive results (Helenius *et al.*, 1978; Massen and Terhorst, 1981). Using immunological, morphological, and biochemical techniques, it was demonstrated that SFV binds preferentially to the major histocompatibility antigens HLA-A and -B on JY cells and H2-K and -L antigens in peripheral murine lymphocytes (Helenius *et al.*, 1978). It remains unclear whether this interaction is relevant in terms of productive infectivity of SFV; these cells express a large amount of HLA and H2 antigens, but they do not serve as efficient host cells. Oldstone *et al.* (1980) have subsequently shown that cells devoid of H2 antigens can be efficiently infected with SFV. In the Sindbis virus study (Massen and Terhorst, 1981), it was shown that the viral spike glycoproteins could be cross-linked chemically to an M_r 90,000 cell-surface protein in JY and Daudi cells.

In conclusion, one can state that the binding sites are very numerous on the surface of many cells, they contain a protein component, and it is very likely that the receptors can be endocytosed (see below). The wide variability in binding efficiency seems to rule out a totally nonspecific binding mechanism, but the receptors for productive infection need to be identified. The molecular nature of alphavirus binding thus remains unclear.

III. ENDOCYTOTIC UPTAKE AND PENETRATION

A. Cell Biology of Endocytosis

Endocytosis is the general term for the processes whereby cells internalize extracellular substances by vesicles formed through plasma membrane invagination (for reviews, see Silverstein *et al.*, 1977; Goldstein *et al.*, 1979; Steinman *et al.*, 1983). The uptake of large particles is called *phagocytosis*, and it is triggered by the attachment of the particle to the cell. Phagocytic uptake is usually associated with professional phagocytic cells (i.e., macrophages and neutrophils), but it can also be seen in modified forms in other cell types. It relies on the formation of a network of microfilaments in the cytoplasm around the forming phagosome, and it is thus sensitive to inhibition by cytochalasins. *Pinocytosis*, the uptake of fluid, solutes, and small particles (<200 nm in diameter), is, in contrast, an ongoing constitutive process, independent of the presence of ligands. It is responsible for continuous and usually highly voluminous endocytotic traffic in the cell. Pinocytosis does not depend on microfilaments, but, like phagocytosis, it requires metabolic energy.

Pinocytosis is responsible for the uptake of fluid and receptor–ligand complexes as well as for clearing the cell surface of adsorbed opportunistic ligands such as antibodies, lectins, and viruses. Most of these functions seem to occur via coated pits. These are specialized microdomains on the cell surface that trap the cell-surface material to be internalized, invaginate, and pinch off to form coated endocytotic vesicles. These primary endocytotic vesicles are usually about 100–180 nm in diameter and have a characteristic bristle coat on their cytoplasmic surface composed of clathrin and associated proteins. Receptor-mediated endocytosis via coated vesicles constitutes an important physiological mechanism for mediating and modulating hormonal signals, internalizing essential nutrients, and controlling membrane homeostasis.

The life-span of an endocytotic coated vesicle is relatively short; within minutes, the internalized material is found within larger uncoated vacuoles (endosomes) in the peripheral cytoplasm of the cell (see Helenius *et al.*, 1983; Pastan and Willingham, 1983). These usually consist of a central vacuolar body and several long radiating membrane tubules. Aside from occasional membrane inclusions, which give them a multivesicular appearance, they usually appear empty of contents. The endosomes play a central role in the endocytotic pathway; they recycle pinocytosed membrane back to the plasmalemma, they dissociate incoming ligands from their receptors, thus allowing receptor reutilization, they modulate receptor expression, and they direct the transport of solutes and membrane components to other target organelles, most prominently the lysosomes. Many of the functions of endosomes depend on their acid pH (pH 5.0–6.0), which is generated by ATP-driven proton pumps located in their membrane (Tycko and Maxfield, 1982; Galloway *et al.*, 1983).

For most incoming ligands, the lysosomes are the final station in their intracellular route. These hydrolytic organelles are usually concentrated in the perinuclear cytoplasm, where they fuse with incoming endosomes and rapidly digest the endosomal contents. The pH of lysosomes is lower still than that of endosomes (pH 4.7–5.0) (Ohkuma and Poole, 1978), and they contain in their membranes high concentrations of negatively charged glycoproteins and possibly glycolipids that may serve to protect the membrane against hydrolysis (Henning *et al.*, 1973; Lewis *et al.*, 1985). The time-course of the pathway followed by a typical ligand in receptor-mediated endocytosis at 37°C is as follows: 3–15 min on the cell surface, 1–2 min in coated vesicles, 15–90 min in endosomes, and finally rapid (<5 min) degradation in lysosomes.

B. Endocytosis of Viruses

Morphological and biochemical experiments have demonstrated that endocytosis of virus particles is quite an efficient process and that it occurs in most cell and virus systems. Whether it is actually needed for productive infection remains unclear in most cases. However, with alphaviruses, the evidence for its obligatory role is exceptionally strong. The studies have used SFV and to a lesser extent Sindbis virus. Quantitative assays have been developed to measure several key steps in the pathway: (1) endocytotic uptake of viruses from the cell surface (Helenius et al., 1980a; Marsh and Helenius, 1980), (2) delivery of viruses into acidic compartments (Helenius et al., 1985; Kielian et al., 1986), (3) transfer from endosomes into lysosomes (Marsh et al., 1983a), (4) lysosomal degradation (Marsh and Helenius, 1980; Helenius et al., 1980a; Talbot and Vance, 1980); (5) release of viral RNA into the cytosolic compartment (Helenius et al., 1982), and (6) onset of viral RNA replication (Miller and Lenard, 1980, 1981; Helenius et al., 1982). The studies have been aided, moreover, by the availability of inhibitors, ionic conditions, and temperature shifts that block and modify the pathway at various points (see below). Extensive morphological data using fluorescence microscopy and transmission and scanning electron microscopy are also available. On the basis of this information, it is possible to reconstruct a fairly detailed step-by-step itinerary for the viral progression through the entry pathway. In the SFV–BHK-21 cell system, the studies have been greatly aided by the high infectivity of the isolated virus (2–3 particles/plaque-forming unit), the high specific activity of metabolically labeled virus (1×10^3 [35S]methionine-labeled and 2×10^4 [^{3}H]uridine-labeled viruses per cpm) and the high efficiency of RNA delivery into the cytosol (45–75% of cell-associated viral RNA is delivered) (Marsh et al., 1983a). Many of the objections frequently raised against biochemical experiments on virus entry at high particle/cell ratios are not valid for these studies, since the experiments have routinely been done at virus/cell ratios as low as 1 (Marsh and Helenius, 1980). In several cases, the biochemical assays have been performed at high as well as low multiplicities, and comparable results have been recorded over huge multiplicity ranges (from 1 to 4×10^4 viruses/cell) (Helenius et al., 1980a; Marsh and Helenius, 1980; Marsh et al., 1983a). In particular, in the multiplicity range of 1×10^2 to 6×10^4 viruses per cell, the delivery efficiency remains virtually constant (Marsh et al., 1983a). The objection that a large portion of viruses might be engaging in cell interactions different from the pathway of productive infection therefore does not apply in this case, since the cells deal with a high number of virus particles in a manner similar to low-multiplicity infection.

After binding to the cell surface, the SFV particles are endocytosed with a half-life of about 8–10 min in BHK-21 cells (Fig. 3). During this time, the viruses are translocated along the membrane, trapped in coated pits, and internalized by coated vesicles. Electron-microscopic studies indicate that coated pits and coated vesicles constitute the main, if not the only, form of pinocytosis in BHK-21 cells (Fig. 4). The process of uptake is faster than that observed for influenza virus and simian virus

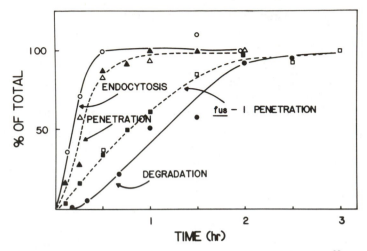

FIGURE 3. Endocytosis, uncoating, and degradation of SFV in BHK-21 cells. [^{35}S]methionine-and [^{3}H]uridine-labeled SFV were allowed to bind to the cells in the cold, and unbound viruses were removed. The cells were then incubated at 37°C for the indicated times and analyzed for virus internalization, release of RNA into the cytosolic compartment, and degradation of viral protein. Of the internalized viral RNA, 45% penetrated; in the figure, the values have been normalized. The RNA-delivery of a viral mutant, *fus-1*, has been included (Kielian *et al.*, 1984, unpublished results).

40, but slower than the rates seen for most physiological ligand–receptor complexes (Steinman *et al.*, 1983).

The capacity of uptake is quite impressive; 1500–3000 virus particles/min per cell can be internalized. As a result of this massive uptake, a transient reduction in the continuous fluid-phase endocytosis of the cells can be recorded, indicating that the virus is displacing some of the fluid within pinocytotic vesicles (Marsh and Helenius, 1980). There are two important corollaries to this finding: (1) The virus is entering by an ongoing constitutive pathway rather than inducing the formation of additional endocytic vesicles and (2) fluid-phase uptake in the BHK-21 cells also occurs to a large extent in coated vesicles.

The process of SFV uptake is remarkably similar to receptor-mediated endocytosis of physiological ligands such as serum low-density lipoprotein. Uptake is not blocked by the addition of cytochalasin B (a microfilament-disrupting agent), colcemid (a microtubule-disrupting agent), or lysosomotropic weak bases or carboxylic ionophores (which elevate the pH in endosomes and lysosomes) (Marsh and Helenius, 1980; Marsh *et al.*, 1982). Phagocytosis thus can be ruled out as a mechanism in the uptake.

C. Role of Endosomes in Virus Penetration

After uptake, the viruses spend only a few minutes in the primary endocytotic vesicles before being delivered by membrane fusion to the

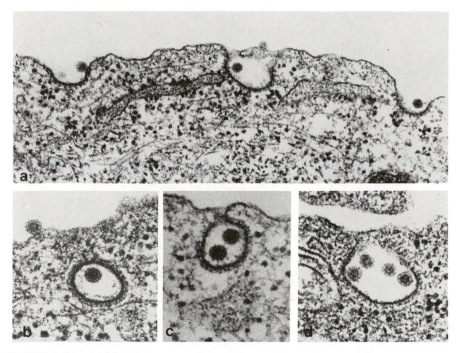

FIGURE 4. Uptake of SFV into coated pits. The virus particles are endocytosed through coated pits and coated vesicles. Although single viruses are usually seen in these endocytotic domains of the cell surface, viral particles do occasionally occur together (C, D).

next organelles in the pathway: the endosomes. As already discussed, these are acidic, and there is rather conclusive evidence that they constitute the main site of SFV penetration. One line of evidence is kinetic: The conversion of the spike glycoproteins into their fusion-active conformation (see below), the release of viral RNA into the cytosol, and infection all occur only a few minutes after uptake, well before the viruses have reached the lysosomes (Marsh *et al.*, 1983a). Moreover, penetration occurs efficiently at 20°C when endocytosed viruses are not delivered to lysosomes due to a temperature-induced block in endosome–lysosome fusion (Marsh *et al.*, 1983a). Furthermore, electron-microscopic analysis has shown that fusion can take place in endosomes (Fig. 5) (Helenius, 1984).

For final confirmation of the role of endosomes in virus entry, studies will be needed that directly demonstrate that the viral nucleocapsids are released from these organelles. It will also be important to determine why only about 50% or so of the incoming viruses are able to release their genome into the cytosol although liposome fusion activity is nearly 100%. Part of the loss may be explained by the observation that virus particles tend to fuse with membrane vesicles inside multivesicular bodies (see Fig. 5). Such "misdirected" penetration would not lead to pro-

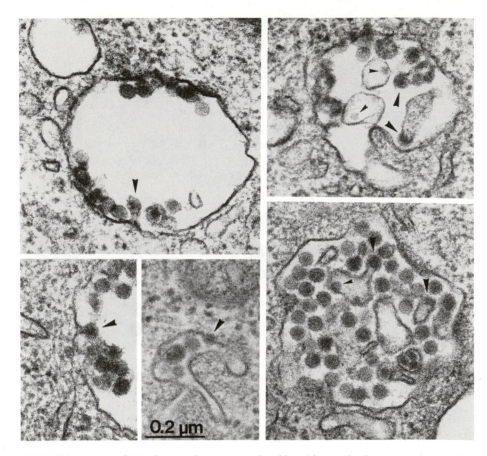

FIGURE 5. Fusion of SFV from endosomes. To be able to detect the fusion reaction in the endosomes morphologically, an experimental design was used that utilized 20°C incubation and NH₄Cl to synchronize the penetration step (Helenius, 1984). Some of the endosomal virus particles are seen in the process of fusion with the limiting membrane of the endosomal compartment; others fuse with the included vesicles present in the multivesicular endosomes.

ductive RNA delivery. Electron micrographs also show virus particles preferentially distributed in the lumen of the larger central endosomes, and within the tubular elements as well. While fusion can apparently occur at either site, numerous apparently intact virus particles remain unfused in these vacuoles.

It is well established that the pH in all endosomes is not equally acidic and that incoming ligands encounter decreasing pH with increasing time in the endosomal compartment (Merion *et al.*, 1983; Murphy *et al.*, 1984; Tanasugarn *et al.*, 1984). The passage of individual viruses through the endocytotic pathway rapidly becomes asynchronous after the initial wave of endocytotic uptake. Thus, some virus particles are delivered to lysosomes as soon as 15–20 min after uptake in BHK cells and others

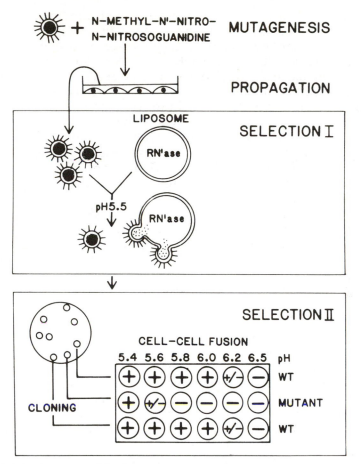

FIGURE 6. Selection protocol for *fus*-1. SFV's *in vitro* fusion activity was used to select for virus mutants with a more acidic pH threshold for membrane fusion. Virus was mutagenized with *N*-methyl-*N'*-nitro-*N*-nitrosoguanidine and propagated at low multiplicity. Progeny virus was concentrated and mixed with RNase-loaded liposomes. The mixture was adjusted to pH 5.5 for 20 min at 37°C, resulting in virus–liposome fusion and a 1×10^{-3} reduction in virus titer. The propagation and selection with liposomes were repeated twice. Isolated plaques of the resulting virus mixture were screened in a cell–cell fusion assay. *fus*-1-Infected cells had a threshold pH of 5.6 compared to 6.2 for *wt* virus (Kielian *et al.*, 1984).

only after 2.5 hr (see Fig. 3). On average, a virus particle spends about 60 min in the endosome compartment (Fig. 3). Analysis of the time–course of uncoating shows that 50% of the wild-type SFV releases its RNA into the cytoplasm within the first 15 min after uptake. Thus, the pH of this "early" endosome compartment is acidic enough (pH 6.2 or lower) to trigger wild-type fusion. In contrast, a mutant of SFV termed *fus*-1 that fuses only at pH values below 5.5 is uncoated with an average delay of 45 min. This mutant was selected for a lower pH threshold by *in vitro* fusion with RNase-containing liposomes (Fig. 6) (Kielian *et al.*, 1984).

Since both viruses are endocytosed at equivalent rates and delivered together into the lysosomal compartment, it can be concluded that viruses that fuse at lower pH are triggered at a later stage of the endosomal route. The kinetics suggest that the penetration of the mutant may occur in perinuclear endosomes, while the wild type fuses in peripheral endosomes (Kielian et al., 1986).

At this point, it may be important to consider why we feel confident that penetration does not occur at the plasma membrane. Considerable work has gone into investigating this possibility. (1) Biochemical studies show that virtually all the virus particles associated with the cell are rapidly endocytosed (see above). (2) Infection occurs efficiently under conitions in which extracellular, plasma-membrane-bound viruses have been removed (Helenius et al., 1982). (3) Virus fusion with the plasma membrane is never seen by electron microscopy unless artificially triggered by low-pH treatment (White et al., 1980). (4) Spike protein antigens from infecting viruses cannot be detected immunologically in the plasma membrane after virus entry (Fan and Sefton, 1978). (5) Penetration is inhibited by agents that increase the pH in endosomes and lysosomes and at concentrations that correlate with the measured pH elevation and with the pH threshold for virus fusion (see below). (6) When penetration is artificially induced at the plasma membrane by lowering the medium pH below physiological levels, infection is not inhibited by these agents, suggesting that their effect is not on cell-surface events (Helenius et al., 1980a; White et al., 1980). (7) The fusion activity of SFV requires a pH well below that normally present in the extracellular medium (see below).

Could lysosomes serve as the site of entry of alphaviruses? They are certainly acidic enough to trigger fusion, and viruses are efficiently delivered to them as a final step in the endocytotic pathway. The main reasons for considering them less important as sites of entry are kinetic; judging by the rate of RNA release into the cytosol, penetration occurs well before virus delivery into lysosomes (see Fig. 3) (Marsh et al., 1983a). Moreover, it is clear that efficient penetration is possible at 20°C, at which temperature traffic from endosomes to lysosomes is inhibited (Dunn et al., 1979; Marsh et al., 1983a). One cannot exclude the possibility that a small fraction of virus particles are rapidly delivered into lysosomes, escape degradation, and penetrate from them. However, viruses may have evolved with fusion activity in the pH 5–6 range, so that penetration will occur before they have reached the hydrolytic environment of lysosomes. Whether viruses that have had time to reach the lysosomes under artificial conditions (such as in the presence of lysosomotropic inhibitors) are able to fuse from this compartment when the inhibitor is removed remains an open question.

IV. INHIBITION OF ENTRY

Alphavirus entry can be blocked at various stages by the addition of inhibitors, by changes in temperature, and by modification of the ionic composition of the medium. Some of the effects have been mentioned above, and they have been quite valuable in defining the route of entry. The lack of interference by other cellular inhibitors and conditions has also been important in characterizing the uptake pathway.

A. Lysosomotropic Weak Bases

It has been known for some time that ammonia and alkylamines inhibit infection by influenza virus and other enveloped viruses in tissue culture (Eaton and Scala, 1961; Jensen *et al.*, 1961) and that they affect the early stages of virus–cell interaction (see Helenius *et al.*, 1982). In a series of studies on SFV and Sindbis virus, the mechanism of inhibition was elucidated using five different weakly basic amines: ammonium chloride, methylamine, tributylamine, chloroquine, and amantadine (Helenius *et al.*, 1980a, 1982; Talbot and Vance, 1980; Miller and Lenard, 1981; Kielian *et al.*, 1984). The studies led to the suggestion that the critical factor in the inhibitory process is the common effect of these agents on the pH of lysosomes, endosomes, and other acidic organelles in the cell.

De Duve, Ohkuma, Poole, and co-workers have shown previously that the amines listed above, as well as numerous other weak bases, accumulate in acidic intracellular compartments and elevate their pH in a concentration-dependent manner (De Duve *et al.*, 1974; Ohkuma and Poole, 1978). Entering in their uncharged form, they become protonated, acquiring a charge that decreases their diffusion out through the membrane and leads to accumulation and pH neutralization. Back-diffusion of the protonated base results in a proton leak, also raising vacuolar pH, and thus inhibiting many lysosomal hydrolases that have acidic pH optima. The high amine concentration frequently causes extensive osmotic swelling of acidic compartments (Ohkuma and Poole, 1981). Although these effects of the lysosomotropic agents were first described in lysosomes, it is now clear that the amines affect other acidic compartments in the same way, including endosomes (Maxfield, 1982), secretory vacuoles (reviewed in Rudnick, 1985), and possibly elements of the Golgi complex (Glickman *et al.*, 1983; Robbins *et al.*, 1984).

The characteristics of alphavirus inhibition by weak bases suggested that the mechanism of inhibition was via elevation of endosome and lysosome pH (Helenius *et al.*, 1980a; Talbot and Vance, 1980; Miller and Lenard, 1981). Amines of highly variable structure and molecular properties but with a common weakly basic and pH-elevating character were all found to inhibit virus entry. The concentration required to inhibit virus infection, block the onset of viral RNA synthesis, and prevent the release of the parental RNA into the cytoplasmic compartment varied, depending on the amine, from 0.1 to 15 mM (Miller and Lenard, 1981; Helenius *et al.*, 1982). Chloroquine was the most efficient, followed by

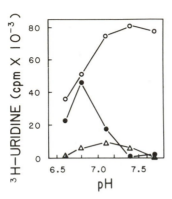

FIGURE 7. pH sensitivity of virus infection, inhibition by NH₄Cl, and inhibition by low-Na medium. Cells were infected with SFV at the indicated pH in control medium (○), medium containing 15 mM ammonium chloride (●), or low-sodium medium (△) (10 mM Na and 100 mM K) (Helenius *et al.*, 1985). Infection was measured as actinomycin-D-resistant [³H]uridine incorporation (Helenius *et al.*, 1982). Infection by virus is optimal at neutral or slightly basic pH. NH₄Cl, 15 mM, blocked infection at pH 7.4; the block is not complete at pH 7.2. Inhibition by low-sodium medium appears virtually pH-independent.

tributylamine, amantadine, methylamine, and ammonium chloride, and the concentration dependence corresponded to the effectiveness in elevating the pH in lysosomes and endosomes (Ohkuma and Poole, 1978; Poole and Ohkuma, 1981; Maxfield, 1982). The inhibitory concentration of amine is very sensitive to the pH of the extracellular medium; a concentration that inhibits totally at pH 7.4 gives incomplete inhibition if the pH is allowed to drop below 7.2 (Fig. 7). It is well known that the elevation in pH of intracellular organelles by the lysosomotropic agents is similarly dependent on medium pH, which affects the active, unprotonated base concentration (De Duve, 1974; Poole and Ohkuma, 1981; Yoshimura *et al.*, 1982). Furthermore, viruses with a high threshold pH for fusion required higher concentrations of a given amine for inhibition than viruses with a lower pH threshold. The *fus*-1 mutant of SFV (see Fig. 6), selected on the basis of its low fusion pH (5.5 as compared to 6.2 for the wild type), was inhibited by 2- to 3-fold less chloroquine, amantadine, or ammonium chloride than the wild type (Kielian *et al.*, 1984). The correlation between inhibitor sensitivity and the pH threshold of fusion is also observed in the comparison of different virus families (Matlin *et al.*, 1981, 1982; Helenius *et al.*, 1982; Yoshimura *et al.*, 1982) or influenza strains of differing pH thresholds (White *et al.*, 1983; Daniels *et al.*, 1985).

To be effective, the amines have to be added no later than a few minutes after the virus has been allowed to enter and must be present continuously thereafter for at least 1 hr (Helenius *et al.*, 1980a; 1982). If added later, the weak bases usually have little effect on the course of the infection. This is consistent with the rate of SFV penetration from endosomes in BHK cells (see Fig. 3). The binding of the virus to the cell surface, the endocytotic uptake of prebound viruses, and the delivery of the virus particles into endosomes are virtually unaffected by the agents (Marsh and Helenius, 1980; Coombs *et al.*, 1981; Miller and Lenard, 1981). These results are similar to those reported for receptor-mediated endocytotic systems in which the first uptake of prebound ligand is not affected (Gonzalez-Noriega *et al.*, 1980). Disturbances in the biosynthetic stages of Sindbis virus infection, RNA synthesis, and the processing of newly synthesized viral proteins have been reported as a result of chloroquine addition at later time points (Cassell *et al.*, 1984; Coombs *et al.*, 1981). Although irrelevant from the point of view of entry, these effects are interesting because they may suggest a role for acidic compartments in the exocytotic pathway. A mutant cell line defective in vacuolar acidification has recently been found to have a similar defect in the post-translational processing of Sindbis glycoproteins (Robbins *et al.*, 1984).

Inhibition of entry is reversed within 2 min following the removal of the agents from the medium (Helenius, 1984). The results of Ohkuma and Poole (1978) have shown that intravacuolar pH returns to near the normal acidic levels within 1 min of chase. It is noteworthy that the vacuolar concentration of amines drops only slowly (Wibo and Poole 1974), suggesting that the pH, rather than the amine concentration *per se*, is crucial for inhibition.

In the presence of inhibitors, SFV does not encounter acidic pH in endosomes or lysosomes, and viral RNA is not delivered into the cytoplasmic compartment. This is not a direct effect of the agents on SFV's fusion activity, since it can be demonstrated by *in vitro* fusion assays that the presence of extreme amine concentrations has no effect provided that the medium is properly acidified (see below). Also, weak bases block the change in trypsin sensitivity of the E1 polypeptide characteristic of the acid-induced conformational change (Kielian *et al.*, 1986), while in the absence of inhibitor, the change takes place soon after the virus enters host cells (see Section V). Finally, degradation of viral proteins is partially inhibited by the agents (Helenius *et al.*, 1980a), in keeping with proteolysis by lysosomal hydrolases with acidic pH optima.

Inhibition by lysosomotropic agents could be bypassed by inducing fusion of viruses with the plasma membrane artificially using brief extracellular acid treatment (Helenius *et al.*, 1980a; White *et al.*, 1981), demonstrating that the agents had no direct effect on fusion. Infection thus apparently proceeded normally if the pH conditions for fusion were fulfilled at the cell surface. It is important to note that medium acidification was performed with cells to which the virus had previously bound in the cold, and the acidification time was only 90 sec (to ensure that no viruses were present in endosomes during the acid treatment). Thus, artifacts resulting from the effect of medium pH on weak-base uptake were eliminated.

The proposition that the elevation in pH of endosomes and lysosomes results in the inhibition of virus entry is thus quite strongly supported by the experimental evidence for SFV and for Sindbis virus. The inhibiting action of other agents that result in the loss of endosome–lysosome acidity (see below) and the common effects on other viruses tend to strengthen the argument. It is apparent, however, that not all investigators in the field are in agreement with the interpretation presented here.

For a different opinion, the reader is referred to a recent paper of Cassell *et al.* (1984) and to Chapter 7.

B. Carboxylic Ionophores

The pH of acidic intracellular compartments can be elevated by a different group of compounds, the carboxylic ionophores, that exchange protons for potassium and sodium across membranes. The carboxylic ionophores integrate into the lipid bilayer and eliminate the pH gradient between the vacuole and the cytoplasm by allowing the exit of protons.

Two carboxylic ionophores, monensin and nigericin, have been shown to block the entry of SFV into BHK cells at concentrations above 10 and 6 μM, respectively (Marsh *et al.*, 1982). Virus binding, endocytosis, and intracellular routing are virtually normal. The general properties of the inhibitory effect are quite similar to those of lysosomotropic weak bases. The mechanism of inhibition is also the same; the viruses do not encounter acidic pH in the endosomes and therefore fail to penetrate. Since these agents, unlike lysosomotropic weak bases, are not sensitive to medium pH, we have found them more reliable than the lysosomotropic agents as entry inhibitors for routine use.

Independent of their effect on entry, the carboxylic ionophores have effects on the biosynthetic stages of the replication cycle (Johnson and Schlesinger, 1980; Kääriäinen *et al.*, 1980). These involve a block in the processing of the newly synthesized spike glycoproteins in the Golgi complex. The monensin concentration needed to obtain these effects is generally about ten-fold lower than that required for the penetration block.

C. Monovalent Cations

The entry of SFV into BHK cells can also be inhibited by changing the cation composition of the extracellular medium (Helenius *et al.*, 1985). Normal media contain high concentrations of sodium and relatively low potassium. When the concentrations are reversed (i.e., 10 mM NaCl and 100 mM KCl), the viruses bind to the cells and are endocytosed, but do not penetrate. The mechanism of this inhibitory effect is not yet fully clarified. It is clear, however, that the viruses in this case do encounter acidic pH in the endosomes, since E1 converts into the acidic conformation (see below), but a productive fusion reaction with the endosomal membranes does not occur. This is rather surprising, since it is clear from *in vitro* fusion experiments that monovalent cations other than protons are not critical and that fusion can easily be induced at the cell surface in low-sodium medium. Not only do the cell-surface-bound viruses fuse with the plasma membrane under these conditions, but also the cells are efficiently infected.

Further studies have shown that the effect correlates with the depolarization of the cell (Helenius *et al.*, 1985). It is not yet known what general effects depolarization may have on the properties of intracellular membranes such as the endosome. The block in SFV entry suggests that the fusion reaction in the endosome can be affected by factors that we cannot reproduce in the simplified *in vitro* assays. The effect could stem indirectly from a rise in cytoplasmic pH or a change in intracellular free calcium.

D. Temperature Effects

Two stages in the receptor-mediated endocytic pathway are differentially temperature-dependent. Initial endocytotic vesicle formation is efficiently blocked at temperatures below 10°C, and it is well known that no penetration or infection of cells with alphaviruses is possible. The delivery of endocytosed material from the endosome to the lysosome is inhibited below 20°C. This leaves a temperature window (10–20°C) in which viruses are internalized but not delivered to lysosomes (Marsh *et al.*, 1983a). As described above, however, they do reach the acidic environment of endosomes, penetrate, and infect the cells.

V. ACID-DEPENDENT MEMBRANE FUSION

A. Properties of Fusion

The membrane fusion activity responsible for the penetration reaction is triggered by mildly acidic pH. Since it was described in 1980 (Helenius *et al.*, 1980a), the properties of this activity have been studied in considerable detail in the SFV system. Together with influenza virus fusion, it now constitutes the best-understood biological membrane fusion activity. A variety of quantitative and qualitative assays are available to monitor the fusion of the viral envelope with natural and artificial membranes, to demonstrate fusion of reconstituted viral membranes (virosomes) with cellular membranes, and to detect conformational changes in the spike glycoproteins induced by acid treatment. Since the results have been recently reviewed in the general context of viral fusion proteins (White *et al.*, 1983), we will concentrate here only on the most pertinent aspects and some recent findings.

Three types of assay systems have been developed to study the overall properties of alphavirus fusion activity. The first employs liposomes as target membranes (White and Helenius, 1980) and the other two the plasma membranes of tissue-culture cells (White *et al.*, 1980) or erythrocytes (Väänanen and Kääriäinen, 1979; Lenard and Miller, 1981). When plasma membranes are used, the fusion can be induced either *from with-*

out (i.e., by the addition of extracellular viruses or reconstituted viral membranes) or *from within*, by biosynthetic incorporation of the fusogenic viral proteins into the cell membrane. A special form of fusion from without is the fusion of two cellular membranes with each other by the bridging action of added viruses (Väänanen and Määriänen, 1980; White *et al.*, 1981; Väänanen *et al.*, 1981, Mann *et al.*, 1983).

The fusion activity is a function of the spike glycoproteins. If integrated in a lipid bilayer membrane, the spike glycoproteins of SFV are alone sufficient. This has been shown by studies with virosomes (Marsh *et al.*, 1983b) and with eukaryotic cells that express the glycoprotein genes (Kondor-Koch *et al.*, 1983). However, no virosomes have been obtained that would match the fusion observed with the intact virus. The proteins may have become partially inactivated during solubilization and reconstitution, or it is possible that the spikes in the reconstituted vesicles are organized in a fashion less advantageous for fusion. The lipid composition of the viral membrane may also be important for the expression of optimal fusion activity. Isolated lipid-free spike glycoproteins are not fusogenic (Marsh *et al.*, 1983b).

The fusion activity is triggered by acid, and it is rapid and highly efficient. The threshold pH for SFV and Sindbis virus is 6.2, with half-maximal fusion at pH 6.0. Judging by the quantitative analysis of SFV's fusion with liposomes, virtually every virus in an isolated preparation is fusogenic. When cellular membranes are used as targets, the fraction of viruses fusing is lower, 45–86%. It is not clear why natural membranes are less effective as targets, but the lower values are certainly consistent with the measured efficiency of delivery of viral RNA to the cytosol discussed above (45–75%). The kinetics of fusion depend on whether the viruses have been prebound to the target membranes, in which case full activity is observed after 5 sec at pH 5 (White *et al.*, 1980), or whether the viruses and the target membranes first have to find each other, in which case fusion is less rapid (White and Helenius, 1980). Cell–cell fusion induced by the presence of viral fusion factors after a brief acidification period is detectable by light microscopy only after 10–20 min of further incubation in neutral medium (White *et al.*, 1981; Mann *et al.*, 1983). The apparent delay is probably not caused by slower fusion, but by the limited resolution of light microscopy; the fusion sites may not be visible before massive cytoplasmic rearrangement and nuclear relocation have occurred.

In terms of phospholipid requirements, the fusion has a very nonspecific character. Not only are natural membranes of a variety of origins compatible with fusion, but also liposomes with artificial phospholipid compositions (White and Helenius, 1980). Unlike the lack of phospholipid specificity, alpha viruses have an absolute sterol requirement. The first indication of this came from lipid-binding studies at low pH; Mooney *et al.* (1975) observed that Sindbis virus attachment to liposomes required the presence of cholesterol. Subsequent studies showed that the effect reflected a cholesterol requirement for fusion (White and Helenius, 1980; Kielian and Helenius, 1984). No fusion could be observed with liposomes lacking the sterol, and optimal activity was obtained when there were one or more cholesterol molecules per 2 phospholipid molecules. A variety of cholesterol analogues including coprostanol, androstanol, and dihydrocholesterol were found to support fusion at variable levels (40–90% relative to cholesterol-containing controls). Thus, major distortions in the steroid ring structure, omission of the 5,6 double bond, and the lack of an isooctyl side chain were tolerated. In contrast, the slightest modification in the 3-β-hydroxyl group rendered steroids incompatible. Epicholesterol (which has the hydroxyl group in the axial 3-α-conformation), 5α-cholestane, 5α-cholestan-3-one, cholesterol methyl ester, and cholesterol acetate were inactive. The structural requirements revealed by such analogue studies did not correlate with the known physical effects of steroids on lipid bilayers (such as the condensation effect or the effects on membrane fluidity). This suggested that the effect is not dependent on a modification of the bilayer properties, but rather, as we discuss below, may involve a more specific interaction between sterol and the viral fusion factor.

When virus binding instead of fusion was measured quantitatively, it was found that binding to liposomes was strictly acid-dependent and that cholesterol-containing liposomes bound virtually all the virus (Mooney et al., 1975; Kielian and Helenius, 1984). Epicholesterol-containing liposomes bound, surprisingly, as much as one fifth the number of viruses in an acid-dependent fashion. This suggested that some binding can occur at low pH without necessarily resulting in fusion.

Both in vitro and in vivo assays have shown that the fusion reaction between SFV and a target membrane is not lytic; i.e. it does not lead to rupture in either of the fusing membranes before, during, or after the fusion reaction. As far as we can tell, there is no exchange of macromolecules or ions between the fusing compartments and the external "bulk" compartment. The first evidence came from SFV–liposome experiments which showed that external RNase does not enter the fusing compartments (White and Helenius, 1980; Kielian and Helenius, 1984). The lack of increased conductance of artificial bilayers showed, subsequently, that no detectable change in the ion permeability—even transiently—accompanied the fusion reaction (Young et al., 1983).

The lack of leakiness is of importance in considering the penetration reaction. It is well known that isolated alphavirus nucleocapsids undergo a dramatic change when exposed to pH below about 6.2 (Söderlund et al., 1972). SFV nucleocapsids shrink, their sedimentation coefficient increases, and they become hydrophobic and insoluble in aqueous solutions. They will partition with the spike proteins into the detergent phase during Triton X-114 partitioning. The conversion to a hydrophobic form depends on a change in the 20-kilodalton proteinase K fragment of the capsid protein (Kielian et al., 1986). Given these dramatic pH-induced changes, it has been tempting to speculate that the alteration reflects a modification taking place in the endosomes and that it could be involved in uncoating. Two lines of evidence from our laboratory argue against this possibility. First, when intact viruses are treated with acid buffers in solution, the spike glycoproteins undergo the characteristic conformational change, but the capsid does not convert into the acid form. Second, capsid does not convert into the acid form during virus entry into BHK cells, as judged by the lack of generation of hydrophobic nucleocapsids. Taken together, these data lead us to conclude that the viral membrane remains tight throughout the entry process and does not allow acidification of the nucleocapsid (Kielian *et al.*, 1986).

B. Acid-Induced Conformational Change in the Viral Spike Glycoproteins

At present, the best-characterized acid-triggered fusion activities are those of influenza A viruses and alhpaviruses. Like alphaviruses, influenza enters cells via endocytosis, and there is increasing evidence that it penetrates by fusion in endosomes (Matlin *et al.*, 1981; Yoshimura and Ohnishi, 1984). Although similar in the general entry pathway, the two viruses appear to have evolved distinct mechanisms of fusion.

The molecular analysis of influenza fusion has reached a higher level of sophistication than the alphaviruses due mainly to the availability of X-ray structure data on the trimeric ectodomain of the hemagglutinin (HA) (the influenza fusion factor) (Brand and Skehel, 1972; Wilson *et al.*, 1981). Acid treatment induces a major, irreversible conformational change in the molecule (Skehel *et al.*, 1982; Daniels *et al.*, 1983; Webster *et al.*, 1983; Yewdell *et al.*, 1983; Doms *et al.*, 1985). The change can be monitored by alterations in the accessibility of groups in the molecule to protease and reducing agents, by changes in the antigenic epitopes, and by modifications in the solubility and spectral properties, as well as by morphology. Mutant hemagglutinins, obtained by selection (Daniels *et al.*, 1985) or site-specific mutagensis (Gething *et al.*, 1986), have also been important in characterizing the role of the various molecular domains. The composite data suggest the following chain of events during influenza fusion: (1) Low pH causes protonation of one or more strategic groups in HA. (2) A conformational change takes place whereby the trimeric molecule opens up and exposes a hydrophobic moiety consisting, at least in part, of the N-terminal peptide segments of the HA2 subunits. (3) These moieties serve to bind the virus hydrophobically to the target membrane and facilitate the close apposition of the two membranes. (4) The interaction leads to fusion of the membranes. While many of the details of the process remain unclear, it seems very likely that the acid HA becomes an integral component of both fusing membranes, with the HA2 C-terminus anchoring the protein to the viral membrane and the HA2 N-ter-

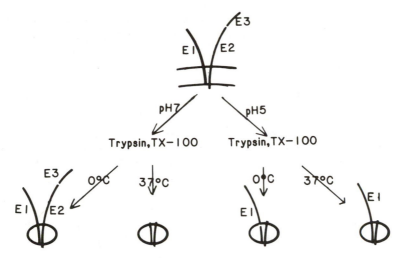

FIGURE 8. Effects of low pH on the trypsin sensitivity of the SFV E1 and E2 glycoproteins. Intact SFV is treated at pH 5 or 7 at 37°C, neutralized, dissolved in Triton X-100, and treated with 100 μg/ml trypsin at either 0 or 37°C for 10 min. At 0°C, both E1 and E2 are resistant to trypsin in the pH 7 virus, while E2 is digested in the pH 5 virus. When trypsinization is carried out at 37°C, E2 is completely digested in either the pH 5 or the pH 7 form. E1, however, is resistant to trypsin in the pH 5 form.

minus attaching it to the target membrane. The HA may in this way bring about the intimate contact between the membranes and the rearrangement in bilayer structure required for fusion.

The molecular changes that accompany fusion in alphaviruses are less well understood. The amino acid sequence data, now available for three alphaviruses, show extensive homologies between the glycoprotein sequences, but there are no obvious similarities with HA or the fusion proteins of other virus families (Garoff et al., 1980; Rice and Strauss, 1981; Dalgarno et al., 1983). Unlike influenza HA, the F-proteins of Paramyxoviridae, and the glycoprotein of murine mammary tumor virus, there is no hydrophobic N-terminal peptide in the glycopolypeptides that could play a role similar to that described for the HA2 N terminus above. Hydrophobicity plots for glycoprotein sequences of alphaviruses also show no internal sequences that would be particularly hydrophobic in character outside the known membrane-spanning regions of E1 and E2 (Dalgarno et al., 1983). An uncharged and extremely conserved peptide region comprising 17 amino acids (from residue 80 to 96 in SFV) in the ectodomain of E1 has been suggested to play a role in fusion (Garoff et al., 1980).

Acid treatment results in irreversible changes in the spike structure. Electron microscopy after negative staining shows that the spikes become disorganized and the virus particles less symmetrical and less homogenous in shape. Acid-treated viruses appear to sediment more slowly in sucrose gradients (Edwards et al., 1983). Major changes occur in the protease sensitivity of both the major glycopolypeptides, E1 and E2 (Fig. 8).

Sindbis virus E2 becomes more sensitive to trypsin (Edwards *et al.*, 1983), and in SFV, it becomes sensitive to trypsin, chymotrypsin, and pronase (Kielian and Helenius, 1985). The alteration is irreversible, and it occurs within 5 sec of acid treatment (Kielian and Helenius, 1985). It is very likely that the change in E2 reflects a conversion involved in fusion.

Changes in SFV E1 can also be observed after trypsin digestion, provided that the viruses are solubilized in Triton X-100 prior to digestion and that the digestion is performed at 37°C (Fig. 8). The neutral form of the E1 is fully digested under these conditions, whereas acid-treated E1 is largely resistant. The conversion to the resistant form is rapid and irreversible, and the pH dependence mirrors that of the fusion activity (Kielian and Helenius, 1985). Again, it seems likely that the change reflects the conformational change in the spikes responsible for fusion activity because viruses entering BHK cells during infection undergo the same change in E1 (Kielian *et al.*, 1986). This intracellular conversion in E1 is blocked by monensin, which, as discussed above, elevates endosomal pH and inhibits virus penetration.

The observation that both E1 and E2 undergo changes in their sensitivity to proteases need not necessarily mean that *both* undergo changes in conformation due to acid effects. Since they form a complex in the viral membrane, a change in one could modify the accessibility of the other. However, it has recently become possible to study the responses of E1 and E2 independently; we observed that SFV digestion with proteinase K at 4°C in the presence of nonionic detergent results in the cleavage of the ectodomains of both proteins from the C-terminal anchoring peptides (Fig. 9) (Kielian and Helenius, 1985). The fragments obtained (called E1* and E2*) are 3000–6000 daltons shorter than the intact proteins. Unlike E1 and E2, they are apparently not complexed to each other, and they are water-soluble in the absence of detergent. Having lost their hydrophobic C-terminal anchor peptides, they correspond to the bromelain fragments of influenza HA, which have proven very useful in studying influenza fusion and spike structure. When isolated E1* and E2* were subjected to low-pH treatment (Kielian and Helenius, 1985), E2* underwent a shift to a trypsin-sensitive form in the same way as intact E2. This suggested that the ectodomain of E2 changes conformation independently of E1. Partioning in Triton X-114 (Bordier, 1981) showed that the acid-treated E2* (pH < 5.6) was reproducibly more hydrophobic than its neutral counterpart. Since no attachment of E2* to liposomes could be observed, the relevance of the change remains unclear. Exposure to fusion-activating pH values apparently revealed hydrophobic surfaces in the ectodomain of E2 large enough for detergent binding, but not sufficient for attaching the spike to lipid bilayers.

The change in E1* to a trypsin-resistant conformation was comparable to that of E1, indicating that the conversion of E1 at acid pH is independent of its interaction with E2. A very interesting finding was that the conversion of E1* to the trypsin-resistant form was dependent not

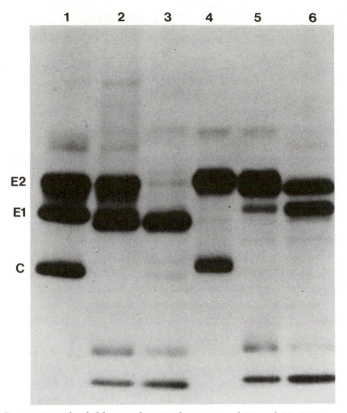

FIGURE 9. Generation of soluble ectodomain fragments of E1 and E2. SFV was dissolved in Triton X-100 and digested with 100 μg/ml proteinase K at 0°C for varying times, precipitated with trichloroacetic acid, and run in sodium dodecyl sulfate–polyacrylamide gel electrophoresis with (lanes 4–6) and without (lanes 1–3) prior reduction and alkylation. Lanes 1 and 4 show virus incubated in the absence of proteinase K. After 5 min of digestion (lanes 2 and 5), E1 is cleaved to yield E1*; after 60 min of digestion (lanes 3 and 6), E2 and E1 have both been cleaved to yield E2* (~41 kd) and E1* (~48 kd).

only on pH but also on the presence of cholesterol. Thus, if isolated E1* was subjected to acid treatment in the absence of added liposomes or in the presence of liposomes containing epicholesterol (3-α-OH), no conversion to the resistant form could be detected. If the E1* was acidified in the presence of cholesterol-containing liposomes, the major fraction became trypsin-resistant. This result is probably very significant in view of the requirement for cholesterol (or other sterols with a 3-β-OH group) in fusion of alphaviruses. It strongly suggests that the cholesterol dependence of fusion is a function of E1. It implies, moreover, that the E1 polypeptide has a central role in the fusion activity of alphaviruses.

E1 and E2 thus appear to respond independently to low pH. Whether both responses are necessary for fusion is unclear. It is hoped that molecular cloning, expression, and sequencing studies in progress on the fusion mutant *fus*-1 will allow a more definite assignment of the roles of the two proteins.

C. Mechanism of Fusion

In view of the relatively limited body of data at present available on alphavirus activation at low pH, it may be premature to speculate about the molecular mechanism of fusion. However, we can make some generalizations and comparisons with the influenza system. Influenza virus and the togaviruses possess in their envelopes a large number of the specific proteins responsible for triggering membrane fusion. In both, acid pH causes profound changes, resulting in the exposure of groups and sites with a potential to interact with membranes and membrane components. The major difference appears to be that the alphaviruses expose specific sites for interaction with cholesterol, whereas influenza HA expresses a hydrophobic peptide moiety that can interact with the target membrane in a more nonspecific manner. The end effect may be the same; a tight association of the two membranes and a disorganization of the local bilayer structure. Future work must focus on the lipid–protein interactions, which no doubt contain the key for understanding the mechanism of any biological fusion event. The viral spike glycoproteins responsible for fusion are interesting, because no cellular proteins with similar activities are yet known. They are also obvious targets for future antiviral strategies, because they are responsible for the only steps in the entry pathway that are virus- and not host-cell-mediated.

VI. CONCLUSION

The entry process of alphaviruses into tissue-culture cells described in this chapter depends to a very large part on constitutive cell functions. It seems clear that viruses exploit the cell's machinery from the very first stages of their interaction. There are two steps at which the viruses need to display their own activities, attachment and membrane fusion, relying on the cell for endocytosis and transport into acidic intracellular compartments. There are many central questions that remain open at this point: What are the advantages of using the endocytotic pathway rather than simply fusing at the cell surface as some other viruses do? Do the viruses also follow the endocytotic pathway elucidated in tissue-culture cells in the cells of the infected animal? What is the significance of the sterol dependence when viewed in light of the fact that the insect vectors of alphaviruses may contain cholesterol only in conjunction with a blood meal? Could the highly specific requirement for 3-β-OH-containing sterol groups be exploited for the design of new antiviral strategies? Have the viral glycoproteins evolved from cellular fusion proteins, and do they reflect general principles of fusion proteins? To what extent can cells inhibit virus entry by the endosomal pathway, and to what extent does the virus adapt itself to the conditions in the endocytotic pathway of specific cells? Many of these questions are amenable to experimental approaches.

ACKNOWLEDGMENTS. We thank Jürgen Kartenbeck for providing Fig. 4, Pam Ossorio for photographic assistance, Bruce Granger for reading the manuscript, and Barbara Longobardi and Joy White for typing.

REFERENCES

Birdwell, C. R., and Strauss, J. H., 1974, Distribution of the receptor sites for Sindbis virus on the surface of chicken and BHK cells, *J. Virol.* **14**:672–678.

Bordier, C., 1981, Phase separation of integral membrane proteins in Triton X-114 solution, *J. Biol. Chem.* **256**:1604–1607.

Brand, C., and Skehel, J., 1972, Crystalline antigen from the influenza virus envelope, *Nature (London) New Biol.* **238**:145–147.

Cassell, S., Edwards, J., and Brown, D. T., 1984, Effects of lysosomotropic weak bases on infection in BHK-21 cells by Sindbis virus, *J. Virol.* **52**:857–864.

Clarke, D. H., and Casals, J., 1958, Techniques for hemagglutination and hemagglutination inhibition with arthropod-borne viruses, *Am. J. Trop. Med. Hyg.* **7**:561–573.

Coombs, K., Mann, E., Edwards, J., and Brown, D. T., 1981, Effects of chloroquine and cytochalasin B on the infection of cells by Sindbis virus and vesicular stomatitis virus, *J. Virol.* **37**:1060–1065.

Dalgarno, L., Rice, C., and Strauss, J., 1983, Ross River virus 26S RNA: Complete nucleotide sequence and deduced sequence of the encoded structural proteins, *Virology* **129**:170–187.

Daniels, R. S., Douglas, A. R., Skehel, J. J., and Wiley, D. C., 1983, Analysis of the antigenicity of influenza hemagglutinin at the pH optimum for virus-mediated membrane fusion, *J. Gen. Virol.* **64**:1657–1662.

Daniels, R. S., Downie, J. C., Hay, A. J., Knossow, M., Skehel, J. J., Wang, M. L., and Wiley, D. C., 1985, Fusion mutants of the influenza virus haemagglutinin glycoprotein, *Cell* **40**:431–439.

De Duve, C., de Barsy, T., Poole, B., Trouet, A., Tulkens, P., and van Hoof, F., 1974, Lysosomotropic agents, *Biochem. Pharmacol.* **23**:2495–2531.

Dimmock, N. J., 1982, Initial stages in infection with animal viruses, *J. Gen. Virol.* **59**:1–22.

Doms, R. W., Helenius, A., and White, J., 1985, Membrane fusion activity of the influenza virus hemagglutinin: The low pH-induced conformational change, *J. Biol. Chem.* **260**:2973–2981.

Dunn, W. A., Hubbard, A. L., and Aronson, N. N., 1979, Low temperature selectively inhibits fusion between pinocytic vesicles and lysosomes during heterophagy of ^{125}I-asialofetuin by the perfused rat liver, *J. Biol. Chem.* **255**:5971–5978.

Eaton, M. D., and Scala, A. R., 1961, Inhibitory effect of glutamine and ammonia on replication of influenza virus in ascites tumor cells, *Virology* **13**:300–307.

Edwards, J., Mann, E., and Brown, D. T., 1983, Conformational changes in Sindbis virus envelope proteins accompanying exposure to low pH, *J. Virol.* **45**:1090–1097.

Fan, D. P., and Sefton, B. M., 1978, The entry into host cells of Sindbis virus, vesicular stomatitis virus and Sendai virus, *Cell* **15**:985–992.

Fries, E., and Helenius, A., 1979, Binding of Semliki Forest virus and its isolated glycoproteins to cells, *Eur. J. Biochem.* **97:**213–220.

Gallaher, W. R., and Howe, C., 1976, Identification of receptors for animal viruses, *Immunol. Commun.* **5**(6):535–552.

Galloway, C. J., Dean, G. E., Marsh, M., Rudnick, G., and Mellman, I., 1983, Acidification of macrophage and fibroblast endocytic vesicles *in vitro, Proc. Natl. Acad. Sci. U.S.A.* **80:**3334–3338.

Garoff, H., Frischauf, A.-M., Simons, K., Lehrach, H.,, and Delius, H., 1980, Nucleotide sequence for cDNA coding of Semliki Forest virus membrane glycoproteins, *Nature* (*London*) **288:**236–241.

Gething, M.-J., Doms, R. W., York, D., and White, J., 1986, Studies on the mechanism of membrane fusion: Site-specific mutagenesis of the hemagglutinin of influenza virus, *J. Cell. Biol.* (in press).

Glickman, J., Croen, K., Kelly, S., and Al-Awqati, Q., 1983, Golgi membranes contain an electrogenic H^+ pump parallel to a chloride conductance, *J. Cell Biol.* **97:**1303–1308.

Goldstein, J. L., Anderson, R. G., and Brown, M. S., 1979, Coated pits, coated vesicles and receptor-mediated endocytosis, *Nature* (*London*) **279:**679–685.

Gonzalez-Noriega, A., Grubb, J. H., Talkad, V., and Sly, W. S., 1980, Chloroquine inhibits lysosomal enzyme pinocytosis and enhances lysosomal enzyme secretion by impairing receptor recycling, *J. Cell Biol.* **85:**839–852.

Hahon, N., and Cooke, K. O., 1967, Primary virus–cell interactions in the immunofluorescence assay of Venezuelan equine encephalomyelitis virus, *J. Virol.* **1**(2):317–326.

Helenius, A., 1984, Semliki Forest virus penetration from endosomes: A morphological study, *Biol. Cell.* **51:**181–186.

Helenius, A., Morrein, B., Fries, E., Simons, K., Robinson, P., Schirrmacher, V., Terhorst, C., and Strominger, J. L., 1978, Human (HLA-A and -B) and murine (H2-K and -D) histocompatibility antigens are cell surface receptors for Semliki Forest virus, *Proc. Natl. Acad. Sci. U.S.A.* **75:**3846–3850.

Helenius, A., Kartenbeck, J., Simons K., and Fries, E., 1980a, On the entry of Semliki Forest virus into BHK-21 cells, *J. Cell Biol.* **84:**404–420.

Helenius, A., Marsh, M., and White, J., 1980b, The entry of viruses into animal cells, *Trends Biochem. Sci.* **5:**104–106.

Helenius, A., Marsh, M., and White, J., 1980c, Virus entry into animal cells, in: *Leukaemias, Lymphomas and Papillomas: Comparative Aspects,* (P. A. Bachmann, ed.), Munich Symposia on Microbiology, pp. 57–63, Taylor and Francis, London.

Helenius, A., Marsh, M., and White, J., 1982, Inhibition of Semliki Forest virus penetration by lysosomotropic weak bases, *J. Gen. Virol.* **58:**47–61.

Helenius, A., Mellman, I., Wall, D., and Hubbard, A., 1983, Endosomes, *Trends Biochem. Sci.* **8:**245–250.

Helenius, A., Kielian, M., Wellsteed, J., Mellman, I., and Rudnick, G., 1985, Effects of monovalent cations on Semliki forest virus entry into BHK-21 cells, *J. Biol. Chem.* **260:**5691–5697.

Henning, R., Plattner, H., and Stoffel, W., 1973, Nature and localization of acidic groups on lysosomal membranes, *Biochim. Biophys. Acta* **330:**61–75.

Hilfenhaus, J., 1976, Propagation of Semliki Forest virus in various human lymphoblastoid cell lines, *J. Gen. Virol.* **33:**539–542.

Jensen, E. M., Force, E. E., and Unger, J. B., 1961, Inhibitory effect of ammonium ions on influenza virus in tissue culture, *Proc. Soc. Exp. Biol. Med.* **107:**447–451.

Johnson, D. C., and Schlesinger, M. J., 1980, Vesicular stomatitis virus and Sindbis virus glycoprotein transport to the cell surface is inhibited by ionophores, *Virology* **103:**407–424.

Johnston, R. E., and Faulkner, P., 1978, Reversible inhibition of Sindbis virus penetration in hypertonic medium, *J. Virol.* **25:**436–438.

Kääriäinen, L., Hashimoto, K., Saraste, J., Virtanen, I., and Pentinen, K., 1980, Monensin and FCCP inhibit the intracellular transport of alphavirus membrane glycoproteins, *J. Cell Biol.* **87**:783–791.

Kielian, M. C., and Helenius, A., 1984, Role of cholesterol in fusion of Semliki forest virus with membranes, *J. Virol.* **52**:281–283.

Kielian, M. C., and Helenius, A., 1985, pH-induced alterations in the fusogenic spike protein of Semliki Forest virus, *J. Cell Biol.* **101**:2284–2291.

Kielian, M., Keränen, S., Kääriäinen, L., and Helenius, A., 1984, Membrane fusion mutants of Semliki Forest virus, *J. Cell Biol.* **98**:139–145.

Kielian, M., Marsh, M., and Helenius, A., 1986, Endosome acidification detected by virus fusion and fusion activation (in prep).

Kondor-Koch, C., Burke, B., and Garoff, H., 1983, Expression of Semliki Forest virus proteins from cloned cDNA. I. The fusion activity of the spike glycoprotein, *J. Cell Biol.* **97**:644–651.

Lenard, J., and Miller, D. K., 1981, pH-dependent hemolysis by influenza, Semliki Forest virus, and Sendai virus, *Virology* **110**:479–482.

Lenard, J., and Miller, D., 1982, Uncoating of enveloped viruses, *Cell* **28**:5–6.

Lewis, V., Green, S. A., Marsh, M., Vihko, P., Helenius, A., and Mellman, I., 1985, Glycoproteins of the lysosomal membrane *J. Cell Biol.* **100**:1839–1847.

Lonberg-Holm, K., and Philipson, L., 1974, Early interactions betwen animal viruses and cells, in: *Monographs in Virology* Vol. 9 (J. L. Melnick, ed.), pp. 1–148, S. Karger, Basel.

Maeda, T., and Ohnishi, S., 1980, Activation of influenza virus by acidic media causes hemolysis and fusion of erythrocytes, *FEBS Lett.* **122**:283–287.

Mann, E., Edwards, J., and Brown, D. T., 1983, Polycaryocyte formation mediated by Sindbis virus glycoproteins, *J. Virol.* **45**:1083–1089.

Marker, S. C., Connelly, D., and Jahrling, P. B., 1977, Receptor interaction between Eastern equine encephalitis virus and chicken embryo fibroblasts, *J. Virol.* **21**(3):981–985.

Marsh, M., 1984, The entry of enveloped viruses into cells by endocytosis, *Biochem. J.* **218**:1–10.

Marsh, M., and Helenius, A., 1980, Adsorptive endocytosis of Semliki Forest virus, *J. Mol. Biol.* **142**:439–454.

Marsh, M., Wellsteed, J., Kern, H., Harms, E., and Helenius, A., 1982, Monensin inhibits Semliki Forest virus penetration into baby hamster kidney (BHK-21) cells, *Proc. Natl. Acad. Sci. U.S.A.* **79**:5297–5301.

Marsh, M., Bolzau, E., and Helenius, A., 1983a, Penetration of Semliki Forest virus from acidic prelysosomal vacuoles, *Cell* **32**:931–940.

Marsh, M., Bolzau, E., White, J,. and Helenius, A., 1983b, Interactions of Semliki Forest virus spike glycoprotein rosettes and vesicles with cultured cells, *J. Cell Biol.* **96**:455–461.

Massen, J. A., and Terhorst, C., 1981, Identification of a cell-surface protein involved in the binding site of Sindbis virus on human lymphoblastic cell lines using a heterobifunctional cross-linker, *Eur. J. Biochem.* **115**:153–158.

Matlin, K., Reggio, H., Helenius, A., and Simons, K., 1981, The infective entry of influenza virus into MDCK-cells, *J. Cell Biol.* **91**:601–613.

Matlin, K., Reggio, H., Simons, K., and Helenius, A., 1982, The pathway of vesicular stomatitis entry leading to infection, *J. Mol. Biol.* **156**:609–631.

Maxfield, F. R., 1982, Weak bases and ionophores rapidly and reversibly raise the pH of endocytic vesicles in cultured mouse fibroblasts, *J. Cell Biol.* **95**:676–681.

Meager, A., and Hughes, R. C., 1977, Virus receptors, in: *Receptors and Recognition*, Series A, Vol. 4 (P. Cuatrecasas and M. F. Greaves, eds.), pp. 143–195, Chapman and Hall, London.

Merion, M., Schlesinger, P., Brooks, R. M., Moehring, J. M., Moehring, T. J., and Sly, W. S., 1983, Defective acidification of endosomes in Chinese hamster ovary cell mutants "cross-resistant" to toxins and viruses, *Proc. Natl. Acad. Sci. U.S.A.* **80**:5315–5319.

Miller, D. K., and Lenard, J., 1980, Inhibition of vesicular stomatitis virus infection by spike glycoprotein, *J. Cell Biol.* **84**:430–437.

Miller, D. K., and Lenard, J., 1981, Antihistaminics, local anesthetics and other amines as antiviral agents, *Proc. Natl. Acad. Sci. U.S.A.* **78:**3605–3609.

Mooney, J. J., Dalrymple, J. M., Alving, C. R., and Russell, P. K., 1975, Interaction of Sindbis virus with liposomal model membranes, *J. Virol.* **15:**225–231.

Murphy, R. F., Powers, S., and Cantor, C. R., 1984, Endosome pH measured in single cells by dual fluorescence flow cytometry: Rapid acidification of insulin to pH 6, *J. Cell Biol.* **98:**1757–1762.

Ohkuma, S., and Poole, B., 1978, Fluorescence probe measurement of the intralysosomal pH in living cells and the perturbation of pH by various agents, *Proc. Natl. Acad. Sci. U.S.A.* **75:**3327–3331.

Ohkuma, S., and Poole, B., 1981, Cytoplasmic vacuolation of mouse peritoneal macrophages and the uptake into lysosomes of weakly basic substances, *J. Cell Biol.* **90:**656–664.

Oldstone, M. B. A., Tishon, A., Dutko, F., Kennedy, S. I. T., Holland, J. J., and Lampert, P. W., 1980, Does the major histocompatibility complex serve as a specific receptor for Semliki Forest virus?, *J. Virol.* **34:**256–265.

Pastan, I., and Willingham, M. C., 1983, Receptor-mediated endocytosis: Coated pits, receptosomes and the Golgi, *Trends Biochem. Sci.* **8:**250–254.

Poole, B., and Ohkuma, S., 1981, Effect of weak bases on the intralysosomal pH in mouse peritoneal macrophages, *J. Cell Biol.* **90:**665–669.

Redmond, S., Peters, G., and Dickson, C., 1984, Mouse mammary tumor virus can mediate cell fusion at reduced pH, *Virology* **133:**393–402.

Rice, C., and Strauss, J., 1981, Nucleotide sequence of the 26S mRNA of Sindbis virus and deduced sequence of the encoded virus structural proteins, *Proc. Natl. Acad. Sci. U.S.A.* **78:**2062–2066.

Robbins, A. R., Oliver, C., Bateman, J. L., Krag, S. S., Galloway, C. J., and Mellman, I., 1984, A single mutation in Chinese hamster ovary cells impairs both Golgi and endosomal functions, *J. Cell Biol.* **99:**1296–1308.

Rudnick, G., 1985, Acidification of intracellular organelles: Mechanism and function, in: *Physiology of Membrane Disorders* (T. Andreoli, D. D. Fanestil, J. F. Hoffman, and S. G. Schultz, eds.), pp. 409–422, Plenum Press, New York.

Silverstein, S. C., Steinman, R. M., and Cohn, Z. A., 1977, Endocytosis, *Annu. Rev. Biochem.* **46:**669–722.

Skehel, J., Bayley, P., Brown, E., Martin, S., Waterfield, M., White J., Wilson, I., and Wiley, D., 1982, Changes in the conformation of influenza virus hemagglutinin at the pH optimum of virus-mediated membrane fusion, *Proc. Natl. Acad. Sci. U.S.A.* **79:**968–972.

Smith, A. L., and Tignor, G. H., 1980, Host cell receptors for two strains of Sindbis virus, *Arch. Virol.* **66**(1)**:**11–26.

Söderlund, H., Kääriainen, L., Von Bonsdorff, C.-H., and Weckstein, P., 1972, Properties of Semliki Forest virus nucleocapsid II: An irreversible contraction by acid pH, *Virology* **47:**753–760.

Steinman, R. M., Mellman, I. S., Muller, W. A., and Cohn, Z. A., 1983, Endocytosis and the recycling of plasma membrane, *J. Cell Biol.* **96:**1–27.

Talbot, P. J., and Vance, D. E., 1980, Sindbis virus infects BHK-cells via a lysosomal route, *Can. J. Biochem.* **58:**1131–1137.

Tanasugarn, L., McNeil, P., Reynolds, G. T., and Taylor, D. L., 1984, Microspectrofluorometry by digital image processing: Measurement of cytoplasmic pH, *J. Cell Biol.* **98:**717–724.

Tycko, B., and Maxfield, F. R., 1982, Rapid acidification of endocytic vesicles containing α_2-macroglobulin, *Cell* **28:**643–651.

Väänanen, P., and Kääriäinen, L., 1979, Hemolysis by two alphaviruses: Semliki Forest virus and Sindbis virus, *J. Gen. Virol.* **43:**593–601.

Väänanen, P., and Kääriäinen, L., 1980, Fusion and haemolysis of erythrocytes caused by three togaviruses: Semliki Forest, Sindbis and rubella, *J. Gen. Virol.* **46:**467–475.

Väänanen, P., Gahmberg, C. G., and Kääriäinen, L., 1981, Fusion of Semliki Forest virus with red cell membranes, *Virology* **110:**366–374.

Webster, R. G., Brown, L. E., and Jackson, D. C., 1983, Changes in the antigenicity of the hemagglutinin molecule of H3 influenza virus at acidic pH, *Virology* **126:**587–599.

White, J., and Helenius, A., 1980, pH-dependent fusion between the Semliki Forest virus membrane and liposomes, *Proc. Natl. Acad. Sci. U.S.A.* **77:**3273–3277.

White, J., Kartenbeck, J., and Helenius, A., 1980, Fusion of Semliki Forest virus with the plasma membrane can be induced by low pH, *J. Cell Biol.* **87:**264–272.

White, J., Matlin, K., and Helenius, A., 1981, Cell fusion by Semliki Forest, influenza and vesicular stomatitis virus, *J. Cell Biol.* **89:**674–679.

White, J., Kielian, M., and Helenius, A., 1983, Membrane fusion proteins of enveloped animal viruses, *Q. Rev. Biophys.* 16:151–195.

Wibo, M., and Poole, B., 1974, Protein degradation in cultured cells. II. The uptake of chloroquine by rat fibroblasts and the inhibition of cellular protein degradation and cathepsin B_1, *J. Cell Biol.* **63:**430–440.

Wilson, I., Skehel, J., and Wiley, D., 1981, Structure of the hemagglutinin membrane glycoprotein of influenza virus at 3 Å resolution, *Nature (London)* **289:**366–373.

Yewdell, J. W., Gerhard, W., and Bachi, T., 1983, Monoclonal anti-hemagglutinin antibodies detect irreversible antigenic alterations that coincide with the acid activation of influenza virus A/PR/834-mediated hemolysis, *J. Virol.* **48:**239–248.

Yoshimura, A., and Ohnishi, S.-I., 1984, Uncoating of influenza virus in endosomes, *J. Virol.* **51:**497–504.

Yoshimura, A., Kuroda, K., Yamashina, S., Maeda, T., and Ohnishi, S.-I., 1982, Infectious cell entry mechanism of influenza virus, *J. Virol.* **43:**284–293.

Young, J. D.-E., Young, G. P. H., Cohn, Z. A., and Lenard, J., 1983, Interaction of enveloped viruses with planar bilayer membranes: Observations of Sendai, influenza, vesicular stomatitis and Semliki Forest viruses, *Virology* **128:**186.

CHAPTER 5

Formation and Assembly of Alphavirus Glycoproteins

Milton J. Schlesinger and Sondra Schlesinger

I. GLYCOPROTEIN STRUCTURES AND DOMAINS

Among the variety of enveloped viruses that exist in nature, the Toga-viridae family is generally considered to be among the simplest with regard to structure and composition of the virus. The virion structure is examined in detail in Chapter 22. In this chapter, we focus on the major protein components of the virion envelope and describe modifications of these proteins that occur as they mature from nascent polypeptides to fully mature and functional glycoproteins that form the spikes of the infectious particle. Most of our information about togavirus glycoproteins comes from biochemical studies of the two alphaviruses, Semliki Forest virus and Sindbis virus (Garoff *et al.*, 1982). These viruses have provided valuable models for the analysis of membrane proteins.

The spike of Semliki Forest and Sindbis viruses consists of two major proteins (E1 and E2) that are embedded in the lipid bilayer surrounding the virion nucleocapsid (Fig. 1). A smaller glycoprotein, called E3, is associated with the E1–E2 spike of Semliki Forest virions. E3 and E2 are initially covalently linked in a precursor form, denoted p62 or PE2, in virus-infected cells (Schlesinger and Schlesinger, 1972). Both E1 and E2 glycoproteins have their major domains (designated 1 in Fig. 1) external to the membrane, and hydrophobic segments of their polypeptides (domain 2) near the carboxy termini extend through the lipid bilayer. The distribution of amino acids in the particular domains is given in Table I; about 25 hydrophobic amino acids anchor the protein in the membrane (Garoff and Söderlund, 1978; Rice *et al.*, 1982). E2 has an additional ~30

Milton J. Schlesinger and Sondra Schlesinger • Department of Microbiology and Immunology, Washington University School of Medicine, St. Louis, Missouri 63110.

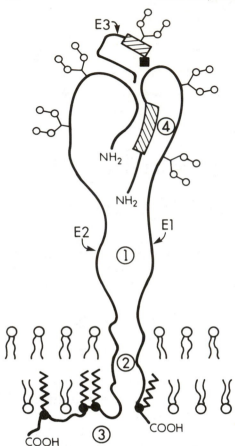

FIGURE 1. Glycoprotein spike of alphaviruses. Domains noted are hydrophilic globular portion (1), membrane-spanning region (2), cytoplasmic fragment (3), and hydrophobic area not in the membrane (4). (oligosaccharide symbol) Oligosaccharides; (■) acetylated amino terminus; (fatty acid symbol) covalent fatty acids; (hatched symbol) hydrophobic regions.

residues on the internal side of the bilayer (domain 3 of Fig. 1), and these are postulated to interact with regions of the capsid polypeptide during assembly (Garoff and Simons, 1974). Two arginine residues are at the internal domain of E1. At least one fatty acid is covalently bound to E1 and three to E2 (Schmidt *et al.*, 1979). Their precise sites of attachment

TABLE I. Domains of Alphavirus Glycoproteins[a]

	Sindbis virus			Semliki Forest virus		
Domain	E1	E2	E3	E1	E2	E3
Hydrophilic, globular (outside)	395	364	45	396	367	47
Membrane-spanning	26	26	—	24	24	—
Cytoplasmic (inside)	2	33	—	2	31	—
Hydrophobic (outside)	16 (80–96)	—	19	16 (80–96)	—	19

[a] Numbers of amino acids in a domain are listed. The numbers in parentheses indicate the positions of the amino acids in the protein's sequence.

TABLE II. Glycosylation Sites on Alphavirus
Glycoproteins[a]

	Protein		
Virus	E1	E2	E3
Sindbis	139, 245	196, 318	14
Semliki Forest	141	200, 262	13, 58
Ross River	141	200, 262	11, 58

[a] Numbers are the positions in the amino acid sequence where oligosaccharides can be attached.

are not known, but probably are at a serine or cysteine in the membrane and cytoplasmic domains of E1 and E2, respectively (Magee et al., 1984). Protein E3 has an acetyl group blocking its amino terminus (Bell et al., 1982). Both E3 and E1 have additional hydrophobic domains; there are 19 amino acids in E3 that function as the signal sequence for binding the 26 S messenger RNA (mRNA) translation complex to the membrane. The E1 hydrophobic sequences that do not span the membrane are located between amino acids 80 (valine) and 96 (cysteine) (domain 4 of Fig. 1) and are highly conserved. This region may be important for the fusion activity of this protein (see Chapter 4).

The potential sites for glycosylation on the glycoproteins have been identified for Sindbis, Semliki Forest, and Ross River viruses by sequencing the complementary DNA (cDNA) of the mRNA for the structural genes (Table II) (Garoff et al., 1980; Rice and Strauss, 1981). Asparagine residues that are located in the tripeptide Asn-X-Ser/Thr serve as potential glycosylation sites. The actual sites of glycosylation on the glycoproteins of Sindbis have been determined experimentally (Hsieh et al., 1983a; Mayne et al., 1985). E1 of Sindbis virus has two glycosylation sites at asparagine residues 139 and 245. Semliki Forest virus has only one glycosylation site at Asn-141. This same site is also a glycosylation site in the E1 of Ross River virus (Dalgarno et al., 1983). The E2 proteins of the three viruses have two glycosylation sites. Sindbis virus has a single glycosylation site on E3. Both Semliki Forest virus and Ross River virus have two potential glycosylation sites on this protein. In Semliki Forest virus, Asn-13 appears to be glycosylated (Simons and Warren, 1984).

II. BIOSYNTHESIS OF POLYPEPTIDES

There is a highly ordered program for the translation of the polycistronic subgenomic mRNA encoding the structural proteins of the alphaviruses that are arranged, from the 5' end of the RNA, as capsid, p62, and E1 (Wirth et al., 1971; refer to Chapter 3 for details). The single initiation site for translation of the 26 S RNA codes for the amino-terminal methionine of the nascent viral capsid (Cancedda et al., 1974).

Immediately following the sequences for the carboxy-terminal tryptophan of the capsid are codons for the amino-terminal serine of the p62
precursor glycoprotein. Thus, for the p62 glycoprotein to be formed, the
capsid sequences must first be translated and released from the nascent
chain by a proteolytic cleavage—believed to be caused by the capsid itself
(Aliperti and Schlesinger, 1978). The first 19 amino acids of p62 comprise
a hydrophobic signal sequence that is recognized by the cellular signal
recognition particle (Walter and Blobel, 1981, 1982). The complex of nascent polypeptide, mRNA, ribosomes, and signal recognition particle binds
to receptors on the endoplasmic reticulum membrane, and transport of
p62 through the membrane begins (Garoff et al., 1978). Glycosylation of
p62 occurs as the nascent polypeptide appears in the lumen of the vesicle,
prior to completion of polypeptide synthesis (Sefton, 1977). We will discuss later the consequences of either blocking the addition of these groups
or preventing the initial processing of the oligosaccharide—the removal
of the glucose residues.

Synthesis of p62 is followed by translation of a segment of the 26 S
RNA encoding a small hydrophobic "linker" polypeptide of 55 amino
acids that appears to act as a discrete signal sequence for E1 (Welch and
Sefton, 1980). A cotranslational proteolytic cleavage releases p62 from
this linker polypeptide and is probably needed to initiate transfer of E1
across the membrane. At some point in the E1 synthesis, an additional
proteolytic activity, possibly the signal peptidase, releases the linker. This
putative E1 signal polypeptide is unusually long, and it is not clear how
it functions and what its ultimate fate is. Evidence that E1 has a signal
sequence distinct from p62 comes from studies with a temperature-sensitive (ts) mutant of Semliki Forest virus that at nonpermissive temperature accumulates a polypeptide consisting of capsid–p62, probably due
to a block in the capsid self-cleavage (Hashimoto et al., 1981). The p62–
E1 cleavage occurs to a limited extent, however, and glycosylated E1 is
detected in the endoplasmic reticulum.

Shortly after E1 biosynthesis is completed, a modification occurs that
is detected in sodium dodecyl sulfate (SDS)–polyacrylamide gels as a decrease in molecular weight from 53,000 to 50,000 (Bonatti and Cancedda,
1982). Conversion is independent of glycosylation and cannot be attributed to disulfide formation or exchange, since all analyses were with
reduced and alkylated preparations. This E1 change is not detected in
SDS–polyacrylamide gels containing 7 M urea, and the authors propose
that E1 alters its conformation so as to increase SDS binding. Consistent
with this interpretation were results of peptide analyses that failed to
show a primary structure difference between the two forms.

Additional changes have been detected in the E1 structure during its
maturation. Two monoclonal antibodies raised against the E1 glycoprotein of Sindbis virus bound more strongly to more mature forms of this
protein than to newly made molecules (Roehrig et al., 1982). Four other

epitopes on E1 of Sindbis virus were identified on the protein in infected cells, but were cryptic on isolated virions (Schmaljohn *et al.*, 1983).

The first identifiable modifications of these glycoproteins are ones that affect the structure of the oligosaccharide and, in turn, the conformation of the protein itself. This set of alterations is detailed below. In adition most of the P62 molecules are acetylated at the amino terminus (Bell and Strauss, 1981), but it is not known when this modification occurs.

III. GLYCOSYLATION

A. Synthesis and Structure of the Oligosaccharide Chains

Asparagine-linked oligosaccharides are synthesized by the *en bloc* transfer of the oligosaccharide $Glc_3Man_9 GlcNAc_2$ from the lipid carrier dolichol phosphate to the nascent polypeptide (R. Kornfeld and S. Kornfeld, 1985). This large oligosaccharide is then processed as the polypeptide moves from the rough endoplasmic reticulum through the Golgi stacks to the plasma membrane (Fig. 2). The removal of glucose and several mannose residues occurs in the rough endoplasmic reticulum. Subsequent processing occurs in the Golgi cisternae. The progress of a glycoprotein through the cellular membranes can be traced by the extent to which the oligosaccharides are processed. Thus, the enzyme *N*-acetylglucosaminyltransferase I is localized to the medial Golgi stack (Dunphy *et al.*, 1985), and a glycoprotein must reach that Golgi stack to obtain the peripheral GlcNAc residues. The galactosyl- and sialyltransferases reside in the trans-Golgi stack, and the addition of galactose or sialic acid to an asparagine-linked oligosaccharide indicates that the particular glycoprotein has reached the trans-Golgi membranes (Griffiths *et al.*, 1982, 1983; Rothman *et al.*, 1984). Viral glycoproteins have served as important tools in identifying the steps in processing; the first description of this pathway came from studies with Sindbis virus (Robbins *et al.*, 1977) and vesicular stomatitis virus (Kornfeld *et al.*, 1978; Tabas *et al.*, 1978; Hunt *et al.*, 1978).

The oligosaccharides most commonly found on the glycoproteins of Sindbis and Semliki Forest viruses are the high-mannose and complex structures (Fig. 3), but the Sindbis glycoproteins were also shown to contain hybrid-type oligosaccharides (Davidson and Hunt, 1985b). These structures arise from the same intermediate (see Fig. 2), and the type that is found depends on (1) the host cell in which the virus is grown and (2) the specific site in the polypeptide chain.

The composition of the oligosaccharide chains on the alphavirus glycoproteins can vary depending on the host cell in which the virus is grown (Keegstra *et al.*, 1975; Hsieh *et al.*, 1983a; Davidson and Hunt, 1983). Differences reported among host cells, however, may not be due only to the origin of the cell. Mayne *et al.* (1985) did not find the same differences

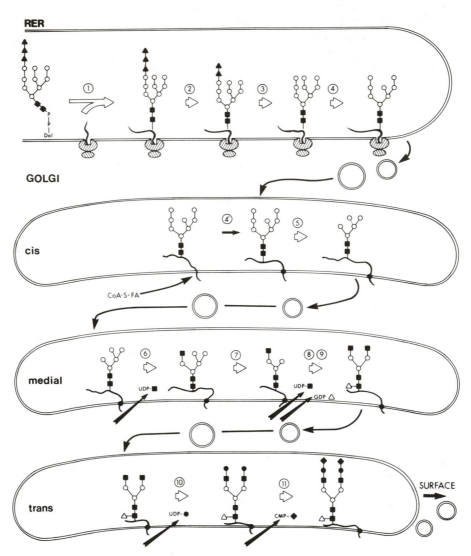

FIGURE 2. Schematic pathway of oligosaccharide processing on newly synthesized glyco-proteins. The reactions are catalyzed by the following enzymes: (1) oligosaccharyltransfer-ase; (2) α-glucosidase I; (3) α-glucosidase II; (4) endoplasmic reticulum α1,2-mannosidase; (4') fatty acyl CoA-protein transferase (postulated); (5) Golgi α-mannosidase I; (6) N-ace-tylglucosaminyltransferase I; (7) Golgi α-mannosidase II; (8) N-acetylglucosaminyltransfer-ase II; (9) fucosyltransferase; (10) galactosyltransferase; (11) sialyltransferase. (RER) Rough endoplasmic reticulum. (■) N-acetylglucosamine, (○) mannose; (▲) glucose; (△) fucose; (●) galactose; (◆) sialic acid. Based on R. Kornfeld and S. Kornfeld (1985) with permission from Annual Reviews, Inc.

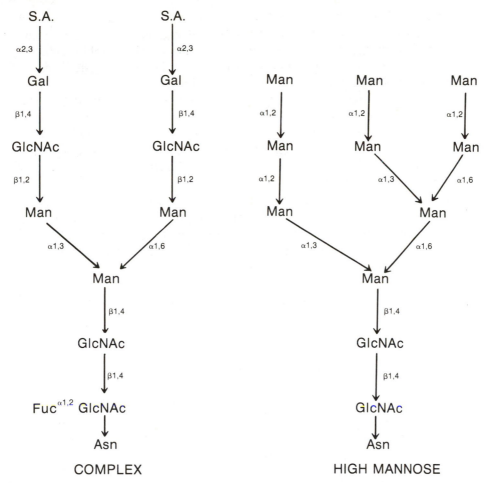

FIGURE 3. Structure of the two major oligosaccharides on alphavirus glycoproteins. (SA) Sialic acid.

between baby hamster kidney (BHK) cells and chicken cells for one of the sites on E1 as had been reported previously (Burke and Keegstra, 1979; Hsieh *et al.*, 1983a). Furthermore, changes in cell growth may be a significant factor. Hakimi and Atkinson (1980) found differences in the oligomannosyl structures depending on the stage of growth of the chicken cells in which Sindbis virus was grown. Glucose deprivation of Chinese hamster ovary cells also affects the oligosaccharides; under this condition, they are not fully processed into acidic-type structures (Davidson and Hunt, 1985a).

The other major source of variation in oligosaccharide chains can be attributed to differences between glycosylation sites. Thus, for Sindbis virus grown in chicken cells, the Asn-196 site in glycoprotein E2 contains a complex oligosaccharide structure, whereas the structure on Asn-318 is high-mannose (Hsieh *et al.*, 1983a). Asn-139 in glycoprotein E1 contains a complex oligosaccharide; the structure at Asn-245 is variable. One explanation for these variations is that the extent of processing depends on the accessibility of a particular site to the processing enzymes. Hsieh *et al.* (1983b) investigated the accessibility of the glycosylation sites in the intact virus. Their strategy was to grow Sindbis virus in a ricin-resistant cell variant that processes oligosaccharides only to $Man_5GlcNAc_2$. This structure, but not the complex structures, will be cleaved between the two GlcNAc residues by the enzyme endo-β-N-acetylglucosaminidase H (endo H). They determined the susceptibility of each of the glycosylation sites to this enzyme and found that the sites most sensitive to endo H were the same as those that were most highly processed in wild-type cells. These results support the hypothesis that the sites with high-mannose structures are less accessible to the processing enzymes than the sites with complex oligosaccharide structures. Pollack and Atkinson (1983) have compiled a list of proteins with both high-mannose and complex oligosaccharide chains. The complex chains are usually the ones closest to the amino terminus.

B. Role of Glycosylation in Virion Formation

Two approaches are being taken to examine the role of oligosaccharides in the localization and biological activity of viral glycoproteins. One approach is the use of mutants, both virus and cell; the other is the use of inhibitors of glycosylation.

One of the first studies with cell mutants was with the ricin-resistant Chinese hamster ovary cell line, described above, that processes asparagine-linked oligosaccharides only as far as $Man_5GlcNAc_2$. Sindbis virus grows normally in these cells (S. Schlesinger *et al.*, 1976) as well as in *Aedes albopictus* mosquito cells, in which the predominant oligosaccharide structure on the virion glycoproteins is $Man_3GlcNAc_2$ (Hsieh and Robbins, 1984). These results establish that the peripheral sugars, GlcNAc, Gal, and sialic acid, are not required for virion formation or for infectivity. In contrast, if carbohydrate is removed from intact Semliki Forest virions using a mixture of endoglycosidases, there is a loss of infectivity due to aggregation (Kennedy, 1974).

Two different ricin-resistant mutants of L cells are unable to replicate either Sindbis or Semliki Forest virus, although they are both hosts for vesicular stomatitis virus (Gottlieb *et al.*, 1979). One of the mutants is unable to process the precursor oligosaccharide beyond $Man_8GlcNAc_2$ (Tabas and Kornfeld, 1978). This mutant is defective in its ability to bind Sindbis virus. Semliki Forest virus does infect these cells, which then produce the viral structural proteins, but the cleavage of p62 to E2 does not occur. The other mutant has an increase in the level of sialylated glycoproteins and is also unable to carry out the cleavage of p62 to E2. For neither of these mutants is it known what the primary defect is or whether the specific change in oligosaccharide structure causes the defect in virus growth. These results demonstrate, however, that cells can be altered in a specific way, presumably in only one or a very few biochemical steps, such that they acquire resistance only to certain viruses.

The mutants of alphaviruses altered in their glycosylation patterns may help us in understanding the host-range specificities of these viruses. Two mutants have been isolated from persistently infected mosquito cells. One mutant of Sindbis virus contains a hyperglycosylated E2 (Durbin and Stollar, 1984). This mutant is temperature-sensitive, but grows normally in mosquito cells at 34.5°C. It grows poorly in vertebrate cells at this temperature. Most important, the phenotypes of hyperglycosylation and the host-specific defect in growth revert together, indicating that a single mutation is responsible for the two effects. This mutant is derived from a strain of Sindbis virus that normally has three glycosylation sites on E2 (Cancedda *et al.*, 1981). Thus, it will be necessary to determine whether it is the addition of a fourth oligosaccharide chain that causes these phenotypes or whether it is glycosylation at a particular site that leads to the defect. Furthermore, it will be important to know at what stage in the virus life cycle the block occurs.

The second mutant is derived from Western equine encephalitis virus (Simizu *et al.*, 1983). This mutant has lost a glycosylation site in E3 (p62). It also shows a decreased ability to grow in vertebrate cells compared to growth in mosquito cells.

The second approach for studying the role of glycosylation—the use of inhibitors of the glycosylation pathway—became feasible with the discovery that there are drugs able to interfere with specific steps in glycosylation. One of the most widely used drugs is tunicamycin, an antibiotic known to block the addition of GlcNAc to the lipid carrier dolichol phosphate (Struck and Lennarz, 1980). Studies with both Semliki Forest virus and Sindbis virus demonstrated that treatment of cells with this drug inhibits virion formation (Schwarz *et al.*, 1976; Leavitt *et al.*, 1977). Viral glycoproteins are synthesized in the presence of tunicamycin, but are not glycosylated. Although the precursor p62 is not cleaved to E2, the glycoproteins can be detected on the cell surface (Scheefers *et al.*, 1980; Mann *et al.*, 1983). They must be in an abnormal conformation, since they are not labeled by iodination, a procedure that does label the glycosylated proteins (Leavitt *et al.*, 1977). An even more convincing argument that the nonglycosylated proteins are not correctly localized on the cell surface is that infected cells treated with tunicamycin and exposed to low pH do not fuse (Mann *et al.*, 1983). Under conditions of infection in which the viral glycoproteins are glycosylated, even in the absence of cleavage of p62 to E2, fusion of the cells occurs on exposure to low pH.

Tunicamycin has been used to inhibit the addition of carbohydrate to a wide spectrum of membrane and secreted glycoproteins. The results have been variable; some proteins are affected, but many are not (Gibson et al., 1980). This variability can be explained by assuming that some proteins require the covalent attachment of oligosaccharides to achieve the correct conformation, whereas other proteins can fold into a stable structure in the absence of carbohydrate. The requirement for oligosaccharides during folding was demonstrated most clearly for the G protein of vesicular stomatitis virus (Gibson et al., 1979, 1980). The folding of this protein becomes temperature-sensitive when it lacks oligosaccharide chains, but G proteins with different amino acid sequences vary in their oligosaccharide requirement. For some proteins, as has been shown for the hemagglutinin of influenza virus, the oligosaccharides protect the protein from proteolytic degradation (Schwarz et al., 1976).

Several compounds that inhibit the removal of glucose residues from the precursor oligosaccharide, $Glc_3Man_9GlcNAc_2$, have been described (Schwarz and Datema, 1984). Two of these compounds, 1-deoxynojirimycin and castanospermine, inhibit $\alpha 1,2$-glucosidase I, the enzyme that removes the terminal glucose residue (Saunier et al., 1982; Pan et al., 1983). The two inner glucose residues are removed by $\alpha 1,3$-glucosidase II. Bromoconduritol inhibits the release of the innermost glucose residue (Datema et al., 1982). All three of these compounds can inhibit the growth of Sindbis virus (Datema et al., 1984; S. Schlesinger et al., 1985).

The studies with deoxynojirimycin and castanospermine demonstrated that inhibition is much more pronounced at 37 than at 30°C, that the cleavage of p62 to E2 is blocked, and that the proteins are able to migrate to the cell surface (S. Schlesinger et al., 1985). It is well established that complex sugars are not required for virion formation. As described above, Sindbis virus grows well in a cell mutant that processes oligosaccharides only to $Man_5GlcNAc_2$ (S. Schlesinger et al., 1976; Hsieh et al., 1983b) and in mosquito cells in which the oligosaccharides are processed to $Man_3GlcNAc_2$ (Hsieh and Robbins, 1984). Thus, it must be the removal of glucose residues from the glycoproteins that is essential for them to function in virion formation.

The effects of retaining glucose residues on the asparagine-linked oligosaccharides of a number of membrane and secreted glycoproteins are reminiscent of those obtained with tunicamycin; i.e., different proteins are affected differently. These differences suggest that the removal of glucose residues during the processing of oligosaccharides is critical only for some proteins to achieve a functional conformation, while others will fold correctly even if the glucose residues are retained. The existence of a single pathway for processing of asparagine-linked oligosaccharides requires that processing occur to meet the structural needs of the most sensitive glycoproteins. Thus, a major role for oligosaccharides and for the initial steps in processing may be in protein folding and stability.

IV. FATTY ACID ACYLATION

At about 10–15 min after completion of polypeptide synthesis, both glycoproteins of the alphaviruses become acylated with long-chain fatty acids—predominantly palmitic acid—when examined in chicken fibroblasts or BHK cells (Schmidt *et al.*, 1979; Schmidt, 1982). The reaction appears to proceed via a fatty acyl-CoA transferase that has been identified and partially purified from liver microsomes (Berger and Schmidt, 1984). The acceptor sites on the glycoproteins reside around the transmembrane domain of the proteins, probably involving the cysteines of the p62 and E1, though transfer to hydroxy amino acids such as serine or threonine may occur later. The precise organelle where acylation occurs has not been rigorously identified; it clearly must be in a late (kinetically) site of the endoplasmic reticulum or immediately on entry to the cis-Golgi. Acylation precedes release of high-mannose chains, since fatty acids are detected on endoglycosidase-H-sensitive forms of the proteins (Schmidt and Schlesinger, 1980). Failure to acylate the proteins does not affect their intracellular transport, but nonacylated proteins are defective in the final stages of virus assembly (M. J. Schlesinger and Malfer, 1982). The basis for this is not known. One theory for a role of fatty acid acylation postulates that acylation blocks inadvertent disulfide formation, which leads to oligomeric aggregates cross-linked by interchain cystine bonds. Such complexes can be found when Sindbis glycoproteins are reacted with hydroxylamine to release lipid (Magee *et al.*, 1984). Further support for this theory comes from observations with different serotypes of vesicular stomatitis virus; the New Jersey form of this virus lacks fatty acid in the G glycoprotein, whereas the Indiana type has lipid bound to G. A major difference between the two proteins is a cysteine residue in the cytoplasmic domain of Indiana G that is absent in the New Jersey type (Gallione and Rose, 1983). Both serotypes are highly infectious in tissue culture and in nature. *In vitro* mutagenesis of Indiana G cDNA that led to conversion of the cytoplasmic cysteine to a serine resulted in loss of fatty acid acylation (Rose *et al.*, 1984). This altered nonacylated G is unaffected in its structure and distribution in the cell's plasma membrane, but not all of its functions, particularly with respect to virion assembly, have been tested.

Acylation could also influence the lipid composition of the membranes in the vicinity of the virus embedded proteins, possibly sparing cholesterol, and protein-bound fatty acids could affect a fusogenic activity known to be associated with most virus glycoproteins. Viruses with transmembranal glycoproteins lacking fatty acids could retain this latter activity by increasing the amounts of sterol and unsaturated fatty acids in the virion envelope.

V. PROTEOLYTIC FORMATION OF E2

The E2 glycoprotein is synthesized initially as a larger polypeptide (p62) with ~65 amino acids at the amino terminus that must be cleaved from the precursor for virion formation. Very shortly after synthesis, this portion of the molecule is glycosylated at the asparagine, and the amino-terminal serine is acetylated (Bell *et al.*, 1982). The amino-terminal region of the E2 precursor has a hydrophobic region that closely resembles the signal sequences utilized for initial targeting to the endoplasmic reticulum membrane and vectorial transport into the vesicle. Unlike most signal sequences, however, p62 is not cleaved by signal peptidase (Bonatti *et al.*, 1979). About 20 min after peptide synthesis is completed (the time varies with the type of cell, temperature, and virus growth conditions), p62 is converted to E2 by a tryptic-like cleavage of the Arg–Ser bond adjacent to basic amino acids (position 64 for Sindbis, 66 for Semliki Forest virus). This cleavage site is analogous to that present in glycoproteins of other viruses as well as in many secreted and membrane-bound proteins made in normal cells; thus, the protease is probably encoded by a cellular rather than a virus gene. The enzyme believed to be responsible for similar cleavages in prohormones and proalbumin is a cathepsin-B-like thiometalloprotease (Steiner *et al.*, 1984), which recognizes the consensus sequence in alphavirus proteins of Arg-X-Arg(Lys)-Arg-X (Dalgarno *et al.*, 1983). This cathepsin-type protease is distinct from that normally associated with cell lysosomes, and, in fact, purified lysosomal cathepsin B does not act on proinsulin (Docherty *et al.*, 1982).

Proteolytic processing of cellular proteins does not occur until the molecules have been transported to the trans-Golgi and possibly into post-Golgi vesicles. The protease could well be in the vesicle, sorting with the substrate polyproteins during budding from the Golgi. A carboxypeptidase B activity that cleaves the terminal basic residue from the newly generated C terminus of many virus glycoproteins (including Sindbis and Semliki Forest viruses) and cellular proteins is present in these vesicles and is also a zinc metalloenzyme (Steiner *et al.*, 1984). Relatively little additional information exists about these proteases, but in pancreatic islet granule fractions, two polypeptides could be labeled with the affinity reagent chloromethyl ketone-Tyr-ala-Lys-Arg, which had been iodinated at tyrosine. A 31.5-kilodalton (kd) and a 39-kd protein were labeled, and a preenzyme of 44 kd was also noted (Docherty *et al.*, 1982). Localization of these enzymes to the secretory vesicle provides a mechanism for regulating the function of a protein that might injure the cell if the protein were active in ER or Golgi membranes.

The cellular site for alphavirus glycoprotein proteolysis has not been rigorously identified. Very little precursor is detected on infected cell surfaces, and no sialic acid is found on this protein (Hakimi and Atkinson, 1982; Bonatti and Cancedda, 1982), leading several investigators to propose that proteolytic processing occurs in an internal organelle, most likely a trans-Golgi type vesicle. Earlier reports (Bracha and Schlesinger, 1976b; Ziemiecki et al., 1980) utilizing antibodies added to the cell surface were interpreted to indicate that precursor was at the surface immediately prior to cleavage. Furthermore, surface perturbations of a nonspecific nature such as treatment with low levels of trypsin will also block precursor conversion (Adams and Brown, 1982).

There are believed to be significant alterations in the structure of the p62 glycoprotein after proteolysis. For example, the interactions between E2 and E1 are stronger than those between p62 and E1, on the basis of analyses utilizing cleavable cross-linking agents such as dimethylsuberimidate and antibodies specific for the individual polypeptides. In Sindbis-virus-infected cells, very low levels of cross-linked p62–E1 molecules could be isolated after a short pulse of radioactive amino acids (Rice and Strauss, 1982). After a chase period in which conversion of radioactive p62 to E2 occurred, a significant amount of cross-linked E1–E2 heterodimers was detected. In addition, a monoclonal antibody that recognizes Semliki Forest virus E2 binds much more strongly to E2 than to p62 (Burke et al., 1983).

E2 formation from p62 does not appear to alter the cytoplasmic domain (topology) of p62, on the basis of results of Ziemiecki and Garoff (1978), who prepared inside-out vesicles of Semliki Forest virus-infected cells and showed that E2 retained a cytoplasmic tail equivalent to its precursor.

Results of many studies show that proteolytic conversion of p62 to E2 is an essential step in the final assembly process for alphaviruses. These include use of protease inhibitors (Bracha and Schlesinger, 1976a), antibodies against the glycoproteins (Bracha and Schlesinger, 1976b), drugs that affect processing and transport through organelles (Johnson and Schlesinger, 1980; Kääriäinen et al., 1980), ts virus mutants (Bracha and Schlesinger, 1976b; Erwin and Brown, 1980), and host cells altered in oligosaccharide processing (Gottlieb et al., 1979).

The cleavage of precursor does not require the presence of a nucleocapsid, since oocytes injected with 26 S virus mRNA make and process p62 to E2, but make no viron RNA (Huth et al., 1984) (see the more detailed discussion below). However, a correct conformation of E1 does seem essential, since expression of Semliki Forest virus cDNA coding for p62 and E1 showed that aberrantly cleaved forms of E2 appeared when the E1 gene was improperly expressed (Konder-Koch et al., 1983). Adams and Brown (1982) noted that trypsin treatment of infected BHK cells blocked the p62 cleavage and inhibited the association of nucleocapsid to membranes. The small E3 glycoprotein remains associated with the Semliki Forest virus spike, but in Sindbis-infected cells, E3 initially appears in the cell and is rapidly secreted into the culture fluid (Mayne *et al.*, 1984).

VI. LOCALIZATION OF GLYCOPROTEINS TO INTRACELLULAR ORGANELLES

The localization of virus structural proteins in the organelles of infected cells has been elegantly elucidated by immunocytochemistry at the resolution of the electron microscope as well as by light microscopy (Green *et al.*, 1981; Griffiths *et al.*, 1983; Saraste and Kaismanen, 1984). For the latter, *ts* mutants of Semliki Forest virus were employed, and the proteins were allowed to accumulate at a specific site when the nonpermissive temperature was imposed on the infection (Kääriänien *et al.*, 1980). Organelle-specific lectins were utilized in double fluorescence microscopy to identify regions that stained with antibodies specific to the virus glycoproteins. For example, in cells infected with the Semliki Forest virus *ts*-1 mutant at the nonpermissive temperature, the glycoproteins were distributed in a pattern resembling that labeled with concanavalin A, which labels mannose residues and, predominantly, the endoplasmic reticulum. On shift to the permissive temperature in the presence of cycloheximide, immunofluorescence is found in Golgi and plasma membranes. The Golgi is visualized with wheat germ agglutinin. When the latter experiment was carried out in the presence of monensin, a sodium ionophore that blocks transport after Golgi, the virus glycoproteins could be seen in a Golgi-complex structure. In the presence of carbonylcyanide-p-trifluoromethoxyphenylhydrazone, an uncoupling agent of oxidative phosphorylation, glycoproteins are arrested in the endoplasmic reticulum. When thin sections of infected BHK cells are prepared and labeled with affinity-purified antibodies directed against the virus glycoproteins followed by protein A–gold conjugates, it is possible to detect the virus antigens (glycoproteins) in the Golgi organelles and to follow movement of the virus glycoproteins out of the Golgi organelles (Green *et al.*, 1981). From quantitation of the material in thin-frozen sections of infected BHK cells, it was determined that the spike proteins take about 15 min to traverse the endoplasmic reticulum and Golgi membranes. From the rates of synthesis (120,000 spikes/min), some 1.8×10^6 proteins are calculated to be in the membranes. On the basis of the dimensions of these organelles and the estimated "dwell time" of 15 min in the membrane, the density of each spike protein is 90 and $750/\mu m^2$ for endoplasmic reticulum and Golgi membranes, respectively (Griffiths *et al.*, 1984; Quinn *et al.*, 1984). The virion has about 22,000 spikes/μm^2, leading to a 240-fold enrichment of the proteins from endoplasmic reticulum to virus. Quinn *et al.* (1984) also measured the total level of endogenous BHK-cell proteins and the Semliki Forest virus E1 in membrane subfractions. From these data, virus glycoproteins account for about 0.3% of the endoplasmic reticulum mem-

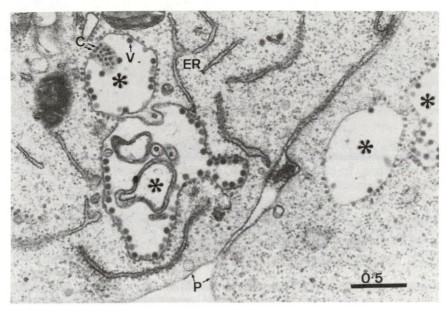

FIGURE 4. Intracellular budding of Semliki Forest virus from vesicles accumulated in the presence of monensin, which blocks transport of glycoproteins beyond trans-Golgi. (ER) Endoplasmic reticulum; (C) nucleocapsids; (V) virions; (P) plasma membrane of BHK cells; (*) intracellular capsid-binding membranes ×50,000. Courtesy of G. A. Warren; from Griffiths *et al.* (1983) with permission from Rockefeller Press.

brane and 2.3% of the Golgi membrane protein. It is postulated that the concentrations of virus glycoproteins in these intracellular membranes are too low to initiate an intracellular budding event. However, when blocks in transport such as that effected by monensin occur, spike proteins accumulate in vesicles, and, indeed, budding of particles into the vesicle cisternae is observed (Fig. 4) (Johnson and Schlesinger, 1980; Griffiths *et al.*, 1983; Quinn *et al.*, 1983).

Shifts to low temperatures when coupled with *ts* mutants to initiate a "synchronous" movement of glycoproteins have also revealed information about movement of the virus glycoproteins through cellular organelles (Saraste and Kaismanen, 1984). With cells infected with Semliki Forest virus *ts*-1 at 39°C, the p62–E1 complex accumulates in endoplasmic reticulum. After a shift down in the presence of cycloheximide to 15°C, the complex moves into a pre-Golgi "vacuolar element." If the shift is to 20°C, the complex moves into and through the Golgi, but accumulates in a trans-post-Golgi structure. Only if the shift is to 28°C do the proteins move to the surface and p62 cleavage is detected. Double immunofluorescence with the antivirus glycoprotein antibodies and with wheat germ agglutinin or concanavalin A led to an assignment of the organelle.

The cell cytoskeleton does not appear to play a significant role in intracellular transport of alphavirus structural proteins. Agents that dissociate microtubules (colchicine) or microfilaments (dibucaine) did not block transport of Semliki Forest virus envelopes proteins to the cell surface (Richardson and Vance, 1982). Cytocholasin B, which affects microfilaments, had no effect on intracellular transport of Semliki Forest virus glycoprotein, although vinblastine inhibited transport by 50% (Kääriäinen *et al.*, 1980).

VII. EXPRESSION OF VIRUS GLYCOPROTEINS FROM COMPLEMENTARY DNAS

A most important development in defining those domains of virus glycoproteins that are critical for transport and assembly is the application of recombinant DNA technology, which has permitted the specific modification of virus genes. Thus far, the technique has been successfully applied to the Semliki Forest virus structural genes. A cDNA containing all the nucleotide sequences of the 26 S mRNA was initially constructed and subsequently modified such that it could be inserted into a vector capable of expressing these genes in an appropriate cell line. Two types of constructs have been made and their expression analyzed. The genes that encode the E1 and E2 glycoproteins were inserted in a (SV 40) virus vector that was microinjected into BHK-cell nuclei (Konder-Koch *et al.*, 1983). Glycoproteins could be detected by immunofluorescence some 6 hr later. Only the p62 was formed, and the E1 was detected by immunoblotting as a somewhat smaller polypeptide, apparently missing portions of the hydrophobic membrane-spanning domain segment at the carboxy terminus. A second recombinant was constructed to eliminate possible aberrant splicing, and expression of this cDNA in BHK cells produced normal E1 and E2 that were localized to the cell surface (Konder-Koch *et al.*, 1983). Cells containing these glycoproteins were capable of cell–cell fusion after treatment for brief times at low pH, conditions previously established for fusogenic activity of Semliki Forest virus glycoproteins (see Chapter 4). Cells missing E1 on their surfaces lacked fusogenic activity; thus, this glycoprotein appears to be the virus fusogen. Yamamoto *et al.* (1981) have shown that purified E1 from Western equine encephalitis virus is a hemolysin when inserted into liposomes.

The cDNA for glycoproteins was further manipulated to produce deletions in those domains of the E2 protein that spanned the membrane and extended into the cytoplasm (Garoff *et al.*, 1983). Exonuclease was used to remove various amounts of nucleotides, and stop codons in all three reading frames were introduced to ensure prompt termination after any frame shift. No E1 was formed when these constructs were expressed in cells. Four variants were analyzed; they all had changes in the original 31 amino acids at the C terminus. They included polypeptides with 8 new amino acids, 7 old plus 7 new amino acids, 12 old plus 7 new ones, and 15 old plus 7 new ones. All these proteins were expressed as p62 and aberrant forms of a cleaved E2. On the basis of immunofluorescence, they all appeared at the surface membrane of the cell with transport rates similar to that of the unmodified protein. The intracellular immunofluorescence was also similar to that of the normal p62/E2. These results imply that the cytoplasmic domain of the protein plays no role in the intracellular transport process; only a basic Arg-Ser-Lys carboxy-terminus sequence adjacent to the bilayer inner surface is common to these four different proteins. The cytoplasmic domain of the polypeptide is believed to have sites for fatty acid acylation; thus, these data suggest further that acylation does not serve as a signal for transport.

Another Semliki Forest virus cDNA that had the entire membrane-spanning and cytoplasmic segment deleted expressed a truncated p62 that remained in the endoplasmic reticulum of transfected BHK cells (Reidel, 1985). This result differs from similar kinds of altered virus glycoproteins that are secreted if their membrane-anchor polypeptide segments are removed. Intracellular transport of Semliki Forest virus p62 was restored by creating chimeric cDNAs in which the p62 polypeptide was fused to membrane domains of either the vesicular stomatitis virus G or the influenza virus hemagglutinin. Both chimeric proteins appeared on surfaces of transfected cells expressing the cDNAs. Immunoblots probed with anti-E2 antiserum showed some formation of lower-molecular-weight proteins in these cells, suggesting partial proteolytic processing of the p62 chimeras during transport. It was not clear, however, whether the proper form of E3 was generated from these polypeptides.

Complementary DNAs have also been constructed for the Sindbis virus structural protein genes and have been inserted into vaccinia virus, where they are expressed during vaccinia infection of cells and animals (C. Rice, personal communication). Expression of virus glycoproteins has also been detected in yeast transformed with a plasmid containing the Sindbis 26 S cDNA (D. Wen and M. J. Schlesinger, unpublished data).

Alphavirus structural gene expression has been studied in the absence of a full infection and virus replication by microinjecting into oocytes a preparation of virus 26 S mRNA isolated from infected cells (Huth et al., 1984). In these experiments, antibodies monospecific for the E1 and E2 glycoproteins were used to detect virus-specific radioactive polypeptides synthesized 24 hr after microinjection of the mRNA. All virus proteins, including the capsid, were formed and apparently processed correctly. The glycoproteins were detected at the surface of the oocyte—about 10% of the total virus products were at this location—and they were not secreted into the extracellular fluid. Pulse–chase studies confirmed the initial synthesis and subsequent proteolytic conversion of the p62 precursor to the E2 and E3 proteins, and antibodies specific to the E2 reacted with material on the cell surface. When tunicamycin was used to block glycosylation in the oocyte, transport of the envelope proteins to the surface occurred, and unglycosylated p62 appeared to be converted to E2; however the latter results were ambiguous, since the bands assigned to unglycosylated precursor and product were not rigorously identified. The ability of these proteins to function correctly in the absence of oligosaccharide may be attributed to the low temperature (25°C) at which these experiments were carried out (see above).

VIII. ASSEMBLY AND BUDDING OF VIRUS

A. Nucleocapsid Formation

Release of virus capsid from the nascent polyprotein is attributed to an autoproteolytic activity encoded within the capsid sequences (Aliperti and Schlesinger, 1978). The amino acid sequence of Gly-Asp-Ser-Gly that is in many serine proteases can be detected in the sequence of Semliki Forest and Sindbis virus capsid (positions 216–220 or 212–216, respectively). Very little information is available on the fate of released capsids; one report shows binding of newly made capsid to the large ribosomal subunit immediately after synthesis (Glanville and Ulmanen, 1976; Söderland and Ulmanen, 1977; Wengler and Wengler, 1984). In a 5-min pulse, no free capsids are detectable, but they are found in virus core particles. Partially completed or empty nucleocapsids are not detected in alphavirus-infected cells. Although reconstituted alphavirus cores, prepared from isolated virions shorn of their envelopes, can bind a wide variety of nucleic acids (Wengler et al., 1982, 1984), from transfer RNAs to fd phage RNA, there is a strict selectivity in vivo, and very little virus-specific 26 S mRNA is ever incorporated into capsid. Neutron diffraction studies of virus cores indicate that the proteins do not form a shell around the nucleic acids; rather, protein and nucleic acid are uniformly distributed (noted in Simons and Warren, 1984). This arrangement implies a concomitant assembly of protein and RNA.

One complementation group (group C) of ts mutants has defects in the capsid. All mutants have the phenotype of an inability to release capsid from the nascent chain at nonpermissive temperature, and high-molecular-weight polypeptides that cannot be subsequently cleaved to yield glycoproteins accumulate in cells infected with these mutants. The phenotype of this class of mutants is also detected when the 26 S mRNA from the mutant is translated in vitro at the nonpermissive temperature; an unprocessed polyprotein of molecular weight 140,000 forms and is not matured (Cancedda et al., 1974). No oligosaccharides appear on these polyproteins.

Assembled nucleocapsids appear rapidly in virions released from the infected cell; however, a large amount of the capsid protein never exits the cell, and it is not clear what the fate of this material is.

B. Interactions at the Plasma Membrane

For effective virion formation in vertebrate cells, the glycoproteins must move from the internal vesicles to the cell surface. A p62–E1 complex appears to be essential for E1 to be transported, since a Semliki Forest virus mutant that is defective in forming p62 can make a normal E1, but the protein does not move from the endoplasmic reticulum vesicles (Hashimoto *et al.*, 1981). In contrast, p62 can move through intracellular organelles in the absence of E1, as shown by studies with cDNA copies of the glycoprotein sequences in which the E1 protein failed to be synthesized in a native form. However, *ts* mutations in Sindbis E1 can affect the p62–E1 complex to the extent that intracellular transport of both p62 and E1 is inhibited at the nonpermissive temperature; both glycoproteins are detectable in smooth membranes (Golgi), but not at the cell surface (Smith and Brown, 1977; Erwin and Brown, 1980). A *ts* mutant of Western equine encephalitis virus with altered E1 also shows a defect in transport of p62 to the cell surface (Hashimoto and Simizv, 1982). Extrusion of virions from infected cells involves a stoichiometric reaction between an E1–E2 heterodimer spike with a capsid polypeptide (Simons and Garoff, 1980). Early electron micrographs of virus-infected cells (Acheson and Tamm, 1967; Brown and Waite, 1972) suggested that the host-cell plasma membrane wrapped around the nucleocapsid and a fusion of the membrane released virus. Few molecular details of this process are known, but the model proposed by Garoff and Simons (1974) accounts reasonably well for the events in virus budding. According to this proposal, an initial interaction occurs between a few glycoprotein spikes at the cell surface and nucleocapsid. Additional virus glycoproteins move into this "nucleation" patch, displacing host-cell membrane proteins. As the cytoplasmic domains of additional E2 proteins bind to more capsid chains, the membrane is pulled around the spherical nucleocapsid, ultimately leading to fusion of the bilayer and release of enveloped virions. The driving force is the protein–protein interactions between E2 and C. An E2–C association in Semliki Forest virus can be demonstrated by cross-linking reagents (Ziemiecki and Garoff, 1978). Furthermore, most of the Semliki Forest virus glycoproteins remain attached to the nucleocapsid when all lipids are removed from virion with octylglucoside at neutral pH (Helenius and Kartenbeck, 1980). Reversible detachment occurs when pH and salt concentrations are varied. The E2 precursor, p62, can also bind nucleocapsids at the cell surface (Brown and Smith, 1975), but envelopment does not proceed, possibly because p62–E1 dimers do not allow for the tight packing required for assembly.

Intracellular budding is generally not seen with alphaviruses and is attributed to (1) low concentration of glycoproteins on internal vesicles, (2) late (post-Golgi) proteolytic conversion of p62 and restructuring of E1–E2 complex, and (3) masking of E2 cytoplasmic sequences by cellular proteins during transport.

There are data indicating that interactions between alphavirus membrane-embedded glycoproteins and nucleocapsids commence before the former emerge at the cell surface. In studies designed to measure lateral mobilities of the glycoproteins on the surfaces of the infected cell, almost all the glycoproteins were in an immobilized state, and the amount of the immobilized material increased with increasing amounts of virus budding (Johnson et al., 1981). Only in the case of a ts mutant examined at the nonpermissive temperature, at which budding was blocked, was there a large mobile fraction of the glycoproteins on the surface. If p62 converts to E2 in a late Golgi vesicle, then the strong E2–E1 complex described earlier should form and create the membrane "patch" recognizable by nucleocapsid. Only fusion of membrane would be needed and may require a specific interaction of the preformed virus complex with the plasma membrane. Budding from intracellular membranes has been seen with a ts mutant of Semliki Forest virus that, at the nonpermissive temperature, fails to have its glycoproteins transported to the surface. Presumably, the latter accumulate inside and, when the temperature is lowered, resume a native conformation that permits budding detectable initially at the internal membranes (Saraste et al., 1980). Intracellular budding is also detected in vesicles that accumulate in cells treated with monensin at concentrations that block transport from Golgi to the plasma membrane (Fig. 4) (Griffiths et al., 1983; Johnson and Schlesinger, 1980).

The question of a selective region of budding from infected cells is not easily answered, although morphological studies suggest a nonrandom location (Brown and Waite, 1972). In chicken fibroblasts, budding virus was detected preferentially in long cell processes (Birdwell et al., 1973). Aberrant assembly of virions in which multiple nucleocapsids bud from the cell within a single vesicle membrane has been found with a ts mutant of Sindbis virus (Strauss et al., 1977).

Another phenomenon, as yet unexplained, has been described for the budding of alphaviruses in chicken embryo cells. In a hypotonic medium, viruses grown in chicken embryo fibroblasts—but not in BHK cells—are blocked in the final stages of budding (Waite and Pfefferkorn, 1970). Precursor nucleocapsid and envelope glycoproteins continue to accumulate, however, and raising the salt concentration produces an immediate burst of virions into the medium. The basis for this peculiar effect is unknown; it is not a block in glycoprotein processing (Bell et al., 1979; Mayne et al., 1984). Furthermore, release from this hypotonic block does not require energy, since it proceeds in the presence of inhibitors such as cyanide, fluoride, azide, and iodoacetic acid. The effect has been found to depend on the strain of virus and the culture conditions of the cells and may involve the interaction between E2 and E1 and organization of the host-cell membrane lipid (Strauss et al., 1980).

Relatively little is known about the role of lipid in assembly and budding. The lipid composition of alphaviruses reflects in general that of the host-cell plasma membrane (reviewed in Lenard, 1980; Hirschberg and Robbins, 1974), and it is generally believed that viral lipids are incorporated from those of the host cell's plasma membrane during budding. It would not be surprising, however, to find some degree of selectivity based on the glycoprotein's composition, particularly in light of the presence of fatty acids on these proteins. A selective lipid assortment distinct from the host cell has been reported for Rous sarcoma virus and vesicular stomatitis virus envelope (Pessin and Glaser, 1980).

The interactions between the alphavirus transmembrane glycoprotein and nucleocapsid must involve a high degree of specificity. Almost no pseudotypes have been found in which alphavirus genomes contain other virus glycoproteins in their membranes, even though alphavirus glycoproteins are found in the membranes of several other enveloped viruses (Zavadova *et al.*, 1977). Pseudotypes are formed in cells coinfected with closely related alphaviruses (Burge and Pfferkorn, 1966; Strauss *et al.*, 1983).

IX. CONCLUSIONS AND PERSPECTIVES

The formation of functional membrane glycoproteins consists of many steps, beginning with synthesis of a polypeptide chain and including a variety of modifications of that polypeptide during its intracellular transport and localization to specific cellular membranes. Investigations of alphavirus glycoprotein synthesis have played an important role in identifying and elucidating some of these steps. The studies on *in vitro* translation of these viral mRNAs were instrumental in identifying cryptic initiation sites for translation and autoprotease activity for processing nascent polyproteins. In this chapter, we have focused on events occurring at the next stages of glycoprotein biosynthesis, which involve posttranslational alterations and migration of the protein through intracellular organelles. Here, too, alphavirus glycoproteins have provided useful models for following a protein from its insertion as a nascent polypeptide into the membrane of the rough endoplasmic reticulum to its mature form on reaching the plasma membrane. There are, however, areas that remain to be explored. For example, we are only beginning to identify conformational changes that must be crucial in the formation of a functional transmembrane protein. For alphavirus glycoproteins, these changes include those that occur in the individual polypeptides as well as those in the heterodimer that result from conversion of p62 to E2. Viral glycoproteins have an additional dimension of complexity, for these proteins must not only reach an appropriate site in cellular membranes but also participate in the assembly and release of the enveloped virion. These latter involve interactions with the nucleocapsid and with the lipid bilayer. We can anticipate that alphavirus glycoproteins will also serve as valuable tools for investigating these events.

ACKNOWLEDGMENTS. Research described in this chapter that was carried out in the authors' laboratories was supported by Public Health Service Grants AI 11377 (to S.S.) and AI 19494 (to M.J.S.) from the National Institute of Allergy and Infectious Diseases.

REFERENCES

Acheson, N. H., and Tamm, I., 1967, Replication of Semliki Forest virus: An electron microscopic study, *Virology* **32**:128.

Adams, R. H., and Brown, D. T. 1982, Inhibition of Sindbis virus maturation after treatment of infected cells with trypsin, *J. Virol.* **41**:692.

Aliperti, G., and Schlesinger, M. J., 1978, Evidence for an autoprotease of Sindbis virus capsid protein, *Virology* **90**:336.

Bell, J. R., and Strauss, J. H., 1981, *In vivo* N-terminal acetylation of Sindbis virus proteins, *J. Biol. Chem.* **256**:8006.

Bell, J. W., Jr., Garry, R. F., and Waite, M. R. F., 1979, Effect of low-NaCl medium on the envelope glycoproteins of Sindbis virus, *J. Virol.* **25**:764.

Bell, J. R., Rice, C. M., Hunkapiller, M. W., and Strauss, J. H., 1982, The N-terminus of PE2 in Sindbis virus-infected cells, *Virology* **119**:255.

Berger, M., and Schmidt, M. F. G., 1984, Cell-free fatty acid acylation of Semliki Forest viral polypeptide with microsomal membranes from eukaryotic cells, *J. Biol. Chem.* **259**:7245.

Birdwell, C. R., Strauss, E. G., and Strauss, J. H., 1973, Replication of Sindbis virus. III. An electron microscopic study of virus maturation using the surface replica technique, *Virology* **56**:429.

Bonatti, S., and Cancedda, F. D., 1982, Postranslational modifications of Sindbis virus glycoproteins: Electrophoretic analysis of pulse–chase-labeled infected cells, *J. Virol.* **42**:64.

Bonatti, S., Cancedda, R., and Blobel, G., 1979, Membrane biogenesis, *in vitro* cleavage, core glycosylation and integration into microsomal membranes of Sindbis virus glycoproteins, *J. Cell Biol.* **80**:219.

Bracha, M., and Schlesinger, M. J., 1976a, Inhibition of Sindbis virus replication by zinc ions, *Virology* **72**:272.

Bracha, M., and Schlesinger, M. J., 1976b, Defects in RNA+ temperature sensitive mutants of Sindbis virus and evidence for a complex of PE2–E1 viral glycoproteins. *Virology* **74**:441.

Brown, D. T., and Smith, J. F., 1975, Morphology of BHK-21 cells infected with Sindbis virus temperature sensitive mutants in complementation groups D and E, *J. Virol.* **15**:1262.

Brown, D. T., and Waite, M. R. F., 1972, Morphology and morphogenesis of Sindbis virus as seen with freeze–etching techniques, *J. Virol.* **10**:524.

Burge, B. W., and Pfefferkorn, E. R., 1966, Phenotypic mixing between group A arboviruses, *Nature (London)* **210**:1397.

Burke, D., and Keegstra, K., 1979, Carbohydrate structure of Sindbis virus glycoprotein E2 from virus grown in hamster and chicken cells, *J. Virol.* **29**:546.

Burke, B., Walter, C., Griffith, G., and Warren, G., 1983, Viral glycoproteins at different stages of intracellular transport can be distinguished using monoclonal antibodies, *Eur. J. Cell Biol.* **31**:315.

Cancedda, R., Swanson, R., and Schlesinger, M. J., 1974, Viral proteins formed in a cell free rabbit reticulocyte system programmed with RNA from a temperature-sensitive mutant of Sindbis virus, *J. Virol.* **14**:664.

Cancedda, R., Villa-Komaroff, L., Lodish, H. F., and Schlesinger, M., 1975, Initiation sites for translation of Sindbis virus 42S and 26S messenger RNAs, *Cell* **6**:215.

Cancedda, R., Bonatti, S., and Leone, A., 1981, One extra oligosaccharide chain of the high-mannose class in the E2 protein of a Sindbis virus isolate, *J. Virol.* **38**:8.

Dalgarno, L., Rice, C. M., and Strauss, J. H., 1983, Ross river virus 26S RNA: Complete nucleotide sequence and deduced sequence of the encoded structural proteins, *Virology* **129**:170.

Datema, R., Romero, P. A., Legler, G., and Schwarz, R. T., 1982, Inhibition of formation of complex oligosaccharides by the glucosidase inhibitor bromoconduritol, *Proc. Natl. Acad. Sci. U.S.A.* **79**:6787.

Datema, R., Romero, P. A., Rott, R., and Schwarz, R. T., 1984, On the role of oligosaccharide trimming in the maturation of Sindbis and influenza virus, *Arch. Virol.* **81**:25.

Davidson, S. K., and Hunt, L. A., 1983, Unusual neutral oligosaccharides in mature Sindbis virus glycoproteins are synthesized from truncated precursor oligosaccharides in Chinese hamster ovary cells, *J. Gen. Virol.* **64**:613.

Davidson, S. K., and Hunt, L. A. 1985a, Sindbis virus glycoproteins are abnormally glycosylated in Chinese hamster ovary cells deprived of glucose, *J. Gen. Virol.* **66**:1457.

Davidson, S. K., and Hunt, L. A., 1985b, Hazelhurst vesicular stomatitis virus G and Sindbis virus E1 glycoproteins undergo similar host cell-dependent variation in oligosaccharide processing, *Biochem. J.* **229**:47–55.

Docherty, K., Carroll, R. J., and Steiner, D. F., 1982, Conversion of proinsulin to insulin: Involvement of a 31,500 molecular weight thiol protease, *Proc. Natl. Acad. Sci. U.S.A.* **79**:4613.

Dunphy, W. G., Brands, R., and Rothman, J. E., 1985, Attachment of terminal N-acetylglucosamine to asparagine-linked oligosaccharides occurs on the central cisternae of the Golgi Stack, *Cell* **40**:463.

Durbin, R. K., and Stollar, V., 1984, A mutant of Sindbis virus with a host-dependent defect in maturation associated with hyperglycosylation of E2, *Virology* **135**:331.

Erwin, C., and Brown, D. T., 1980, Intracellular distribution of Sindbis virus membrane proteins in BHK-21 cells infected with wild-type virus and maturation defective mutants, *J. Virol.* **36**:775.

Gallione, C. J., and Rose, J. K., 1983, Nucleotide sequence of a cDNA clone encoding the entire glycoprotein from the New Jersey serotype of vesicular stomatitis virus, *J. Virol.* **46**:162.

Garoff, H., and Simons, K., 1974, Location of the spike glycoproteins in the Semliki Forest virus membrane, *Proc. Natl. Acad. Sci. U.S.A.*, **71**:3988.

Garoff, H., and Söderlund, H., 1978, The amphiphilic membrane glycoproteins of Semliki Forest virus are attached to the lipid bilayer by their COOH-terminal ends, *J. Mol. Biol.* **124**:535.

Garoff, H., Simons, K., and Dobberstein, B., 1978, Assembly of the Semliki Forest virus membrane glycoproteins in the membrane of the endoplasmic reticulum *in vitro*, *J. Mol. Biol.* **124**:587.

Garoff, H., Frischauf, A.-M., Simons, K., Lehrach, H., and Delius, H., 1980, Nucleotide sequence of cDNA coding for Semliki Forest virus membrane glycoproteins, *Nature (London)* **288**:236.

Garoff, H., Kondor-Koch, C., and Riedel, H., 1982, Structure and assembly of alphaviruses, *Curr. Top. Microbiol. Immunol.* **99**:1.

Garoff, H., Kondor-Koch, C., Petterson, R., and Burke, B., 1983, Expression of Semliki Forest virus proteins from cloned complementary DNA. II. The membrane-spanning glycoprotein E2 is transported to the cell surface without its normal cytoplasmic domain, *J. Cell Biol.* **97**:652.

Gibson, R., Schlesinger, S., and Kornfeld, S., 1979, The nonglycosylated glycoprotein of vesicular stomatitis virus is temperature-sensitive and undergoes intracellular aggregation at elevated temperatures, *J. Biol. Chem.* **254**:3600.

Gibson, R., Kornfeld, S., and Schlesinger, S., 1980, A role for oligosaccharides in glycoprotein biosynthesis, *Trends Biochem. Sci.* **5**:290.

Glanville, N., and Ulmanen, J., 1976, Biological activity of *in vitro* synthesized protein:

Binding of Semliki Forest virus capsid protein to the large ribosomal subunit, *Biochem. Biophys. Commun.* **71**:393.

Gottlieb, C., Kornfeld, S., and Schlesinger, S., 1979, Restricted replication of two alphaviruses in ricin-resistant mouse L cells with altered glycosyltransferase activities, *J. Virol.* **29**:344.

Green, J., Griffith, G., Louvard, D., Quinn, P., and Warren, G., 1981, Passage of viral membrane glycoproteins through the Golgi complex, *J. Mol. Biol.* **152**: 663.

Griffiths, G., Brands, R., Burke, B., Lowcard, D., and Warren, G., 1982, Viral membrane proteins acquire galactose in trans Golgi cisternae during intracellular transport, *J. Cell Biol.* **95**:781.

Griffiths, G., Quinn, P., and Warren, G., 1983, Dissection of the Golgi complex, 1. Monensin inhibits the transport of viral membrane proteins from medial to trans Golgi cisternae in baby hamster kidney cells infected with Semliki Forest virus, *J. Cell Biol.* **96**:835.

Griffiths, G., Warren, G., Quinn, P., Mathieu-Costello, and Hoppeler, H., 1984, Density of newly synthesized plasma membrane proteins in intracellular membranes. I. Sterological studies, *J. Cell Biol.* **98**:7133.

Hakimi, J., and Atkinson, P. H., 1980, Growth-dependent alterations in oligomannosyl glycopeptides expressed in Sindbis virus glycoproteins, *Biochemistry* **19**:5619.

Hakimi, J., and Atkinson, R. H., 1982, Glycosylation of intracellular Sindbis virus glycoproteins, *Biochemistry* **21**:2140.

Hashimoto, K., and Simizu, B., 1982, A temperature-sensitive mutant of Western equine encephalitis virus with an altered envelope protein E1 and a defect in the transport of envelope glycoproteins, *Virology* **119**:276.

Hashimoto, K., Erdel, S., Keranen, S., Saraste, J., and Kääriäinen, L., 1981, Evidence for a separate signal sequence for the carboxy-terminal envelope glycoprotein E1 of Semliki Forest virus, *J. Virol.* **38**:34.

Helenius, A., and Kartenbeck, J., 1980, The effects of octylglucoside on the Semliki Forest virus membrane: Evidence for a spike-protein–nucleocapsid interaction, *Eur. J. Biochem.* **106**:613.

Hirschberg, C. B., and Robbins, P. W., 1974, The glycolipids and phospholipids of Sindbis virus and their relation to the lipids of the host cell plasma membrane, *Virology* **61**:602.

Hsieh, P., and Robbins, P. W., 1984, Regulation of asparagine-linked oligosaccharide processing, *J. Biol. Chem.* **259**:2375.

Hsieh, P., Rosner, M. R., and Robbins, P. W., 1983a, Host-dependent variation of asparagine-linked oligosaccharides at individual glycosylation sites of Sindbis virus glycoproteins, *J. Biol. Chem.* **258**:2548.

Hsieh, P., Rosner, M. R., and Robbins, P. W., 1983b, Selective cleavage by endo-β-N-acetylglucosaminidase H at individual glycosylation sites of Sindbis virion envelope glycoproteins, *J. Biol. Chem.* **258**:2555.

Hunt, L. A., Etchison, J. R., and Summers, D. F., 1978, Oligosaccharide chains are trimmed during synthesis of the envelope glycoprotein of vesicular stomatitis virus, *Proc. Natl. Acad. Sci. U.S.A.* **75**:754.

Huth, A., Rapoport, T. A., and Kääriäinen, L., 1984, Envelope proteins of Semliki Forest virus synthesized in *Xenopus* oocytes are transported to the cell surface, *EMBO J.* **3**:767.

Ishida, I., Simizu, B., Koizumi, S., Oya, A., and Yamada, M., 1981, Nucleoside triphosphate phosphohydrolase produced in BHK-cells infected with WEE is probably associated with an 82 k dalton non-structural protein, *Virology* **108**:13–20.

Johnson, D. C., and Schlesinger, M. J., 1980, Vesicular stomatitis virus and Sindbis virus glycoprotein transport to the cell surface is inhibited by ionophores, *Virology* **103**:407.

Johnson, D. C., Schlesinger, M. J., and Elson, E. L., 1981, Fluorescence photobleaching recovery measurements reveal differences in envelopment by Sindbis and vesicular stomatitis virus, *Cell* **23**:423.

Johnston, R. E., 1983, Requirement for host replication of Sindbis virus, *J. Virol.* **45**:200.

Kääriäinen, L., Hashimoto, K., Saraste, J., Virtanen, I., and Penttinen, K., 1980, Monensin and FCCP inhibit the intracellular transport of alphavirus membrane glycoproteins, *J. Cell Biol.* **87**:783.

Kaluza, G., and Pauli, G., 1980, The influence of intramolecular disulfide bonds on the structure and function of Semliki Forest virus membrane glycoproteins, *Virology* **102**:300.

Kaluza, G., Roh, R., and Schwarz, R. T., 1980, Carbohydrate-induced conformational changes of Semliki Forest virus glycoproteins determine antigenicity, *Virology* **102**:286.

Keegstra, K., Sefton, B., and Burke, D., 1975, Sindbis virus glycoproteins; Effect of the host cell on the oligosaccharides, *J. Virol.* **16**:613.

Kennedy, S. I. T., 1974, The effect of enzymes on structural and biological properties of Semliki Forest virus, *J. Gen. Virol.* **23**:129.

Konder-Koch, C., Burke, B., and Garoff, H., 1983, Expression of Semliki Forest virus proteins from cloned complementary DNA. 1. The fusion activity of the spike glycoprotein, *J. Cell Biol.* **97**:644.

Kornfeld, R., and Kornfeld, S., 1985, Assembly of asparagine-linked oligosaccharides, *Annu. Rev. Biochem.* **54**:631.

Kornfeld, S., Li, E., and Tabas, I., 1978, The synthesis of complex-type oligosaccharides. II. Characterization of the processing intermediates in the synthesis of the complex oligosaccharide units of the vesicular stomatitis virus G protein, *J. Biol. Chem.* **253**:7771.

Leavitt, R., Schlesinger, S., and Kornfeld, S., 1977, Tunicamycin inhibits glycosylation and multiplication of Sindbis and vesicular stomatitis virus, *J. Virol.* **21**:375.

Lenard, J., 1980, Lipids of alphaviruses, in: *The Togaviruses* (R. W. Schlesinger, ed), pp. 335–341, Academic Press, New York.

Magee, A. I., Koyama, A. H., Malfer, C., Wen, D., and Schlesinger, M. J., 1984, Release of fatty acids from virus glycoproteins by hydroxylamine, *Biochim. Biophys. Acta* **798**,156.

Mann, E., Edwards, J., and Brown, D. T., 1983, Polycaryocyte formation mediated by Sindbis virus glycoproteins, *J. Virol.* **45**:1083.

Mayne, J. T., Rice, C. M., Strauss, E. G., Hunkapiller, M. W., and Strauss, J. H., 1984, Biochemical studies of the maturation of the small Sindbis virus glycoprotein E2, *Virology* **134**:338.

Mayne, J. T., Bell, J. R., and Strauss, J. H., 1985, Pattern of glycosylation of Sindbis virus envelope proteins synthesized in hamster and chicken cells, *Virology* (in press).

Pan, Y. T., Hori, H., Saul, R., Sanford, B. A., Molyneux, R. J., and Elbein, A. D., 1983, Castanospermine inhibits the processing of the oligosaccharide portion of the influenza viral hemagglutinin, *Biochemistry* **22**:3975.

Pessin, J. E., and Glaser, M., 1980, Budding of Rous sarcoma virus and vesicular stomatitis virus from localized regions in the plasma membrane of chicken embryo fibroblasts, *J. Biol. Chem.* **255**:9044.

Pollack, L., and Atkinson, P. H., 1983, Correlation of glycosylation forms with position in amino acid sequence, *J. Cell Biol.* **97**:293.

Quinn, P., Griffiths, G., and Warren, G., 1983, Dissection of the Golgi complex, II. Density separation of specific Golgi functions in virally infected cells treated with monensin, *J. Cell Biol.* **96**:851.

Quinn, P., Griffiths, G., and Warren, G., 1984, Density of newly synthesized plasma membrane proteins in intracellular membranes. II. Biochemical studies, *J. Cell Biol.* **98**:2147.

Reidel, H., 1985, Different membrane anchors allow the Semliki Forest virus spike subunit E2 to reach the cell surface, *J. Virol.* **54**:224.

Rice, C. M., and Strauss, J. H., 1981, Nucleotide sequence of the 26S mRNA of Sindbis virus and deduced sequence of the encoded virus structural proteins, *Proc. Natl. Acad. Sci. U.S.A.* **78**:2062.

Rice, C. M., and Strauss, J. H., 1982, Association of Sindbis virion glycoproteins and their precursors, *J. Mol. Biol.* **154**:325.

Rice, C. M., Bell, J. R., Hunkapiller, M. W., Strauss, E. G., and Strauss, J. H., 1982, Isolation and characterization of the hydrophobic COOH-terminal domains of the Sindbis virion glycoproteins, *J. Mol. Biol.* **154**:355.

Richardson, C. D., and Vance, D. E., 1978, The effect of colchicine and dibucaine on the morphogenesis of Semliki Forest virus, *J. Biol. Chem.* **253**:4584.

Robbins, P. W., Hubbard, S. C., Turco, S. J., and Wirth, D. F., 1977, Proposal for a common oligosaccharide intermediate in the synthesis of membrane glycoproteins, *Cell* **12**:893.

Roehrig, J. T., Gorski, D., and Schlesinger, M. J., 1982, Properties of monoclonal antibodies directed against the glycoproteins of Sindbis virus, *J. Gen. Virol.* **59**:421.

Rose, J. K., Adams, G. A., and Gallione, C., 1984, The presence of cysteine in the cytoplasmic domain of the vesicular stomatitis virus glycoprotein is required for palmitate addition, *Proc. Natl. Acad. Sci. U.S.A.* **81**:2050.

Rothman, J. E., Miller, R. L., and Urbane, L. J., 1984, Intercompartmental transport in the Golgi complex is a dissociative process: Facile transfer of membrane protein between two Golgi populations, *J. Cell Biol.* **99**:260.

Saraste, J., and Kaismanen, E., 1984, Pre- and post-Golgi vacuoles operate in the transport of Semliki Forest virus membrane glycoproteins to the cell surface, *Cell* **38**:535.

Saraste, J., VonBonsdorff, C. H., Hashimoto, K., Kääriäinen, L., and Keränen, S., 1980, Semliki Forest virus mutants with temperature-sensitive transport defect of envelope proteins, *Virology* **100**:229.

Saunier, B., Kilker, R. D., Jr., Tkacz, J. S., Quaroni, A., and Herscovics, A., 1982, Inhibition of N-linked complex oligosaccharide formation by 1-deoxynojirimycin, an inhibitor of processing glucosidases, *J. Biol. Chem.* **259**:14155.

Scheefers, H., Scheefers-Borchel, U., Edwards, J., and Brown, D. T., 1980, Distribution of virus structural proteins and protein–protein interactions in plasmid membrane of baby hamster kidney cells infected with Sindbis or vesicular stomatitis virus, *Proc. Natl. Acad. Sci. U.S.A.* **77**:7277.

Schlesinger, M. J., and Malfer, C., 1982, Cerulenin blocks fatty acid acylation of glycoproteins and inhibits vesicular stomatitis and Sindbis virus particle formation, *J. Biol. Chem.* **257**:9887.

Schlesinger, S., and Schlesinger, M. J., 1972, Formation of Sindbis virus proteins: Identification of a precursor for one of the envelope proteins, *J. Virol.* **10**:925.

Schlesinger, S., Gottlieb, C., Feil, P., Gelb, N., and Kornfeld, S., 1976, Growth of enveloped RNA viruses in a line of Chinese hamster ovary cells with deficient N-acetyl-glucosaminyltransferase activity, *J. Virol.* **17**:239.

Schlesinger, S., Malfer, C., and Schlesinger, M. J., 1984, The formation of vesicular stomatitis virus (San Juan strain) becomes temperature-sensitive when glucose residues are retained on the oligosaccharides of the glycoprotein, *J. Biol. Chem.* **259**:7597.

Schesinger, S., Koyama, A. H., Malfer, C., Gee, S. L., and Schlesinger, M. J., 1985, The effects of inhibitors of glucosidase I on the formation of Sindbis virus, *Virus Res.* **2**:139.

Schmaljohn, A. L., Kokabun, K. M., and Cole, G. A., 1983, Protective monoclonal antibodies define maturational and pH-dependent antigenic changes in Sindbis virus E1 glycoprotein, *Virology* **130**:144.

Schmidt, M. F. G., 1982, Acylation of viral spike glycoproteins, a feature of envelope RNA viruses, *Virology* **116**:327.

Schmidt, M. F. G., and Schlesinger, M. J., 1980, Relation of fatty acid attachment to the translation and maturation of vesicular stomatitis and Sindbis virus membrane glycoproteins, *J. Biol. Chem.* **255**:3334.

Schmidt, M. F. G., Bracha, M., and Schlesinger, M. J., 1979, Evidence for covalent attachment of fatty acids to Sindbis virus glycoproteins, *Proc. Natl. Acad. Sci. U.S.A.* **76**:1687.

Schwarz, R. T., and Datema, R., 1984, Inhibitors of trimming: New tools in glycoprotein research, *Trends Biochem. Sci.* **9**:32.

Schwarz, R. T., Rohrschneider, J. M., and Schmidt, M. F. G., 1976, Suppression of glycoprotein formation of Semliki Forest influenza, and avian sarcoma virus by tunicamycin, *J. Virol.* **19**:782.

Sefton, B. M., 1977, Immediate glycosylation of Sindbis virus membrane proteins, *Cell* **10**:659.

Simizu, B., Hashimoto, K., and Ishida, I., 1983, A varient of Western equine encephalitis virus with nonglycosylated E3 protein, *Virology* **125**:99.

Simons, K., and Garoff, H., 1980, The budding mechanism of enveloped animal viruses, *J. Gen. Virol.* **50**:1.

Simons, K., and Warren, G., 1984, Semliki Forest virus: A probe for membrane traffic in the animal cell, *Adv. Protein Chem.* **36**:79.

Smith, J. F., and Brown, D. T., 1977, Envelopment of Sindbis virus: Synthesis and organization of proteins in cells infected with wild type and maturation defective mutants, *J. Virol.* **22**:662.

Söderlund, H., and Ulmanen, I., 1977, Transient association of Semliki Forest virus capsid protein with ribosomes, *J. Virol.* **24**:907.

Steiner, D. F., Docherty, K., and Carroll, R., 1984, Golgi/granule processing of peptide hormone and neuropeptide precursors: A minireview, *J. Cell. Biochem.* **24**:121.

Strauss, E. G., Birdwell, C. R., Lenches, E. M., Staples, S. E., and Strauss, J. H., 1977, Mutants of Sindbis virus. II. Characterization of a maturation-defective mutant *ts*103, *Virology* **82**:122.

Strauss, E. G., Lenches, E., and Stamreich-Martin, M., 1980, Growth and release of several alphaviruses in chick and BHK cells, *J. Gen. Virol.* **49**:297.

Strauss, E. G., Tsukeda, H., and Simizu, B., 1983, Mutants of Sindbis virus. IV. Heterotypic complementation and phenotypic mixing between temperature-sensitive mutants and wild type Sindbis and Western equine encephalitis virus, *J. Gen. Virol.* **64**:1581.

Struck, D. K., and Lennarz, W. J., 1980, The function of saccharide-lipids in synthesis of glycoproteins, in: *The Biochemistry of Glycoproteins and Proteoglycans* (W. J. Lennarz, ed.), pp. 35–83, Plenum Press, New York.

Tabas, I., and Kornfeld, S., 1978, The synthesis of complex-type oligosaccharides, *J. Biol. Chem.* **253**:779.

Tabas, I., Schlesinger, S., and Kornfeld, S. 1978, Processing of high mannose oligosaccharides to form complex type oligosaccharides on the newly synthesized polypeptide of the vesicular stomatitis virus G protein and the IgG heavy chain, *J. Biol. Chem.* **253**:716.

Ulmanen, I., Söderlund, H., and Kääriäinen, L., 1979, Role of protein synthesis in the assembly of Semliki Forest virus nucleocapsid, *Virology* **99**:265.

Waite, M. R. F., and Pfefferkorn, E. R., 1970, Inhibition of Sindbis virus production by media of low ionic strength: Intracellular events and requirements for reversal, *J. Virol.* **5**:60.

Walter, P., and Blobel, G., 1981, Translocation of proteins across the endoplasmic reticulum. III. Signal recognition protein (SRP) causes signal-dependent and site-specific arrest of chain elongation that is released by microsomal membranes, *J. Cell Biol.* **91**:557.

Walter, P., and Blobel, G., 1982, Signal recognition particle contains a 7S RNA essential for protein translocations across the endoplasmic reticulum, *Nature (London)* **299**:691.

Welch, W. J., and Sefton, B. M., 1980, Characterization of a small, nonstructural viral polypeptide present late during infection of BHK cells by Semliki Forest virus, *J. Virol.* **33**:230.

Wengler, G., Boege, U., Wengler, G., Bischoff, H., and Wahn, K., 1982, The core protein of the alphavirus Sindbis virus assembles into core-like nucleoproteins with the viral genome RNA and with other single-stranded nucleic acids *in vitro*, *Virology* **118**:401.

Wengler, G., Wengler, G., Boege, U., and Wahn, K., 1984, Establishment and analysis of a system which allows assembly and disassembly of alphavirus core-like particles under physiological conditions *in vitro*, *Virology* **132**:401.

Wengler, G., and Wengler, G., 1984, Identification of a transfer of viral core protein to cellular ribosomes during the early stages of alphavirus infection, *Virology* **134**:435.

Wirth, D. F., Katz, F., Small, B., and Lodish, H. F., 1971, How a single Sindbis virus mRNA directed the synthesis of one soluble protein and two integral membrane glycoproteins, *Cell* **10**:253.

Yamamoto, K., Suzuki, K., and Simizu, B., 1981, Hemolytic activity of the envelope glycoproteins of western equine encephalitis virus in reconstitution experiments, *Virology* **109**:452.

Zavadova, Z., Zavada, J., and Weiss, R., 1977, Unilateral phenotypic mixing of envelope antigens between togaviruses and vesicular stomatitis virus or avian RNA tumor virus, *J. Gen. Virol.* **37**:557.

Ziemiecki, A., and Garoff, H., 1978, Subunit composition of the membrane glycoprotein complex of Semliki Forest virus, *J. Mol. Biol.* **122**:259.

Ziemiecki, A., Garoff, H., and Simons, K., 1980, Formation of the Semliki Forest virus membrane glycoprotein complexes in the infected cells, *J. Gen. Virol.* **50**:111.

CHAPTER 6

Defective RNAs of Alphaviruses

SONDRA SCHLESINGER AND BARBARA G. WEISS

I. INTRODUCTION

Most viruses, when passaged at high multiplicity in cultured cells, accumulate deletion mutants characterized by their ability to interfere with the replication of the standard virus. These mutants are defined as defective interfering (DI) particles (Huang and Baltimore, 1970; Perrault, 1981). One of their hallmarks is the specificity of their inhibition; they interfere only with the replication of homologous or closely related viruses. Why study DI particles? What can they tell us about the standard virus or about virus–host interactions? The following points attempt to answer these questions and provide the framework for this chapter.

1. DI genomes contain deletions of the standard virus genome, rendering them incapable of self-replication. To be propagated, they must conserve specific sequences or structures that can be recognized by the coinfecting helper virus proteins. Therefore, identification of sequences retained by DI genomes should help to define specific regulatory sequences in the standard genome.

2. DI particles may be a factor in establishing or maintaining persistent viral infections. This idea was proposed by Huang and Baltimore (1970) and spurred an interest in the role of DI particles in persistent infections, particularly in cultured cells.

3. DI particles may serve as a probe for analyzing viral RNA replication and detecting host factors involved in virus replication.

Although these points apply to all virus systems, we discuss them almost entirely in the context of alphaviruses as is befitting a chapter in

SONDRA SCHLESINGER AND BARBARA G. WEISS • Department of Microbiology and Immunology, Washington University School of Medicine, St. Louis, Missouri 63110.

this book. The two alphaviruses, Sindbis virus and Semliki Forest virus, have been studied in most detail, and almost all the analyses have been with DI particles derived from these viruses.

II. IDENTIFICATION OF SEQUENCES RETAINED IN THE DEFECTIVE INTERFERING GENOMES OF SINDBIS AND SEMLIKI FOREST VIRUSES

A. Characterization of the Defective Interfering RNAs

The overall properties of the defective genomes of Sindbis and Semliki Forest viruses have been described (Stollar, 1979, 1980; Holland *et al.*, 1980; Perrault, 1981) and are summarized here only briefly. The generation and amplification of the DI RNAs of Sindbis and Semliki Forest viruses show many parallels. Depending on the cells and the conditions of passaging, the first DI RNAs to be detected after about 3–5 high-multiplicity passages are about one half the size of the virion RNA (Guild and Stollar, 1975; Stark and Kennedy, 1978). These molecules soon disappear on subsequent passaging and are replaced by molecules one fourth to one fifth the size of the original genome.

One interesting aspect of the analysis of these DI particles was the finding that the defective particles and the envelope-free nucleocapsids are either similar in size and density or heavier than the infectious particles (Shenk and Stollar, 1973; Bruton and Kennedy, 1976; Kennedy *et al.*, 1976). These results suggest that a definite quantity of RNA is required for capsid stability and that multiple copies of specific sizes of DI RNAs are encapsidated.

The initial studies on the structure of the DI genomes of alphaviruses used solution hybridization (Guild and Stollar, 1977) and T₁ RNase-resistant oligonucleotide mapping (Kennedy, 1976; Dohner *et al.*, 1979) to demonstrate that sequences near the 3' and 5' termini of the viral genomes are retained in the DI RNAs. These data suggested a model for the generation of DI RNAs in which successive internal deletions of genomic RNA produced DI molecules of decreasing size, with a limit size being reached due to packaging constraints (Stark and Kennedy, 1978). The DI RNAs, however, were soon found to be more complex than had originally been thought.

B. Sequence Organization of Defective Interfering RNAs of Sindbis and Semliki Forest Viruses

The first direct demonstration of the complexity of the alphavirus DI genomes came from the work of Pettersson (1981), who demonstrated that the 5'-terminal sequences of DI RNAs of Semliki Forest virus are

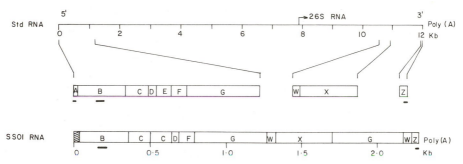

FIGURE 1. A comparison between the deduced sequence of the standard Sindbis HR virus RNA and the SS01 DI RNA. The top line represents the total 12 kilobases (Kb) of standard virion RNA; the regions of homology with the DI RNA are expanded below. The bold underlines represent the conserved regions of the genome (see the text). Regions A–G are derived from the 5' two thirds of the standard genome, which codes for the nonstructural proteins. Regions W–Z are derived from the 3' one third of the standard genome, which codes for the structural proteins. The hatched area in SS01 RNA is the tRNA^{Asp}. Modified from Monroe and Schlesinger (1984) with permission from the American Society of Microbiology.

heterogeneous and different from that of the standard virus. This observation was followed by the sequence determination of the cloned complementary DNA (cDNA) of a DI RNA of Semliki Forest virus (Lehtovaara *et al.*, 1982). This sequence consists of three repeated units derived from the 5' two thirds of the viral genome and a region derived from the 3' terminus. A second cloned DI cDNA from this DI RNA population also shows repeats and rearrangements (Söderlund *et al.*, 1981; Lehtovaara *et al.*, 1982). Sequence analysis of DI RNAs of Sindbis virus demonstrated that these molecules have similar features (Fig. 1) (Monroe *et al.*, 1982; Monroe and Schlesinger, 1984). In this case, the peculiar structure of the genome was initially overshadowed by the discovery that the 5' terminus of the DI RNA is almost identical to the sequence of a rat transfer (t)RNA^{Asp} (Monore and Schlesinger, 1983). The 5' termini of two independently generated DI RNAs were sequenced and were found to contain all but the 9 nucleotides at the 5' terminus of the tRNA^{Asp} (Fig. 2). The 3' C-C-A terminus of the tRNA is covalently attached near the 5' terminus of the DI RNA. In one DI RNA, the tRNA replaces the 5'-terminal 31 nucleotides of the virion RNA; in the other DI RNA, 23 nucleotides from the 49 S viral RNA are replaced.

We considered two ways in which the covalent linkage between tRNA^{Asp} and viral RNA sequences may have formed: There could have been a copy-choice event during synthesis of the negative strand such that the replicase switched templates from DI RNA to tRNA^{Asp}, or the tRNA could have served as a primer for the transcription of plus-strand DI RNA (Fig. 3). The suggestion that tRNA^{Asp} may serve as a primer for DI RNA synthesis implies that it might also serve a similar role in stan-

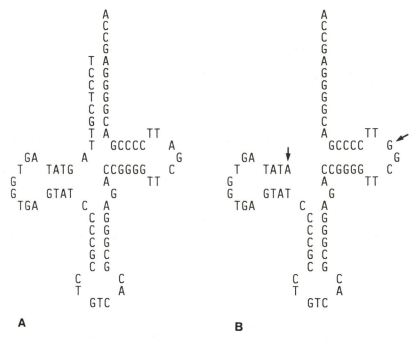

FIGURE 2. (A) Sequence of tRNA^Asp (Sekiya *et al.*, 1981). (B) Sequence of the 67 nucleotides at the 5′ terminus of two independently derived DI RNAs (Monroe and Schlesinger, 1983). The arrows indicate the two bases that differ from the tRNA^Asp sequence.

dard viral RNA synthesis. In the latter case, however, the tRNA primer would be removed before the RNA was encapsidated.

There is precedence for the utilization of cellular RNA sequences in the synthesis of viral RNA. Influenza virus messenger RNA (mRNA) synthesis is initiated with primers derived from the 5′ ends of cellular mRNAs (Plotch *et al.*, 1981). The recognition of the cellular RNAs does not involve base-pairing with the viral RNA templates, and influenza

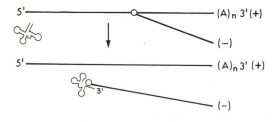

FIGURE 3. Two models to depict how tRNA^Asp might be added to the 5′–terminal region of Sindbis DI RNAs.

virus mRNAs found in infected cells contain 10–15 nucleotides at the 5′ ends, including the cap, that are nonviral sequences. There is also precedence for a cellular tRNA functioning as a primer; it does so in the synthesis of retrovirus cDNA (Varmus and Swanstrom, 1982). Avian retrovirus RNAs contain 18–19 bases 100–180 nucleotides from their 5′ end that are complementary to the 3′ sequences of their respective tRNA primers and include the C-C-A terminus. In this regard, 13 of 14 nucleotides at the 5′ end of the truncated tRNA^{Asp} are homologous to the standard Sindbis virion RNA beginning at nucleotide 892. Even if these sequences (on the negative strand) are able to bind tRNA^{Asp}, it is difficult to imagine how a site at this location can be involved in initiation of plus-strand RNA synthesis unless in the replication complex this site is brought into close proximity to the proper initiation site. Furthermore, we have not detected sequences in the negative strand of virion RNA that could provide a binding site for the 3′ terminus of the tRNA.

Although there is no evidence to support a role for tRNA^{Asp} in Sindbis RNA replication, there is reason to believe that the incorporation of the tRNA into the DI RNAs was not a random event. Three of the four times that DI RNAs were generated in chicken embryo fibroblasts, the 5′ terminus was tRNA^{Asp} (Monroe and Schlesinger, 1984). The predominant species of tRNA^{Asp} would be expected to constitute only about 5% of the total cellular tRNA, making it improbable that the same tRNA would be selected at random.

The Strausses and their collaborators have sequenced the entire genome of Sindbis virus (Rice and Strauss, 1981a,b; Strauss *et al.*, 1984), and a direct comparison of the infectious and a DI genome is shown in Fig. 1 and Table I. The DI population from which the cDNA was cloned is heterogeneous, but the particular clone represents one of the major DI species in that population (Monroe and Schlesinger, 1984). Most of the sequences of the DI genome come from the first 1200 nucleotides of the standard genome. This DI genome, like those of Semliki Forest virus, contains repeated segments as well as large deletions and rearrangements. Although several different DI RNAs of Sindbis virus retain sequences from the structural region of the genome, these sequences do not appear to be essential, since they are absent in some DI genomes (Monroe and Schlesinger, 1984).

The Strausses and their collaborators have obtained considerable sequence information for several related alphaviruses. Their data show that four regions of the alphavirus genomes are conserved: (1) 19 nucleotides at the 3′ terminus, (2) 21 nucleotides spanning the start of a subgenomic 26 S mRNA, (3), 51 nucleotides near the 5′ terminus, and (4) about 40 nucleotides at the 5′ terminus.

Sequence information from the defective genomes of Sindbis and Semliki Forest viruses reveals interesting correlates with respect to these four regions. The 3′ termini of several different DI RNAs from both viruses are identical to those of the standard genomes, demonstrating that

TABLE I. Comparison of Nucleotide Sequence Blocks in a Defective
Interfering and Virion RNA of Sindbis Virus

Segment (see Fig. 1)	Nucleotides in the DI RNA	Nucleotides in 49 S virion RNA
—	1–66 (tRNA^{Asp})	—
A	Deleted	1–30
B	67–384	31–349
C	385–525	350–490
C	526–666	350–490
D	667–722	491–546
E	Deleted	547–646
F	723–821	647–745
G	822–1,302	746–1,226
—	Deleted	1,227–10,238
W	1,303–1,356	10,239–10,291
X	1,357–1,728	10,292–10,665
G	1,729–2,209	746–1,226
W	2,210–2,263	10,238–10,291
Z	2,264–2,312	11,655–11,703

this region is also conserved in defective genomes (Lehtovaara *et al.*, 1982; Monroe *et al.*, 1982). The junction region is missing from all alphavirus DI genomes examined to date. In contrast, the 51-nucleotide region is conserved in these genomes.

The 5′-terminal sequences are the most problematical. They are the least conserved among the four conserved regions of the alphavirus genomes and are not conserved in many of the defective genomes of Sindbis virus. The 5′-terminal sequences of six different DI RNAs generated either in chicken embryo fibroblasts or in baby hamster kidney (BHK) cells have been identified (Tsiang *et al.*, 1985). Each independently generated DI RNA actually represents a population of molecules of DI RNA; however, in all cases but one, we detected only a single 5′ terminus. The most common (3 of 6) and the most unusual is the tRNA^{Asp} terminus, but one of two DI RNAs (DI-1) generated in BHK cells has the 5′ terminus of 26 S RNA with a deletion from nucleotides 24 to 67 of the 26 S RNA (Fig. 4). The other DI RNA population generated in BHK cells has the same 5′ end as the virion RNA. The DI RNA population generated in chicken cells that does not have a tRNA^{Asp} 5′ terminus is the one in which several different 5′ termini were detected. They consisted of rearrangements of regions near the 5′ end. All six DI RNAs are replicated efficiently by the viral enzymes. Although it is possible that these 5′-terminal sequences all form some related stem-loop structure (Ou *et al.*, 1983), it may be that a variety of structures or sequences at the exact 5′ terminus are consonant with replication of the RNA.

Sequences lost from the DI RNAs must be those that are not required for replication or packaging. They are not completely without interest,

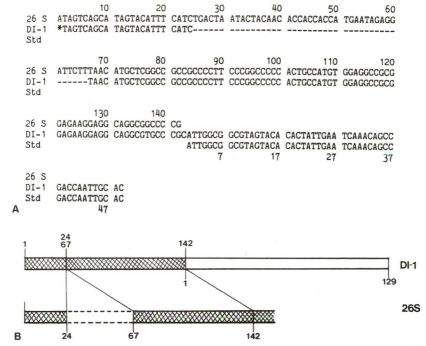

```
                 10         20         30         40         50         60
    26 S  ATAGTCAGCA TAGTACATTT CATCTGACTA ATACTACAAC ACCACCACCA TGAATAGAGG
    DI-1  *TAGTCAGCA TAGTACATTT CATC------ ---------- ---------- ----------
    Std

                 70         80         90        100        110        120
    26 S  ATTCTTTAAC ATGCTCGGCC GCCGCCCCTT CCCGGCCCCC ACTGCCATGT GGAGGCCGCG
    DI-1  ------TAAC ATGCTCGGCC GCCGCCCCTT CCCGGCCCCC ACTGCCATGT GGAGGCCGCG
    Std

                130        140
    26 S  GAGAAGGAGG CAGGCGGCCC CG
    DI-1  GAGAAGGAGG CAGGCGTGCC CGCCATTGGCG GCGTAGTACA CACTATTGAA TCAAACAGCC
    Std                       ATTGGCG GCGTAGTACA CACTATTGAA TCAAACAGCC
                                     7         17         27         37

    26 S
    DI-1  GACCAATTGC AC
    Std   GACCAATTGC AC
  A                47
```

FIGURE 4. (A) Nucleotide sequence of the DI-1 RNA 5′ terminus deduced from the cDNA obtained by primer extension. The 26 S RNA sequence is aligned with that of the 5′ end of DI-1 RNA to show the 42-nucleotide deletion in DI-1 RNA (-----). The 5′-terminal sequence of standard RNA is also aligned with that of DI-1 RNA to show that the 26 S RNA sequence is attached to the standard 5′ end with the intervention of a C residue in the DI-1 RNA. (∗) Nucleotides not determined. (B) A diagram comparing the 5′ terminal sequences of DI-1 and 26 S RNA. Reproduced from Tsiang *et al.* (1985) with permission from the American Society of Microbiology.

however, for they can provide some clues about the generation of the defective molecules. The junction region may be one of the first regions of the molecule to be deleted (Stark and Kennedy, 1978). The sequences directly upstream from the junction region are highly structured (Ou *et al.*, 1982a) and thus may be more readily deleted. The loss of this region could provide a selective advantage for a DI RNA, the negative strand of which would serve as a template only for replication and would not be engaged in transcription of a subgenomic mRNA.

C. Expression of Cloned Defective Interfering Genomes

Cloning and sequencing of DI genomes represent an important first step in using DI molecules to identify sequences essential for replication and packaging. Once this was accomplished for alphavirus DI genomes,

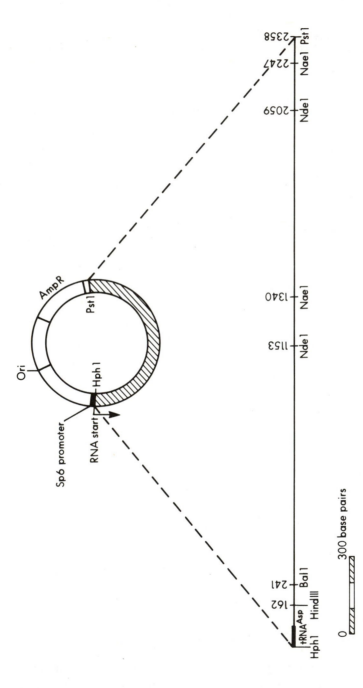

FIGURE 5. Plasmid containing the Sindbis DI cDNA insert. The restriction enzyme sites mentioned in the text are marked. (☐) Bacterial sequences; (▨) promoter for the SP6 polymerase; (▨) Sindbis DI cDNA.

TABLE II. Biological Activity of Defective Interfering RNAs Obtained by
Transcription of Complementary DNA Clones[a]

DI RNA structure	DI RNA detected in passaging	Plaque titer of passage 3 ($\times 10^9$)
A. —	–	15
SSO1[b]	+	0.14
Deletion of first 5 nucleotides at the 5' end	–	22
Inversion of first 21 nucleotides at the 5' end	+	3
Deletion from nucleotides 22 to 162	–	18
Deletion from nucleotides 162 to 241	+	0.16
B. —	–	1.4
Deletion from nucleotides 2248 to 2289 at the 3' end	+	0.19
Deletion from nucleotides 2448 to 2295 at the 3' end	+	0.35
Deletion from nucleotides 2248 to 2300 at the 3' end	–	1.3
Deletion from nucleotides 2248 to 2301 at the 3' end	–	1.5
C. —	–	9
Deletion from nucleotides 1153 to 2059	+	0.4
Deletion from nucleotides 1340 to 2247	+	0.15

[a] The data were obtained from three different experiments. In experiment B, the virus in the medium
was harvested at an earlier time than in the other experiments and therefore the yield was lower.
[b] SS01 refers to the DI RNA sequence, the diagram of which is shown in Fig. 1. The clone is now referred
to as KDI25 (Levis et al., 1986).

the next step of determining which of the sequences retained by DI genomes are required for their biological activity became feasible. Two different approaches to achieve this goal have been taken and are described
below.

1. Transcription of the Complementary DNA and Transfection of the Transcribed RNA

We have inserted a complete cDNA copy of the Sindbis DI genome
into a plasmid to position it directly downstream from the promoter for
the SP6 bacteriophage DNA-RNA polymerase (Fig. 5). The plasmid can
be linearized at a site downstream from the 3' terminus of the DI cDNA
so that the enzyme transcribes the DI genome starting almost exactly at
the 5' end and terminating soon after the polyadenylate stretch. The RNA
transcribed in vitro is transfected into chicken embryo fibroblasts that
are simultaneously infected with helper virus. Only a small fraction of
the cells are transfected with DI RNA, but all the cells are infected with
infectious virus so that every cell receiving a DI RNA will be able to
replicate that molecule. The transfected DI RNA becomes the predominant viral RNA species after 2–3 passages (Fig. 6). In addition, the ac-

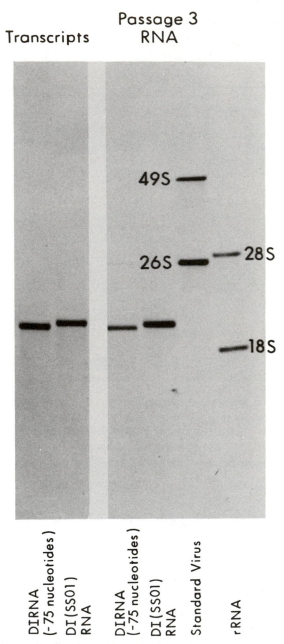

FIGURE 6. Amplification of transcribed DI RNAs after transfection and passaging of Sindbis virus in chicken embryo fibroblasts. Cells were infected with Sindbis virus with or without transfection of DI RNA. The medium harvested 12 hr later (passage 1) was used to infect new cells. The medium was again harvested after 12 hr (passage 2) and was used to infect new cells that were labeled with [³H]uridine in the presence of actinomycin D. RNA, isolated from these cells (passage 3), was denatured and analyzed by electrophoresis in an agarose gel. The RNA transcripts used for the transfections were labeled during synthesis with [³H]-

cumulation of this RNA and DI particles is also evident from the drop in titer of infectious virus by the second or third passage (Table II).

The strategy adopted to obtain the complete DI genome as a cDNA copy correctly positioned next to the SP6 promoter required the insertion of a 21-base-paired fragment corresponding to the first 21 bases of the tRNA^{Asp} segment of the 5' terminus (Levis et al., 1986). These 21 bases form the D stem and loop segment of the tRNA^{Asp} region (see Fig. 2). In addition to obtaining a clone with the correct insertion, a second clone with this fragment inverted was also isolated and is biologically active (Table II). In contrast, a cDNA clone missing the first 5 nucleotides from the 5' terminus and a cDNA clone with a deletion of 140 nucleotides from positions 22 to 162 are transcribed into RNA that is nonfunctional. The latter deletion removes 46 bases of the tRNA as well as additional sequences in the Sindbis genome.

Two regions in the DI genome that are highly conserved in the 49 S genome—the region containing the 19 nucleotides at the 3' terminus and the 51-base region near the 5' terminus—were also subjected to deletion analysis. To introduce deletions at the 3' end, the cDNA was cut at the NaeI site (Fig. 5, nucleotide 2247, 67 nucleotides from the 3' terminus) and treated with Bal31. Deletions that extend downstream from nucleotide 2248 to nucleotide 2295 do not inactivate the biological activity of the DI transcript. Nucleotide 2295 is 20 nucleotides from the 3' terminus; thus, the biological activity of the DI RNA is maintained if the 19 3'-terminal nucleotides (the 3' conserved region) is kept intact. A deletion that extends to nucleotide 2300, however, renders the DI RNA inactive (Table II). The loss of activity supports the proposal of Ou et al. (1981) that the 19-nucleotide sequence represents a recognition sequence for negative-strand RNA replication.

The 51-nucleotide region located at positions 191–241 in the DI RNA was deleted using HindIII and BalI to remove 75 bases flanking and including the 51-base sequence. Surprisingly, deletion of this region has no effect on the biological activity of the DI RNA transcript (Fig. 6). This result demonstrates that this highly conserved region is not essential for replication or packaging of the DI RNA or, presumably, of the virion RNA. There are, however, several ways in which the DI RNA could differ from the virion RNA in its requirement for this conserved region. First, this sequence may be involved in stabilizing a circularized structure in virion RNA. Ou et al. (1983) found that the 51-base 5' conserved region shows significant base-pairing ability with regions near the 3' terminus of the alphavirus genome. There are several such regions near the 3' termini of the virion genomes (Ou et al., 1982b), which are deleted in DI genomes. DI genomes might carry out this function through a mechanism in which

UTP. The DI (SS01) RNA is the RNA transcribed from the plasmid shown in Fig. 5. The DI RNA (−75 nucleotides) is the RNA that has the HindIII/BalI deletion (Fig. 5) and is missing the 51-base conserved region.

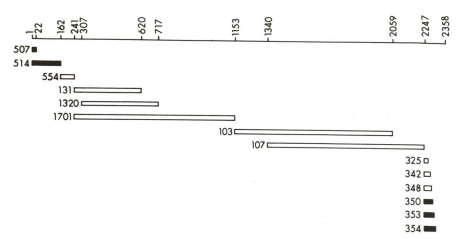

FIGURE 7. A deletion map of the genome of a Sindbis DI RNA. The deletions, which cover the entire DI genome are all indicated on this diagram. The open boxes represent regions which were deleted without causing a loss of biological activity. The closed boxes are regions which when deleted led to a loss of biological activity. The numbers beside each box identify the different cDNA clones. The numbers at the top of the figure refer to the sites at which the DI cDNA was cut by various restriction enzymes to generate the deletions.

multiple DI RNAs interact or through the mediation of the tRNA^{Asp} sequences. Second, this region may be essential for a function unrelated to genome replication or packaging, e.g., control of the level of transcription of 26 S RNA relative to 49 S RNA or control of translation. Such functions would be dispensable for a DI RNA.

Two internal deletions, each removing approximately 900 nucleotides from the DI cDNA, were constructed using the restriction enzymes *Nde*I and *Nae*I (see Fig. 5). These deleted cDNAs are both transcribed into RNAs that are amplified and packaged. Thus, the W, X, and second copy of the G segment (see Fig. 1 and Table I) are not required for biological activity. These findings indicate that the minimum size of DI RNAs observed in nature (about 2×10^3 bases) is not a fundamental "limit size" requirement for packaging (Stark and Kennedy, 1978), but may be related to the way in which deletions are generated.

We have constructed a series of cDNA clones containing deletions covering the entire DI genome (Fig. 7). The biological activity of the DI RNA transcribed from these cDNAs was determined by the transfection and amplification assay described above. Our data show that more than 90% of the sequences in the DI RNA have no specific role in the replication or packaging of the molecule (Levis *et al.*, 1986). Only sequences in the 162 nucleotide region at the 5' terminus and sequences in the 19 nucleotide region at the 3' terminus are specifically required for replication and packaging of the genome (Fig. 7).

2. Transcription of a Defective Interfering Complementary DNA under the Control of Simian Virus 40

The cloned DNA from a DI RNA of Semliki Forest virus has been inserted into the late gene region of Simian virus 40 (SV40) (Jalanko and Söderlund, 1985). The DI RNA is synthesized in monkey kidney cells coinfected with the SV40–Semliki Forest DI hybrid virus and helper SV40 virus. This clone lacks the 5' terminus of the Semliki Forest DI RNA. When cells expressing this DI RNA are superinfected with Semliki Forest virus, the particle yield is reduced about 2-fold. There is no effect on the synthesis of virus-specific RNA, nor is there any evidence that the DI RNA *per se* can be packaged. The data suggest that the DI RNA expressed from the SV40 vector does not contain recognition sites for replication by the Semliki Forest virus enzymes, but that it does retain packaging signals because it interferes with particle yield. Because the DI RNA itself cannot be packaged, all the information required for packaging may not have been retained. This type of study suggests that it would be possible to design an RNA that would more effectively inhibit virion release but would not itself be packaged. Such an RNA could be a model for a new type of antiviral agent.

III. DEFECTIVE INTERFERING PARTICLES OF ALPHAVIRUSES AND PERSISTENT INFECTIONS

The first series of *Comprehensive Virology* included in Vol. 16 excellent chapters on persistent infection (Youngner and Preble, 1980) and DI RNA viruses (Holland *et al.*, 1980). These articles provide the basis for an understanding of this subject, describing the several different virus–cell systems that had been analyzed up to 1980. Three major conclusions emerge from studies on persistent infections in cultured cells:

1. Any of several factors, either alone or in concert, may convert an infection that is normally cytopathic into one that is persistent. Studies with DI particles, temperature-sensitive mutants, and interferon show that each of these can play a role in establishing or maintaining persistence. Which factor predominates may be determined by the host cell.
2. Persistent infections maintained by DI RNAs can often be distinguished from infections maintained by interferon. When DI RNAs are the predominant factor, almost all the cells in the population are infected, whereas in those cells that survive due to the presence of interferon, only a small percentage of the cells show evidence of viral infection.
3. Even the most stable long-term persistent infections are dynamic

systems in which the virus and DI RNAs, when present, are evolving.

Characteristics of persistent infections of mammalian cells by alphaviruses fit very well into the framework described (Holland *et al.*, 1980).

Alphaviruses also set up persistent infections in insect cells. In insect cells, it appears that the cell is a major factor in determining the course of the infection. An analysis of different clones from a population of *Aedes albopictus* cells showed that the response to infection and the subsequent outcome of that infection vary with individual clones (Tooker and Kennedy, 1981). The role of DI particles in these infections is reviewed in Chapter 7.

A. Characteristics of Persistent Infections in Mammalian Systems

Two types of long-term alphavirus infections of mammalian cells have been well characterized, one in L cells and the other in BHK cells. In both examples, DI particles were required in the original inoculum to permit cell survival. With L cells infected with Semliki Forest virus, interferon becomes the predominant factor in cell survival (Meinkoth and Kennedy, 1980). The DI particles are lost, and only a small fraction of the cells show signs of infection. These cultures exhibit periodic crises in which a majority of the cells die, but occasionally the cells are spontaneously cured of infection. In contrast, DI particles appear to be an important factor in maintaining the persistent infection of BHK cells by Sindbis virus (Weiss *et al.*, 1980). Almost all cells in the culture are infected, the cells do not undergo periodic crises, and intracellular DI RNAs can be demonstrated in abundance. This persistent state was studied for over 18 months, during which time both the virus and DI RNAs continually evolved and the majority of cells in the culture remained infected.

B. Evolution of Sindbis Virus and Its Defective Interfering RNAs during Persistent Infection of Baby Hamster Kidney Cells

The adaptation of a virulent virus during a long-term persistent infection requires that it become attenuated, permitting the cells to survive, and that it continue to replicate at a high enough level to survive itself. A long-term persistent infection in cultured cells provides a means to study how a virus evolves to accommodate to this type of environment. As shown by Holland and his collaborators for vesicular stomatitis virus, during the establishment of a carrier state, there is a selection for mutationally altered viruses with reduced cytopathic properties (Horodyski *et al.*, 1983; O'Hara *et al.*, 1984a,b). During the long-term maintenance of such cultures, viruses evolve that are increasingly resistant to those DI particles previously present in the population. Thus, the maintenance of virus in the carrier cultures appears to depend on the selection of mutants that can grow in the presence of preexisting DI RNAs. Replication of these mutants results in the generation of new DI RNAs that can specifically suppress these mutant viruses, and thus a vectorial pattern of coevolution of DI particles and infectious virus is established. This pattern, so well documented for vesicular stomatitis virus, is also observed with Sindbis virus grown in BHK cells.

Within 1–2 months after infection of BHK cells with Sindbis virus, a small plaque variant predominates in the virus population (Weiss *et al.*, 1980). This virus is temperature-sensitive in its ability to synthesize RNA, has reduced cytopathogenicity, and shows decreased susceptibility to the DI particles involved in the initiation of the carrier state (Weiss and Schlesinger, 1981). Resistance to these DI particles is more pronounced in the virus isolated 2 months after infection, and by 16 months, the virus appears to be totally resistant to the original DI particles. During this entire period, the cells remain resistant to superinfection with homologous virus and by several criteria continue to synthesize viral RNA and protein (Weiss *et al.*, 1983). The DI RNAs detected within the cell are extremely heterogeneous in size. Moreover, they have undergone structural alterations such that they no longer can be detected in extracellular particles (Weiss *et al.*, 1983). The loss of packaging signals may be fortuitous, since extracellular release of particles is not a requirement for survival. It is also possible that in a persistent infection, the inability of a DI RNA to be encapsidated can provide a selective advantage. It would do so if those RNA molecules that do not bind to capsid protein would be more available for interaction with replicase and therefore would be replicated to a higher level.

The properties of the DI RNAs and viruses obtained from the persistently infected cultures are especially interesting, since they relate to attenuation of virulence as well as to the mechanisms of DI RNA interference. An analysis of these genomes should help in defining how these properties are expressed.

IV. DEFECTIVE INTERFERING PARTICLES AS A PROBE FOR ANALYSIS OF VIRAL REPLICATION AND IDENTIFICATION OF HOST FACTORS

A major interest in the study of DI particles is to learn more about the replication of the homologous, infectious virus. DI particles of alphaviruses have not yet provided a significant contribution in this realm, but this is not surprising, given the current state of knowledge about the details of transcription and replication of the RNA of these viruses. Some of the data do suggest, however, that analyzing these DI particles may help to uncover differences in host cells and show how DI RNAs interfere with virus growth.

A. Generation and Accumulation of Defective Interfering Particles in Different Host Cells

One of the first reports of the generation of DI particles of Semliki Forest virus described a much greater level of interference in mouse cells than in chicken embryo cells (Levin *et al.*, 1973). This observation was extended by Stark and Kennedy (1978), who compared the generation and amplification of DI RNAs in seven different cultured cell types. They found differences both in the time of first appearance and in the rate of accumulation of DI particles. In one line of HeLa cells, DI particles were not detected even after 200 passages.

The differences among cells could depend on the generation of DI RNAs, on the amplification of the DI RNA, and on the packaging efficiency of the RNA. Stark and Kennedy (1978) showed that if DI particles were added to the one nonproducing HeLa cell line, the DI RNA could be replicated. They concluded that the absence of DI particles was due to their inability to be generated.

There have been several reports showing that DI particles obtained from vertebrate cells are unable to replicate and interfere in *A. albopictus* cells (for reviews, see Stollar, 1979, 1980). Steacie and Eaton (1984) showed that a stock of DI particles generated in chicken embryo fibroblasts replicates, but does not interfere, in mosquito cells. These authors also demonstrated quite clearly that DI particles generated in mosquito cells are heterogeneous with respect to their ability to replicate in vertebrate cells. On the basis of the studies of Tooker and Kennedy (1981), the results with mosquito cells may also be complicated by heterogeneity in the host-cell population.

There are at least two ways we can think about host-cell differences regarding the ability to propagate DI particles of a particular virus. One hypothesis would be that a specific host factor or factors are involved directly in the generation or amplification of DI RNAs. The exact nature of these putative factors would vary from cell type to cell type and would affect the extent to which DI particles accumulated. An alternative explanation is that the differences between host cells are due to more indirect effects. These indirect effects may be related to quantitative differences in the amounts of virus-specific products produced in different cells. For example, if the probability of generating a DI RNA occurs once during a given number of replication events, the generation of DI RNAs would occur more rapidly in cells producing more virion RNA. A second quantitative difference could occur depending on the translational capacity of different cells. The present data suggest that DI RNA competes very effectively for replicase molecules, but may compete poorly for capsid protein (Kääriäinen *et al.*, 1981; Weiss *et al.*, 1983). Thus, the total level of nonstructural or structural proteins produced in a particular cell type could influence the number of passages required to detect DI particles, as well as the extent to which DI particles are able to inhibit the replication of the infectious virus. Some of the differences observed, particularly between vertebrate and mosquito cells, must be due to fundamental differences between these host cells, but the possibility that some of the phenomena observed are of a more indirect nature should be kept in mind.

B. Biological Heterogeneity among Alphavirus Defective Interfering Particles

One conclusion from the observation that only some of the DI RNA species generated in mosquito cells were capable of replicating in other cell types (Steacie and Eaton, 1984) is that DI RNAs of Sindbis virus can be functionally distinct. One of the first clear-cut biological differences among DI RNAs of Sindbis virus was the difference in their ability to induce interferon. Fuller and Marcus (1980) demonstrated that early-passage DI particles of Sindbis virus are capable of inducing interferon, whereas late-passage particles are not. They correlated these differences with the size of the DI RNA and proposed that the DI genomes first generated during high-multiplicity passaging retain the capacity to translate the nonstructural genes. If the viral proteins are functional, they will synthesize some double-stranded RNA that would induce interferon. The late-passage DI genomes lose this translational capacity.

In more recent studies, Barrett and Dimmock (1984c) and Barrett *et al.* (1984a) report that different passages of DI particles of Semliki Forest virus are biologically distinct. They used two different methods to quantitate DI particles, one depending on the reduction in infectious virus yield and the other on the inhibition of viral RNA synthesis, and found differences in the ratio of the two assays. They suggest that the biological differences they are measuring are due to differences in the sequences of the DI RNA species.

Barrett and Dimmock (1984a,b,d) and Barrett *et al.* (1984b) have extended their analysis of the biological heterogeneity of DI particles of Semliki Forest virus to include infection of mice. They examined the ability of DI particles to protect mice against a lethal encephalitis caused by Semliki Forest virus. The DI particle preparations fell into three groups: One of the preparations protected the mice not only from disease, but also from subsequent challenge; a second preparation did not protect the mice from a later challenge; and the third preparation had no protective effect against the virus infection.

As described earlier, DI RNAs of alphavirus generated during high-multiplicity passaging are heterogeneous. Furthermore, differences in the size of the predominant species occur with continued passaging. Now that it is feasible to examine the biological activity of a cloned DI nucleic acid, it is possible to determine whether the various biological differences can each be correlated with a particular RNA structure.

V. CONCLUSION

This chapter began by describing three aspects of DI particles that we believe make them valuable tools for studies of virus replication and virus–host interactions. The studies reviewed here suggest that DI particles of alphavirus are just beginning to play a definitive role in expanding our knowledge of these viruses.

The establishment of persistent infections, in both mammalian and insect cells, has led to the accumulation and characterization of potentially interesting virus mutants. Some of these mutants are altered in their response to DI particle interference (Weiss *et al.*, 1983), but several show a change in host range (Durbin and Stollar, 1984; Simizu *et al.*, 1983). Characterization of mutants like these should help to identify changes that affect virulence and host-cell tropisms.

The site-directed mutagenesis and expression of cloned DI genomes, in particular, will be an important step in defining the sequences required for replication and encapsidation of alphaviruses. The manipulation of DI genomes offers several advantages over studies with intact virion RNAs in this type of analysis. First, DI genomes retain only a subset of standard viral sequences. The DI RNAs of Sindbis and Semliki Forest virus retain 20% or less of the sequences present in the virion RNA. This narrows considerably the number of sequences that must be scanned for functionally important regions. Second, the biological activity of DI RNAs depends only on *cis*-acting sequences that must be recognized by the proteins of the helper virus for replication and encapsidation. Modifications of many regions of the DI RNA are found to be compatible with replication, interfering activity, and encapsidation. Similar alterations in 49 S RNA, however, are more likely to have dramatic effects on infectivity, which represents the sum total of transcriptional and translational events required for virus maturation.

ACKNOWLEDGMENTS.   The work carried out in the authors' laboratory was supported by Public Health Service Grant R01 AI11377 from the National Institute of Allergy and Infectious Disease.

REFERENCES

Barrett, A. D. T., and Dimmock, N. J., 1984a, Modulation of a systemic Semliki Forest virus infection in mice by defective-interfering virus, J. Gen. Virol. **65:**1827.

Barrett, A. D. T., and Dimmock, N. J., 1984b, Properties of host and virus which influence defective-interfering virus mediated protection of mice against Semliki Forest virus lethal encephalitis, *Arch. Virol.* **81**:185.

Barrett, A. D. T., and Dimmock, N. J., 1984c, Variation in homotypic and heterotypic interference by defective-interfering viruses derived from different strains of Semliki Forest virus and from Sindbis virus, *J. Gen. Virol.* **65**:1119.

Barrett, A. D. T., and Dimmock, N. J., 1984d, Modulation of Semliki Forest virus-induced infection of mice by defective-interfering virus, *J. Infect. Dis.* **150**:98.

Barrett, A. D. T., Crouch, C. F., and Dimmock, N. J., 1984a, Defective interfering Semliki Forest virus populations are biologically and physically heterogeneous, *J. Gen. Virol.* **65**:1273.

Barrett, A. D. T., Guest, A. R., Mackenzie, A., and Dimmock, N. J., 1984b, Protection of mice infected with a lethal dose of Semliki Forest virus by defective interfering virus: Modulation of virus multiplication, *J. Gen. Virol.* **65**:1909.

Bruton, C. J., and Kennedy, S. I. T., 1976, Defective-interfering particles of Semliki Forest virus: Structural differences between standard virus and defective-interfering particles, *J. Gen. Virol.* **31**:383.

Dohner, D., Monroe, S., Weiss, B., and Schlesinger, S., 1979, Oligonucleotide mapping studies of standard and defective Sindbis virus RNA, *J. Virol.* **29**:794.

Durbin, R. K., and Stollar, V., 1984, A mutant of Sindbis virus with a host-dependent defect in maturation associated with hyperglycosylation of E2, *Virology* **135**:331.

Fuller, F. J., and Marcus, P. I., 1980, Interferon induction by viruses. IV. Sindbis virus: Early passage defective-interfering particles induce interferon, *J. Gen. Virol.* **48**:63.

Guild, G. M., and Stollar, V., 1975, Defective interfering particles of Sindbis virus. III. Intracellular viral RNA species in chick embryo cell cultures, *Virology* **67**:24.

Guild, G. M., and Stollar, V., 1977, Defective-interfering particles of Sindbis virus. V. Sequence relationships between SV$_{STD}$ 42S RNA and intracellular defective viral RNA, *Virology* **77**:175.

Holland, J. J., Kennedy, S. I. T., Semler, B. L., Jones, C. L., Roux, L., and Grabau, E. A., 1980, Defective interfering RNA viruses and host-cell response, in: *Comprehensive Virology*, Vol. 16 (H. Fraenkel-Conrat and R. R. Wagner eds.), pp. 137–192, Plenum Press, New York.

Horodyski, F. M., Nichol, S. T., Spindler, K. R., and Holland, J. J., 1983, Properties of DI particle resistant mutants of vesicular stomatitis virus isolated from persistent infections and from undiluted passages, *Cell* **33**:801.

Huang, A. S., and Baltimore, D., 1970, Defective viral particles and viral disease processes, *Nature (London)* **226**:325.

Jalanko, A., and Söderlund, H., 1985, The repeated regions of Semliki Forest virus defective-interfering RNA interferes with the encapsidation process of the standard virus, *Virology* **141**:257.

Kääriäinen, L., Pettersson, R. F., Keränen, S., Lehtovaara, P., Söderlund, H., and Ukkonen, P., 1981, Multiple structurally related defective-interfering RNAs formed during undiluted passages of Semliki Forest virus, *Virology* **113**:686.

Kennedy, S. I. T., 1976, Sequence relationships between the genome and the intracellular RNA species of standard and defective-interfering Semliki Forest virus, *J. Mol. Biol.* **108**:491.

Kennedy, S. I. T., Bruton, C. J., Weiss, B., and Schlesinger, S., 1976, Defective-interfering passages of Sindbis virus: Nature of the defective virion RNA, *J. Virol.* **19**:1034.

Lehtovaara, P., Söderlund, H., Keränen, S., Pettersson, R. F., and Kääriäinen, L., 1981, 18S defective-interfering RNA of Semliki Forest virus contains a triplicated linear repeat, *Proc. Natl. Acad. Sci. U.S.A.* **78**:5353.

Lehtovaara, P., Söderlund, H., Keränen, S., Pettersson, R. F., and Kääriäinen, L., 1982, Extreme ends of the genome are conserved and rearranged in the defective-interfering RNAs of Semliki Forest virus, *J. Mol. Biol.* **156**:731.

Levin, J. G., Ramseur, J. M., and Grimley, P. M., 1973, Host effect on arbovirus replication: Appearance of defective-interfering particles in murine cells, *J. Virol.* **12**:1401.

Levis, R., Weiss, B. G., Tsiang, M., Huang, H. and Schlesinger, S., 1986, Deletion mapping of Sindbis virus DI RNAs derived from cDNAs defines the sequences essential for replication and packaging, *Cell* **44**.

Meinkoth, J., and Kennedy, S. I. T., 1980, Semliki Forest virus persistence in mouse L929 cells, *Virology* **100**:141.

Monroe, S. S., and Schlesinger, S., 1983, RNA's from two independently isolated defective-interfering particles of Sindbis virus contain a cellular tRNA sequence at their 5' ends, *Proc. Natl. Acad. Sci. U.S.A.* **80**:3279.

Monroe, S. S., and Schlesinger, S., 1984, Common and distinct regions of defective-interfering RNAs of Sindbis virus, *J. Virol.* **49**:865.

Monroe, S. S., Ou, J.-H., Rice, C. M., Schlesinger, S., Strauss, E. G., and Strauss, J. H., 1982, Sequence analysis of cDNA's derived from the RNA of Sindbis virions and of defective-interfering particles, *J. Virol.* **41**:153.

O'Hara, P. J., Horodyski, F. M., Nichol, S. T., and Holland, J. J., 1984a, Vesicular stomatitis virus mutants resistant to defective-interfering particles accumulate stable 5'-terminal and fewer 3'-terminal mutations in a stepwise manner, *J. Virol.* **49**:793.

O'Hara, P. J., Nichol, S. T., Horodyski, F. M., and Holland, J. J., 1984b, Vesicular stomatitis virus defective-interfering particles can contain extensive genomic sequence rearrangements and base substitutions, *Cell* **36**:915.

Ou, J.-H., Strauss, E. G., and Strauss, J. H., 1981, Comparative studies of the 3'-terminal sequences of several alphavirus RNAs, *Virology* **109**:281.

Ou, J.-H., Rice, C. M., Dalgarno, L., Strauss, E. G., and Strauss, J. H., 1982a, Sequence studies of several alphavirus genomic RNAs in the region containing the start of the subgenomic RNA, *Proc. Natl. Acad. Sci. U.S.A.* **79**:5235.

Ou, J.-H., Trent, D. W., and Strauss, J. H., 1982b, The 3'-non-coding regions of alphavirus RNAs contain repeating sequences, *J. Mol. Biol.* **156**:719.

Ou, J.-H., Strauss, E. G., and Strauss, J. H., 1983, The 5'-terminal sequences of the genomic RNAs of several alphaviruses, *J. Mol. Biol.* **168**:1.

Perrault, J., 1981, Origin and replication of defective interfering particles, *Curr. Top. Microbiol. Immunol.* **93**:151.

Pettersson, R. F., 1981, 5'-Terminal nucleotide sequence of Semliki Forest virus 18S defective-interfering RNA is heterogenous and different from the genomic 42S RNA, *Proc. Natl. Acad. Sci. U.S.A.* **78**:115.

Plotch, S. J., Bouloy, M., Ulmanen, I., and Krug, R. M., 1981, A unique cap (m^{7}GpppXm)-dependent influenza virion endonuclease cleaves capped RNAs to generate the primers that initiate viral RNA transcription, *Cell* **23**:847.

Rice, C. M., and Strauss, J. H., 1981a, Nucleotide sequence of the 26S mRNA of Sindbis virus and deduced sequence of the encoded virus structural proteins, *Proc. Natl. Acad. Sci. U.S.A.* **78**:2062.

Rice, C. M., and Strauss, J. H., 1981b, Synthesis, cleavage and sequence analysis of DNA complementary to the 26S messenger RNA of Sindbis virus, *J. Mol. Biol.* **150**:315.

Sekiya, T., Kuchino, Y., and Nishimura, S., 1981, Mammalian tRNA genes: Nucleotide sequence of rat genes for tRNA^{Asp}, tRNA^{Gly} and tRNA^{Glu}, *Nucleic Acids Res.* **9**:2239.

Shenk, T. E., and Stollar, V., 1973, Defective-interfering particles of Sindbis virus. I. Isolation and some chemical and biological properties, *Virology* **53**:162.

Simizu, B., Hashimoto, K., and Ishida, I., 1983, A variant of western equine encephalitis virus with nonglycosylated E3 protein, *Virology* **125**:99.

Söderlund, H., Keränen, S., Lehtovaara, P., Palva, I., Pettersson, R. F., and Kääriäinen, L., 1981, Structural complexity of defective-interfering RNAs of Semliki Forest virus as revealed by analysis of complementary DNA, *Nucleic Acids Res.* **9**:3403.

Stark, C., and Kennedy, S. I. T., 1978, The generation and propagation of defective-interfering particles of Semliki Forest virus in different cell types, *Virology* **89**:285.

Steacie, A. D., and Eaton, B. T., 1984, Properties of defective-interfering particles of Sindbis virus generated in vertebrate and mosquito cells, *J. Gen. Virol.* **65:**333.

Stollar, V., 1979, Defective-interfering particles of togaviruses, *Curr. Top. Microbiol. Immunol.* **86:**35.

Stollar, V., 1980, Defective-interfering alphaviruses, in: *The Togaviruses; Biology, Structure, Replication* (R. W. Schlesinger, ed.), pp. 427–455, Plenum Press, New York.

Strauss, E. G., Rice, C. M., and Strauss, J. H., 1984, Complete nucleotide sequence of the genomic RNA of Sindbis virus, *Virology* **133:**92.

Tooker, P., and Kennedy, S. I. T., 1981, Semliki Forest virus multiplication in clones of *Aedes albopictus* cells, *J. Virol.* **37:**589.

Tsiang, M., Monroe, S. S., and Schlesinger, S., 1985, Studies of defective-interfering RNAs of Sindbis virus with and without tRNA^{Asp} sequences at their 5' termini, *J. Virol.* **54:**38.

Varmus, H., and Swanstrom, R., 1982, Replication of retroviruses, in: *RNA Tumor Viruses* (R. Weiss, N. Teich, H. Varmus, and J. Coffin, eds.), pp. 369–512, Cold Spring Harbor Laboratory, New York.

Weiss, B., and Schlesinger, S., 1981, Defective-interfering particles of Sindbis virus do not interfere with the homologous virus obtained from persistently infected BHK cells but do interfere with Semliki Forest virus, *J. Virol.* **37:**840.

Weiss, B., Rosenthal, R., and Schlesinger, S., 1980, Establishment and maintenance of persistent infection by Sindbis virus in BHK cells, *J. Virol.* **33,**463.

Weiss, B., Levis, R., and Schlesinger, S., 1983, Evolution of virus and defective-interfering RNAs in BHK cells persistently infected with Sindbis virus, *J. Virol.* **48:**676.

Youngner, J. S., and Preble, O. T., 1980, Viral persistence: Evolution of viral populations, in: *Comprehensive Virology*, Vol. 16 (H. Fraenkel-Conrat and R. R. Wagner eds.), pp. 73–135, Plenum Press, New York.

CHAPTER 7

Replication of Alphaviruses in Mosquito Cells

DENNIS T. BROWN AND LYNN D. CONDREAY

I. INTRODUCTION

The alphaviruses are members of a group of infectious agents once referred to as "arboviruses" (*arthropod-bo*rne viruses). Although alphaviruses are clearly identified as agents of human and animal disease, the
term arbovirus recognizes the active roles of invertebrates in the natural
life cycle of these viruses. Alphaviruses are perpetuated in the wild, in
part, through an interplay between insect and vertebrate hosts and are
transmitted to vertebrates by the bite of infected arthropods, usually mosquitos or ticks. The use of the term "insect vector" to describe the invertebrate counterpart implies a reduced importance, or possibly passive
role, of the insect in the transmission of these agents to vertebrates. It
is very clear today that active replication of alphaviruses in the invertebrate is essential to the perpetuation of the virus in nature. Furthermore, it now appears that the constant participation of the vertebrate
host may not be essential for the maintenance of these viruses in the
wild. Evidence strongly suggests that alphaviruses (as well as other insect-
borne viruses) are transmitted vertically (transovarially) from generation
to generation (for a review, see Leake, 1984).

The development of lines of tissue-cultured cells of insect origin that
can be serially passaged in the laboratory (Singh, 1967) has made possible
comparative studies of the replication of alphaviruses in vertebrate and
invertebrate cells. This system provides a unique opportunity for the
biologist to examine the expression of a single piece of genetic infor-

DENNIS T. BROWN AND LYNN D. CONDREAY • Cell Research Institute and Department of Microbiology, The University of Texas at Austin, Austin, Texas 78713.

mation in two phylogenetically unrelated biochemical and genetic environments.

The alphaviruses are hybrid entities composed of structures that are specified by the host, as well as by the virus. Although the primary amino acid sequence of the virus envelope proteins is virally encoded, the pattern of glycosylation of these proteins is specified by host enzymes; thus, the virus glycoproteins reflect in their carbohydrates the glycosylating capabilities of the host (Stollar et al., 1976; Sarver and Stollar, 1978). The virus membrane, on the other hand, is derived completely from host-specified lipid bilayers, resulting in striking differences in lipid composition (Luukonen et al., 1977) and membrane viscosity (Moore et al., 1976), depending on the cell type in which the virus replicates. Although the composition of the host-specified components of the virion differs from vertebrate to invertebrate host, the biological properties of the virions [e.g., infectivity (Stollar et al., 1976), sensitivity to antiserum (Stollar et al., 1976), low-pH inactivation (Edwards et al., 1983), and ability to fuse cultured cells (Edwards et al., 1983)] are remarkably similar. As will be outlined below, studies of alphavirus replication in cultured insect and vertebrate cells have revealed some dramatic differences, both in the way the virus replicates in the two cell systems and in the response of the particular host to alphavirus infection. These studies have provided important insight into the maintenance of these agents in nature.

II. GROWTH OF ALPHAVIRUSES IN WHOLE INSECTS AND INSECT LARVAE

The mosquito is the primary vector for many alphaviruses. When a female mosquito feeds on an infected vertebrate host, the infectious blood meal is deposited in the posterior part of the mesenteron (i.e., the midgut). This is composed of a single layer of epithelial cells surrounded by a basal lamina. The foregut has a noncellular chitinous lining; thus, it is protected from initial infection. An infection threshold exists for each mosquito; this varies, depending on both virus and vector species. Furthermore, both the number of infected cells within the midgut and the distribution of these cells have been shown to vary (Chamberlain and Sudia, 1961; McLintock, 1978; Hardy et al., 1983). Eastern equine encephalitis (EEE) virus in *Aedes triseriatus* has been shown to infect approximately one in every three mesenteronal cells. The level of infectious virus in the mesenteron appears to first decrease and then rise sharply to reach a peak within a few days, at which point the titer falls until only a low level of virus production is seen (Kramer et al., 1981; Miles et al., 1973; Thomas, 1963; Hardy et al., 1983).

The method of attachment and the nature of specific host receptors (if any) remain unclear. Recently, Houk et al. (1985), using Western equine encephalitis (WEE) virus and three strains of *Culex tarsalis*, ob-

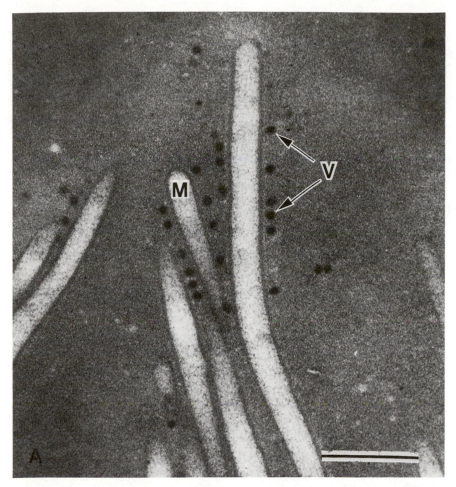

FIGURE 1. Early events in the infection of mosquito mesenteronal epithelial cells by WEE virus *in vivo*. (A) Numerous WEE virions (V) attached to the surface of mesenteronal microvilli (M) at $2\frac{1}{2}$ hr after ingestion of a blood meal. (B) Fusion of the apical plasma membrane of mesenteronal epithelial cell with the envelope of a WEE virion (▼) releasing the nucleocapsid (N) into the cell cytoplasm. (C) Fusion of a WEE viral envelope and the plasma membrane of a microvillus (▶). These experiments were conducted by allowing adult female mosquitos to feed on defibrinated rabbit blood or rabbit serum containing 10^6–10^7 plaque-forming units of WEE per 0.2 ml. The mesenteronal tissues were dissected directly into fixative and prepared for electron microscopy. In these photographs, the tissues are electron-transparent and the gut lumen containing the blood meal is more opaque. Scale bars: 400 nm. Reproduced from Houk *et al.* (1985) with the permission of the authors and Elsevier Publishers, Amsterdam.

served membrane fusion events between the virion membrane and the microvillar surface of the mesenteron in electron micrographs (Fig. 1). In addition, uncoated nucleocapsids were visible at 3 hr postinfection (postingestion of the infected blood meal) in the apical region of the cell. In

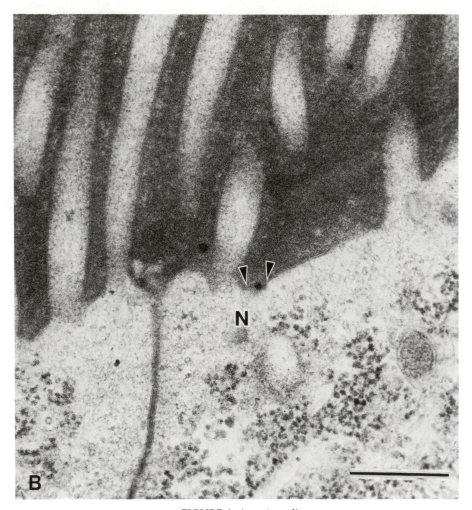

FIGURE 1. (continued)

contrast to some reports of alphavirus penetration in vertebrate cells (Helenius et al., 1980a,b), no viral particles were found in association with
endocytotic or lysosomal vacuoles during the initial period of infection.
Thus, direct fusion of the viral envelope with the host plasma membrane
appears to be the method of penetration and uncoating in the insect cell.

The route of viral maturation appears to vary with both virus and
mosquito species. WEE virus in *Aedes dorsalis* has been observed associated with endoplasmic reticulum and then found in extracellular spaces
between the basal lamina and the plasma membrane of the cell on the
hemocoel side. WEE in *C. tarsalis* is released only into the extracellular
spaces above the basal lamina (Fig. 2). The response of the cells to infection has also been shown to vary. On infection of the Knight's Landing

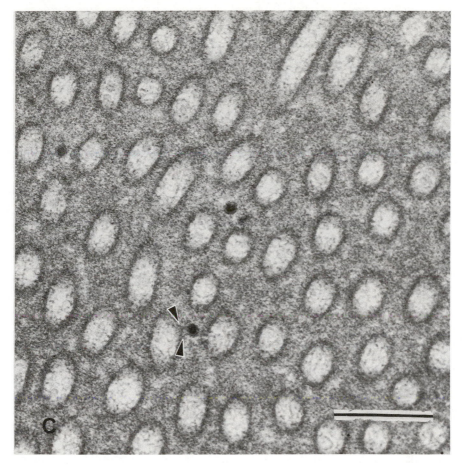

FIGURE 1. *(continued)*

strain of *C. tarsalis* with WEE, no subcellular morphological changes are seen. However, infection of either *C. tarsalis* strain WS-3 or *A. dorsalis* with WEE results in drastic alterations. In both cases, apical accumulations of nucleocapsids embedded in an amorphous matrix, or associated with membranes, were seen by 29 hr postinfection. *Aedes dorsalis* also developed many cytoplasmic vacuoles (Houk *et al.*, 1985). The time required for virus attachment, penetration, replication, and release from the mosquito mesenteron varies with the virus type, virus dosage, mosquito species, and temperature of incubation (Hardy *et al.*, 1983).Thus, the necessity to examine alphavirus infection of mosquitos on an individual basis arises.

A "leaky gut" hypothesis involving an intercellular route has also been suggested for bypassing the mesenteronal epithelial cells entirely, as a way to the hemocoel (Miles *et al.*, 1973; Houk, 1977; Houk and Hardy, 1979). The initial hypothesis arose due to the short lag period

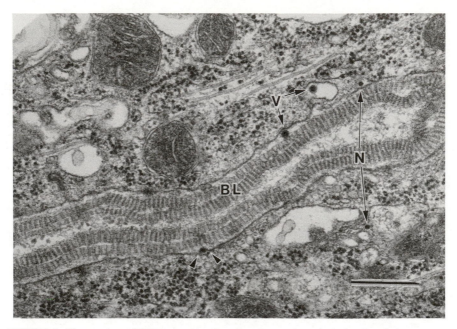

FIGURE 2. Maturation of WEE virus in the mesenteronal epithelial cells of the mosquito
A. dorsalis. Maturation of WEE virus occurs when nucleocapsids (N), lying free within the
cytoplasm, bind to basolateral membranes and bud (▲) into the extracellular spaces. Mature
virions (V) are formed immediately adjacent to the basal lamina (BL) of the mesenteronal
epithelial cells of the mosquito *A. dorsalis* 35 hr after ingestion of an infectious blood meal.
Scale bar: 400 nm. Kindly provided by Edward J. Houk, Naval Biosciences Laboratory, Oak-
land, California.

between ingestion of some arboviruses and the appearance of virions in
the hemocoel (0.5–4 hr with Whataroa in *Aedes australis*) (Miles *et al.*,
1973). Morphological evidence supporting this idea has also been presented. In response to a blood meal, the cellular morphology is altered
from columnar to squamous epithelial (Howard, 1962; Houk, 1977). This
may involve the breakdown and re-formation of zonula continua (occlu-
sive intercellular junctions) and thus allow virus to pass around the epithelial gut lining. This is supported by the findings of Howard (1962)
that after mosquitos had been fed blood, 5-μm polystyrene beads (greater
than the normal size allowed to pass through the junctions) could be
found within a few microns of the basal lamina. Thus, two possible routes
apparently exist: intracellular through the mesenteron cells and extracellular through the dissolved junctions that occur due to cellular morphological changes on feeding.

It has been accepted for some time that at least one barrier exists for
alphavirus replication in insects. This is the mesenteronal infection barrier, and it is both dose- and host-dependent. Higher virus concentrations
are more likely to result in infection, but the dosage of each virus varies
with each host (Hardy *et al.*, 1983). Many factors could influence this

barrier, including both formation time and mean porosity of the peritrophic membrane, digestive enzymes secreted into the mesenteron, and the surface charge and the distribution of these charges on the mesenteronal epithelial (ME) cell surface, as well as various hypothesized receptor sites. A dissemination barrier blocking the spread of virus from an infected mesenteron has also been demonstrated. Recent evidence has shown an accumulation of naked WEE nucleocapsids along a basal edge of ME cells (Houk *et al.*, 1985). This could indicate that the block lies in the maturation of virus. In any event, for the further infection of the mosquito to occur, it is necessary for the virus to pass through the basal lamina of the mesenteron; the mechanism for this passage is unclear.

Generally, amplification of the virus occurs in the mesenteron before the infection of the salivary glands (Hardy *et al.*, 1983). In some cases, a 10-fold amplification of the virus in other cells and tissues (e.g., fat bodies and neural tissues) is seen. In any case, virus released into the hemolymph is most often responsible for salivary gland infection. A salivary gland infection barrier, similar to the midgut barrier in that it is dose-dependent, has been indicated with WEE in *C. tarsalis* (Kramer *et al.*, 1981). This barrier, however, is also time-dependent, decreasing in its effectiveness over longer periods in preventing infection of the salivary gland. Virus apparently traverses the basement membrane of the salivary gland to attach to the plasma membrane of the acinar cells. As a rule, only a few cells in the lateral lobes become infected. The virus appears to pass from the distal to the apical portion of these cells (Whitfield *et al.*, 1973). Electron micrographs show virions in the endoplasmic reticulum lumen and later in the diverticulum. In fact, crystallized aggregates of virus may be seen due to the large concentrations of virus that can occur in the salivary glands. The time from the initial infection of the mosquito to the time when a sufficient concentration of virus to infect a vertebrate host is produced is termed the "extrinsic incubation period." This time has been shown to vary with temperature; the period is longer in lower temperatures (Hardy *et al.*, 1973).

The preceding discussion has dwelt largely on the accepted vector role of the mosquito in the transmission of alphaviruses from vertebrate to invertebrate. It has now been indicated that the vertebrate is not a necessary part of the alphavirus cycle. The hypothesis of transovarial transmission solves many of the problems of the alphavirus life cycle. For example, how are viruses propagated or maintained during interepidemic periods? In the egg stage (the stage most suited to surviving adverse conditions), the virus is protected and could survive winter or dry seasons. During these seasons, vertebrate hosts may not be numerous enough to ensure a meeting with an infected vector. Thus, by parent-to-progeny transmission, the risk of finding a susceptible vertebrate host is eliminated, and the virus could be maintained in the invertebrate host, but not detected in vertebrates in the area. A general acceptance of the concept of transovarial transmission of some arboviruses (bunyaviruses and flaviviruses) exists (Leake, 1984). In alphaviruses, evidence for transovarial transmission has also been presented, including the isolation of Semliki Forest (Stones, 1960) and EEE viruses (Chamberlain and Sudia, 1961) from male mosquitos, EEE virus from naturally infected insect larvae (Chamberlain and Sudia, 1961), and WEE virus from *C. tarsalis* eggs (Thomas, 1963). It is clear that ovarian tissue is susceptible to viral infection, since Thomas (1963) detected WEE virus in the ovaries of *C. tarsalis* at 4–10 days postinfection of the mosquito. It is not clear, however, whether intact virions or infectious viral RNAs are responsible for the actual infection of the egg (Leake, 1984). The anatomy of the ovary dictates that an infecting virus particle must first traverse the numerous ovarian membranes and penetrate the follicular epithelium. Furthermore, the follicle is readily permeable only to particles less than 11 nm in diameter (Anderson and Spielman, 1971). Also, by 36–48 hr after a blood meal, desmosomes form to block interfollicular channels, and the vitelline membrane separates the oocyte from the follicular epithelium. The reported sizes of alphaviruses are all considerably larger than the diameter of the interfollicular channels in *Aedes aegypti*. It seems possible, therefore, that RNA is the agent for spreading virus infection in the ovaries.

III. DEVELOPMENT OF CELL CULTURES FOR USE IN STUDIES OF ALPHAVIRUS–INSECT CELL INTERACTIONS

Prior to the mid 1960s, *in vitro* studies of mosquito-borne arboviruses were restricted to vertebrate cells and the surviving tissue of the insect. While many attempts had been made at mosquito-cell culture, little success had been achieved. In most cases, the tissue would survive for a few weeks, but no growth was seen. This changed in 1967 when Singh (1967) reported a line of *Aedes albopictus* cells derived from larvae that were grown in the medium of Mitsuhashi and Maramorosch (1964) supplemented with heat-inactivated fetal calf serum. The Mitsuhashi and Maramorosch medium is undefined, with amino acids supplied by lactalbumin hydrolysate and yeast hydrolysate. It is interesting that the Singh cells that grow in this medium can be readily adapted to Eagle's medium, which has a much lower osmotic pressure (Stollar, 1980). A report of the establishment of a line of *A. aegypti* cells by Peleg (1968) rapidly followed Singh's. In this case, eggs containing developing embryos were surface-sterilized with copper sulfate and mercuric chloride in ethanol, homogenized, pressed through a grid, cleansed of remaining shell fragments, and seeded into dishes containing Kitamura's medium. Surviving cells were subsequently passaged to produce the lines employed in the studies described below. Because these cells were derived from macerated tissues of whole larvae, it is not possible to establish what particular tissue types are present in these mosquito-cell cultures. The possibility that cells of different tissue origins differ in their ability to produce virus and in their response to virus is underscored in studies on various sublines derived from these original isolates (described below).

The cell lines derived by Peleg and by Singh are both capable of supporting alphavirus growth. As with the intact mosquito, it seems that initially high virus titers may be reached that will then decline to a low level of production. The time required for this varies with both cell line and virus type. Another parallel with intact mosquitos occurs in that a cytopathic effect is generally not found in cultured mosquito cells. Currently, although the Singh and Peleg cell lines are the most common, many mosquito cell lines are available in which alphaviruses can grow. At a molecular level, this has made the study of the virus–host interactions in the mosquito much easier. Thus, many valid comparisons between the infected invertebrate and vertebrate systems may now be drawn with regard to the infectious cycle from a molecular aspect.

IV. EARLY EVENTS IN THE INFECTION OF CULTURED INSECT CELLS WITH ALPHAVIRUSES

Little is known about the molecular mechanism by which alphaviruses attach to and penetrate cultured insect cells. Current explanations for how these events may occur in vertebrate cells are presented, together with supporting data, in Chapter 4.

Available observations suggest that a specific receptor for alphaviruses on the surface of susceptible cells of any type may not exist. The host range of alphaviruses is so broad, indeed spanning phyla, that a strict requirement for a receptor of defined composition and conformation is virtually prohibited. In a laboratory assay, alphaviruses can be demonstrated to fuse with protein-free liposomes, the only requirement being the presence of cholesterol in the artificial membrane (White and Helenius, 1980). A fusion event occurring between the membrane of the alphavirion and a membrane of a susceptible host cell is generally agreed to be an essential event in the penetration of host cells by these viruses. Thus in the aforementioned liposome system, the requisites for attachment that precede the penetration event may have been met. The cellular location where the penetration–fusion event takes place is, however, the subject of some controversy.

Studies of alphavirus penetration of cultured mosquito cells have been conducted, following the experimental protocols devised for similar investigations using vertebrate host cells. Studies of the early events in the interaction of alphaviruses with insect cells in culture are fraught with problems common to these studies, regardless of the source of the host cell. Large amounts of virus must be added to cultured cell populations to bring the number of virus–cell interactions to a level sufficient to allow evaluation by relatively insensitive electron-microscopic or biochemical assays. When large multiplicities of virus are added to cultured cells, many virions may follow a pathway of interaction with the host cell that does not lead to infection, a situation that may be exaggerated by the fact that less than 10% of the virus particles in a preparation can be demonstrated to be infectious. In the electron microscope, large numbers of viruses are found attached to the surface of tissue-cultured insect cells, and, with time, they accumulate in intracellular vesicles. Fusion events occurring between virions and host-cell membranes at the cell surface, or in intracellular compartments, are not detected. Indeed, the only electron-microscopic images that show this fusion event are those found in whole-insect studies (described above). Houk *et al.* (1985) have provided morphological evidence for the fusion of the membranes of invading virions with the surfaces of cells lining the midgut of mosquitos after the ingestion of a blood meal (see Fig. 1). Although the midgut is normally alkaline, it is believed that the pH of the midgut is the pH of the blood the insect has consumed. Since this pH is typically neutral, the fusion event seen by Houk and co-workers has probably occurred without the necessity for exposure to acid pH.

Edwards and Brown (1984) have demonstrated that high multiplicities of Sindbis virus can be used to induce the fusion of cultured *A. albopictus* cells. As is the case in vertebrate-cell fusion studies, the fusion of mosquito cells was found to require a brief exposure to acidic pH, followed by an obligatory return to neutral pH conditions. The conditions for initiating the fusion event in insect cells were found to differ from those in vertebrate cells. Sindbis virus grown in baby hamster kidney (BHK)-21 cells was found to optimally fuse BHK-21 cells after a short exposure to pH 5.3, followed by a return to pH 7.2. The same vertebrate-grown virus was found to mediate fusion of cultured *Aedes* cells only if the pH was first lowered to 4.6. Fusion at any pH with *Aedes* cells was observed only if the cells were first adapted for growth in Eagle's minimal essential medium (the same medium in which the BHK cells are cultured). Cells perpetuated in insect culture media could not be fused with high multiplicities of virus after exposure to any pH from 4.0 to 7.2. Although the insect cells differed in their pH requirement for virus-mediated fusion, it was found that BHK-grown Sindbis virus infects insect cells and BHK cells with equal efficiency. We have also found that Sindbis virus infection of insect cells shows a reduced sensitivity to the antiviral activities of lysosomotropic weak bases (Edwards and Brown, unpublished observation). These observations seem to separate the phenomenon of low-pH virus-mediated cell fusion from the events that lead to the infection of the cell. Sindbis virus from a common origin demonstrates different pH requirements for cell fusion, depending on the cell type to which it is attached, and yet infects both cell types with equal efficiency. These data suggested that in the case of *A. albopictus* cells, the laboratory phenomenon of low-pH-mediated fusion demonstrates the ability of virus proteins in certain conformations to induce fusion (see above); however, some other as yet unknown interaction of the virus with the cell surface may induce conformational changes essential for the fusion event in the absence of fluctuation in pH.

V. DEVELOPMENT OF ALPHAVIRUSES DURING ACUTE INFECTION OF INSECT CELLS

A. Kinetics of Virus Production

The rate at which invertebrate cells produce alphaviruses, as well as the total yield of virus produced, are dependent on the species of insect cell employed. Within a given species of cultured insect cells, clonal isolates that vary in their ability to produce progeny virions can be obtained. Cultured lines of A. albopictus cells that produce yields of virus equivalent to the best virus-producing vertebrate cells at approximately the same kinetic rate at temperatures in the physiological range of the insect cells have been obtained (Stollar et al., 1975; Gliedman et al., 1975). Cell cultures derived from A. aegypti, on the other hand, produce alphaviruses poorly, even though these insects are vectors for many alphaviruses in nature (Peleg, 1968a,b). Surprisingly, cells derived from the silk moth (Bombyx mori) produce Sindbis virus as efficiently as A. albopictus cells, although this nonhematophagous insect is not known to be a natural host for any of the alphaviruses (Scheefers-Borchel et al., 1981).

Within a population of A. albopictus cells known to produce high yields of virus, clones of individual cells that differ in their ability to produce virions have been isolated. Tooker and Kennedy (1981) reported that of 115 clones of A. albopictus cells, 70% produced poor yields of virus, while 30% produced yields equivalent to that obtained in vertebrate cells. Cells that gave poor yields of virus also produced virus at a slower kinetic rate. Tooker and Kennedy demonstrated that the inability of certain clones to produce high virus yields was not due to an inability of the cell to attach virus or to convert incoming viral RNA into double-stranded molecules. In a separate study, Edwards and Brown (1984) demonstrated that Sindbis virus produced from BHK-21 cells generated infectious centers of A. albopictus cells with an efficiency equal to, or slightly better than, that obtained with BHK cells. Thus, it seems that the amount of virus produced per invertebrate cell, and the kinetic rate of that production, are determined by factors that express themselves after the establishment of infection. In the study of Tooker and Kennedy (1981), it was observed that the ability of a cell line to produce high yields of virus was related to the ability of the infecting virus to establish a cytopathic effect in that cell line. Thus, it was concluded that cells that produced large amounts of virus did so in part because of an ability of the virus to compete successfully with the host cell for the necessary biosynthetic machinery to produce virus components. Tooker and Kennedy further observed that cells that produced low yields of virus did not demonstrate cytopathic effect and concluded that these cells coexisted with the production of virus because of some ability to maintain host-cell functions in the presence of virus replication. The implication of these observations for understanding the establishment of persistent infections of invertebrate cells by alphaviruses is described in more detail below.

B. Synthesis of Viral RNA and Protein

The events that result in the production of alphavirus RNA and protein in cultured mosquito cells are in large part poorly understood. Because the biological properties of the virus particle produced from invertebrate cells are very similar to those of viruses produced from vertebrate cells (Stollar et al., 1976), it is not surprising that the types of RNA and protein generated by the virus in these two cell systems are similar (Igarashi and Stollar, 1976; Tooker and Kennedy, 1981). An evaluation of the forms of RNA produced after infection of cultured mosquito cells with alphaviruses revealed the production of double-stranded RNA, 42 S progeny RNA, and 26 S subgenomic messenger RNA, similar in structure and in ratios to each other, as are found in infected vertebrate-cell cultures (Igarashi and Stollar, 1976; Tooker and Kennedy, 1981). The rate of total virus-induced RNA synthesis in invertebrate cells parallels the rate observed in the vertebrate cells under identical conditions (Condreay and Brown, unpublished observations).

Biochemical analysis of the proteins synthesized by alphaviruses in cultured invertebrate cells has indicated that the processing of virus structural proteins is similar to that in cultured vertebrate cells. Peptides PE2, E1, E2, and C are readily detected in virus-infected insect cells at 5 hr post-infection (Scheefers-Borchel et al., 1981). Gillies and Stollar (1981) have demonstrated that extracts of A. albopictus cells are competent to produce some of these proteins in vitro. These authors found an efficient production of polypeptide C (capsid), as well as traces of the 100K precursor to polypeptides PE2 and E1 in these cell-free extracts. These results are similar to those obtained when alphavirus 26 S RNA is used to program an in vitro synthesizing system derived from rabbit reticulocytes (Cancedda and Schlesinger, 1974; Simmons and Strauss, 1974).

Assays directed at determining the percentage of the cell population responding to Sindbis virus infection by production of virus polypeptides have indicated that nearly 100% of all invertebrate cells are virus-positive by immunofluorescent staining with antiviral antiserum at 7–10 hr postinfection (Riedel and Brown, 1977). Individual cell fluorescence is maximal at 18–24 hr postinfection (Riedel and Brown, 1977).

Whereas the route of intracellular processing of virus in cultured vertebrate cells is quite well understood, nothing is known about the

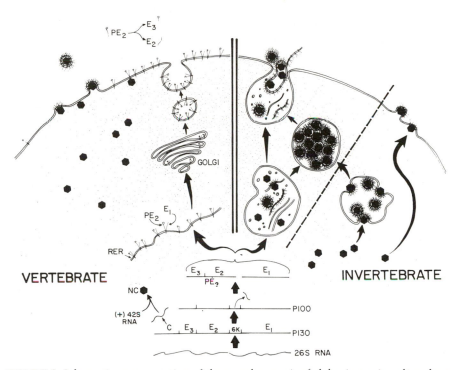

FIGURE 3. Schematic representation of the morphogenesis of alphaviruses in cultured vertebrate and invertebrate cells. The posttranslational proteolytic processing of virus structural proteins after their synthesis from the subgenomic 26 S messenger RNA, illustrated at the bottom of the figure, is believed to be similar in both hosts. In the vertebrate host, nucleocapsids are assembled in the cell cytoplasm from capsid protein and 42 S progeny RNA. The nucleocapsids are mature by envelopment at the plasma membrane by association with regions of the membrane-containing virus glycoproteins. The virus envelope glycoproteins are processed sequentially by host enzymes and transported from their site of synthesis in the rough endoplasmic reticulum through the Golgi membranes to the plasma membrane (a complete description of this process is presented in Chapter 5). The development of virus components in the invertebrate cell is understood only at the level of electron-microscopic morphology. Three profiles of virus maturation have been reported and are described in detail in the text. Some cultured mosquito cells show virus assembly to occur almost exclusively in membrane-limited cytoplasmic "virus factories" (see Fig. 4). Virus is released into the medium by the exocytotic expulsion of this vesicle by fusion of the vesicle membrane with the plasma membrane. In other instances, virus maturation in mosquito cells follows a pathway similar to that observed in vertebrate cells with nucleocapsids assembled in the cell cytoplasm matured at the plasma membrane. The formation of cytoplasmic vesicles containing large numbers of mature virions has also been reported to occur by the maturation of cytoplasmic nucleocapsids by envelopment in the membrane of the vesicle. The release of these virions may also occur by fusion of the vesicle with the cell surface.

intracellular site of the various processing events in the mosquito cell. Because the gene products produced in the invertebrate cell are similar to those seen in the vertebrate cell, it is likely that many of the processing steps take place in environments similar to those of the vertebrate host (reviewed in Chapter 5). In the vertebrate host, structural protein synthesis is initiated on cytoplasmic polyribosomes and, after proteolytic removal of the amino-terminal capsid protein, proceeds on membranes of the rough endoplasmic reticulum. Integration into the rough endoplasmic reticular membranes is accompanied by the cleavage of the terminally situated PE2 polypeptide from E1 with the release of an intervening 6000-dalton peptide (Fig. 3). The proteolytic activity for this cleavage is not identified. Glycosylation of the E1 and PE2 proteins is initiated in the rough endoplasmic reticulum and completed in the Golgi membranes, and the PE2 and E1 peptides are transported to the plasma membrane, where PE2 is cleaved to E2 and E3. This cleavage is mediated by an unidentified protease and is essential for virus maturation.

The observation that *in vitro* translating systems derived from *A. albopictus* cells, like those derived from rabbit reticulocytes, produce capsid protein but not polypeptides PE2 and E1 (Gillies and Stollar, 1981) suggests that in the invertebrate, protein processing beyond the production of capsid protein takes place in cellular membrane compartments. The type of cellular membrane where these events occur is not clear. Virus proteins are glycosylated in insect cells by a pathway that yields glycoproteins of reduced carbohydrate content (Sarver and Stollar, 1978; Luukonen *et al.*, 1977; Hsieh and Robbins, 1984). Virus glycoproteins produced in insect cells are sialic-acid-free, since these cells contain no sialyl transferase activity (Sarver and Stollar, 1978). Determination of the intracellular site of various processing events must await the development of techniques for the fractionation of subcellular components from mosquito cells. Some studies have indicated, however, that certain mosquito-cell populations actively producing Sindbis virus do not express virus antigens at their surface in amounts sufficient to be detected by heamadsorption (Gliedman *et al.*, 1975), specific antibody binding (Gliedman and Brown, unpublished observation), lactoperoxidase radioiodination (Renz and Brown, unpublished observation), or acid-induced, virus-protein-mediated fusion from within (Mann and Brown, unpublished observation). The cellular sites of virus maturation dictate the location at which final events in virus protein processing may take place, and various cell types demonstrate differing locations for the envelopment event (see below).

C. Assembly and Maturation of Alphavirus

The process of assembly of alphaviruses involves extremely complex molecular interactions among the virus structural components. These interactions proceed sequentially and unidirectionally, and the product is mature virus particles that have biological properties that are not affected by the host cell in which the virus is produced (Stollar *et al.*, 1976).

After its release from the 130,000-dalton polyprotein (see Fig. 3), the capsid protein proceeds through a series of macromolecular interactions, each step preparing the developing structure for subsequent stages in assembly. The capsid protein recognizes and binds to 42 S plus-polarity progeny RNA and, subsequent to this binding, folds the RNA into a ribonucleoprotein complex (the nucleocapsid) in which 240 copies of the capsid protein combine with the RNA to produce an icosahedral structure in which all capsid protein is exposed on the surface. This nucleocapsid is capable of binding to the carboxy terminus of the plasma-membrane-associated PE2 polypeptide, which has been processed through a separate pathway, according to the scheme in Fig. 3. This recognition reaction reunites the capsid with the envelope proteins and initiates the terminal stage in virus maturation, the envelopment of the virus particle in the membrane bilayer. In cultured vertebrate cells, this process occurs almost exclusively at the plasma membrane (Brown, 1980).

Most information on the maturation of the alphaviruses in cultured insect cells has been derived from electron-microscopic studies of cells at various times postinfection. These studies have produced observations on the site of virus maturation that suggest that the insect cells themselves differ in the way in which they support virus growth, a conclusion that is supported from data to be described below on the effects of inhibitors on virus replication and cellular responses to alphavirus infection.

Morphological studies have produced three possible pathways for virus maturation in cultured insect cells (see Fig. 3). Stollar *et al.* (1979) found the morphological profile of Sindbis virus growth in mosquito cells to be similar to that observed in cultured vertebrate cells. Nucleocapsids were seen in the cytoplasm of infected cells, and envelopment (budding) was seen at the cell surface. Gliedman *et al.* (1975) found that their cultures of *A. albopictus* cells produced Sindbis virus by a very unique pathway. Nucleocapsids were found in the cytoplasm of infected cells, and virus envelopment at the cell surface was seen in only a very small percentage of cells. In the majority of cells, Sindbis virus was found to mature in restricted areas of the infected cell. These areas were visualized as membrane-limited structures that contained virus components in various stages of development (Fig. 4), together with membranes and what appeared to be ribosomes. These "virus factories" contained particles in various stages of envelopment, and, with time, these vesicles became filled with mature virions. In some instances, the packing of particles was so close that paracrystalline arrays were seen (Fig. 4). Release of the virus particles occurred by an exocytosis-like process when the membranes of the vesicle fused with the cell surface to release virions into the surrounding medium (Fig. 4). The release of these virus factories into

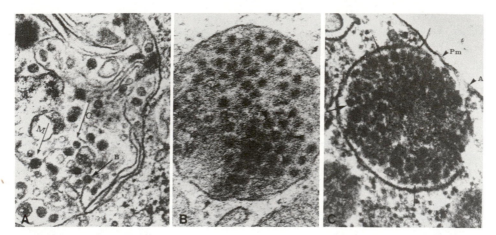

FIGURE 4. Maturation of Sindbis virus in cultured *A. albopictus* cells. (A) A membrane-limited cytoplasmic "virus factory" containing nucleocapsids (C), mature virions (M), and nucleocapsids partly enveloped in membranes within the vesicle (B). (B) A cytoplasmic vesicle containing many mature Sindbis virions. (C) Release of the virions into the surrounding medium by fusion of the vesicle membrane with the plasma membrane (Pm). This fusion is best seen at point A. Mature virions (→) are seen in the electron-dense vesicle. Reproduced from Gliedman *et al.* (1975) with the permission of the authors and the American Society for Microbiology.

the surrounding medium occurred at any time in virus development. Thus, the exocytotic process released membranes with partially enveloped capsids, naked capsid structures, membranes containing numerous capsids, and other components (Fig. 5).

Raghow *et al.* (1973b) observed both the aforedescribed processes in individual mosquito cells infected with Ross River virus. These authors

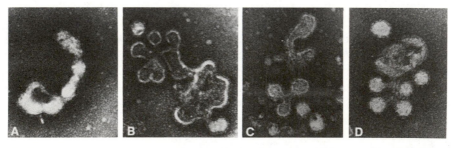

FIGURE 5. Electron micrographs of virus-modified membranes and partially matured virus capsid membrane complexes released into the medium surrounding Sindbis-virus-infected *A. albopictus* cells. (A) A membrane fragment with surface spikes containing serial nucleocapsids. (B) A large fragment of modified membrane with externally located spikes and one or two enveloped nucleocapsids. Some areas of this membrane appear to have partly enveloped nucleocapsids that were lost. (C, D) Extracellular complexes containing modified membranes and maturing virions. In some instances, the viral envelopes are continuous with the modified membranes from which they are developed. Reproduced from Gliedman *et al.* (1975) with the permission of the authors and the American Society of Microbiology.

found capsids in the cell cytoplasm budding into veiscles of the type described by Gliedman and co-workers and virus budding at the cell surface. These authors did not assign any particular importance to the intracellular site of virus maturation. Simizu and Maeda (1981) examined the morphogenesis of WEE virus in the C6/36 clone of *A. albopictus* cells. These authors found cells containing virus inclusions similar to those described by Gliedman *et al.* (1975) and by Raghow *et al.* (1973a,b), as well as virus budding from the cell surface.

It is difficult to establish from these studies a predominant path for morphogenesis of alphaviruses in insect cells. As indicated above, the various profiles of virus development may be due in part to differences in the subclones of mosquito cells used by these authors. This contention is supported by other observations made on the interaction of these viruses with cultured insect cells. *Aedes albopictus* cells employed in the studies of Brown and co-workers could be cured of virus infection by growth in antiviral serum (Riedel and Brown, 1977). Riedel and Brown concluded that curing occurred when the virus factory was released from the cell and that reinfection of the cells by extracellular virus was essential for maintaining the infected state of the cell population. Renz and Brown (1976) found that temperature-sensitive (*ts*) mutants that complement one another readily in tissue-cultured vertebrate cells complemented poorly, or not at all, in invertebrate cells. Recently, Condreay and Brown (unpublished observation) have found that *A. albopictus* cells infected with Sindbis virus do not establish a homologous interference phenomenon through expression of nonstructural virus proteins (as is the case in vertebrate cells). All these observations can be explained if the infecting viruses establish independent sites responsible for the replication of only one virus genome. Riedel and Brown (1977) and Condreay and Brown (unpublished observation) have also observed no cytopathic effects of virus infection on their clone of *A. albopictus* cells, an observation that was explained with the suggestion that compartmentalization of infection protects the host cell from the virus components responsible for the cytotoxic effects of infection.

Both the aforedescribed morphological pathways for virus production may be important in nature. Whole insects show no deleterious effects of virus infection and pass the virus infection transovarially to offspring while showing no obvious indications of the presence of virus in the abdomen and thorax (Leake, 1984). The salivary glands of these insects, however, produce large amounts of virus throughout the life cycle of the insect. In subcloning lines of *A. albopictus* cells that were derived from larvae, individual laboratories may have selected for cell types that are destined to develop into different components of the mature mosquito and that respond to and support virus replication in unique ways. The maturation pattern described by Gliedman *et al.* (1975) may be important in the establishment of cryptic persistent infections, whereas the pathway described by Stollar *et al.* (1979) and Raghow *et al.* (1973a,b) may be important for production of large amounts of extracellular viruses in other tissue, such as the salivary glands of the mature mosquito.

D. Cell Response to Alphavirus Infection

Infection with alphaviruses has profoundly different effects on vertebrate and invertebrate hosts. Alphaviruses can cause serious clinical diseases with high fever, encephalitis, hemorrhagia, and death in man and domestic animals. The mosquito vector, however, demonstrates no deleterious effects of alphavirus infection, and the infection persists throughout the life of the arthropod. These facts have suggested that a comparative study of the replication of alphaviruses in cultured cells of vertebrate and invertebrate origin may provide important information on which aspects of virus infection are responsible for the cytocidal effects and how one host coexists with a persistent infection of the virus while others do so only poorly and only under rather elaborate conditions of infection.

Although it is generally agreed that infection of cultured mosquito cells with alphaviruses leads to the establishment of a culture of persistently infected cells that can be cultured indefinitely, different laboratories have reported varying degrees of cytotoxic effects during the acute phase of the infection. These observed differences in cell response to virus infection, like the differing observations on virus maturation described above, most likely result from subtle, unknown differences in the lines of cells employed in the various studies.

In the most benign situation, Brown and co-workers (Gliedman *et al.*, 1975; Riedel and Brown, 1977, 1979) have reported no cytopathic effects of Sindbis virus infection in their line of *A. albopictus* cells. Riedel and Brown (1977) have shown that monolayers of these cells infected with 50–100 plaque-forming units (PFU) of Sindbis virus/cell are 100% infected, as determined by infectious-center assay and by immunofluorescence determination of the presence of virus antigens. These acutely infected mosquito cells produce about 6000–9000 PFU of virus/cell in the first 24 hr of infection. These cells were found to be indistinguishable from uninfected cells in morphology, rate of cell division, and ability to concentrate neutral red or exclude trypan blue. More recently, Scheefers-Borchel *et al.* (1981) and Condreay and Brown (unpublished observations) have shown that these infected cells incorporate [3H]thymidine or [35S]methionine at kinetic rates equivalent to those in uninfected cells. This noncytopathic cell response to Sindbis virus infection was not altered by culturing these cells in the insect medium of Mitsuhashi and Maramorosch (1964) or in Eagle's (1959) minimal essential medium, and repeated attempts to subclone cells that demonstrate a cytopathic response have been unsuccessful.

Although derived from the same Singh parent line of *A. albopictus* cells as those of Brown and co-workers, the *A. albopictus* cells employed in the studies of Stollar and co-workers, and in the laboratories of other investigators employing cell lines received from Stollar, respond very differently to alphavirus infection than the cell lines employed by Brown and co-workers. Tooker and Kennedy (1981), using *A. albopictus* cells from the laboratory of Stollar, found that they could clone these cells into subpopulations that responded very differently to infection with Semliki Forest virus. Of these clones, 70% produced virus very poorly and demonstrated no observable cytopathic effects. The remaining 30% produced high yields of progeny virus and demonstrated extensive cytopathology, including "almost total cell destruction." A further study revealed that clones showing absence of cytotoxic response to virus infection not only produced less virus but also did so at a greatly reduced kinetic rate.

Stalder *et al.* (1983) also found an extensive cytopathic response of the Igarashi–Stollar *A. albopictus* cell line to Semliki Forest virus infection. These investigators found that only 10% of the insect cells survived the acute phase of infection to go on to produce a persistently infected cell population. In contrast to the report of Tooker and Kennedy, clones derived from this parent line did not demonstrate varied response to Semliki Forest virus infection. All clones responded to infection like the parent line, with approximately 90% of the cells succumbing to the toxic effects of virus infection. As in other investigations, Stalder and co-workers found that the surviving population of cells went on to produce a persistently infected population that could be cultured indefinitely and that was apparently indistinguishable from uninfected cell populations.

The most concerted effort to understand the role that alphaviruses and insect cells individually play in the development of virus-induced cytopathology has been exerted by Stollar and co-workers. These investigators made the first efforts to clone *A. albopictus* cells into sublines that would respond differently to Sindbis virus infection [an approach subsequently employed by Tooker and Kennedy (see above)]. Stollar and co-workers have also made host-range mutants of Sindbis virus that demonstrate differing abilities to grow in vertebrate and invertebrate cells. Sarver and Stollar (1977) produced clones of *A. albopictus* cells that differ in their cytopathic response to Sindbis virus infection. As was subsequently confirmed by Tooker and Kennedy (1981) (with the same cell line), Stollar found that certain lines demonstrated little or no cytopathology while other lines showed extensive cytopathology after Sindbis virus infection. Unlike Stalder *et al.* (1983), Stollar found that the selected phenotype of the clone with respect to cytopathic effect (CPE) was stable, i.e., that 100% of the CPE-positive cells remained so on passage, as did CPE-negative cells. Unlike Tooker and Kennedy, Sarver and Stollar found no correlation between the amount of virus produced from the clones and the extent of the CPE; in addition, some clones were temperature-sensitive with respect to the extent of cytopathology, with the effect greater at 34 than at 28°C. The presence or absence of serum, in addition to the temperature, also affected the cytopathic response of A. *albopictus* cells to infection with vesicular stomatitis virus.

These collected studies demonstrate that the line of A. *albopictus* cells produced from insect larvae by Singh contains a variety of cell types capable of responding differently to alphavirus infection. One line (used in the laboratory of Brown and co-workers) produced high yields of virus with no detectable deleterious effects on the host cell (Riedel and Brown, 1977). The AISC-3 line isolated by Stollar and co-workers (Sarver and Stollar, 1977) from the Singh cell line demonstrated a similar ability to produce large amounts of virus in the absence of CPE. Other sublines produced large amounts of virus with extensive CPE [Tooker and Kennedy (1981) and Stalder *et al.* (1983), as well as the LT-C7 line of Sarver and Stollar (1977)]. Yet another subline produced very low yields of virus without CPE (Tooker and Kennedy, 1981).

The mechanism by which alphaviruses establish a CPE in any cultured host cell is unclear. Alphavirus infection of cultured vertebrate cells is followed by an arrest of host-cell protein and RNA synthesis (Wengler, 1980). Viral messenger RNA (primarily 26 S) forms the majority of polysomal complexes to the exclusion of host message (Wengler and Wengler, 1976). Van Steeg *et al.* (1984) have suggested that this substitution of virus message for host is mediated by virus capsid protein that blocks the formation of initiation complexes containing host message. Garry *et al.* (1979) and Ulug *et al.* (1984) have suggested that an alteration in intracellular salt concentration mediated by modification of the function of plasma-membrane-associated Na^+ favors translation of virus messages. Ulug and Bose (1985) have proposed that the inhibition of ion-pump activity results from the modification of the plasma membrane by insertion of virus glycoprotein. Contreras and Carrasco (1981) have suggested that an alteration in membrane permeability may facilitate the process of alphavirus-induced host-cell shut-off of vertebrate cells. Nozawa and Apostolov (1982) have implicated an alteration in membrane fatty acids in this process. Compartmentalization of virus replication in cultured insect cells of the type described by Gliedman *et al.* (1975) and Riedel and Brown (1977) might prevent either or both of these putative cytotoxic functions of virus structural proteins. The plasma membranes of these cells are not detectably modified by virus glycoprotein, and capsid structures are not detected in the cytoplasm. These cells show no CPE of Sindbis virus infection. Other cultures of insect cells, showing maturation of virus at the cell surface and presence of capsid in the cytoplasm, show varying degrees of CPE from transient rounding of cells to cell death. An understanding of the variation in the cells that permits this broad spectrum of responses to alphavirus infection will certainly prove important to understanding the interaction of alphaviruses with the various host organs in the intact mosquito.

That the whole insect contains cell populations capable of responses different from those seen in cell culture was demonstrated by Kowal and Stollar (1981) and Stollar and Hardy (1984). These investigators have shown that mutants of Sindbis virus can be produced that grow to high titer in chick embryo fibroblast cells (vertebrate cells) but demonstrate very poor growth in the LT-C7 line of *A. albopictus* cells at 34°C. These viruses have an RNA-negative phenotype in the invertebrate host at nonpermissive temperature that is not observed in the vertebrate cells at the same temperature. Although these Sindbis virus mutants grow poorly in cultured *A. albopictus* cells at 34°C, they produce yields equivalent to wild-type virus at 34°C when injected intrathoracically into *Aedes* mosquitos. This result again suggests that the insect contains a variety of cell populations that differ in their ability to support Sindbis virus growth.

VI. ROLE OF INSECT HOST-CELL FUNCTIONS IN THE REPRODUCTION OF ALPHAVIRUSES

Host cells play obvious roles in the replication of alphaviruses. Virus glycoproteins are glycosylated by host enzymes while transported by host vesicles from their site of synthesis on host rough-endoplasmic-reticulum-associated polyribosomes. Virus capsids are enveloped in a membrane bilayer provided by the host in the final stages of virus assembly. The ability of alphaviruses to efficiently exploit existing host biosynthetic pathways during replication becomes more remarkable when one considers the astonishing range of diverse cells and organisms in which these agents replicate. The considerable selective pressure placed on alphaviruses by the diverse range of hosts infers that those biochemical processes, which are unique to vertebrates or invertebrates, cannot play a significant or essential role in the outcome of virus infection. As indicated above, the alphaviruses produced from an acute infection of either vertebrate or invertebrate cells are similar in their biological properties despite differences in glycosylation and lipid content. Stollar *et al.* (1976) demonstrated that Sindbis virus grown in cultured insect cells was equivalent to vertebrate-grown virus in its response to a standard antiserum and in its ability to plaque in BHK-21 cells. Edwards *et al.* (1983) and Mann *et al.* (1983) further demonstrated that insect-grown Sindbis virus was identical to vertebrate-grown virus in its pH requirements for inducing cell fusion in BHK cells and in its kinetic rate of inactivation after exposure to pH 5.3.

Recent studies have shown that host-cell functions participate in the process of virus replication and assembly in subtle, critical ways that are only beginning to be understood. Adams and Brown (1982) have demonstrated that mild treatment of vertebrate cells with a proteolytic enzyme at $1\frac{1}{2}$–3 hr after establishment of an infection with Sindbis virus arrests the synthesis of viral RNA and the production of progeny virions. Vertebrate cells infected with vesicular stomatitis virus, similarly treated, showed no alteration of viral RNA synthesis or virus production. Similar treatment of *A. albopictus* cells infected with Sindbis virus produced no effect on the outcome of virus replication at any concentration of enzyme tested or at any time postinfection at which the experiment was carried out. The time postinfection at which the sensitivity of virus replication to trypsin treatment occurred in the vertebrate cell coincided with the onset of virus-induced arrest of host protein synthesis. These experiments implied that a host-coded function, or structure, which could be perturbed by trypsin treatment, was essential for the replication of Sindbis virus RNA, but not vesicular stomatitis virus RNA. Until host protein synthesis was arrested by virus, the host competence to participate in virus replication could be restored in some part. The insensitivity of Sindbis-virus-infected insect cells to similar treatment implies the lack of an externally situated trypsin-sensitive cell protein that can initiate a transmembrane effect resulting in the perturbation of the putative host function. Alternatively, the inability of Sindbis virus infection to arrest mosquito-cell protein synthesis may allow the insect cell to rapidly reestablish the function if it is disturbed.

That vertebrate host cells provide an essential component for alphavirus RNA synthesis has been further substantiated in a recent study on the expression of homologous interference by Sindbis virus in BHK-21 cells. Johnston *et al.* (1974) have shown that Sindbis-virus-induced homologous interference is expressed within 15 min of virus infection and is a virus-coded function associated with the genes that code for nonstructural proteins. Adams and Brown (1985) have suggested that the establishment of homologous interference requires the interaction of a virus nonstructural protein encoded in the 42 S RNA with an unidentified host component that is present in limited quantities. The complete utilization of this host factor by an infecting virus blocks the replication of a superinfecting virus at a stage after the attachment, penetration, and translation of the 42 S parental RNA and before the synthesis of 26 S RNA or progeny 42 S RNA. Edwards and Brown (unpublished observation) have further shown that a vertebrate cell contains 50–100 copies of this limited host-coded function by demonstrating that at multiplicities less than 50, a Sindbis virus *ts* mutant incapable of RNA synthesis (TS-18) demonstrates a multiplicity-dependent reduced ability to exclude superinfecting virus.

Aedes albopictus cells persistently infected with Sindbis virus also prevent the expression of superinfecting homologous virus. Condreay and Brown (unpublished observation) recently found, however, that this exclusion is not by the same mechanism as that functioning in the vertebrate cell. *Aedes albopictus* cells infected with multiplicities of up to 500 PFU/cell do not exclude the expression of superinfecting virus for up to 24 hr. These cells can be 100% infected with multiplicities of 25–50 PFU/cell (as determined by infectious-center assay and immunofluorescence), and maximum virus production is established in 9–12 hr at 28°C.

The establishment of interference in *A. albopictus* cultures seems to coincide with the production by the infected cell population of an antiviral protein produced as the cells enter the persistent phase of infection. It is possible that the putative, limited host factor responsible for the establishment of interference and initiation of virus RNA synthesis is the function eliminated by trypsin treatment of Sindbis-virus-infected vertebrate cells. Invertebrate cells may continue to produce this entity long after virus infection, since host-cell macromolecular synthesis is not inhibited. This continued production would provide a constant new supply of this material to participate in, and support the expression of, superinfecting virus.

In addition to host participation in virus RNA replication, a product of the insect-cell genome is required for Sindbis virus maturation. Scheefers-Borchel *et al.* (1981) found that production of mature progeny virions from cultured *A. albopictus* cells was prevented by pretreatment of cells for 2 hr with 4 μg/ml of actinomycin D. Actinomycin D treatment of mosquito cells did not affect their ability to produce vesicular stomatitis virus. A similar pretreatment of vertebrate cells causes a slight increase in Sindbis virus or vesicular stomatitis virus yields. When actinomycin D was added to mosquito-cell cultures after infection, 90 min was required for the arrest of virus production. The block in Sindbis virus maturation was not due to trapping of progeny virions in cytoplasmic vesicles. Erwin and Brown (1983) further examined the role of host nuclear functions in Sindbis virus replication in BHK-21 and *A. albopictus* cells by examining production of Sindbis virus in enucleated cells. Enucleation blocked the production of Sindbis virus but not vesicular stomatitis virus in cultured mosquito cells. Enucleation did not affect production of either virus in BHK-21 cells. Nearly 100% of *A. albopictus* cytoplasts infected with Sindbis virus demonstrated a strong, positive fluorescence when stained with fluorescence-conjugated anti-Sindbis virus antiserum, indicating the production of protein in the absence of virus maturation. Likewise, actinomycin-D-treated *A. albopictus* cells infected with Sindbis virus produced virus proteins PE2, E1, E2, and capsid and shed large amounts of these proteins in a nonvirus form into the culture medium. This system demonstrates that conversion of PE2 to E2 can take place in the absence of virus maturation.

The observations summarized above point out striking differences in the way invertebrate cells respond to alphavirus infection compared to vertebrate cells. Studies employing actinomycin D or enucleation show that in tissue-cultured vertebrate cells, alphavirus replication has no requirement for continued expression of the host genome. In cultured insect cells, however, the continued expression of nuclear information seems essential for maturation of alphaviruses, but not of rhabdoviruses. Stollar and co-workers further investigated the interaction of alphaviruses and insect cells by preparation of cell mutants expressing varying sensitivity to a variety of metabolic inhibitors and by examining the effects of drugs on Sindbis virus replication. Mento and Stollar (1978) prepared mutants of *A. albopictus* cells resistant to ouabain, α-amanitin, and bromodeoxyuridine. These investigators examined the replication of Sindbis virus in these cells in the presence and absence of these drugs. It was found that only ouabain treatment affects Sindbis virus production and only in ouabain-sensitive cells. This result showed that the reduced virus production resulted from an effect of the drug on the cell itself, not from a direct effect on the virus synthetic machinery. Malinoski and Stollar (1980, 1981a,b) examined the effect of the guanosine analogue ribavirin on the replication of Sindbis virus in cultured vertebrate and invertebrate cells. Ribavirin reduced Sindbis virus RNA synthesis and progeny production in mosquito cells, but in BHK cells, sensitivity to this drug was very dependent on the physiological condition of the cells themselves. Ribavirin reduced Sindbis virus replication in low-passage-number rather than high-passage-number cells, in subconfluent rather than confluent monolayers, and in the presence rather than in the absence of serum. In both vertebrate and invertebrate cells, ribavirin acts by reducing intracellular GTP levels. In cultured insect cells, the inhibitory effects of ribavirin could be reversed by treatment with a low concentration of actinomycin D (0.2 μg/ml), presumably by reducing host competition for limited amounts of GTP present in the drug-treated cells. Baric *et al.* (1983a,b) have described a sensitivity of Sindbis virus RNA synthesis (but not that of vesicular stomatitis virus) to actinomycin D if the drug was present 18 hr prior to infection, suggesting the participation of a long-lived host function in Sindbis virus RNA synthesis in cultured vertebrate cells. It has not been demonstrated whether similar lengthy treatments with actinomycin D can prevent Sindbis virus RNA synthesis in cultured insect cells.

We have recently demonstrated that alphavirus maturation and RNA synthesis are arrested when infected BHK cells are treated with 0.1 mM chloroquine (Cassell *et al.*, 1984). The mechanism of inhibition of these two virus functions is unknown. Chloroquine at 0.1 mM has no effect on the outcome of Sindbis virus infection of *A. albopictus* cells when present from 1 hr prior to infection to 18 hr postinfection (Coombs *et al.*, 1981). Tenfold higher concentrations of chloroquine are required to produce inhibitory effects on Sindbis replication of mosquito cells similar to those seen in BHK-21 cells.

Recently, Durbin and Stollar (1984) have described a spontaneous variant of Sindbis virus resulting from repeated passage in cultured *A. albopictus* cells. This variant (designated as SVap 15/21) grew normally in insect cells at 34°C, but poorly in BHK cells at any temperature tested. Growth in vertebrate cells was somewhat temperature-dependent, allowing the growth defect to be attributed to a *ts* lesion that, by complementation, fell into group E (an RNA[+] defect in virus maturation). The primary phenotype of SVap 15/21 was an overglycosylation of the PE2 and E2 proteins. Durbin and Stollar concluded that this alteration resulted from the addition of a new glycosylation site in the PE2 polypeptide. The addition of this site had little effect on virus maturation in insect cells at permissive temperature, but it blocked virus assembly in BHK cells. The ability of the infected insect cell to tolerate this change in the primary sequence of the PE2 precursor protein, compared to the inability of the infected BHK cell to mature the same protein into virus, again points to subtle differences in the interaction of virus and host-cell components during virus maturation.

VII. ESTABLISHMENT AND MAINTENANCE OF PERSISTENT ALPHAVIRUS INFECTIONS OF INSECT CELLS

A. Transition from Acute to Persistent Infection

A most striking characteristic of the interaction of alphaviruses with cultured insect cells is the ease with which persistent infections are established and maintained with indefinite passage. Persistent infections of cultured mosquito cells by arthropod-vectored viruses were first described by Rehacek (1968) and Peleg (1969). These initial observations underscored the potential of this virus–cell system for (1) understanding how these viral agents are perpetuated in nature; (2) determining at the molecular level the factors, both cellular and viral, that participate in the establishment and maintenance of a persistent virus infection; and (3) understanding those virus-induced events that result in host-cell death.

The first important observations describing the events that lead to the establishment of persistent infections of cultured insect cells came from the laboratory of Lynn Dalgarno. Davey and Dalgarno (1974) followed the state of infection of *A. albopictus* cells after an initial infection with Semliki Forest virus. They found that initially, all cells were infected as determined by infectious-center and immunofluorescene assays for virus antigens [a result subsequently reproduced by Igarashi *et al.* (1977) and Riedel and Brown (1977)]. More important, these authors showed that with time, there was a "shutdown" of virus-specific RNA synthesis and progeny virus production as the cell population moved from the acute phase of infection to the persistent phase of infection. In the persistently infected cell population, only 2% of the cells could be shown to be virus-producing.

Raghow *et al.* (1973a) conducted a morphological study of Semliki Forest virus maturation in *A. albopictus* cells and Raghow *et al.* (1973b) studied the growth of Ross River virus in these cells in an attempt to determine whether some unique characteristics of virus development could account for the noncytolytic response of the insect cells to virus infection and whether any morphological differences could be detected that could account for the control of virus production during persistence.

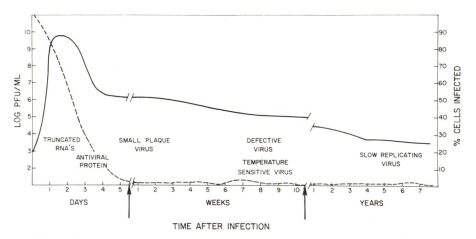

FIGURE 6. Progress of an infection of *A. albopictus* cells with alphavirus. The yield of progeny virus in PFU/ml (———) and the percentage of cells infected (-----) as determined in an infectious-center assay or by immunofluorescence is plotted against time. The points in the course of infection at which virus-induced entities appear are indicated. This figure is a composite of data described in the text.

Electron-microscopic investigation of these infected cells (described in detail above) showed that alphavirus maturation could take place at the cell surface (as occurs in vertebrate cells) or in membrane-limited vesicles. The latter process resulted in the accumulation of cell-associated viruses. It was suggested that the destruction of these vesicles and their contents by lysosomal enzymes might occur as persistence is established and result in the reduction of virus yields and the curing of cells, reducing the number of total cells that are virus-positive in the persistent phase of infection.

Since the pioneering studies of Rehacek, Peleg, and Dalgarno and co-workers, there have been many studies on the establishment of the acute high-titer virus-producing phase of alphavirus infection of mosquito cells and the events that occur during the transition from the acute phase to the low-virus-yielding persistent phase of infection. Many of the events that occur in this process are summarized in Fig. 6. Although some confusion exists in the literature regarding the toxic effects of virus on cultured insect cells during the acute phase of infection [which may in large part have to do with the cell type and culture conditions employed (see above)], there is unanimity in the observation that cells persistently infected with alphaviruses are indistinguishable from uninfected cell populations and that they do not demonstrate the periodic crises seen in many types of persistent infections established in cultured vertebrate cells. In the most benign situations (Gliedman *et al.*, 1975; Riedel and Brown, 1977), the infection of cultured mosquito cells with alphaviruses results in the participation of 100% of the cell population in the production of progeny virions. Yields of virus are high, equivalent to that obtained from cultured vertebrate cells. The cells show no CPE during this acute phase of infection, and all divide at normal rates and survive to enter the persistent phase of infection. As the cell population enters the persistent phase of infection, the production of virus RNA and protein decreases, progeny virus production drops to about 2 PFU/cell per day, and less than 2% of the cell population can be shown to be virus-positive by infectious-center assay or immunofluorescence. The persistently infected cell culture is, however, refractory to superinfection by homologous virus; i.e., yields of virus from the population of cells cannot be increased by attempts to reinfect the cell population by the addition of exogenous virus particles.

B. Role of Temperature-Sensitive Mutants and Defective Interfering Particles in the Establishment of Persistent Infection

In an attempt to understand the critical factors involved in establishing the persistent infection, various laboratories have examined the role played by the virus itself through repeated infection of the cell population by progeny viruses and the generation of ts mutants and defective interfering (DI) particles. The participation of the host cell in the establishment of the persistent infection has been examined by electron-microscopic observations of the cellular sites of virus maturation and by assaying for the production of interferonlike agents or other antiviral compounds produced in response to virus infection.

As the persistent infection, with its characteristic of low virus production, is established, all viruses appearing in the culture medium become very-small-plaque variants (Davey and Dalgarno, 1974; Eaton, 1981; Stalder et al., 1983; Shenk et al., 1974). This plaque-morphology change is detected as early as 3–4 days postinfection, and the small-plaque virus is the only type seen in the supernatant of the infected cell culture by day 8 or 9. As the cells are repeatedly cultured at low temperature (28°C), ts virus appears in the media. The shift to the ts phenotype occurs slowly, with progeny virus showing predominantly ts character at 12–14 weeks after infection. Most of the ts viruses isolated from Sindis-virus-persistently infected A. albopictus cells had RNA$^+$ phenotype (Shenk et al., 1974). These investigators were unable to complement these isolates with other known ts Sindbis virus mutants and concluded that virus from the persistently infected cell cultures contained multiple genetic defects.

Simizu and Maeda (1981) have examined the pattern of growth of ts mutants of WEE virus in BHK cells and mosquito cells. The mutants were isolated from early-passage persistently infected mosquito cells (5–30 days after infection) and from late-passage persistently infected cells (80–170 days after infection). Early-passage mutants were found to be of both RNA$^+$ and RNA$^-$ phenotypes. Late-passage mutants were RNA$^-$.

These investigators found that whereas wild-type WEE and early-passage RNA$^+$ mutants, and some RNA$^-$ mutants, produced CPE in cultured insect cells, late-passage RNA$^-$ mutants produced little or no CPE in the same cells. These late RNA$^-$ *ts* mutants produced more infectious virus than either wild-type or early mutants in cultured insect cells. The authors concluded that late-passage mutants had adapted to mature in mosquito cells more efficiently than wild-type virus.

Condreay and Brown (unpublished observation) found that a culture of *A. albopictus* cells persistently infected with Sindbis virus for 8 years produces a virus that is detectable in plaque assay on BHK-21 cells only if the assay is incubated at 28°C for 5–6 days prior to staining with neutral red. This extended assay yields plaques that are barely detectable by eye, being only 1 mm in diameter. By comparison, wild-type Sindbis virus in BHK-21 cells produces plaques 3–4 mm in cross section in 2 days at 28°C. The medium of this persistently infected cell population contains only 10^3–10^4 of this small-plaque virus, or about 0.0001 PFU/cell. These viruses grow very slowly in BHK-21 cells compared to wild-type virus, but demonstrate a more rapid growth pattern in *A. albopictus* cells when used in a new infection. One or two passages in BHK-21 cells restore a rapid-growth phenotype to the virus for BHK-21 cells. DI particles are not detected in the culture medium of these particular long-term, persistently infected *A. albopictus* cell cultures.

Insect cells persistently infected with alphaviruses generate intracellular truncated RNAs typical of DI particles that are produced during repeated passage of these viruses in cultured vertebrate cells (Tooker and Kennedy, 1981; Stalder *et al.*, 1983). The truncated RNAs are detectable as early as 24 hr after virus infection. The fate of these defective RNAs is not clearly understood. Stalder *et al.* (1983) reported the presence of defective RNAs associated with *A. albopictus* cells persistently infected with Semliki Forest virus using a hybridization assay with a DNA probe prepared from the 3' end of the standard virus genome. These investigators could not detect these RNAs in the media of persistently infected cells and concluded that although they represented the majority population of intracellular RNAs in the culture, they were not efficiently packaged into infectious virions. King *et al.* (1979) found that prolonged culturing of persistently infected *A. albopictus* cells for 48 hr without subculturing resulted in the appearance of DI particles in the medium. The ability of DI particles to interfere with the outcome of an infection of mosquito cells is also unclear.

The production of *ts* phenotypes of virus during persistent infection of *A. albopictus* cells is not considered to play an important role in the establishment of persistence. These viruses appear weeks after the persistent infection is established and likely reflect the loss of selective pressure against the production of *ts* phenotypes by repeated growth at low temperature. Alphaviruses repeatedly passaged at low temperature in cultured chick embryo fibroblast cells or BHK-21 cells also produce *ts* virus, although the pathogenicity of the virus in these vertebrate cells is maintained.

It is at present unclear whether the development of defective RNAs is a significant contributor to the establishment of persistent infections. Brown and Gliedman (1973) found that infection of *A. albopictus* cells for 5 days resulted in the appearance of virus particles that were approximately 80% the size of normal Sindbis virions. These particles possessed serum-blocking capability against anti-Sindbis virus serum, but were not infectious. Johnston *et al.* (1975), and recently Barrett *et al.* (1984), have demonstrated that DI particles of Sindbis virus and Semliki Forest virus produced in BHK cells show a similar reduction in size to 80–84% that of standard virus. Brown and Gliedman did not demonstrate an interfering capability associated with the smaller particles produced by mosquito cells. Assayable amounts of DI particles are not found in the media of persistently infected insect cells until 6–7 weeks after the characteristic persistently infected state is established (Eaton, 1981), much too late to participate in the events leading to the reduction in virus macromolecular synthesis seen in the transition to persistent infection. Truncated alphavirus RNAs have, however, been reported by Tooker and Kennedy (1981) to be produced in one subclone of *A. albopictus* cells that demonstrates high virus-producing characteristics and CPE at 24 hr after infection with standard virus. This subclone of cells went on to produce a persistently infected population after an initial cytopathic crisis. Likewise, Stalder *et al.* (1983) detected the production of truncated, defective Semliki Forest virus RNAs at 24 hr post-infection in their experiments examining the establishment of persistent infections of *A. albopictus* cells. These cells also went on to produce a persistently infected population after the initial death of 90% of the cell population. It is possible that the early presence of DI RNAs was responsible for preventing cell death in a limited number of cells, allowing them to make the transition to the persistent state. In the surviving population of persistently infected cells, these defective RNAs exist as a majority population and may, as such, continue to prevent cell death. The failure to mature these RNAs, reported by Stalder *et al.* (1983), into infectious particles may be of significance in that phase of the life cycle of these viruses in which these agents are transmitted from blood-sucking insects to vertebrate hosts. This particular defective alphavirus RNA may also prove important in that it may lack a sequence of nucleotides essential for nucleocapsid formation.

A major problem in implicating virus DI RNAs or virus *ts* mutants as regulators of alphavirus production during the persistent phase of infection is that under this type of control, the infection should be maintained indefinitely in all cells in the culture and be transmitted to daughter cells during cell division. Riedel and Brown (1977) and Igarashi *et al.* (1977) have shown that insect cells infected with alphaviruses can be cured of virus infection if the reinfection of the cells is prevented by

culturing in antivirus serum. Persistently infected cells "cured" in this
fashion lost the ability to prevent expression of superinfecting homologous virus and, after reinfection with homologous virus, reentered the
acute phase of infection followed by transition to persistent infection, as
was seen with uninfected cell cultures. A possible explanation for how
cells could become cured of virus infection has been provided by electron-
microscopic studies of *A. albopictus* cells infected by alphaviruses. As
described in detail above, alphavirus-infected insect cells that do not demonstrate cytopathic response to virus infection show virus maturation
occurring within restricted areas of the cell cytoplasm, producing vesicular structures containing numbers of mature, and partially assembled,
particles (Gliedman *et al.*, 1975). Riedel and Brown (1977) suggested that
the repeated expulsion of these virus-producing entities from the cell
cytoplasm could ultimately result in curing the cell of the virus infection.
Once these cells were cured, the individual cell would become susceptible
to reinfection by extracellular virus, producing a cell that could then, as
an individual entity, reenter the acute phase and ultimately return to the
persistent phase of low levels of virus production. The pattern of morphogenesis of alphaviruses in the insect-cell cultures employed in the
laboratory of Brown and co-workers differs from that reported in other
laboratories in that the assembly of virus structures free in the cytoplasm
of cells is almost never seen and virus maturation at the cell surface is
a very infrequent event. Likewise, the cells employed in the studies of
Brown and co-workers show no cytopathic or cytotoxic response to Sindbis virus infection, but rather behave during the acute phase of high virus
production as would uninfected cells with respect to their ability to grow
or attach to substrate, their rate of division, and other properties. A possibly important role for this pattern of development of viruses in virus
factories in developing the persistent infection of insect cells is supported
by a number of other observations: (1) That these cells show no cytotoxic
response to virus infection implies that the sequestering of virus maturation into intracellular compartments may protect the host cell from
those virus-produced entities that are toxic to the physiological processes
of the cell itself. (2) The sequestering of individual invading virions to
individual virus-producing entities would prevent DI particles from interfering with standard virus replication in the insect-cell culture population, the defective functions of these particles not being complemented by the functions present in the standard virions. (3) Renz and
Brown have shown that Sindbis virus *ts* mutants that readily complement
one another in cultured vertebrate cells complement each other poorly,
or not at all, in invertebrate-cell populations, implying that the development of the mutants is separated such that proteins required for complementation do not intermix; this could be explained by the sequestering
of individual infecting virions to separate virus-replicating entities in the
cell cytoplasm.

C. Appearance and Function of Antiviral Compounds

Although sequestering of virus replication in restricted areas of cell cytoplasm may provide a mechanism for isolating the virus infection from the host, thus preventing cell cytopathology and allowing for spontaneous curing of the cell population, it does not explain how the rate and quantity of virus production are suppressed during the transition from the acute to the persistent phase of infection. Indeed, Davey and Dalgarno (1974) demonstrated a dramatic reduction in virus RNA synthesis as this transition takes place. Riedel and Brown (1979) have described an entity that could serve to regulate virus replication in persistently infected cells. They have shown that A. albopictus cells persistently infected with Sindbis virus release into the medium a low-molecular-weight (dialyzable) peptide that inhibits Sindbis virus replication in a manner that is both cell- and virus-specific. This protein is very heat-labile and sensitive to selected proteases. It is insensitive to RNase, DNase, and a variety of glycosidases and lipases, and it is insensitive to treatment at pH 12. When infected cells are treated with media containing this peptide for 48 hr, they fail to produce an acute infection when subsequently infected with Sindbis virus at high multiplicities (Riedel and Brown, 1979). Indeed, these cells immediately assume the characteristics of a persistently infected cell population with a low percentage of cells immunofluorescing, a low percentage of cells registering infectious centers, and virus production at about 2 PFU/cell per day. This antiviral protein does not appear to be a virus structural component, since its activity is not affected by treatment with antisera prepared against purified Sindbis virus or against Sindbis-virus-infected BHK-21 cell extracts. This antiviral compound has no effect on Sindbis virus infection in BHK cells or on Semliki Forest virus infection of A. albopictus cells. Recently, Newton and Dalgarno (1983) have identified a similar antiviral activity produced by A. albopictus cells infected with Semliki Forest virus. This antiviral activity efficiently prevents the high-yield infection of A. albopictus cells by Semliki Forest virus, but has no effect on Sindbis virus infection; thus, this antiviral protein is the reciprocal of the one described by Riedel and Brown. The antiviral protein seems to act by inducing an antiviral state. Approximately 48 hr is required to establish this virus-specific refractory state, in which cells seem blocked in their ability to replicate virus RNA. The antiviral protein is detected in the media of cells at the time when cells are undergoing the transition from acute to persistent infections.

The antiviral protein also seems to be responsible for the expression of homologous interference in alphavirus-infected insect cells. The onset of exclusion of homologous virus in A. albopictus cells coincides with the appearance of the antiviral protein and the transition to the persistent phase of infection (Riedel and Brown, 1977; Newton and Dalgarno, 1983).

The role that the antiviral protein plays in persistent infections is not clear. Its ability to suppress virus RNA synthesis and prevent the production of progeny virus when applied to uninfected cells prior to infection suggests that these agents could have an important role in regulating virus replication. The antiviral protein *induces* a virus-resistant state in the mosquito cell. The changes in the cell during this condition are unknown. The antiviral protein is not sensitive to anti-Sindbis virus serum, and it is unclear how treatment of persistently infected cells with antiserum results in curing of the cell population, allowing an "acute" infection on subsequent virus addition. It is possible that in the absence of repeated reinfection by extracellular virus, production of the antiviral compound ceases and the cells revert from the antiviral state to normal phenotype. This possibility is supported by the observation that complete curing of the cell population by growth in virus antiserum can require up to 5 days (Riedel and Brown, 1977).

VIII. CONCLUSIONS AND PERSPECTIVES

As indicated above, the examination of alphavirus infection of cultured invertebrate cells has produced a spectrum of seemingly contradictory and inconsistent observations. Cultured insect cells vary in their degree of cytotoxic response to virus infection, in their morphological aspects of virus development, and in their ability to produce and respond to DI particles. These broad variations in the response of cultured mosquito cells to alphavirus infection may be due (as implied above) to differences in the clones of cells employed in the particular studies. The original culture of *Aedes* cells derived from larvae may have contained a diverse population of partially differentiated cells with subpopulations destined for development into particular organelles in the mature insect. Within the mature insect, the ability of completely differentiated cells to express a high virus-yielding response, or to develop a cryptic virus infection, may explain how the insect is able to sustain production of large amounts of virus in the salivary gland while passing an inapparent infection transovarially to offspring. Further characterization of the cell types in the preparations of insect cells using developmental markers would be important in establishing the potential of particular clonal subsets for virus production. An examination of the growth of viruses and the response of the cells in these more defined clones would prove invaluable in understanding the interaction of alphaviruses with insect vectors.

Continued study of the relationship between the alphavirion and the insect cell during persistent infection at the molecular level will be a fruitful area of investigation. It will be particularly valuable to elucidate the mechanism by which truncated defected RNAs are generated in cells and, in conjunction with structural studies, to determine whether defective RNAs are selected for deletions that differ from those produced in vertebrate cells. The role these truncated RNAs play in preventing cell death, and in the suppression of virus replication as the persistent infection is established, should be a particularly important area of investigation. The ultimate purification and characterization of antiviral proteins produced by alphavirus-infected invertebrate cells may have medical importance in addition to their value in understanding their regulation of virus production in insect cells. Dalgarno and co-workers, Riedel and Brown, and Condreay and Brown have found a remarkable virus, as well as cell specificity for these antiviral proteins. The ability of these compounds to specifically arrest the synthesis of RNA of alphaviruses that induce their production seems to be a critical characteristic for the establishment of persistence. It has been shown previously that the transition from acute to persistent infection is accompanied by a rapid reduction in the rate of virus RNA synthesis and that this reduction occurs temporally as these antiviral proteins are detected at assayable levels in the cultured fluids. It is most important to determine the origin of these antiviral proteins and to establish the molecular events that result in their induced production. A determination of the molecular structure of these proteins may make it possible to alter the specificity of their inhibitory effects for both virus and cell type. The achievement of this latter goal may result in the development of a new battery of therapeutic and chemoprophylactic agents for the control of alphavirus-induced disease.

REFERENCES

Adams, R. H., and Brown, D. T., 1982, Inhibition of Sindbis virus maturation after treatment of infected cells with trypsin, *J. Virol.* **41:**692–702.

Adams, R. H., and Brown, D. T., 1985, BHK cells expressing Sindbis virus induced homologous interference allow the translation of superinfecting virus nonstructural genes, *J. Virol.* **54:**351–357.

Anderson, W. A., and Spielman, A., 1971, Permeability of the ovarian follicle of *Aedes aegypti* mosquitos, *J. Cell Biol.* **50:**201–221.

Baric, R. S., Carlin, L. J., and Johnston, R. E., 1983a, Requirements for host transcription in the replication of Sindbis virus, *J. Virol.* **45:**200–250.

Baric, R. S., Lineberger, D. W., and Johnston, R. E., 1983b, Reduced synthesis of Sindbis virus negative-strand RNA in cultures treated with host transcription inhibitors, *J. Virol.* **47:**46–54.

Barrett, A. D. T., Cubitt, W. D., and Dimmock, N. J., 1984, Defective interfering particles of Semliki Forest virus are smaller than particles of standard virus, *J. Gen. Virol.* **65:**2265–2268.

Brown, D. T., 1980, The assembly of alphaviruses, in: *The Togaviruses: Biology, Structure, Replication* (R. W. Schlesinger, ed.), pp. 473–501, Academic Press, New York.

Brown, D. T., and Gliedman, J. B., 1973, Morphological variants of Sindbis virus obtained from infected mosquito tissue culture cells, *J. Virol.* **12:**1535–1539.

Cancedda, R., and Schlesinger, M. J., 1974, Formation of Sindbis virus capsid protein in mammalian cell-free extracts programmed with viral RNA, *Proc. Natl. Acad. Sci. U.S.A.* **71:**1843–1847.

Cassell, S., Edwards, J., and Brown, D. T., 1984, The effects of lysosomotropic weak bases on the infection of BHK-21 cells by Sindbis virus, *J. Virol.* **52:**857–864.

Chamberlain, R. W., and Sudia, W. D., 1961, Mechanism of transmission of viruses by mosquitos, *Annu. Rev. Entomol.* **6:**371–390.

Contreras, A., and Carrasco, L., 1981, Selective inhibition of protein synthesis in virus-infected mammalian cells, *J. Virol.* **29:**114–122.

Coombs, K., Mann, E., Edwards, J., and Brown, D. T., 1981, Effects of chloroquine and cytochalasin B on the infection of cells by Sindbis virus and vesicular stomatitis virus, *J. Virol.* **37:**1060–1065.

Davey, M. W., and Dalgarno, L., 1974, Semliki Forest virus replication in cultured *Aedes albopictus* cells: Studies on the establishment of persistence, *J. Gen. Virol.* **24:**453–463.

Durbin, R. K., and Stollar, V., 1984, A mutant of Sindbis virus with a host-dependent defect In maturation associated with hyperglycosylation of E_2. *Virology* **135:**331–334.

Eagle, H., 1959, Amino acid metabolism in mammalian cell cultures, *Science* **130:**432–437.

Eaton, B. T., 1981, Viral interference and persistence in Sindbis virus infected *Aedes albopictus* cells, *Can. J. Microbiol* **27:**563–567.

Edwards, J., and Brown, D. T., 1984, Sindbis virus induced fusion of tissue cultured *Aedes albopictus* (mosquito) cells, *Virus Res.* **1:**705–711.

Edwards, J., Mann, E., and Brown, D. T., 1983, Conformational changes in Sindbis virus envelope proteins accompanying exposure to low pH, *J. Virol.* **45:**1090–1097.

Erwin, C., and Brown, D. T., 1983, Requirement of cell nucleus for Sindbis virus replication in cultured *Aedes albopictus* (mosquito) cells, *J. Virol.* **45:**792–799.

Garry, R. F., Bishop, J. M., Parker, S., Westbrook, K., Lewis, G., and Waite, M. R. F., 1979, Na^+ and K^+ concentrations and the regulation of protein synthesis in Sindbis virus-infected chick cells, *Virology* **96:**108–120.

Gillies, S., and Stollar, V., 1981, Translation of vesicular stomatitis and Sindbis virus mRNAs in cell-free extracts of *Aedes albopictus* cells, *J. Biol. Chem.* **256:**13,188–13,192.

Gliedman, J. B., Smith, J. F., and Brown, D. T., 1975, Morphogenesis of Sindbis virus in cultured *Aedes albopictus* cells, *J. Virol.* **16:**913–926.

Hardy, J. L., Houk, E. J., Kramer, L. D., and Reeves, W. C., 1983, Intrinsic factors affecting vector competence of mosquitos for arboviruses, *Annu. Rev. Entomol.* **28:**229–262.

Helenius, A., Kartenbeck, J., Simons, K., and Fries, E., 1980a, On the entry of Semliki Forest virus into BHK-21 cells, *J. Cell Biol.* **84:**404–420.

Helenius, A., Marsh, M., and White, J., 1980b, The entry of viruses into animal cells, *Trends Biochem. Sci.* **5:**104–106.

Houk, E. J., 1977, Midgut ultrastructure of *Culex tarsalis* (Diptera: Culicidae) before and after a bloodmeal, *Tissue Cell* **9**(1):103–118.

Houk, E. J., and Hardy, J. L., 1979, *In vivo* negative staining of the midgut continuous junction in the mosquito *Culex tarsalis* (Diptera: Culicidae), *Acta. Trop.* **36:**267–276.

Houk, E. J., Kramer, L. D., Hardy, J. L., and Chiles, R. E., 1985, Western equine encephalomyelitis virus. *In vivo* infection and morphogenesis in mosquito mesenteronal epithelial cells, *Virus Res.* **2:**123–138.

Howard, L. M., 1962, Studies on the mechanism of infection of the mosquito midgut by *Plasmodium gallicaceum*, *Am. J. Hyg.* **75:**287–300.

Hsieh, P., and Robbins, P. W., 1984, Regulation of asparagine-linked oligosaccharide processing in *Aedes albopictus* mosquito cells, *J. Biol. Chem.* **259:**2375–2382.

Igarashi, A., and Stollar, V., 1976, Failure of defective interfering particles of Sindbis virus produced in BHK or chicken cells to affect viral replication in *Aedes albopictus* cells, *J. Virol.* **19:**398–408.

Igarashi, A., Koo, R., and Stollar, V., 1977, Evolution and properties of *Aedes albopictus* cell cultures persistently infected with Sindbis virus, *Virology* **82:**69–83.

Johnston, R. E., Wan, K., and Bose, H. R., Jr., 1974, Homologous interference induced by Sindbis virus, *J. Virol.* **14:**1076–1082.

Johnston, R. E., Tovell, D. R., Brown, D. T., and Faulkner, P., 1975, Interfering passages of Sindbis virus: Concomitant appearance of interference, morphological variants and truncated viral RNA, *J. Virol.* **16:**951–958.

King, C.-C., King, M. W., Garry, R. F., Wan, K. M.-M., Ulug, E. T., and Waite, M. R. F., 1979, Effect of incubation time on the generation of defective-interfering particles during undiluted serial passage of Sindbis virus in *Aedes albopictus* and chick cells, *Virology* **96**:229–238.

Kowal, K. J., and Stollar, V., 1981, Temperature-sensitive host-dependent mutants of Sindbis virus, *Virology* **114**:140–148.

Kramer, L. D., Hardy, J. L., Presser, S. B., and Houk, E. J., 1981, Dissemination barriers for Western equine encephalomyelitis virus in *Culex tarsalis* infected after ingestion of low viral doses, *Am. J. Trop. Med. Hyg.* **30**(1):190–197.

Leake, C. J., 1984, Transovarial transmission of arboviruses by mosquitos, in: *Vectors in Virus Biology* (M. A. Mayo and K. A. Harrap, eds.), pp. 63–91, Academic Press, London.

Luukonen, A., von Bonsdorff, C.-H., and Renkonen, F., 1977, Characterization of Semliki Forest virus grown in mosquito cells: Comparison with the virus from hamster cells, *Virology* **78**:331–335.

Malinoski, F., and Stollar, V., 1980, Inhibition of Sindbis virus replication in *Aedes albopictus* cells by virazole (ribavirin) and its reversal by actinomycin: A correction, *Virology* **102**:473–476.

Malinoski, F., and Stollar, V., 1981a, Inhibitors of IMP dehydrogenase prevent Sindbis virus replication and reduce GTP levels in *Aedes albopictus* cells, *Virology* **110**:281–291.

Malinoski, F., and Stollar, V., 1981b, Inhibition of Sindbis virus replication by ribavirin: Influence of cultural conditions and of the host cell phenotype, *Antiviral Res.* **1**:287–299.

Mann, E., Edwards, J., and Brown, D. T., 1983, Polycaryocyte formation mediated by Sindbis virus glycoproteins, *J. Virol.* **45**:1083–1089.

McLintock, J., 1978, Mosquito–virus relationships of American encephalitides, *Annu. Rev. Entomol.* **23**:17–37.

Mento, S. J., and Stollar, V., 1978, Effect of ouabain on Sindbis virus replication in oubain-sensitive and oubain-resistant *Aedes albopictus* cells (Singh), *Virology* **87**:58–65.

Miles, J. A. R., Pillai, J. S., and Maguire, T., 1973, Multiplication of Whataroa virus in mosquitos, *J. Med. Entomol.* **10**:176–185.

Mitsuhashi, J., and Maramorosch, K., 1964, Leafhopper tissue culture: Embryonic, nymphal and imaginal tissues from aseptic insects, *Contrib. Boyce Thompson Inst.* **22**:435–460.

Moore, N. F., Barenholz, Y., and Wagner, R. R., 1976, Microviscosity of togavirus membranes studied by fluorescence depolarization: Influence of envelope proteins and the host cell, *J. Virol.* **19**:126–135.

Newton, S. E., and Dalgarno, L., 1983, Antiviral activity released from *Aedes albopictus* cells persistently infected with Semliki Forest virus, *J. Virol.* **47**:652–655.

Nozawa, C. M., and Apostolov, K., 1982, Association of the cytopathic effect of Sindbis virus with increased fatty acid saturation, *J. Gen. Virol.* **59**:219–222.

Peleg, J., 1968a, Growth of arboviruses in monolayers from subcultured mosquito embryo cells, *Virology* **35**:617–619.

Peleg, J., 1968b, Growth of arboviruses in primary tissue culture of *Aedes aegypti* embryos, *Am. J. Trop. Med. Hyg.* **17**:219–223.

Peleg, J., 1969, Inapparent persistent virus infection in continuously grown *Aedes aegypti* mosquito cells, *J. Gen. Virol.* **5**:463–471.

Raghow, R. S., Davey, M. W., and Dalgarno, L., 1973a, The growth of Semliki Forest virus in cultured mosquito cells: Ultrastructural observations, *Arch. Gesamte Virusforsch.* **43**:165–168.

Raghow, R. S., Grace, T. D. C., Filshie, B. K., Bartly, W., and Dalgarno, L., 1973b, Ross River virus replication in cultured mosquito and mammalian cells: Virus growth and correlated ultrastructural changes, *J. Gen. Virol.* **21**:109–122.

Rehacek, J., 1968, The growth of arboviruses in mosquito cells *in vitro*, *Acta Virol.* **12**:241–246.

Renz, D., and Brown, D. T., 1976, Characteristics of Sindbis virus temperature-sensitive mutants in cultured BHK-21 and *Aedes albopictus* (mosquito) cells, *J. Virol.* **19**:775–781.

Riedel, B., and Brown, D. T., 1977, Role of extracellular virus in the maintenance of the persistent infection induced in *Aedes albopictus* (mosquito) cells by Sindbis virus, *J. Virol.* **23**:554–561.

Riedel, B., and Brown, D. T., 1979, Novel antiviral activity found in the media of Sindbis virus-persistently infected mosquito (*Aedes albopictus*) cell cultures, *J. Virol.* **29**:51–60.

Sarver, N., and Stollar, V., 1977, Sindbis virus-induced cytopathic effect in clones of *Aedes albopictus* (Singh) cells, *Virology* **80**:390–400.

Sarver, N., and Stollar, V., 1978, Virazole prevents production of Sindbis virus and virus-induced cytopathic effect in *Aedes albopictus* cells, *Virology* **91**:267–282.

Scheefers-Borchel, U., Scheefers, H., Edwards, J., and Brown, D. T., 1981, Sindbis virus maturation in cultured mosquito cells is sensitive to actinomycin D, *Virology* **110**:292–301.

Shenk, T. E., Koshelnyk, K. A., and Stollar, V., 1974, Temperature-sensitive virus from *Aedes albopictus* cells chronically infected with Sindbis virus, *J. Virol.* **13**:439–447.

Simizu, B., and Maeda, S., 1981, Growth patterns of temperature-sensitive mutants of Western equine encephalitis virus in cultured *Aedes albopictus* (mosquito) cells, *J. Gen. Virol.* **56**:349–361.

Simmons, D. T., and Strauss, J. H., 1974, Translation of Sindbis virus 26S RNA and 49S RNA in lysates of rabbit reticulocytes, *J. Mol. Biol.* **86**:397–409.

Singh, K. R. P., 1967, Cell cultures derived from larvae of *Aedes albopictus* (Skuse) and *Aedes aegypti* (L.), *Curr. Sci.* **36**:506–508.

Stalder, J., Reigel, F., and Koblet, H., 1983, Defective viral RNAs in *Aedes albopictus* C6/36 cells persistently infected with Semliki Forest virus, *Virology* **129**:247–254.

Stollar, V., 1980, Togaviruses in cultured arthropod cells, in: *The Togaviruses: Biology, Structure, Replication* (R. W. Schlesinger, ed.), pp. 584–621, Academic Press, New York.

Stollar, V., and Hardy, J. L., 1984, Host dependent mutants of Sindbis virus whose growth is restricted in cultured *Aedes albopictus* cells produce normal yields of virus in intact mosquitos, *Virology* **134**:177–183.

Stollar, V., Shenk, T. E., Koo, R., Igarashi, A., and Schlesinger, W., 1975, Observations on *Aedes albopictus* cell cultures persistently infected with Sindbis virus, *Ann. N. Y. Acad. Sci.* **266**:214–231.

Stollar, V., Stollar, B. D., Koo, R., Harrap, K. A., and Schlesinger, R. W., 1976, Sialic acid contents of Sindbis virus from vertebrate and mosquito cells: Equivalence of biological and immunological viral properties, *Virology* **69**:104–115.

Stollar, V., Harrap, K. A., Thomas, V., and Sarver, N., 1979, Observations related to cytopathic effect in *Aedes albopictus* cells infected with Sindbis virus, in: *Arctic and Tropical Arboviruses* (E. Kurstak, ed.), pp. 277–296. Academic Press, New York.

Stones, P. B., 1960, Symposium on the evolution of arbovirus diseases. *Trans. R. Soc. Trop. Med. Hyg.* **54**:132.

Thomas, L. A., 1963, Distribution of the virus of Western equine encephalomyelitis in the mosquito vector *Culex tarsalis*, *Am. J. Hyg.* **78**:150–165.

Tooker, P., and Kennedy, S. I. T., 1981, Semliki Forest virus multiplication in clones of *Aedes albopictus* cells, *J. Virol.* **37**:589–600.

Ulug, E. T., and Bose, H. R., Jr., 1985, Effect of tunicamycin on the development of the cytopathic effect in Sindbis virus-infected avian fibroblasts, *Virology* **143**:546–557.

Ulug, E. T., Garry, R. F., Waite, M. R. F., and Bose, H. R., Jr., 1984, Alterations in monovalent cation transport in Sindbis virus-infected chick cells, *Virology* **132**:118–130.

Van Steeg, H., Kaspelaitis, M., Voorma, H. O., and Beene, R., 1984, Infection of neuroblastoma cells by Semliki Forest virus: The interference of viral capsid protein with the binding of host messenger RNAs into initiation complexes is the cause of shut-off of host protein synthesis, *Eur. J. Biochem.* **138**:473–478.

Wengler, G., 1980, Effects of alphaviruses on host cell macromolecular synthesis, in: *The*

Togaviruses: Biology, Structure, Replication (R. W. Schlesinger, ed.), pp. 459–472, Academic Press, New York.

Wengler, G., and Wengler, G., 1976, Protein synthesis in BHK-21 cells infected with Semliki Forest virus, *J. Virol.* **17:**10–19.

White, J., and Helenius, A., 1980, pH dependent fusion between the Semliki Forest virus membrane and liposomes, *Proc. Natl. Acad. Sci. U.S.A.* **77:**3273–3277.

White, J., Kartenbeck, J., and Helenius, A., 1980, Fusion of Semliki Forest virus with the plasma membrane can be induced by low pH, *J. Cell Biol.* **87:**264–272.

Whitfield, S. G., Murphy, F. A., and Sudia, W. D., 1973, St. Louis encephalitis virus: An ultrastructural study of infection in a mosquito vector, *Virology* **56:**70–87.

CHAPTER 8

Alphavirus Pathogenesis and Immunity

DIANE E. GRIFFIN

I. INTRODUCTION

Most alphavirus infections in nature occur in wild birds, rodents, and occasionally reptiles and amphibians. Some of these natural hosts sustain a relatively asymptomatic, prolonged, high-titered viremia, facilitating transmission of the virus by mosquitoes (Chamberlain, 1980). Although there is undoubtedly mortality and morbidity in the wild due to alphavirus infections, many native species have inapparent infection, and epizootic, epidemic, or endemic disease is usually recognized after mosquito-mediated transmission to a domestic host (e.g., horses, pigeons, captive pheasants) or to man. Affected individuals develop a wide range of symptoms, including fever, skin rashes, arthritis, myositis, and/or encephalitis. Depending on the virus and on the host, the infection may be clinically inapparent or cause fulminant disease and death (Shope, 1980).

Investigation of this closely related group of viruses has proven to be a fruitful ground for studies of viral pathogenesis. Over the years, investigators have sought to identify the parameters of virus virulence and host defense that determine the outcome of infection. This pursuit is facilitated in alphavirus infections by the comparatively simple structure of the virion and extensive knowledge of the genome and synthesis of viral polypeptides gained from work in many laboratories. Most pathogenesis studies have utilized animal models of alphavirus-induced disease in which experimental parameters are more easily manipulated than they are in man. Immunological studies have been encouraged by the ability of alphaviruses to cause diseases in mice analogous to those seen in man.

DIANE E. GRIFFIN • Departments of Medicine and Neurology, The Johns Hopkins University School of Medicine, Baltimore, Maryland 21205.

A number of strains of mice that are well characterized genetically and immunologically are available for experimentation, making this an ideal species for the study of pathogenesis in general and immune responses in particular. This chapter will be devoted primarily to a synthesis of information gained from studies of alphavirus infections in laboratory animals, with the addition of relevant data from other species when available. Only selected information from the use of monoclonal antibodies will be covered, since Chapter 9 is devoted to this subject.

II. DISEASE PATHOGENESIS

A. Eastern Equine Encephalitis

1. Natural Infection

Eastern equine encephalitis (EEE) virus is among the most virulent of the alphaviruses, causing severe epidemic encephalitis in man, horses, pigeons, and pheasants. The case fatality rate in man is 50–75% in all age groups (Farber et al., 1940; Hart et al., 1964; Feemster, 1957), up to 90% in horses (Farber et al., 1940), and 50–70% in pheasants (Sussman et al., 1958; Eisner and Nusbaum, 1983).

Birds vary in their susceptibility, with pheasants, pigeons, crows, owls, and a variety of small birds developing disease, while many other avian species show no morbidity or mortality even though they may maintain a prolonged viremia (Fothergill and Dingle, 1938; Davis, 1940; Kissling et al., 1954a,b; Satriano et al., 1957; Jungherr et al., 1957). Pheasant deaths are caused by encephalitis (Jungherr et al., 1957), while in young chickens, death appears to be due to myocarditis (Tyzzer and Sellards, 1941). In susceptible wild birds, encephalitis is also unusual (Jungherr et al., 1957). Virus persists in the feather follicles of infected pheasants, and secondary transmission among penned pheasants can occur through feather picking and cannibalism (Satriano et al., 1957).

Man is a highly susceptible species. Serological surveys suggest that there are approximately 23 inapparent infections for every case of recognized encephalitis, but this declines to only 8 to 1 for children under 4 (Goldfield et al., 1968). The encephalitis tends to be fulminant and associated with fever, headache, altered consciousness, and seizures (Farber et al., 1940; Hart et al., 1964). Histopathology demonstrates a diffuse meningoencephalitis with widespread neuronal destruction, perivascular cuffing, which includes polymorphonuclear as well as mononuclear leukocytes early, and vasculitis with vessel occlusion (Farber et al., 1940). In survivors, paralysis, seizures, and mental retardation are common sequelae (Farber et al., 1940).

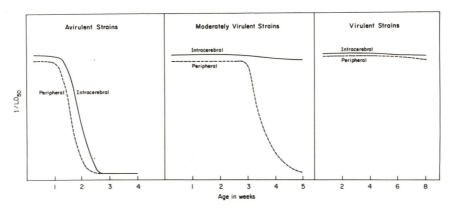

FIGURE 1. Schematic representation of the change with age in LD_{50} of avirulent, moderately virulent, and virulent strains of alphaviruses given by intracerebral and peripheral routes of inoculation. Examples of avirulent viruses are Sindbis, Ross River, and attenuated strains of WEE. Examples of moderately virulent viruses are EEE, WEE, Chikungunya, and attenuated Semliki Forest virus (SFV). An example of a virulent virus is SFV.

2. Laboratory Infection

Limited laboratory studies indicate a similar neurovirulence of EEE for monkeys, mice, and hamsters when virus is given by either peripheral or intracerebral inoculation (Wyckoff and Tesar, 1939; Liu *et al.*, 1970; Brown and Officer, 1975; Dremov *et al.*, 1978). The virus replicates in both neurons and glia of the central nervous system (CNS), causing extensive lysis of these cells in all species examined (Murphy and Whitfield, 1970; Liu *et al.*, 1970; Bastian *et al.*, 1975). Hamsters also exhibit necrosis of hepatocytes and evidence of infection of lymphatic tissue (Dremov *et al.*, 1978). Cellular destruction is accompanied by a local infiltration of mononuclear leukocytes.

At 3–4 weeks of age, mice become relatively resistant to peripheral, but not intracerebral, inoculation of EEE (Morgan, 1941; Lennette and Koprowski, 1944; Brown *et al.*, 1975) (Fig. 1). Hamsters, however, remain susceptible to peripheral inoculation. Attenuated strains of EEE have been described that are less likely than wild-type to cause encephalitis after peripheral inoculation of mice (Brown and Officer, 1975; Brown *et al.*, 1975). In hamsters, attenuated strains replicate primarily in lymphoid tissue, with CNS infection occurring in only a small proportion of animals (Dremov *et al.*, 1978).

B. Western Equine Encephalitis

1. Natural Infection

Western equine encephalitis (WEE) virus causes encephalitis in horses and in man in widespread areas of North America, but the case fatality rate of 10% in man (Reeves and Hammon, 1962) and 20–40% in horses (Doby *et al.*, 1966) is lower than that observed for EEE virus infections. As with other viral encephalitides, WEE is associated with fever, headache, irritability, tremors, and seizures, along with signs and symptoms of meningitis such as nuchal rigidity and photophobia (Kokernot *et al.*, 1953). Severe disease, fatal encephalitis, and significant sequelae are more likely to occur in infants and young children than in older children and adults (Kokernot *et al.*, 1953; Finley *et al.*, 1955; Longshore *et al.*, 1956; Earnest *et al.*, 1971). Transplacental transmission can occur (Shinefield and Townsend, 1953; Copps and Giddings, 1959). Pathological examination of brains from fatal cases demonstrates early perivascular extravasation of blood followed by endothelial hyperplasia, perivascular mononuclear and polymorphonuclear inflammation, and parenchymal necrosis (Noran and Baker, 1945).

2. Laboratory Infection

The main target organs of WEE in most laboratory animals are muscles and the CNS (Aguilar 1970; Zlotnick *et al.*, 1972b). With increasing age, mice become relatively resistant to fatal infection after peripheral, but not intracerebral, inoculation (Lennette and Koprowski, 1944; Aguilar, 1970) (Fig. 1), while hamsters remain susceptible by all routes (Zlotnick *et al.*, 1972b). Neonatal mice develop acute inflammation and necrosis in skeletal and smooth muscle, cartilage, and bone marrow (Aguilar, 1970). After peripheral inoculation of weanling mice, the virus replicates in skeletal muscles, causing a viremia (Fig. 2). During the viremia, the heart and brain become infected (Liu *et al.*, 1970; Aguilar, 1970; Monath *et al.*, 1978). Viral antigen in the CNS is localized first to choroid plexus and ependyma before appearing in neurons of the brain and spinal cord, suggesting that choroidal infection during the viremia seeds the cerebrospinal fluid (CSF), resulting in spread of virus throughout the CNS (Liu *et al.*, 1970). In animals with encephalitis, the brain shows multifocal areas of necrosis and widespread lymphocytic infiltration of the leptomeninges and perivascular regions of the brain parenchyma (Aguilar, 1970). The heart shows a necrotizing, inflammatory myocarditis associated with electrocardiographic changes that suggest that lesions are pathophysiologically significant (Monath *et al.*, 1978). Infiltration of mononuclear leukocytes into skeletal muscle and interstitial areas of lung, liver, and brown fat also occurs. Attenuated strains of WEE (B628 clones) have been described that are avirulent after intracerebral inoculation of weanling mice (H. N. Johnson, 1963) and hamsters (Zlotnick *et al.*, 1972b), but these have not been studied further.

C. Sindbis Virus

1. Natural Infection

Sindbis virus is among the least virulent of the alphaviruses. Serological surveys suggest that infection is relatively common in certain

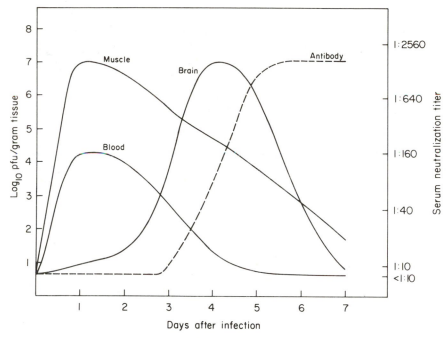

FIGURE 2. Schematic representation of the replication and systemic spread of alphaviruses that infect muscle and brain (e.g., WEE, Sindbis, Semliki Forest virus) occurring after peripheral inoculation of virus. The development of virus-neutralizing antibody is also depicted.

regions, but clinically apparent disease is unusual (Taylor *et al.*, 1955; Doherty, 1977). Fever, myalgias, tendonitis, arthritis, and/or rash may occur. The rash is notable in that it is sometimes vesicular or hemorrhagic and may involve the palms and soles as well as other skin surfaces (Malherbe *et al.*, 1963; McIntosh *et al.*, 1964; Doherty *et al.*, 1969; Guard *et al.*, 1982). A syndrome of headache, fever, and malaise may occur a few days prior to the onset of the rash. Although protracted disease occurs occasionally, neither fatal infections nor encephalitis has been recognized (Malherbe *et al.*, 1963; McIntosh *et al.*, 1964). A closely related virus, Ockelbo, has recently been associated with a similar disease complex in Sweden (Espmark and Niklasson, 1984; Niklasson *et al.*, 1984), and serological studies suggest that a Sindbislike virus also caused recent outbreaks in Finland and Russia (Shope, 1985).

2. Laboratory Infection

Infection of mice either peripherally or intracerebrally with the wild-type strain AR339 or AR86 causes encephalitis in which the mortality is age-dependent (Taylor *et al.*, 1955; Weinbren *et al.*, 1956; Reinarz *et al.*, 1971) (see Fig. 1). Between the ages of 1 and 3 weeks, mice abruptly become completely resistant to fatal encephalitis, although virus still

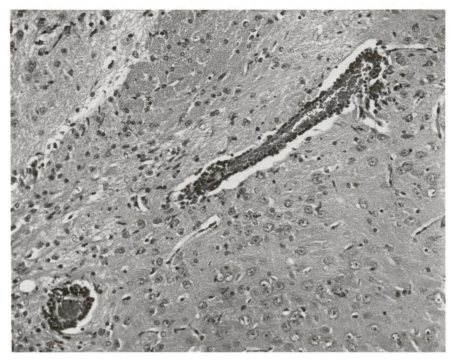

FIGURE 3. Mononuclear perivascular inflammatory reaction in the brain of a weanling mouse infected with Sindbis virus. This infiltrate begins to develop 3–4 days after infection and is indicative of a cellular immune response to the virus.

grows in the brain and encephalitis is still evident histologically (R. T. Johnson *et al.*, 1972; Hackbarth *et al.*, 1973; Griffin, 1976). After peripheral inoculation of mice at any age, Sindbis virus replicates in skeletal muscle, producing a viremia and secondary infection of the brain and meninges (R. T. Johnson, 1965, 1966; Griffin, 1976) (Fig. 2). In newborn mice, invasion of the CNS occurs by infection of capillary endothelial cells (R. T. Johnson, 1965, 1966). The site of CNS invasion in older mice has not been established. Virus replication in the brain is focal, involves all neural cell types, and is accompanied by a mononuclear cell inflammatory response in the perivascular regions of the brain, the leptomeninges, and the CSF (R. T. Johnson, 1971; Griffin, 1976, 1981) (Figs. 3 and 4). Ocular infection has also been demonstrated (Carreras *et al.*, 1982).

Strains of Sindbis virus that show altered virulence for mice have been derived. One (SB-RL) has decreased neurovirulence for newborn mice (Olmsted *et al.*, 1984), and others (NSV, HR-SMB-10, and HR-WMB-14) have increased neurovirulence for adult mice (Weinbren *et al.*, 1956; Boone and Brown, 1976; Griffin and Johnson, 1977) after intracerebral inoculation. The reasons for altered virulence are not yet clear, but decreased neurovirulence for newborn mice *in vivo* correlates *in vitro* with an accelerated penetration of virus into baby hamster kidney (BHK)-21

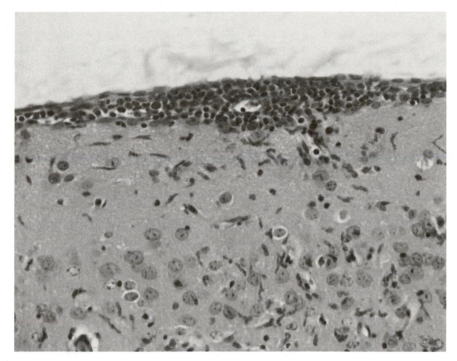

FIGURE 4. Meningitis in a mouse infected with Sindbis virus. This cellular infiltrate is also composed of mononuclear cells appearing in the meninges as a part of the cellular immune response.

cells and with a probable conformational change in the E2 glycoprotein (Baric *et al.*, 1981; Olmsted *et al.*, 1984). Increased neurovirulence for adult mice is associated with replication to higher titer in the brain and spinal cord of infected mice (Griffin and Johnson, 1977) and conformational changes in the neutralizing and hemagglutinating epitopes of the E2 glycoprotein (Stanley *et al.*, 1985). Taken together, these observations suggest an important role for E2 in determining Sindbis virus virulence.

D. Venezuelan Equine Encephalitis

1. Natural Infection

Venezuelan equine encephalitis (VEE) virus causes encephalitis in horses and in man. During epidemics, the infection in horses is frequently fatal (40–80%) and associated with leukopenia and a high-titered, relatively prolonged viremia that probably serves to amplify the infection for mosquito-mediated transmission during epidemics (Kissling *et al.*, 1956; Henderson *et al.*, 1971; Sudia and Newhouse, 1975). In contrast to WEE and EEE, the disease in horses is not consistently an encephalitis, and virus is shed in nasal, eye, and mouth secretions as well as in urine and milk. Transmission in horses can occur by the respiratory route as well as by mosquitoes. The pathology in horses includes cellular depletion of bone marrow, spleen, and lymph nodes, pancreatic necrosis, and, in cases with encephalitis, swelling of vascular endothelial cells, edema, and a modest mononuclear infiltration in the brain (Kissling *et al.*, 1956).

A number of species of wild birds become infected with VEE, but they develop little evidence of disease and develop immunity to reinfection. The viremia in birds is relatively short (1–2 days) and of lower titer than in many mammals, but sufficient to infect mosquitoes, conferring on birds a potential role in endemic disease (Chamberlain *et al.*, 1956).

Infection in man can occur by the respiratory route as well as by mosquito inoculation, as has been demonstrated by a number of laboratory-acquired infections (Lennette and Koprowski, 1943). The disease in adults is usually a mild illness associated with fever, headache, myalgias, and pharyngitis (mortality <1%). Virus can be recovered from the pharynx as well as the blood during the early phases of illness (Lennette and Koprowski, 1943; Scherer *et al.*, 1972). More severe disease, including fulminant reticuloendothelial system infection and encephalitis, may occur in young children (Ehrenkranz and Ventura, 1974). Fetal abnormalities have been reported with infection during pregnancy in both man (Wenger, 1977) and monkeys (London *et al.*, 1977). Children with encephalitis may be left with profound neurological deficits (Leon *et al.*, 1975). Spread to the CNS is most likely via the bloodstream, although studies in primates suggest that during respiratory infection, the olfactory route may also play a role (Danes *et al.*, 1973). At autopsy, cerebral edema, perivascular cuffing in the brain, cellular depletion of lymphatic tissue, and occasionally myocarditis and hepatitis are found (K. M. Johnson *et al.*, 1968).

VEE virus has considerable regional variation, so many natural field isolates as well as laboratory-adapted strains are available for experimental studies. Some field strains were isolated during large epizootics, others from cases of endemic disease, and still others by *in vitro* passage of wild-type strains (Hearn, 1960; Henderson *et al.*, 1971; Monath *et al.*, 1974; Walder and Bradish, 1975). Epizootic strains can be transmitted by various species of mosquitoes and are generally more virulent for man, horses, and experimentally infected monkeys (Henderson *et al.*, 1971; Ehrenkranz and Ventura, 1974; Monath *et al.*, 1974) than are the enzootic strains, which are transmitted in a cycle primarily involving *Culex* mosquitoes and rodents (Grayson and Galindo, 1968; Sudia *et al.*, 1971).

2. Laboratory Infection

Experimental infection of laboratory animals has included mice, guinea pigs, hamsters, rats, and rabbits. Different patterns of infection and disease are produced in different species (Table I). For guinea pigs and

TABLE I. Alphavirus-Induced Disease

Virus	Host	Disease
EEE	Man	Encephalitis
	Horses	Encephalitis
	Pheasants	Encephalitis
	Chicks	Myocarditis
	Monkeys	Encephalitis
	Mice	Encephalitis
	Hamsters	Encephalitis, hepatitis
WEE	Man	Encephalitis
	Horses	Encephalitis
	Monkeys	Encephalitis
	Mice	Encephalitis, myocarditis
	Hamsters	Encephalitis
Sindbis	Man	Skin rash, arthritis
	Mice	Encephalitis
VEE	Man	Pharyngitis, myocarditis, encephalitis, leukopenia, hepatitis
	Horses	Leukopenia, encephalitis, pancreatitis
	Monkeys	Encephalitis
	Mice, rats	Myocarditis, encephalitis, pancreatitis, late demyelination, placentitis
	Guinea pigs, rabbits	Hepatitis, lymphadenitis, leukopenia
	Hamsters	Leukopenia, lymphadenitis, encephalitis, pancreatitis
Semliki Forest	Man	Encephalitis
	Mice	Encephalitis, late demyelination, encephalopathy, placentitis
Chikungunya	Man	Skin rash, arthritis, encephalitis
	Mice	Encephalitis
Ross River	Man	Skin rash, arthritis
	Mice	Myositis, myocarditis, ependymitis, late demyelination

rabbits, the primary target organs for VEE virus appear to be the liver, lymphoid, and myeloid tissue. Peripheral inoculation of VEE produces necrosis in lymph nodes, spleen, thymus, intestinal and conjunctival lymphoid tissue, liver, and bone marrow (Victor *et al.*, 1956; Gleiser *et al.*, 1962; Airhart *et al.*, 1969). Numbers of peripheral white blood cells decrease within 48 hr, and death after infection with virulent strains occurs in 3–4 days. Virulence for guinea pigs may correlate with virulence for horses (Scherer and Chin, 1977).

After subcutaneous inoculation of virulent strains of VEE virus, hamsters, like guinea pigs and rabbits, show widespread involvement of myeloid and lymphoid tissue (Austin and Scherer, 1971; Jahrling and Scherer, 1973a,b; Walker *et al.*, 1976) in addition to evidence of impaired mitosis in bone marrow (Pruslin and Rodman, 1978). However, hamsters also develop evidence of encephalitis, with focal hemorrhagic lesions in the brain, and of pancreatitis, with necrosis of both acinar and islet cells (Gorelkin and Jahrling, 1974). Early deaths may be due in part to secondary bacterial infections associated with virus-induced ileal necrosis and impaired function of the reticuloendothelial system (Gorelkin and Jahrling, 1975). Avirulent strains of VEE produce encephalitis in hamsters after peripheral inoculation, with particular localization to the olfactory bulbs and tracts (Dill *et al.*, 1973). Temperature-sensitive (*ts*) mutants of VEE have been produced in which attenuation of virulence in hamsters is associated with restricted growth in spleen and bone marrow (Krieger *et al.*, 1979).

Experimentally infected rats and mice have more limited destruction of lymphoid and myeloid tissues than guinea pigs, rabbits, and hamsters. Initial replication still occurs in thymus, lymph nodes, and spleen, but is followed by viremia (Victor *et al.*, 1956; Tasker *et al.*, 1962; Gleiser *et al.*, 1962) and spread of virus to the CNS. Death is usually caused by encephalitis. Virus enters the CNS from the blood by infection of the endothelial cells of cerebral capillaries (Gorelkin, 1973). Infection causes neuronal necrosis, severe mononuclear cell cuffing of cerebral vessels, and meningitis (Jahrling *et al.*, 1978; Garcia-Tamayo *et al.*, 1979; Gleiser *et al.*, 1962). Oligodendroglial cells also become infected (Gorelkin, 1973), which may be relevant to the later appearance of demyelinating lesions in the white matter of the brain and spinal cord (Dal Canto and Rabinowitz, 1981). In pregnant animals, virus infects the placenta and may result in fetal death (Spertzel *et al.*, 1972; Garcia-Tamayo *et al.*, 1981). Neonatal mice are susceptible to infection by peripheral or intracerebral inoculation with evidence of extensive replication in many tissues including brain, myocardium, and pancreas (Kundin *et al.*, 1966; Garcia-Tamayo, 1973).

Studies of the biological differences between virulent and avirulent strains of VEE have shown that virulent virus (Trinidad donkey strain) is cleared more slowly from the blood of susceptible animals than avirulent virus (TC-83) (Jahrling and Scherer, 1973b; Jahrling *et al.*, 1977). As described for virulence variants of Sindbis virus (Olmsted *et al.*, 1984), virulent virus (63U2) adsorbs to and grows more slowly in BHK-21 cells than avirulent virus (TC-83) (Scherer *et al.*, 1971). Biochemical studies comparing the hamster-virulent 68U201 strain to an avirulent mutant *ts* 126 identified differences in surface characteristics associated with a change in the charge on the El glycoprotein (Emini and Wiebe, 1981). Similar comparisons of the virulent Trinidad donkey strain and its vaccine derivative, TC-83, suggest a decrease from three to two glycosylation sites on the E2 protein (Mecham and Trent, 1982) and a less rapid production of infectious virus by the avirulent strain in Vero cells (Mecham and Trent, 1983).

E. Semliki Forest Virus

Semliki Forest virus (SFV), Chikungunya virus, and Ross River virus (RRV) are closely related viruses of moderate virulence for man. SFV is generally considered to be of low pathogenicity for man, although one laboratory-acquired infection has been associated with fatal encephalitis (Willems *et al.*, 1979). The disease in mice is more predictable and more severe. After intraperitoneal or intramuscular inoculation, SFV replicates in skeletal muscle and fibroblasts, giving rise to a viremia and secondary infection of the CNS (Murphy *et al.*, 1970; Grimley and Friedman, 1970; Pusztai *et al.*, 1971) (see Fig. 2). Virus is reported to enter the brain from the blood by transport across capillary endothelial cells without actual infection of the endothelial cells (Pathak and Webb, 1974).

A large number of SFV isolates are available for study that vary in their virulence for mice, rabbits, and guinea pigs (Bradish *et al.*, 1971). All are capable of replicating in brain as well as muscle. Virulent strains (e.g., V13, L10) produce a fatal encephalitis with high titers of virus in the brain after peripheral as well as intracerebral inoculation in all ages of mice (Bradish *et al.*, 1971; Bradish and Allner, 1972). Mice with fatal infection show inflammation in the brain and CSF and acute degenerative lesions with glial and neuronal necrosis in the brain (Zlotnick and Harris, 1970; Bradish *et al.*, 1971; Doherty, 1973; Mackenzie *et al.*, 1978).

Lethal SFV infection can be modified in certain strains of mice by the administration of UV-inactivated tissue-culture fluid containing defective interfering (DI) particles intraperitoneally or intranasally simultaneously with the infecting inoculum (Dimmock and Kennedy, 1978; Barrett and Dimmock, 1984a,b). Virus replication in the brain is reduced, the brain appears normal histopathologically, and mortality decreases from 100 to 15%. This effect appears to be due to DI-particle interference with standard virus replication, rather than to an alteration in the immune response to the virus (Dimmock and Kennedy, 1978; Crouch *et al.*, 1982).

Avirulent strains of SFV (e.g., A7) exhibit an age-dependent induction of disease. Neonatal mice uniformly succumb to a fatal encephalitis after either peripheral or intracerebral inoculation (Bradish *et al.*, 1971) with evidence of widespread viral replication in the CNS (Murphy *et al.*, 1970). Between 2 and 3 weeks of age, mice become relatively resistant to fatal infection after peripheral inoculation, but remain susceptible to intracerebral inoculation (Bradish *et al.*, 1971) (see Fig. 1). Decreased susceptibility with age is associated with decreased yields of virus from muscle early after infection (Grimley and Friedman, 1970), lower-titered viremias, and decreased viral replication in the CNS. Avirulent strains of SFV infect the brain after peripheral or intracerebral inoculation and infect the placenta to cause fetal infection (Pusztai *et al.*, 1971; Atkins *et al.*, 1982; Milner and Marshall, 1984). Infection is widespread in the CNS, including the retina (Illavia *et al.*, 1982), and is associated pathologically with a perivascular mononuclear inflammatory response and microcystic changes that appear within 5 days. Focal spongiform lesions irregularly distributed in brain and spinal cord develop that are associated with astrocytic hypertrophy and lysosomal glycosidase activity (Bradish *et al.*, 1971; Suckling *et al.*, 1976; Mackenzie *et al.*, 1978). Mice do not appear clinically ill. An inflammatory response to infection is evident in CSF as well as in the brain parenchyma (Doherty, 1973; Parsons and Webb, 1982).

Instead of complete recovery after infection with avirulent strains of SFV, a number of late pathological changes in the CNS have been described. A late degenerative encephalopathy has been noted histopathologically (Zlotnick et al., 1972a) and a proportion (≈20%) of infected mice will develop focal areas of demyelination in the cerebellum, optic nerve, brainstem, and spinal cord 2–4 weeks after infection (Chew-Lim et al., 1977a; Suckling et al., 1978b; Illavia et al., 1982). Demyelinating lesions are associated with increased levels of acid proteinase thought to be indicative of macrophage activity in the brain (Suckling et al., 1978a).

In addition to the SFV field isolates that vary in virulence, a number of mutant viruses with decreased virulence compared to wild-type virus have been derived from the L10 virulent strain (Barrett et al., 1980). Two of these avirulent mutants invade the CNS after peripheral inoculation (as do the avirulent field strains) and produce demyelination. Morphological studies on mice infected with these strains suggest that demyelination is secondary to preferential lytic replication of the mutant virus in oligodendrocytes. The parent virulent strain replicates lytically in both neurons and oligodendroglia, causing fatal encephalitis (Sheahan et al., 1981; Atkins and Sheahan, 1982). Thus, it is postulated that the mutants' decreased tropism for neurons allows demyelinating lesions caused by oligodendroglial infection to become apparent. This mechanism is further suggested by the preferential in vitro replication of mutant viruses in oligodendroglial cell lines (Atkins and Sheahan, 1982) and by the fact that similar biological differences have been shown for the naturally occurring avirulent strains of SFV that are also associated with demyelination (Atkins, 1983).

F. Chikungunya and O'nyong-nyong Viruses

1. Natural Infection

Chikungunya ("that which bends up", so-called because of the pain caused by infection) and O'nyong-nyong viruses in Africa and Asia (Robinson, 1955; Haddow et al., 1960) cause fever and acute polyarthritis in man. The onset of the disease is sudden, without prodromal symptoms, and is manifested by fever, severe joint and muscle pains, and moderate headache. A pruritic maculopapular rash and generalized lymphadenopathy appear 3–4 days later (Robinson, 1955; Thiruvengadam et al., 1965; Carey et al., 1969). Virus can be isolated from the blood early in the course of disease (Williams and Woodall, 1961; Carey et al., 1969). Pain, swelling, and morning stiffness involving both large and small joints may continue in a chronic form for months to years after infection, but joint deformity does not occur (Robinson, 1955; Doherty et al., 1961; Kennedy et al.,

1980). Chikungunya may cause encephalitis in young children, and fatal disease and cases with neurological sequelae have been reported (Carey et al., 1969).

2. Laboratory Infection

Chikungunya virus causes a fatal encephalitis in suckling mice and in weanling mice after intracerebral inoculation, but fails to invade the CNS after peripheral inoculation (Igarashi et al., 1971; Suckling et al., 1978a). The encephalitis is severe, with necrosis and leptomeningeal and perivascular inflammation (Suckling et al., 1978a).

G. Ross River Virus

1. Natural Infection

RRV in Australia and the Pacific region (Anderson et al., 1961; Doherty et al., 1964; Doherty, 1977) causes fever and acute polyarthritis in man, a disease in many ways similar to the diseases caused by Chikungunya and O'nyong-nyong viruses in Africa and Asia. Approximately 50 subclinical infections occur for each clinically recognized case (Aaskov et al., 1981d). The onset of disease is sudden, with fever, joint and muscle pain, headache, and rash (Doherty et al., 1961; Aaskov et al., 1981b). Virus can be isolated from blood early in the course of disease (Rosen et al., 1981; Tesh et al., 1981), and transplacental transmission during the viremia can occur (Aaskov et al., 1981c). Arthritic symptoms may persist for months (Doherty et al., 1961; Rosen et al., 1981). Adults are more likely to be symptomatic than children (Aaskov et al., 1981d). Fatal cases or encephalitis have not been recognized.

2. Laboratory Infection

In neonatal mice, RRV causes an acute paralytic disease after peripheral inoculation. The virus infects striated, smooth, and cardiac muscle, producing necrosis of muscle fibers (Murphy et al., 1973). The paralytic disease is age-dependent, with limited virus replication and no evidence of clinical disease occurring in weanling mice (Seay et al., 1981) (see Fig. 1). Neonatal mice infected with the less virulent NB5092 strain develop minimal CNS disease. CNS infection is primarily ependymal, leading to late-onset aqueductal stenosis and hydrocephalus (Mims et al., 1973). Infection with the more virulent T48 strain causes more extensive CNS disease in addition to myositis, lymphoid infection (Mims et al., 1973; Seay et al., 1981), and fetal infection and death in pregnant animals (Aaskov et al., 1981a; Milner and Marshall, 1984). The majority of CNS viral antigen is found in the cerebellum, brainstem, and spinal cord. Perivascular inflammation with both mononuclear and polymorphonuclear leukocytes is followed by the appearance of focal areas of demyelination in mice infected at 1 week of age (Seay and Wolinsky, 1982). Necrotic lesions in brown fat, myocardium, and brain tend to become calcified within 10–14 days after infection (Murphy *et al.*, 1973; Seay and Wolinsky, 1982).

III. DEVELOPMENT OF THE IMMUNE RESPONSE

Most alphaviruses replicate rapidly after intracerebral, subcutaneous, intravenous, or intraperitoneal inoculation into a susceptible host, and peak virus titers are reached in the organ of initial replication within 1–2 days after infection with even a small amount of virus (see Fig. 2). The initial period of replication results in a plasma viremia that may allow spread of virus to other target organs such as the brain, pancreas, skin, or joints (Fig. 2). Most alphaviruses cause more severe disease in younger individuals than in older ones (see Fig. 1), but differ in whether the most severe manifestation of disease is the myositis, the encephalomyelitis, the arthritis, or the reticuloendothelial system necrosis (see Table I). In addition, a few of these viruses may be associated with the late onset of a degenerative or demyelinating CNS disease. During all these infections of animals, nonspecific (natural) as well as immunologically specific host defenses are induced. Each of these will be discussed in turn.

A. Natural Immunity

The earliest virus-specific immune response after primary infection is not detectable until the 3rd or 4th day after infection when both immunoglobulin M (IgM) antibody and the mononuclear inflammatory response begin to appear. Prior to this, virus is replicating rapidly in the primary site of infection, a viremia has been established, and spread to other organ systems has begun (Fig. 2). Several host resistance factors, however, are capable of mounting early nonspecific antiviral activity that may limit viral replication and virus spread prior to the appearance of virus-specific immunity. Three of these factors have been studied in some detail for one or another of the alphaviruses: complement, interferon, and natural killer cells.

1. Complement

One function of the complement system is amplification of antibody activity by way of the classic pathway of activation of the complement cascade. Classic pathway activation usually requires binding of the first component of complement (C1) to the Fc portion of antibody complexed with antigen. This binding initiates the complement cascade, resulting in opsonization, chemotactic factor production, enhanced neutralization, and/or lysis of virions or cells. In the absence of antibody, the complement cascade can be activated by an alternative pathway that is initiated by the binding of the third component of complement (C3b) to the surface of an activating particle (Hirsch, 1982; Sissons *et al.*, 1982). Sindbis virus (Hirsch *et al.*, 1980b), but not RRV (Aaskov *et al.*, 1985), is capable of activating complement by such an antibody-independent mechanism. The capacity to activate complement may be important during infection, since a deficiency of complement, either genetic (Hirsch *et al.*, 1980a) or pharmacologically induced (Hirsch *et al.*, 1978), results in a prolonged viremia and increased infection of the CNS. The late complement components (C5–9) (Hirsch *et al.*, 1980a), but not C3 (Hirsch *et al.*, 1978), may also contribute to clearance of virus from tissue.

RRV virions do not activate complement, but instead inhibit the cleavage of C3 by either the classic or the alternative pathway of activation. It is not clear what role this inhibition plays *in vivo*, but it may contribute to RRV's success as a human pathogen, since this property would be predicted to retard viral clearance from blood (Aaskov *et al.*, 1985).

Antibody-independent activation of complement by Sindbis virions or Sindbis-virus-infected cells is dependent on the sialic acid content of the envelope glycoproteins (Hirsch *et al.*, 1981), since sialic acid interferes with C3 cleavage (Fearon, 1979). The sialic acid content of Sindbis virus is dependent on the host cell in which the virus is grown (Stollar *et al.*, 1976; Keegstra and Burke, 1977; Hsieh *et al.*, 1983). Virus grown in mosquito cells has essentially no sialic acid and activates complement by the alternative pathway more efficiently, and is cleared from the blood more rapidly, than virus grown in BHK-21 cells (Hirsch *et al.*, 1981). Since the sialic acid content of glycoproteins is genetically determined for individuals within a species (Nydegger *et al.*, 1978), it is possible that the amount of glycoprotein sialic acid may be partly responsible for individual variation in susceptibility to alphavirus infection. This possibility has been tested in outbred mice. Higher levels of cellular sialic acid correlate with a higher incidence and level of viremia after peripheral inoculation with Sindbis virus. These data suggest that the amount of sialic acid in glycosylated proteins of an individual host influences the host's ability to limit infection with Sindbis virus and possibly other alphaviruses as well (Hirsch *et al.*, 1983).

2. Interferon

Alphaviruses induce the production of type I (α and β) interferon both *in vivo* (Vilcek, 1964; Postic *et al.*, 1969) and *in vitro* (Schleupner *et al.*, 1969; Blackman and Morris, 1984). Interferon produces in host cells an antiviral state that limits replication and infection of new cells in the first few days after infection (Mayer *et al.*, 1973; Tazulakhova *et al.*, 1973). Interferon activity can be identified in blood and tissues within a few hours of virus infection (Baron and Buckler, 1963; Bradish and Allner, 1972; Hackbarth *et al.*, 1973). The amount of interferon induced generally correlates with the amount of virus being produced by the infected host, so that virulent infections are associated with more interferon than avirulent infections (Vilcek, 1964; Grimley and Friedman, 1970; Bradish and Allner, 1972; Hackbarth *et al.*, 1973; Bradish *et al.*, 1975). Exogenously administered interferon or stimulation with interferon inducers reduces mortality in SFV infections if given after virus challenge (Worthington and Baron, 1971) and in VEE and SFV virus infections if given before virus challenge (Finter, 1966; Rabinowitz and Adler, 1973; Bradish and Titmuss, 1981).

3. Natural Killer Cells

In addition to the direct effects of interferon on virus replication, interferon also induces the activity of natural killer (NK) cells (Gidlund *et al.*, 1978). NK cells can kill virus-infected cells (Rager-Zisman and Bloom, 1982) and tumor cells (Herberman and Holden, 1978) in an immunologically nonspecific fashion. The induction of such cytolytic activity during viral infections was first described in SFV infection of mice (Rodda and White, 1976; Macfarlan *et al.*, 1977). Cells with NK activity are detectable in the blood and spleen within 24 hr after infection. NK cells are also present in the CSF within 3 days after intracerebral inoculation with Sindbis virus and even earlier in athymic nude mice (Griffin and Hess, 1986). Although potentially of considerable importance for the early elimination of virus-infected cells, no essential role for NK cells in recovery from alphavirus infection has been demonstrable (Hirsch, 1981; Gates *et al.*, 1984).

B. Virus-Specific Immunity

Virus-specific immune responses are mediated either by antiviral antibody or by T lymphocytes that are virus-specific. Other cells, particularly monocytes and macrophages, may be important participants in immunologically specific immune responses, but are directed in these activities either by antibody, which is produced by mature B lymphocytes, or by lymphokines, which are produced by mature T lymphocytes of the helper or inducer subsets. The virus-specific immune response may play an important role in a number of phases of infection. One is termination of viremia, which is necessary to prevent continued seeding of secondary sites of infection. Another is elimination of virus from the tissue, a step that is necessary for eventual complete recovery from infection. A third is the prevention of reinfection.

In addition to these beneficial features, the immune response has the potential to contribute to the tissue damage during infection above and beyond that which is caused by the virus itself (immunopathological injury). Virus-infected cells may be damaged by immunologically activated cells or antibody (Gilden et al., 1972; E. D. Johnson and Cole, 1975). Immune complexes composed of viral antigens and nonneutralizing antibody may be formed, and these complexes may cause damage at distant sites such as the renal glomerulus or the choroid plexus (Oldstone, 1984). Antibody and cells may be expected to participate to varying degrees in each of these aspects of the immune response, and the effect of immunosuppression could be either to delay recovery (more severe disease) or to prevent damage (less severe disease).

1. Antibody

Using sensitive methods, virus-specific antibody can usually be detected in serum 3–4 days after virus infection (Griffin and Johnson, 1973; Hackbarth et al., 1973; Doherty, 1973; Monath et al., 1974; Griffin, 1976; Rabinowitz, 1976; Rodda and White, 1976; Jagelman et al., 1978; Suckling et al., 1978a; Bradish et al., 1979; Seay et al., 1981) (see Figs. 2 and 5). This initial antibody is of the IgM class (Griffin and Johnson, 1977; Park et al., 1980) and capable of neutralizing virus. Its appearance correlates with termination of viremia (Griffin and Johnson, 1977; Seay et al., 1981) (Fig. 2). By 7–8 days after infection, antibody formation switches to antibody predominantly of the IgG class (Burns et al., 1975; Griffin and Johnson, 1977; Hirsch and Griffin, 1979; Park et al., 1980). The virus-specific antibody response peaks in 10–14 days, then plateaus and can be boosted by reexposure to the same virus (Griffin and Johnson, 1973) (Fig. 5). Significant differences have not been detected in the rapidity or amount of antibody produced by animals of different ages (Griffin, 1976).

The biological function of antibody is dependent on its antigenic specificity and the antibody isotype. Isotype is determined by the heavy chain (e.g., μ, $\gamma 1$, $\gamma 2$, $\gamma 3$, α) of the immunoglobulin molecule, which in turn determines biological properties such as complement fixation, attachment to cells, placental transfer, and others (Spiegelberg, 1974). Antibodies with a number of different biological functions of potential importance to recovery from infection and protection from reinfection are produced. These include neutralizing, complement-fixing, fusion-inhibiting, and hemagglutination-inhibiting antibodies and antibodies that direct the lysis of virus-infected cells through the action of either complement or cells.

a. Neutralization

Clearance of virus from the blood could occur by antibody that neutralizes viral infectivity, participates in complement-mediated lysis of

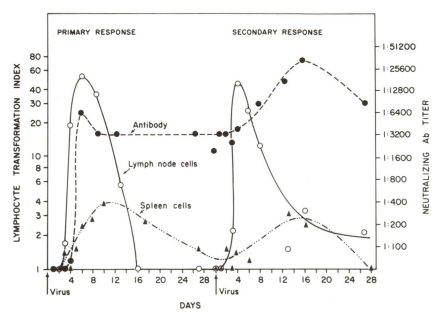

FIGURE 5. Development of virus-specific serum neutralizing antibody and proliferative responses of cells from the draining lymph node and spleen at various times after primary subcutaneous inoculation of Sindbis virus and after rechallenge (Griffin and Johnson, 1973).

virions (Stollar, 1975), or mediates enhanced physical clearance of virions by the liver and spleen (Jahrling *et al.*, 1983). Classic experiments using polyclonal antibody prepared against purified viral proteins localized the neutralizing epitopes to the E2 glycoproteins of Sindbis virus (Dalrymple *et al.*, 1976) and VEE virus (Pedersen and Eddy, 1974; France *et al.*, 1979). Additional studies with polyclonal antibodies (Symington *et al.*, 1977) and more recent studies with monoclonal antibodies have demonstrated that for Sindbis virus (Schmaljohn *et al.*, 1982, 1983; Chanas *et al.*, 1982; Stanley *et al.*, 1985, SFV (Boere *et al.*, 1984), and possibly VEE virus (Roehrig *et al.*, 1982), antibodies against both the E1 and E2 glycoproteins can effectively neutralize infectivity. IgM, IgA, and IgG antibodies can all mediate neutralization. The mechanism of alphavirus neutralization by antibody to either protein is unknown.

b. Hemagglutination Inhibition

Isolated E1 proteins from SFV (Helenius *et al.*, 1976) and Sindbis virus (Dalrymple *et al.*, 1976) can hemagglutinate goose erythrocytes, and polyclonal antibody to E1 inhibits hemagglutination (Dalrymple *et al.*, 1976). However, for these same viruses, monoclonal antibodies directed to either the E1 or the E2 glycoprotein can inhibit hemagglutination (Schmaljohn *et al.*, 1983; Boere *et al.*, 1984; Stanley *et al.*, 1985). In contrast, for VEE virus, only polyclonal (Pedersen and Eddy, 1974; France *et al.*, 1979) and monoclonal antibodies (Roehrig *et al.*, 1982) made to the E2 protein are capable of inhibiting hemagglutination.

c. Cell Lysis

Antibody-dependent complement-mediated lysis of virus-infected cells requires integration of viral antigen into the lipid bilayer of the infected cell (Morein *et al.*, 1979) and antibody that is both specific for an accessible portion of either the E1 or the E2 glycoprotein (Schmaljohn *et al.*, 1982, 1983; Stanley *et al.*, 1986) and of an isotype capable of binding complement. Smith *et al.* (1985) report that polyclonal antibody to the capsid (C) protein can also mediate lysis of infected cells, but monoclonal antibodies to C with this capability have not been identified (Stanley *et al.*, 1986). Since IgM is efficient at fixing complement, antibody that mediates lysis of infected cells is often present early after infection (Rodda and White, 1976). Antibody can also direct cytolysis of virus-infected cells mediated by macrophages or killer cells independent of complement (Macfarlan *et al.*, 1977) and may represent an important recovery mechanism.

d. Protection from Infection

Antibody administered passively prior to or at the time of challenge with a number of virulent alphaviruses has been shown to protect animals from fatal disease (Cremer *et al.*, 1966; Igarashi *et al.*, 1971; Seamer *et al.*, 1971; Zlotnick *et al.*, 1972a; Pedersen *et al.*, 1973; Rabinowitz and Adler, 1973; Brown and Officer, 1975; Griffin and Johnson, 1977; Kraaijeveld *et al.*, 1979a). Use of monoclonal antibodies has shown that this protective capacity includes, but is not limited to, antibodies with neutralizing activity and includes antibodies to both the E1 and the E2 glycoprotein (Schmaljohn *et al.*, 1982; Mathews and Roehrig, 1982; Boere *et al.*, 1983, 1984; Hunt and Roehrig, 1985; Stanley *et al.*, 1986). Even though less efficient than neutralizing anti-E2 antibodies (Mathews and Roehrig, 1982; Boere *et al.*, 1984), the ability of nonneutralizing anti-E1 antibodies to protect from death after a normally fatal challenge may explain the previously puzzling observations that cross protection by related alphaviruses occurs in the absence of production of cross-neutralizing antibodies (Casals, 1963; Hearn and Rainey, 1963; Cole and McKinney, 1971; Brown and Officer, 1975).

Antibody to the E2 protein is more virus-specific, while antibody to the E1 protein tends to be more cross-reactive (Dalrymple *et al.*, 1976; Roehrig *et al.*, 1982; Chanas *et al.*, 1982). This is consistent with the greater homology observed in the E1 region compared to the E2 region of alphavirus genomes (Bell *et al.*, 1984). E1-region homology is probably associated with the cross-reactive, antibody-dependent, complement-mediated cytolysis seen between SFV and Sindbis virus antibody and infected cells, which does not correlate with virus neutralization (King *et al.*, 1977; Wolcott *et al.*, 1982a,b). These E1 epitopes are more accessible on the surface of virus-infected cells than they are on mature virions, which is probably indicative of a conformational change in the protein associated with virus budding (Kaluza *et al.*, 1980; Gates *et al.*, 1982; Schmaljohn *et al.*, 1983; Wolcott *et al.*, 1984; Hunt and Roehrig, 1985).

e. Promotion of Recovery

In studies of WEE, neurovirulent Sindbis, VEE, and SFV, antibodies can also promote recovery from an otherwise fatal infection when given 24–48 hr after infection (Howitt, 1932; Zichis and Shaughnessy, 1940; Olitsky *et al.*, 1943; Berge *et al.*, 1961; Seamer *et al.*, 1971; Griffin and Johnson, 1977). The SFV, VEE, and WEE experiments used a peripheral route of inoculation, and antibody appeared to delay CNS infection (Olitsky *et al.*, 1943; Berge *et al.*, 1961; Seamer *et al.*, 1971). The normal production of antibody to SFV and WEE was also delayed, resulting in a chronic encephalitis (Olitsky *et al.*, 1943; Seamer *et al.*, 1971). In the Sindbis virus experiments, the virus was inoculated intracerebrally, normal antibody production was not delayed, and recovery was complete (Griffin and Johnson, 1977). Promotion of recovery after Sindbis virus infection can be mediated by neutralizing E2 antibodies and by neutralizing and nonneutralizing E1 antibodies (Stanley *et al.*, 1986).

The mechanism of protection when antibody is given either before or after virus challenge is not yet clear. To date, all Sindbis virus protective antibodies studied have either neutralized virus, mediated complement-dependent lysis of infected cells, or bound to live virus-infected cells (Schmaljohn *et al.*, 1982; Stanley *et al.*, 1986). Antibody digestion and cytotoxic drug regimens have indicated that a host cell along with both the Fc and the Fab portion of the antibody molecule are necessary for antibody-mediated protection (Cremer *et al.*, 1966; Griffin and Johnson, 1977; Hirsch *et al.*, 1979), suggesting that a mechanism such as antibody-dependent cell-mediated cytotoxicity may be important. Complement does not appear to be necessary *in vivo*, suggesting that complement-mediated cell lysis is not the only mechanism of protection (Hirsch *et al.*, 1979).

f. Enhancement of Infectivity

Antibody to the E1 glycoprotein of Sindbis virus enhances infectivity of this virus for macrophagelike cell lines (Chanas *et al.*, 1982). The importance of this activity for virus replication *in vivo* is not clear, since there is no evidence that Sindbis virus can replicate in primary macrophage cultures or in macrophages *in vivo* (Lagwinska *et al.*, 1975; Mokhtarian *et al.*, 1982). Such a mechanism could potentially be of more im-

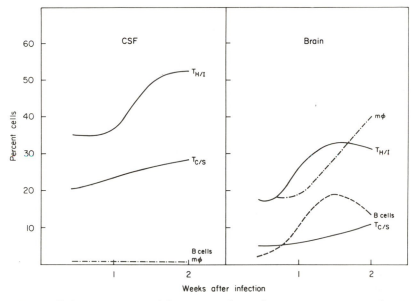

FIGURE 6. Cellular composition of the mononuclear inflammatory response in the CSF and brain at various times after intracerebral inoculation of Sindbis virus. The cells identified are T lymphocytes of the helper/inducer ($T_{H/I}$) and cytotoxic/suppressor ($T_{C/S}$) phenotypes, B lymphocytes, and macrophages (mϕ). Adapted from Moench and Griffin (1984).

portance in VEE virus infections, since VEE virus, but not WEE or EEE virus, does replicate in monocytes (Levitt *et al.*, 1979).

g. Local Antibody Production

In addition to the production of systemic antibody, there is evidence that antibody is also produced locally at the sites of virus replication. Studies of WEE (Schlesinger *et al.*, 1944; Schlesinger, 1949) and Sindbis (Griffin, 1981) virus encephalitis have demonstrated that antibody is produced locally in the brain after CNS infection. This antibody is presumably produced by alphavirus-committed B cells that migrate into the perivascular cuff areas of the brain parenchyma beginning 7–8 days after virus infection (Moench and Griffin, 1984) (Fig. 6). These B cells are thought to differentiate into plasma cells, which then produce virus-specific antibody at the site of virus replication. Locally produced antibody can readily be measured in the CSF and includes antibodies of the IgA, as well as the IgG, class (Griffin, 1981) (Fig. 7). Virus-specific antibody can be measured in the CSF beginning approximately 1 week after infection at a time when virus has usually been cleared from the brain. Antibody synthesis continues for several weeks after recovery (Schlesinger, 1949; Griffin, 1981). The role or importance of IgA or IgG antibodies in the CNS for recovery from infection has not been determined. However, the presence of antiviral antibody in the CNS is important for sub-

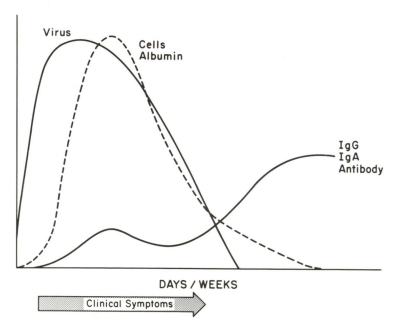

FIGURE 7. Schematic representation of the magnitude of the changes occurring within the CNS during active infection and during recovery from alphavirus-induced encephalitis. The course of the disease and recovery period may extend over days to weeks.

sequent resistance to intracerebral virus challenge in experimental animals (Morgan *et al.*, 1942).

h. Effect of Virus Infection on Nonviral Immune Responses

In addition to the generation of an antibody response to viral polypeptides, infection with the avirulent strain of VEE virus affects antibody responses to other antigens. Mice and guinea pigs challenged with a heterologous antigen within 2 weeks after VEE virus infection show more rapid clearance of, and increased antibody production to, the challenge antigen (Craig *et al.*, 1969; Howard *et al.*, 1969; Hruskova *et al.*, 1972). This phenomenon is associated with proliferation of reticuloendothelial cells and enhanced phagocytosis (Staab *et al.*, 1970). If VEE virus infection occurs shortly after antigenic challenge, antibody responses are decreased (Airhart *et al.*, 1969; Hruskova *et al.*, 1972).

2. Cellular Immunity

Cellular immune responses have been more complex to analyze than antibody responses, since they encompass the activities of a variety of T cells (helper/inducer, cytotoxic, and suppressor T lymphocytes). In turn, these T cells are necessary for B cells to mature into antibody-producing cells and for monocytes to enter sites of viral replication and mature into macrophages. All these responses are restricted by antigens of the major histocompatibility complex (MHC), which must be present on the interacting cell populations. Since MHC recognition is necessary in addition to specific antigen recognition for T-lymphocyte activation, these cells are incapable of recognizing and interacting with free virus as antigen and can recognize viral antigen only on the surface of cells that also display MHC antigens. Thus, the cellular immune response is designed for controlling virus replication at tissue sites of replication, rather than cell-free virus in the blood, CSF, or interstitial fluid (Doherty, 1984).

a. Lymphoproliferation

Alphavirus-specific lymphoproliferative responses are detectable in cells from lymph nodes draining the site of infection or in spleen cells 3–4 days after virus inoculation, at approximately the same time that antibody responses appear (Adler and Rabinowitz, 1973; Griffin and Johnson, 1973; Hirsch and Griffin, 1979; Milner *et al.*, 1984) (see Fig. 5). Virus-specific lymphoproliferation is dependent on macrophages for antigen presentation, but primarily reflects the recognition of antigen by the helper subset of T cells. Lymphoproliferation is demonstrable in lymphocytes from mice of all ages (Griffin, 1976). The lymphoproliferative response is transient, returning completely to baseline levels in mice within 2–3 weeks after infection, but can be promptly restimulated by rechallenge with the same virus (Griffin and Johnson, 1973; Adler and Rabinowitz, 1973) (Fig. 5). This time–course is more prolonged in man, in whom proliferative responses in peripheral blood lymphocytes can be measured for weeks after infection with live VEE virus vaccine (Marker and Ascher, 1976).

b. T-Cell Cytotoxicity

Cytotoxic T cells are generated after infection with most alphaviruses. The time–course for the appearance (3–4 days after infection), disappearance (2–3 weeks after infection), and restimulation of these cells is similar to that for the lymphoproliferative response (Rodda and White, 1976; Peck *et al.*, 1979; McFarland, 1974; Wolcott *et al.*, 1982b; Milner *et al.*, 1984; Blackman and Morris, 1984). The specific viral antigens on the surface of infected cells that are recognized by cytotoxic T cells have not been definitively identified. It is known, however, that cytotoxic T lymphocytes are cross-reactive among Sindbis virus, Bebaru virus, and SFV (Peck *et al.*, 1979; Mullbacher *et al.*, 1979; Wolcott *et al.*, 1982b), while neutralizing antibodies are not. These observations suggest that one of the antigens is likely to be the E1 protein, since this protein is known to contain cross-reacting, protective epitopes (Wolcott *et al.*, 1982a).

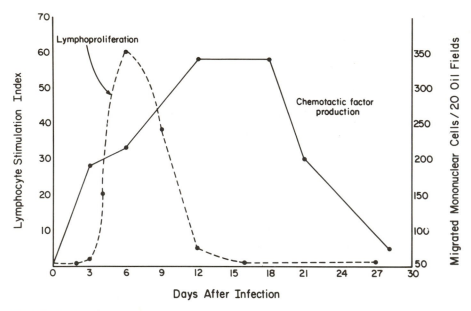

FIGURE 8. Lymphoproliferation and the production of a lymphokine mononuclear chemotactic factor *in vitro*, by lymphocytes taken from the draining lymph node at various times after subcutaneous infection with Sindbis virus (Griffin *et al.*, 1983).

c. Lymphokine Production

Helper/inducer T cells are capable of producing a number of biologically active soluble factors (lymphokines) after stimulation by antigen (Oppenheim, 1981). These include macrophage inhibiting factor, γ-interferon, B-cell proliferation and differentiation factors, interleukin 2, mononuclear chemotactic factor, and others. Production of mononuclear chemotactic factor (Mokhtarian *et al.*, 1982) and γ-interferon (Glasgow, 1966; Blackman and Morris, 1984) has been specifically demonstrated in response to alphavirus stimulation, but the others are undoubtedly produced as well. The time–course of T-cell lymphokine production that is demonstrable *in vitro* by cells from infected mice is more prolonged than lymphoproliferation or cytotoxicity, lasting for 4–5 weeks after virus infection (Fig. 8) (Mokhtarian *et al.*, 1982; Griffin *et al.*, 1983).

d. Generation of Inflammation

The mononuclear inflammatory response that is first detectable at 3–4 days after infection (Berge *et al.*, 1961; R. T. Johnson, 1971; Doherty, 1973; Chew-Lim, 1975; Suckling *et al.*, 1978a; Anders *et al.*, 1979) (see Figs. 3 and 4) in virus-infected tissues is another manifestation of the virus-specific cellular immune response (McFarland *et al.*, 1972). Inflammatory cells also appear in the CSF (Doherty, 1973; Griffin, 1981). In the

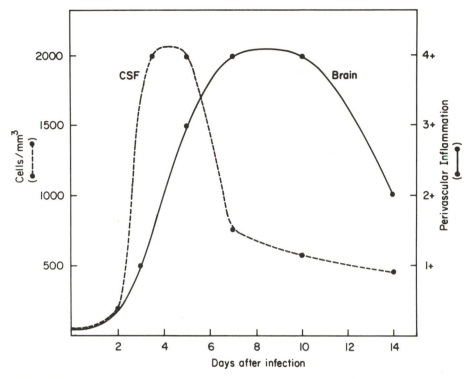

FIGURE 9. Time–course of the development of the mononuclear inflammatory response in the CSF and brains of mice inoculated intracerebrally with Sindbis virus. Adapted from Moench and Griffin (1984).

CNS during Sindbis-virus-induced encephalitis, the cell populations in the perivascular regions of the brain parenchyma are more varied than the cells present in the CSF (see Fig. 6). The time–course of appearance and resolution of inflammation at these two sites is also distinct (Moench and Griffin, 1984). The inflammatory infiltrate in CSF peaks and resolves before that of the brain parenchyma (Fig. 9). The CSF inflammatory cells are primarily T lymphocytes (Moench and Griffin, 1984) and NK cells (Griffin and Hess, 1986). Inflammatory cells in the brain parenchymal perivascular cuffs are a mixture of T lymphocytes, B lymphocytes, and macrophages. NK cell activity has not been assessed, but may contribute to the unidentified cells early. T cells of the helper/inducer phenotype predominate over those of the cytotoxic/suppressor phenotype. The percentage of B cells and macrophages increases at later time points (Moench and Griffin, 1984).

The generation of the mononuclear inflammatory response in the CNS is complicated by the presence of the blood–brain barrier. It has been suggested that the earliest-arriving virus-specific helper T lymphocytes are stimulated by antigen present in infected tissue to produce a variety of lymphokines. Some of these soluble substances may stimulate mast cells to open endothelial tight junctions (Mokhtarian and Griffin, 1984), while others attract monocytes from the peripheral blood (Mokhtarian *et al.*, 1982) and stimulate the appearance and local differentiation of B lymphocytes (Griffin, 1984). The B lymphocytes that accumulate locally are likely to be the source of the immunoglobulin that appears in the CSF beginning 7–8 days after infection (Griffin, 1981) (see Fig. 7). This mononuclear inflammatory reaction is analogous in cell type and time-course to the classic delayed-type hypersensitivity reaction. A footpad-swelling test has been used as a measure of delayed-type hypersensitivity in mice infected with SFV (Kraaijeveld *et al.*, 1979b).

e. Role of Cell-Mediated Immunity in Virus Clearance and Recovery from Infection

Identification of a role for cellular immune responses distinct from antibody is always fraught with difficulty, since antibody is produced by cells (B lymphocytes) and a subpopulation of T cells (helper/inducer cells) is necessary for optimal production of antibody to most viral antigens. Two types of experiments have been done to assess the relative roles of antibody and cellular immunity in recovery from infection: (1) infection with avirulent strains of virus of animals deficient, either genetically or by experimental manipulation, in immune function and (2) passive transfer of various components of the immune response to animals infected with virulent alphavirus strains prior to the onset of the normal immune response. In both, the effect of the experimental manipulation on the outcome of infection can be analyzed. A second issue that is addressed directly or indirectly by many of these experiments is whether any aspect of the disease produced by infection is actually due to the immune response.

For avirulent infections such as that caused by Sindbis virus, there is little evidence that the inflammatory response or the cellular immune response in general plays an important role in recovery from infection. Athymic nude mice, which lack mature T lymphocytes, have no detectable inflammation in the brain, and virus is cleared from the CNS in a manner indistinguishable from that of normal mice (Hirsch and Griffin, 1979). Sindbis-virus-infected athymic nude mice do not, however, develop cross-protective immunity to SFV (Latif *et al.*, 1979). IgM antibody is produced normally (Hirsch and Griffin, 1979; Latif *et al.*, 1979), and NK cells appear in the CSF (Griffin and Hess, 1986). As with Sindbis virus, athymic nude mice infected with avirulent strains of VEE virus clear infection from blood and brain similarly to their normal littermates. Athymic mice infected with avirulent strains of SFV clear virus from the blood normally and remain asymptomatic, but do not clear virus from the brain (Jagelman *et al.*, 1978; Bradish *et al.*, 1979). Depletion of T cells by thymectomy and irradiation also eliminates the inflammatory response without altering virus clearance when compared with control mice (Park *et al.*, 1980; Griffin *et al.*, 1983). In contrast, virulent strains of VEE virus and SFV kill athymic nude mice faster than normal mice, although amounts of virus in target organs are not identifiably different (Doherty, 1973; LeBlanc *et al.*, 1978).

Treatment with cyclophosphamide delays virus clearance from the brain, but does not otherwise alter the benign character of infection with Sindbis virus or avirulent strains of SFV and WEE (Thind and Price, 1969; McFarland *et al.*, 1972; Bradish *et al.*, 1975; Suckling *et al.*, 1977; Berger, 1980). Cyclophosphamide-induced immunosuppression abrogates both the inflammatory response and the antibody response. Both return together 12–14 days after a single dose of cyclophosphamide, at which time virus is cleared (McFarland *et al.*, 1972; Berger, 1980). Similar results have been obtained in mice using other forms of immunosuppression such as irradiation (Smillie *et al.*, 1973; Chew-Lim *et al.*, 1977b).

Cyclophosphamide-treated hamsters infected with WEE or VEE viruses have delayed virus clearance, but also a higher mortality due to infection (Zlotnick *et al.*, 1972b; Jahrling *et al.*, 1974). In addition, hamsters that survive acute WEE virus infection develop subacute or chronic sequelae localized in the olfactory cortical regions of the brain (Zlotnick *et al.*, 1972b).

For the virulent neuroadapted strain of Sindbis virus, cyclophosphamide has little effect on the mortality or the mean day of death (Hirsch *et al.*, 1979). Passive transfer of splenic lymphocytes from immune animals does not protect these infected mice, but antiserum alone will (Griffin and Johnson, 1977). In studies with virulent strains of VEE virus, passive transfer of splenic T lymphocytes or immune serum protects mice from the fatal infection (Rabinowitz and Adler, 1973). Sensitized cells from spleen or lymph nodes inhibit the growth of VEE and Sindbis viruses *in vitro* (Rabinowitz and Proctor, 1974; McFarland, 1974). The effector cell in these *in vitro* systems was a macrophage, but for VEE virus, the antiviral activity of the macrophage was dependent on the presence of immune T cells (Rabinowitz and Proctor, 1974). In contrast, antithymocyte serum delayed slightly (≈ 2 days) the deaths of mice caused by VEE, possibly due to depletion of cells for viral growth (Woodman *et al.*, 1975).

Studies with SFV and RRV have also addressed the role of the immune response in the late demyelinating lesions. A number of studies using various modes of immunosuppression and the avirulent A7 strain of SFV suggest that demyelination is immunologically mediated (Jagelman *et al.*, 1978; Suckling *et al.*, 1978b; Berger, 1980; Fazakerley *et al.*, 1983). Chew-Lim (1979), however, found increased evidence of demyelination in athymic nude mice compared to immunocompetent mice, suggesting that virus replication is primarily responsible for the demyelination. Immunosuppression with cyclophosphamide does not alter the demyelination induced by RRV (Seay and Wolinsky, 1982), and extensive demyelination is induced by the SFV mutant virus (M9) despite immunosuppression (Gates *et al.*, 1984). Resolution of this controversy may come from more detailed studies combining virus titration, pathology, and documented immunosuppression.

IV. SUMMARY

Alphaviruses cause a wide range of diseases in animals and man, but generally tend to infect muscle, brain, reticuloendothelial system tissue, and/or joints as target tissues. Disease may be so mild as to be clinically inapparent or may cause death in a few days. Infections are associated with the prompt development of antibodies to all viral proteins, and the antibodies display multiple biologically important properties (e.g., neutralization, protection, complement-fixation, hemagglutination-inhibition). Infection is also associated with a mononuclear inflammatory response indicative of the cellular immune response to the virus. Recovery from infection correlates best with development of antiviral antibody, rather than with virus-specific cellular immunity.

ACKNOWLEDGMENTS.   Work from this laboratory was supported by Research Grants NS-18596 and NS-15721 from the National Institutes of Health.

REFERENCES

Aaskov, J. G., Davies, C. E. A., Tucker, M., and Dalglish, D., 1981a, Effect on mice of infection during pregnancy with three Australian arboviruses, *Am. J. Trop. Med. Hyg.* **30**:198–203.

Aaskov J. G. Mataika, J. U., Lawrence, G. W. Rabukawaga, V., Tucker, M.M. Miles, J. A. R., and Daglish, D. A. 1981b, An epidemic of Ross River virus infection in Fiji, 1979, *Am. J. Trop. Med. Hyg.* **30**:1053–1059.

Aaskov, J. G., Nair, K., Lawrence, G. W., Dalglish, D. A., and Tucker, M., 1981c, Evidence for transplacental transmission of Ross River virus in humans, *Med. J. Aust.* **2**:20–21.

Aaskov, J. G., Ross, P., Davies, C. E. A., Innis, M. D., Guard, R. W., Stallman, N. D., and Tucker, M., 1981d, Epidemic polyarthritis in northeastern Australia, 1978–1979, *Med. J. Aust.* **2**:17–19.

Aaskov, J. G., Hadding, U., and Bitter-Suermann, 1985, Interaction of Ross River virus with the complement system, *J. Gen. Virol.* **66**:121–129.

Adler, W. H., and Rabinowitz, S. G., 1973, Host defenses during primary Venezuelan equine encephalomyelitis virus infection in mice. II. *In vitro* methods for the measurement and qualitation of the immune response, *J. Immunol.* **110**:1354–1362.

Aguilar, M. J., 1970, Pathological changes in brain and other target organs of infant and weanling mice after infection with non-neuroadapted Western equine encephalitis virus, *Infect. Immun.* **2**:533–542.

Airhart, J. W., Trevino, G. S., and Craig, C. P., 1969, Alterations in immune responses by attenuated Venezuelan equine encephalitis vaccine. II. Pathology and soluble antigen localization in guinea pigs, *J. Immunol.* **102**:1228–1234.

Anders, E. M., Miller, J. F. A. P., and Gamble, J., 1979, A radioisotopic technique for measuring the mononuclear inflammatory response in Sindbis virus-induced encephalitis of mice, *J. Immunol. Methods* **29**:167–171.

Anderson, S. G., Doherty, R. L., and Carley, J. G., 1961, Epidemic polyarthritis: Antibody to a group A arthropod-borne virus in Australia and the island of New Guinea, *Med. J. Aust.* **1**:273–276.

Atkins, G. J., 1983, The avirulent A7 strain of Semliki Forest virus has reduced cytopathogenicity for neuroblastoma cells compared to the virulent L10 strain, *J. Gen. Virol.* **64**:1401–1404.

Atkins, G. J., and Sheahan, B. J., 1982, Semliki Forest virus neurovirulence mutants have altered cytopathogenicity for central nervous system cells, *Infect. Immun.* **36**:333–341.

Atkins, G. J., Carter, J., and Sheahan, B. J., 1982, Effect of alphavirus infection on mouse embryos, *Infect. Immun.* **38**:1285–1290.

Austin, F. J., and Scherer, W. F., 1971, Studies of viral virulence. I. Growth and histopathology of virulent and attenuated strains of Venezuelan encephalitis virus in hamsters, *Am. J. Pathol.* **62**:195–209.

Baric, R. S., Trent, D. W., and Johnston, R. E., 1981, A Sindbis virus variant with a cell-determined latent period, *Virology* **110**:237–242.

Baron, S., and Buckler, C. E., 1963, Circulating interferon in mice after intravenous injection of virus, *Science* **144**:1061–1063.

Barrett, A. D. T., and Dimmock, N. J., 1984a, Modulation of a systemic Semliki Forest virus infection in mice by defective interfering virus, *J. Gen. Virol.* **65**:1827–1831.

Barrett, A. D. T., and Dimmock, N. J., 1984b, Properties of host and virus which influence defective interfering virus mediated-protection of mice against Semliki Forest virus lethal encephalitis, *Arch. Virol.* **81**:185–188.

Barrett, P. N., Sheahan, B. J., and Atkins, G. J., 1980, Isolation and preliminary characterization of Semliki Forest virus mutants with altered virulence, *J. Gen. Virol.* **49**:141–147.

Bastian, F. O., Wende, R. D., Singer, D. B., and Zeller, R. S., 1975, Eastern equine encephalomyelitis: Histopathologic and ultrastructural changes with isolation of the virus in a human case, *Am. J. Clin. Pathol.* **64**:10–13.

Bell, J. R., Kinney, R. M., Trent, D. W., Strauss, E. G., and Strauss, J. H., 1984, An evolutionary tree relating eight alphaviruses, based on amino-terminal sequences of their glycoproteins, *Proc. Natl. Acad. Sci. U.S.A.* **81**:4702–4706.

Berge, T. O., Gleiser, C. A., Gochenour, W. S., Jr., Miesse, M. L., and Tigertt, W. P., 1961, Studies on the virus of Venezuelan equine encephalomyelitis. II. Modification by specific immune serum of response of central nervous system of mice, *J. Immunol.* **87**:509–517.

Berger, M. L., 1980, Humoral and cell-mediated immune mechanisms in the production of pathology in virulent Semliki Forest virus encephalitis, *Infect. Immun.* **30**:244–253.

Blackman, M. J., and Morris, A. G., 1984, Gamma interferon production and cytotoxicity of spleen cells from mice infected with Semliki Forest virus. *J. Gen. Virol.* **65**:955–961.

Boere, W. A. M., Benaissa-Trouw, B. J., Harmsen, M., Kraaijeveld, C. A., and Snippe, H., 1983, Neutralizing and non-neutralizing monoclonal antibodies to the E_2 glycoprotein of Semliki Forest virus can protect mice from lethal encephalitis, *J. Gen. Virol.* **64**:1405–1408.

Boere, W. A. M., Harmsen, T., Vinje, J., Benaissa-Trouw, B. J., Kraaijeveld, C. A., and Snippe, H., 1984, Identification of distinct antigenic determinants on Semliki Forest virus by using monoclonal antibodies with different antiviral activities, *J. Virol.* **52**:575–582.

Boone, L. R., and Brown, A., 1976, Variants of the HR strain of Sindbis virus lethal for mice, *J. Gen. Virol.* **31**:261–263.

Bradish, C. J., and Allner, K., 1972, The early responses of mice to respiratory or intraperitoneal infection by defined virulent and avirulent strains of Semliki Forest virus, *J. Gen. Virol.* **15**:205–218.

Bradish, C. J., and Titmuss, D., 1981, The effects of interferon and double-stranded RNA upon the virus–host interaction: Studies with togavirus strains in mice, *J. Gen. Virol.* **53**:21–30.

Bradish, C. J., Allner, K., and Maber, H. B., 1971, The virulence of original and derived strains of Semliki Forest virus for mice, guinea pigs and rabbits, *J. Gen. Virol.* **12**:141–160.

Bradish, C. J., Allner, K., and Fitzgeorge, R., 1975, Immunomodification and the expression of virulence in mice by defined strains of Semliki Forest virus: The effects of cyclophosphamide, *J. Gen. Virol.* **28**:225–237.

Bradish, C. J., Fitzgeorge, R., Titmuss, D., and Baskerville, A., 1979, The responses of nude-athymic mice to nominally avirulent togavirus infections, *J. Gen. Virol.* **42**:555–566.

Brown, A., and Officer, J. E., 1975, An attenuated variant of Eastern encephalitis virus: Biological properties and protection induced in mice, *Arch. Virol.* **47**:123–138.

Brown, A., Vosdingh, R., and Zebovitz, E., 1975, Attenuation and immunogenicity of ts mutants of Eastern encephalitis virus for mice, *J. Gen. Virol.* **27**:111–116.

Burns, W. H., Billups, L. C., and Notkins, A. L., 1975, Thymus dependence of viral antigens, *Nature (London)* **256**:654–655.

Carey, D. E., Myers, R. M., DeRainitz, C. M., Jadhav, M., and Reuben, R., 1969, The 1964 Chikungunya epidemic at Vellore, South India, including observations on concurrent dengue, *Trans. R. Soc. Trop. Med. Hyg.* **63**:434–445.

Carreras, B., Griffin, D. E., and Silverstein, A. M., 1982, Sindbis virus-induced ocular immunopathology, *Invest. Ophthalmol. Vis. Sci.* **22**:571–578.

Casals, J., 1963, Relationships among arthropod-borne animal viruses determined by cross-challenge tests, *Am. J. Trop. Med. Hyg.* **12**:587–596.

Chamberlain, R. W., 1980, Epidemiology of arthropod-borne togaviruses: The role of arthropods as hosts and vectors and of vertebrate hosts in natural transmission cycles, in: *The Togaviruses: Biology, Structure, Replication* (R. W. Schlesinger, ed.), pp. 175–227, Academic Press, New York.

Chamberlain, R. W., Kissling, R. E., Stamm, D. D., Nelson, D. B., and Sikes, R. K., 1956, Venezuelan equine encephalomyelitis in wild birds, *Am. J. Hyg.* **63**:261–273.

Chanas, A. C., Gould, E. A., Clegg, J. C. S., and Varma, M. G. R., 1982, Monoclonal antibodies to Sindbis virus glycoprotein E1 can neutralize, enhance infectivity, and independently inhibit haemagglutination or haemolysis, *J. Gen. Virol.* **58**:37–46.

Chew-Lim, M., 1975, Mouse encephalitis induced by avirulent Semliki Forest virus, *Vet. Pathol.* **12**:387–393.

Chew-Lim, M., 1979, Brain viral persistence and myelin damage in nude mice, *Can. J. Comp. Med.* **43**:39–43.

Chew-Lim, M., Suckling, A. J., and Webb, H. E., 1977a, Demyelination in mice after two or three infections with avirulent Semliki Forest virus, *Vet. Pathol.* **14**:67–72.

Chew-Lim, M., Webb, H. E., and Jagelman, S., 1977b, The effect of irradiation on demyelination induced by avirulent Semliki Forest virus, *Br. J. Exp. Pathol.* **58**:459–464.

Cole, F. E., Jr., and McKinney, R. W., 1971, Cross-protection in hamsters immunized with group A arbovirus vaccines, *Infect. Immun.* **4**:37–41.

Copps, S. C., and Giddings, L. E., 1959, Transplacental transmission of Western equine encephalitis: Report of a case, *Pediatrics* **24**:31–33.

Craig, C. P., Reynolds, S. L., Airhart, J. W., and Staab, E. V., 1969, Alterations in immune responses by attenuated Venezuelan equine encephalitis vaccine. I. Adjuvant effect of VEE virus infection in guinea pigs, *J. Immunol.* **102**:1220–1227.

Cremer, N. E., Lennette, E. H., Hagens, S. J., and Fujimoto, F. Y., 1966, Difference in mechanism of viral neutralization under *in vitro* and *in vivo* conditions, *J. Immunol.* **96**:284–288.

Crouch, C. F., Mackenzie, A., and Dimmock, N. J., 1982, The effect of defective-interfering Semliki Forest virus on the histopathology of infection with virulent Semliki Forest virus in mice, *J. Infect. Dis.* **146**:411–416.

Dal Canto, M. C., and Rabinowitz, S. G., 1981, Central nervous system demyelination in

Venezuelan equine encephalomyelitis infection: An experimental model of virus-induced myelin injury, *J. Neurol. Sci.* **49:**397–418.

Dalrymple, J. M., Schlesinger, S., and Russell, P. K., 1976, Antigenic characterization of two Sindbis envelope glycoproteins separated by isoelectric focusing, *Virology* **69:**93–103.

Danes, L., Kufner, J., Hruskova, J., and Rychterova, V., 1973, The role of the olfactory route on infection of the respiratory tract with Venezuelan equine encephalomyelitis virus in normal and operated *Macaca rhesus* monkeys. I. Results of virological examination, *Acta Virol.* **17:**50–56.

Davis, W. A., 1940, Study of birds and mosquitoes as hosts for virus of Western equine encephalomyelitis, *Am. J. Hyg. Sect. C* **32:**45–59.

Dill, G. S., Pederson, C. E., and Stookey, J. L., 1973, A comparison of the tissue lesions produced in adult hamsters by two strains of avirulent Venezuelan equine encephalomyelitis virus, *Am. J. Pathol.* **72:**13–24.

Dimmock, N. J., and Kennedy, S. T., 1978, Prevention of death in Semliki Forest virus-infected mice by administration of defective interfering Semliki Forest virus, *J. Gen. Virol.* **39:**231–242.

Doby, P. B., Schnurrenberger, P. R., Martin, R. J., Hanson, L. E., Sherrick, G. W., and Schoenholz, W. K., 1966, Western encephalitis in Illinois horses and ponies, *J. Am. Vet. Med. Assoc.* **148:**422–427.

Doherty, P. C., 1973, Quantitative studies of the inflammatory process in fatal viral meningoencephalitis, *Am. J. Pathol.* **73:**607–622.

Doherty, P. C., 1984, Clearance of experimental viral infections of the central nervous system, in: *Serono Symposia*, Vol. 12, *Neuroimmunology* (P. O. Behan and F. Spreafico, eds.), pp. 301–310, Raven Press, New York.

Doherty, R. L., 1977, Arthropod-borne viruses in Australia, *Aust. J. Exp. Biol. Med. Sci.* **55:**103–130.

Doherty, R. L., Anderson, S. G., Aaron, K., Farnworth, J. K., Knyvett, A. F., and Nimmo, D., 1961, Clinical manifestations of infection with group A arthropod-borne viruses in Queensland, *Med. J. Aust.* **1:**276–279.

Doherty, R. L., Gorman, B. M., Whitehead, R. H., and Carley, J. G., 1964, Studies of epidemic polyarthritis: The significance of three group-A arboviruses isolated from mosquitoes in Queensland, *Aust. Ann. Med.* **13:**322–327.

Doherty, R. L., Bodey, A. S., and Carew, J. S., 1969, Sindbis virus infection in Australia, *Med. J. Aust.* **2:**1016–1017.

Dremov, D. P., Solyanik, R. G., Miryutova, T. L., and Laptokova, L. M., 1978, Attenuated variants of Eastern equine encephalomyelitis virus: Pathomorphological, immunofluorescence and virological studies of infection in Syrian hamsters, *Acta Virol.* **22:**139–145.

Earnest, M. P., Goolishian, H. A., Calverley, J. R., Hayes, R. O., and Hill, H. R., 1971, Neurologic, intellectual and psychologic sequelae following western encephalitis: A follow-up of 35 cases, *Neurology* **21:**969–974.

Ehrenkranz, N. J., and Ventura, A. K., 1974, Venezuelan equine encephalitis virus infection in man, *Annu. Rev. Med.* **25:**9–14.

Eisner, R. J., and Nusbaum, S. R., 1983, Encephalitis vaccination of pheasants: A question of efficacy, *J. Am. Vet. Med. Assoc.* **183:**280–281.

Emini, E. A., and Wiebe, M. E., 1981, An attenuated mutant of Venezuelan encephalitis virus: Biochemical alterations and their genetic association with attenuation, *Virology* **110:**185–196.

Espmark, A., and Niklasson, B., 1984, Ockelbo disease in Sweden: Epidemiological, clinical and virological data from the 1982 outbreak, *Am. J. Trop. Med. Hyg.* **33:**1203–1211.

Farber, S., Hill, A., Connerly, M. L., and Dingle, J. H., 1940, Encephalitis in infants and children caused by the virus of the Eastern variety of equine encephalitis, *J. Am. Med. Assoc.* **114:**1725–1731.

Fazakerley, J. K., Amor, S., and Webb, H. E., 1983, Reconstitution of Semliki Forest virus infected mice induces immune mediated pathological changes in the CNS, *Clin. Exp. Immunol.* **52:**115–120.

Fearon, D. T., 1979, Activation of the alternative complement pathway, *Crit. Rev. Immunol.* **1:**1–32.

Feemster, R. F., 1957, Equine encephalitis in Massachusetts, *N. Engl. J. Med.* **257:**701–704.

Finley, K. H., Longshore, W. A., Palmer, R. J., Cook, R. E., and Riggs, N., 1955, Western equine and St. Louis encephalitis: Preliminary report of a clinical followup study in California, *Neurology* **5:**223–235.

Finter, N. B., 1966, Interferon as an antiviral agent *in vivo:* Quantitative and temporal respects of the protection of mice against Semliki Forest virus, *Br. J. Exp. Pathol.* **47:**361–369.

Fothergill, L. D., and Dingle, J. H., 1938, Fatal disease of pigeons caused by the virus of the Eastern variety of equine encephalitis, *Science* **88:**549–550.

France, J. K., Wyrick, B. C., and Trent, D. W., 1979, Biochemical and antigenic comparisons of the envelope glycoproteins of Venezuelan equine encephalomyelitis strains, *J. Gen. Virol.* **44:**725–740.

Garcia-Tamayo, J., 1973, Venezuelan equine encephalomyelitis virus in the heart of newborn mice, *Arch. Pathol.* **96:**294–297.

Garcia-Tamayo, J., Carreno, G., and Esparza, J., 1979, Central nervous system alterations in sequelae of Venezuelan equine encephalitis virus infection in the rat, *J. Pathol.* **128:**87–91.

Garcia-Tamayo, J., Esparza, J., and Martinez, J., 1981, Placental and fetal alterations due to Venezuelan equine encephalitis virus in rats, *Infect. Immun.* **32:**813–821.

Gates, M. D., Brown, A., and Wust, C. J., 1982, Comparison of specific and cross reactive antigens of alphaviruses on virions and infected cells, *Infect. Immun.* **35:**248–255.

Gates, C., Sheahan, B. J., and Atkins, G. J., 1984, The pathogenicity of the M9 mutant of Semliki Forest virus in immunocompromised mice, *J. Gen. Virol.* **65:**73–80.

Gidlund, M., Orn, A., Wigzell, H., Senik, A., and Gresser, I., 1978, Enhanced NK cell activity in mice injected with interferon and interferon inducers, *Nature (London)* **273:**759–761.

Gilden, D. H., Cole, G. A., and Nathanson, N., 1972, Immunopathogenesis of acute central nervous system disease produced by lymphocytic choriomeningitis virus. II. Adoptive immunization of virus carriers, *J. Exp. Med.* **135:**874–889.

Glasgow, L. A., 1966, Leukocytes and interferon in the host response to viral infections. II. Enhanced interferon response of leukocytes from immune animals, *J. Bacteriol.* **91:**2185–2191.

Gleiser, C. A., Cochenour, W. S., Berge, T. O., and Tigertt, W. D., 1962, The comparative pathology of experimental Venezuelan equine encephalomyelitis infection in different animal hosts, *J. Infect. Dis.* **110:**80–97.

Goldfield, M., Welsh, J. N., and Taylor, B. F., 1968, The 1959 outbreak of Eastern encephalitis in New Jersey. 5. The inapparent infection:disease ratio, *Am. J. Epidemiol.* **87:**32–38.

Gorelkin, L., 1973, Venezuelan equine encephalomyelitis in an adult animal host: An electron microscopy study, *Am. J. Pathol.* **73:**425–434.

Gorelkin, L., and Jahrling, P. B., 1974, Pancreatic involvement by Venezuelan equine encephalomyelitis virus in the hamster, *Am. J. Pathol.* **75:**349–362.

Gorelkin, L., and Jahrling, P. B., 1975, Virus-initiated septic shock: Acute death of Venezuelan encephalitis virus-infected hamsters, *Lab. Invest.* **32:**78–85.

Grayson, M. A., and Galindo, P., 1968, Epidemiologic studies of Venezuelan equine encephalitis virus in Almirante, Panama, *Am. J. Epidemiol.* **88:**80–96.

Griffin, D. E., 1976, Role of the immune response in age-dependent resistance of mice to encephalitis due to Sindbis virus, *J. Infect. Dis.* **133:**456–464.

Griffin, D. E., 1981, Immunoglobulins in the cerebrospinal fluid: Changes during acute viral encephalitis in mice, *J. Immunol.* **126:**27–31.

Griffin, D. E., 1984, The inflammatory response to acute viral infections, in: *Concepts in*

Viral Pathogenesis (A. L. Notkins and M. B. A. Oldstone, eds.), pp. 46–52, Springer-Verlag, New York.

Griffin, D. E., and Hess, J. L., 1986, Cells with natural killer activity in the CSF of normal and athymic nude mice with acute Sindbis virus encephalitis *J. Immunol.* (in press).

Griffin, D. E., and Johnson, R. T., 1973, Cellular immune response to viral infection: *In vitro* studies of lymphocytes from mice infected with Sindbis virus, *Cell. Immunol.* **9:**426–434.

Griffin, D. E., and Johnson, R. T., 1977, Role of the immune response in recovery from Sindbis virus encephalitis in mice, *J. Immunol.* **118:**1070–1075.

Griffin, D. E., Mokhtarian, F., Park, M. M., and Hirsch, R. L., 1983, Immune responses to acute alphavirus infection of the central nervous system: Sindbis virus encephalitis in mice, in: *Progress in Brain Research,* Vol. 59, *Immunology of Nervous System Infections* (P. O. Behan, V. ter Meulen, and F. C. Rose, eds.), pp. 11–21, Elsevier, New York.

Grimley, P. M., and Friedman, R. M., 1970, Arboviral infection of voluntary striated muscles, *J. Infect. Dis.* **122:**45–52.

Guard, R. W., McAuliffe, M. J., Stallman, N. D., and Bramston, B. A., 1982, Haemorrhagic manifestations with Sindbis virus: Case report, *Pathology* **14:**89–90.

Hackbarth, S. A., Reinerz, A. B. G., and Sagik, B. P., 1973, Age-dependent resistance of mice to Sindbis virus infection: Reticuloendothelial role, *J. Reticuloendothel. Soc.* **14:**405–425.

Haddow, A. J., Davies, C. W., and Walker, A. J., 1960, O'nyong-nyong fever: An epidemic virus disease in East Africa. I. Introduction, *Trans. R. Soc. Trop. Med. Hyg.* **54:**517–522.

Hart, K. L., Keen, D., and Belle, E. A., 1964, An outbreak of Eastern equine encephalomyelitis in Jamaica, West Indies. I. Description of human cases, *Am. J. Trop. Med.* **13:**331–334.

Hearn, H. J., Jr., 1960, A variant of Venezuelan equine encephalomyelitis virus attenuated for mice and monkeys, *J. Immunol.* **84:**626–629.

Hearn, H. J., and Rainey, C. T., 1963, Cross-protection in animals infected with group A arbovirus, *J. Immunol.* **90:**720–724.

Helenius, A., Fries, E., Garoff, H., and Simons, K., 1976, Solubilization of the Semliki Forest virus membrane with sodium deoxycholate, *Biochim. Biophys. Acta* **436:**319–334.

Henderson, B. E., Chappel, W. A., Johnston, J. G., Jr., and Sudia, W. D., 1971, Experimental infection of horses with three strains of Venezuelan equine encephalitis virus. I. Clinical and virological studies, *Am. J. Epidemiol.* **93:**194–205.

Herberman, R. B., and Holden, H. T., 1978, Natural cell-mediated cytotoxicity, *Adv. Cancer Res.* **27:**305–377.

Hirsch, R. L., 1981, Natural killer cells appear to play no role in the recovery of mice from Sindbis virus infection, *Immunology* **43:**81–89.

Hirsch, R. L., 1982, The complement system: Its importance in the host response to viral infection, *Microbiol. Rev.* **46:**71–85.

Hirsch, R. L., and Griffin, D. E., 1979, The pathogenesis of Sindbis virus infection in athymic nude mice, *J. Immunol.* **123:**1215–1218.

Hirsch, R. L., Griffin, D. E., and Winkelstein, J. A., 1978, The effect of complement depletion on the course of Sindbis virus infection in mice, *J. Immunol.* **121:**1276–1278.

Hirsch, R. L., Griffin, D. E., and Johnson, R. T., 1979, Interactions between immune cells and antibody in protection from fatal Sindbis virus encephalitis, *Infect. Immun.* **23:**320–324.

Hirsch, R. L., Griffin, D. E., and Winkelstein, J. A., 1980a, The role of complement in viral infections. IV. The participation of the terminal complement components (C5–9) in recovery of mice from Sindbis virus infection, *Infect. Immun.* **30:**899–901.

Hirsch, R. L., Winkelstein, J. A., and Griffin, D. E., 1980b, The role of complement in viral infections. III. Activation of the classical and alternative pathways by Sindbis virus, *J. Immunol.* **124:**2507–2510.

Hirsch, R. L., Griffin, D. E., and Winkelstein, J. A., 1981, Host modification of Sindbis virus sialic acid content influences alternative complement pathway activation and virus clearance, *J. Immunol.* **127:**1740–1743.

Hirsch, R. L., Griffin, D. E., and Winkelstein, J. A., 1983, Natural immunity to Sindbis virus is influenced by host tissue sialic acid content, *Proc. Natl. Acad. Sci. U.S.A.* **80:**548–550.

Howard, R. J., Craig, C. P., Trevino, G. S., Dougherty, S. F., and Mergenhagen, S. E., 1969, Enhanced humoral immunity in mice infected with attenuated Venezuelan equine encephalitis virus, *J. Immunol.* **103:**699–707.

Howitt, B. F., 1932, Equine encephalomyelitis, *J. Infect. Dis.* **51:**493–510.

Hruskova, J., Rychterova, V., and Kliment, V., 1972, The influence of infection with Venezuelan equine encephalomyelitis virus on antibody response against sheep erythrocytes. I. Experiments on mice, *Acta. Virol.* **16:**115–124.

Hsieh, P., Rosner, M. R., and Robbins, P. W., 1983, Host-dependent variation of asparagine-linked oligosaccharides at individual glycosylation sites of Sindbis virus glycoproteins, *J. Biol. Chem.* **258:**2548–2554.

Hunt, A. R., and Roehrig, J. T., 1985, Biochemical and biological characteristics of epitopes on the E1 glycoprotein of Western equine encephalitis virus, *Virology* **142:**334–346.

Igarashi, A., Fukuoka, T., and Fukai, K., 1971, Passive immunization of mice with rabbit antisera against Chikungunya virus and its components, *Biken J.* **14:**353–355.

Illavia, S. J., Webb, H. E., and Pathak, S., 1982, Demyelination induced in mice by avirulent Semliki Forest virus. I. Virology and effects on optic nerve, *Neuropathol. Appl. Neurobiol.* **8:**35–42.

Jagelman, S., Suckling, A. J., Webb, H. E., and Bowen, E. T. W., 1978, The pathogenesis of avirulent Semliki Forest virus infections in athymic nude mice, *J. Gen. Virol.* **41:**599–607.

Jahrling, P. B., and Scherer, W. F., 1973a, Histopathology and distribution of viral antigens in hamsters infected with virulent and benign Venezuelan encephalitis viruses, *Am. J. Pathol.* **72:**25–38.

Jahrling, P. B., and Scherer, W. F., 1973b, Growth curves and clearance rates of virulent and benign Venezuelan encephalitis viruses in hamsters, *Infect. Immun.* **8:**456–462.

Jahrling, P. B., Dindy, E., and Eddy, G. A., 1974, Correlates to increased lethality of attenuated Venezuelan encephalitis virus vaccine for immunosuppressed hamsters, *Infect. Immun.* **9:**924–930.

Jahrling, P. B., Hilmas, D. E., and Heard, C. D., 1977, Vascular clearance of Venezuelan equine encephalomyelitis viruses as a correlate to virulence for rhesus monkeys, *Arch. Virol.* **55:**161–164.

Jahrling, P. B., DePaoli, A., and Powanda, M. C., 1978, Pathogenesis of a Venezuelan encephalitis virus strain lethal for adult white rats, *J. Med. Virol.* **2:**109–116.

Jahrling, P. B., Hesse, R. A., Anderson, A. O., and Gangemi, J. D., 1983, Opsonization of alphaviruses in hamsters, *J. Med. Virol.* **12:**1–16.

Johnson, E. D., and Cole, G. A., 1975, Functional heterogeneity of lymphocytic choriomeningitis virus-specific T lymphocytes. I. Identification of effector and memory subsets, *J. Exp. Med.* **141:**866.

Johnson, H. N., 1963, Selection of a variant of Western encephalitis virus of low pathogenicity for study as a live virus vaccine, *Am. J. Trop. Med. Hyg.* **12:**604–610.

Johnson, K. M., Shelokov, A., Peralta, P. H., Dammin, G. J., and Young, N. A., 1968, Recovery of Venezuelan equine encephalomyelitis virus in Panama, *Am. J. Trop. Med. Hyg.* **17:**432–440.

Johnson, R. T., 1965, Virus invasion of the central nervous system: A study of Sindbis virus infection in the mouse using fluorescent antibody, *Am. J. Pathol.* **46:**929–943.

Johnson, R. T., 1966, The incubation period of viral encephalitis, NINDB Monogr. 2, *Slow, Latent and Temperate Virus Infections,* National Institute of Neurological Diseases and Blindness Monogr. 2 (D. C. Gajdusek, C. J. Gibbs, and M. Alpers, eds.), U. S. Govt. Printing Office, Washington, D. C., pp. 119–124.

Johnson, R. T., 1971, Inflammatory response to viral infection, in: *Immunological Disorders of the Nervous System, Res. Publ. Assoc. Res. Nerv. Ment. Dis.* **69:**305–312.

Johnson, R. T., McFarland, H. F., and Levy, S. E., 1972, Age-dependent resistance to viral encephalitis: Studies of infections due to Sindbis virus in mice, *J. Infect. Dis.* **125:**257–262.

Jungherr, E. L., Helmboldt, C. F., Satriano, S. F., and Luginbuhl, R. E., 1957, Investigation of Eastern equine encephalomyelitis. III. Pathology in pheasants and incidental observations in feral birds, *Am. J. Hyg.* **67:**10–20.

Kaluza, G., Rott, R., and Schwarz, R. T., 1980, Carbohydrate-induced conformational changes of Semliki Forest virus glycoproteins determine antigenicity, *Virology* **102:**286–299.

Keegstra, K., and Burke, D., 1977, Comparison of the carbohydrate of Sindbis virus glycoprotein with the carbohydrate of host glycoproteins, *J. Supramol. Struct.* **7:**371–379.

Kennedy, A. C., Fleming, J., and Solomon, L., 1980, Chikungunya viral arthropathy: A clinical description, *J. Rheumatol.* **7:**231–236.

King, B., Wust, C. J., and Brown, A., 1977, Antibody-dependent complement mediated homologous and cross cytolysis of togavirus-infected cells, *J. Immunol.* **119:**1289–1292.

Kissling, R. E., Chamberlain, R. W., Sikes, R. K., and Eidson, M. E., 1954a, Studies on the North American arthropod-borne encephalitides. III. Eastern equine encephalitis in wild birds, *Am. J. Hyg.* **60:**251–265.

Kissling, R. E., Eidson, M. E., and Stamm, D. D., 1954b, Transfer of maternal neutralizing antibodies against Eastern equine encephalomyelitis virus in birds, *J. Infect. Dis.* **95:**179–181.

Kissling, R. E., Chamberlain, R. W., Nelson, D. B., and Stamm, D. D., 1956, Venezuelan equine encephalomyelitis in horses, *Am. J. Hyg.* **63:**274–287.

Kokernot, R. H., Shinefield, H. R., and Longshore, W. A., 1953, The 1952 outbreak of encephalitis in California, *Calif. Med.* **79:**73–77.

Kraaijeveld, C. A., Harmsen, M., and Boutahar-Trouw, B. K., 1979a, Cellular immunity against Semliki Forest virus in mice, *Infect. Immun.* **23:**213–218.

Kraaijeveld, C. A., Harmsen, M., and Boutahar-Trouw, B. K., 1979b, Delayed type hypersensitivity against Semliki Forest virus in mice, *Infect. Immun.* **23:**219–223.

Krieger, J. N., Scherer, W. F., Wiebe, M. E., Pancake, B. A., and Harsayi, Z. P., 1979, A hamster-attenuated, temperature-sensitive mutant of Venezuelan encephalitis virus, *Infect. Immun.* **25:**873–879.

Kundin, W. D., Liu, C., and Rodina, P., 1966, Pathogenesis of Venezuelan equine encephalomyelitis virus. I. Infection in suckling mice, *J. Immunol.* **96:**39–48.

Lagwinska, E., Stewart, C. C., Adles, C., and Schlesinger, S., 1975, Replication of lactic dehydrogenase virus and Sindbis virus in mouse peritoneal macrophages: Induction of interferon and phenotypic mixing, *Virology* **65:**204–214.

Latif, Z., Gates, D., Wust, C. J., and Brown, A., 1979, Cross protection among togaviruses in nude mice and littermates, *J. Gen. Virol.* **45:**89–98.

LeBlanc, P. A., Scherer, W. F., and Susdorf, D. H., 1978, Infections of congenitally athymic (nude) and normal mice with avirulent and virulent strains of Venezuelan encephalitis virus, *Infect. Immun.* **21:**779–785.

Lennette, E. H., and Koprowski, H., 1943, Human infection with Venezuelan equine encephalomyelitis virus: A report on eight cases of infection acquired in the laboratory, *J. Am. Med. Assoc.* **123:**1088–1095.

Lennette, E. H., and Koprowski, H., 1944, Influence of age on the susceptibility of mice to infection with certain neurotropic viruses, *J. Immunol.* **49:**175–191.

Leon, C. A., Jaramillo, R., Martinez, S., Fernandez, F., Tellez, H., Lasso, B., and de Guzman, R., 1975, Sequelae of Venezuelan equine encephalitis in humans: A four year follow-up, *Int. J. Epidemiol.* **4:**131–140.

Levitt, N. H., Miller, H. V., and Edelman, R., 1979, Interaction of alphaviruses with human peripheral leukocytes: *In vitro* replication of Venezuelan equine encephalomyelitis virus in monocyte cultures, *Infect. Immun.* **24:**642–646.

Liu, C., Voth, D. W., Rodina, P., Shauf, L. R., and Gonzalez, G., 1970, A comparative study of the pathogenesis of Western equine and Eastern equine encephalomyelitis viral infections in mice by intracerebral and subcutaneous inoculations, *J. Infect. Dis.* **122:**53–63.

London, W. T., Levitt, N. H., Kent, S. G., Wong, V. G., and Sever, J. L., 1977, Congenital cerebral and ocular malformations induced in rhesus monkeys by Venezuelan equine encephalitis virus, *Teratology* **16:**285–296.

Longshore, W. A., Stevens, I. M., Hollister, A. C., Gittelsohn, A., and Lennette, E. H., 1956, Epidemiologic observations on acute infectious encephalitis in California, with special reference to the 1952 outbreak, *Am. J. Hyg.* **63:**69–86.

Macfarlan, R. I., Burns, W. H., and White, D. O., 1977, Two cytotoxic cells in the peritoneal cavity of virus-infected mice: Antibody-dependent macrophages and non-specific killer cells, *J. Immunol.* **119:**1569–1574.

Mackenzie, A., Suckling, A. J., Jagelman, S., and Wilson, A. M., 1978, Histopathological and enzyme histochemical changes in experimental Semliki Forest virus infection in mice and their relevance to scrapie, *J. Comp. Pathol.* **88:**335–344.

Malherbe, H., Strickland-Cholmley, M., and Jackson, A. L., 1963, Sindbis virus infection in man: Report of a case with recovery of virus from skin lesions, *S. Afr. Med. J.* **37:**547–552.

Marker, S. C., and Ascher, M. S., 1976, Specific *in vitro* lymphocyte transformation with Venezuelan equine encephalitis virus, *Cell. Immunol.* **23:**32–38.

Mathews, J. H., and Roehrig, J. T., 1982, Determination of the protective epitopes on the glycoproteins of Venezuelan equine encephalomyelitis virus by passive transfer of monoclonal antibodies, *J. Immunol.* **129:**2763–2767.

Mayer, V., Ibrahim, A. H., and Gajdosova, E., 1973, Viral infection and resistance in immunosuppressed host. III. Intracerebral challenge with Sindbis virus in immunized or interferon inducer given mice, *Acta Virol.* **17:**29–40.

McFarland, H. F., 1974, *In vitro* studies of cell-mediated immunity in an acute viral infection, *J. Immunol.* **113:**173–180.

McFarland, H. F., Griffin, D. E., and Johnson, R. L., 1972, Specificity of the inflammatory response in viral encephalitis. I. Adoptive immunization of immunosuppressed mice infected with Sindbis virus, *J. Exp. Med.* **136:**216–226.

McIntosh, D. M., McGillivrary, G. M., Dickinson, D. B., and Malherbe, H., 1964, Illness caused by Sindbis and West Nile viruses in South Africa, *S. Afr. Med. J.* **38:**291–294.

Mecham, J. O., and Trent, D. W., 1982, Glycosylation patterns of the envelope glycoproteins of an equine-virulent Venezuelan encephalitis virus and its vaccine derivative, *J. Gen. Virol.* **63:**121–129.

Mecham, J. O., and Trent, D. W., 1983, A biochemical comparison of the *in vitro* replication of a virulent and an avirulent strain of Venezuelan encephalitis virus, *J. Gen. Virol.* **64:**1111–1119.

Milner, A. R., and Marshall, I. D., 1984, Pathogenesis of *in utero* infections with abortigenic and non-abortigenic alphaviruses in mice, *J. Virol.* **50:**66–72.

Milner, A. R., Marshall, I. D., and Mullbacher, A., 1984, Effect of pregnancy on stimulation of alphavirus immunity in mice, *J. Virol.* **50:**73–76.

Mims, C. A., Murphy, F. A., Taylor, W. P., and Marshall, I. D., 1973, Pathogenesis of Ross River virus infection in mice. I. Ependymal infection, cortical thinning, and hydrocephalus, *J. Infect. Dis.* **127:**121–128.

Moench, T. R., and Griffin, D. E., 1984, Immunocytochemical identification and quantitation of mononuclear cells in cerebrospinal fluid, meninges, and brain during acute viral encephalitis, *J. Exp. Med.* **159:**77–88.

Mohktarian, F., and Griffin, D. E., 1984, Role of mast cells in virus-induced CNS inflammation in the mouse, *Cell. Immunol.* **86:**491–500.

Mokhtarian, F., Griffin, D. E., and Hirsch, R. L., 1982, Production of mononuclear chemotactic factors during Sindbis virus infection in mice, *Infect. Immun.* **35:**965–973.

Monath, T. P., Calisher, C. H., Davis, M., Bowen, G. S., and White, J., 1974, Experimental studies of rhesus monkeys infected with epizootic and enzootic subtypes of Venezuelan equine encephalitis virus, *J. Infect. Dis.* **129:**194–200.

Monath, T. P., Kemp, G. E., Cropp, C. B., and Chandler, F. W., 1978, Necrotizing myocarditis in mice infected with Western equine encephalitis virus: Clinical, electrocardiographic, and histopathologic correlations, *J. Infect. Dis.* **138:**59–66.

Morein, B., Barz, D., Koszinowski, U., and Schirrmacher, V., 1979, Integration of a virus membrane protein into the lipid bilayer of target cells as a prerequisite for immune cytolysis, *J. Exp. Med.* **150:**1383–1398.

Morgan, I. M., 1941, Influence of age on susceptibility and on immune response of mice to Eastern equine encephalomyelitis virus, *J. Exp. Med.* **74:**115–132.

Morgan, I. M., Schlesinger, R. W., and Olitsky, P. K., 1942, Induced resistance of the central nervous system to experimental infection with equine encephalomyelitis virus. I. Neutralizing antibody in the central nervous system in relation to cerebral resistance, *J. Exp. Med.* **76:**357–369.

Mullbacher, A., Marshall, I. D., and Blanden, R. V., 1979, Crossreactive cytotoxic T cells to alphavirus infection, *Scand. J. Immunol.* **10:**291–296.

Murphy, F. A., and Whitfield, S. G., 1970, Eastern equine encephalitis virus infection: Electron microscopic studies of mouse central nervous system, *Exp. Mol. Pathol.* **13:**131–146.

Murphy, F. A., Harrison, A. K., and Collin, W. K., 1970, The role of extra-neural arbovirus infection in the pathogenesis of encephalitis: An electron microscopic study of Semliki Forest virus infection in mice, *Lab. Invest.* **22:**318–328.

Murphy, F. A., Taylor, W. P., Mims, C. A., and Marshall, I. D., 1973, Pathogenesis of Ross River virus infection in mice. II. Muscle, heart and brown fat lesions, *J. Infect. Dis.* **127:**129–138.

Niklasson, B., Espmark, A., LeDuck, J. W., Gargan, T. P., Ennis, W. A., Tesh, R. B., and Main, A. J., Jr., 1984, Association of a Sindbis-like virus with Ockelbo disease in Sweden, *Am. J. Trop. Med. Hyg.* **33:**1212–1217.

Noran, H. H., and Baker, A. B., 1945, Western equine encephalitis: The pathogenesis of the pathological lesions, *J. Neuropathol. Exp. Neurol.* **4:**269–276.

Nydegger, U. E., Fearon, D. T., and Austen, K. F., 1978, Autosomal locus regulates inverse relationship between sialic content and capacity of mouse erythrocytes to activate human alternative complement pathway, *Proc. Natl. Acad. Sci. U.S.A.* **75:**6078–6083.

Oldstone, M. B. A., 1984, Virus-induced immune complex formation and disease: Definition, regulation, importance, in: *Concepts in Viral Pathogenesis* (A. L. Notkins and M. B. A. Oldstone, eds.), pp. 201–209, Springer-Verlag, New York.

Olitsky, P. K., Schlesinger, R. W., and Morgan, I. M., 1943, Induced resistance of the central nervous system to experimental infection with equine encephalomyelitis virus. II. Serotherapy in Western virus infection, *J. Exp. Med.* **77:**359–375.

Olmsted, R. A., Baric, R. S., Sawyer, B. A., and Johnston, R. E., 1984, Sindbis virus mutants selected for rapid growth in cell culture display attenuated virulence in animals, *Science* **225:**424–426.

Oppenheim, J. J., 1981, Lymphokines, in: *Cellular Functions in Immunity and Inflammation* (J. J. Oppenheim, D. L. Rosenstreich and M. Potter, eds.), pp. 259–282, Elsevier, New York.

Park, M. M., Griffin, D. E., and Johnson, R. T., 1980, Analysis of the role of immune responses in recovery from acute Sindbis virus encephalitis: Studies in adult thymectomized lethally irradiated mice, *Infect. Immun.* **34:**306–309.

Parsons, L. M., and Webb, H. E., 1982, Virus titres and persistently raised white cell counts in cerebrospinal fluid in mice after peripheral infection with demyelinating Semliki Forest virus, *Neuropathol. Appl. Neurobiol.* **8:**395–401.

Pathak, S., and Webb, H. E., 1974, Possible mechanisms for the transport of Semliki Forest virus into and within mouse brain: An electron microscopic study, *J. Neurol. Sci.* **23:**175–184.

Peck, R., Brown, A., and Wust, C. J., 1979, In vitro heterologous cytotoxicity by T effector cells from mice immunized with Sindbis virus, J. Immunol. 123:1763–1766.

Pedersen, C. E., Jr., Slocum, D. R., and Eddy, G. A., 1973, Immunological studies on the envelope component of Venezuelan equine encephalomyelitis virus, Infect. Immun. 8:901–906.

Pedersen, C. E., Jr., and Eddy, G. A., 1974, Separation, isolation, and immunological studies of the structural proteins of Venezuelan equine encephalomyelitis virus, J. Virol. 14:740–744.

Postic, B., Schleupner, C. J., Armstrong, J. A., and Ho, M., 1969, Two variants of Sindbis virus which differ in interferon induction and serum clearance. I. The phenomenon, J. Infect. Dis. 120:339–347.

Pruslin, F. H., and Rodman, T. C., 1978, Venezuelan encephalitis virus: In vivo induction of a chromosomal abnormality in hamster bone marrow cells, Infect. Immun. 19:1104–1106.

Pusztai, R., Gould, E. A., and Smith, H., 1971, Infection patterns in mice of an avirulent and virulent strain of Semliki Forest virus, Br. J. Exp. Pathol. 52:669–677.

Rabinowitz, S. G., 1976, Host immune responses after administration of inactivated Venezuelan equine encephalomyelitis virus vaccines. II. Kinetics of neutralizing antibody responses in donors and adoptively immunized recipients, J. Infect. Dis. 134:39–47.

Rabinowitz, S. G., and Adler, W. H., 1973, Host defenses during primary Venezuelan equine encephalomyelitis virus infection in mice. I. Passive transfer of protection with immune serum and immune cells, J. Immunol. 110:1345–1353.

Rabinowitz, S. G., and Proctor, R. A., 1974, In vitro study of antiviral activity of immune spleen cells in experimental Venezuelan equine encephalomyelitis infection in mice, J. Immunol. 112:1070–1077.

Rager-Zisman, B., and Bloom, B. R., 1982, Natural killer cells in resistance to virus-infected cells, Springer Semin. Immunopathol. 4:397–414.

Reeves, W. C., and Hammon, W., Mc.D., 1962, Epidemiology of the Arthropod-Borne Viral Encephalitides in Kern County, California, 1943–1952, University of California Publications in Public Health, Vol. 4, University of California Press, Berkeley.

Reinarz, A. B. G., Broome, M. G., and Sagik, B. P., 1971, Age-dependent resistance of mice to Sindbis virus infection: Viral replication as a function of host age, Infect. Immun. 3:268–273.

Robinson, M. C., 1955, An epidemic of virus disease in Southern province, Tanganyika Territory, in 1952–1953. I. Clinical features, Trans. R. Soc. Trop. Med. Hyg. 49:28–32.

Rodda, S. J., and White, D. O., 1976, Cytotoxic macrophages: A rapid nonspecific response to viral infection, J. Immunol. 117:2067–2072.

Roehrig, J. T., Day, J. W., and Kinney, R. M., 1982, Antigenic analysis of the surface glycoproteins of a Venezuelan equine encephalitis virus (TC-83) using monoclonal antibodies, Virology 118:269–278.

Rosen, L., Gubler, D. J., and Bennett, P. H., 1981, Epidemic polyarthritis (Ross River) virus infection in the Cook Islands, Am. J. Trop. Med. Hyg. 30:1294–1302.

Satriano, S. F., Luginbuhl, R. E., Wallis, R. C., Jungherr, E. L., and Williamson, L. A., 1957, Investigation of Eastern equine encephalomyelitis. IV. Susceptibility and transmission studies with virus of pheasant origin, Am. J. Hyg. 67:21–34.

Scherer, W. F., and Chin, J., 1977, Responses of guinea pigs to infections with strains of Venezuelan encephalitis virus, and correlations with equine virulence, Am. J. Trop. Med. Hyg. 26:307–312.

Scherer, W. F., Ellsworth, C. A., and Ventura, A. K., 1971, Studies of viral virulence. II. Growth and adsorption curves of virulent and attenuated strains of Venezuelan encephalitis virus in cultured cells, Am. J. Pathol. 62:211–219.

Scherer, W. F., Campillo-Sainz, C., Mucha-Macias, J. de, Dickerman, R. W., Wong Chia, C., and Zarate, M. L., 1972, Ecologic studies of Venezuelan encephalitis virus in Southeastern Mexico. VII. Infection of man, Am. J. Trop. Med. Hyg. 21:79–85.

Schlesinger, R. W., 1949, The mechanism of active cerebral immunity to equine encephalomyelitis virus. II. The local antigenic booster effect of the challenge inoculum, *J. Exp. Med.* **89**:507–527.

Schlesinger, R. W., Olitsky, P. K., Morgan, I. M., 1944, Induced resistance of the central nervous system to experimental infection with equine encephalomyelitis virus. III. Abortive infection with Western virus and subsequent interference with the action of heterologous virus, *J. Exp. Med.* **80**:197–211.

Schleupner, C. J., Postic, B., Armstrong, J. A., Atchison, R. W., and Ho, M., 1969, Two variants of Sindbis virus which differ in interferon induction and serum clearance. II. Virological characterization, *J. Infect. Dis.* **120**:348–355.

Schmaljohn, A. L., Johnson, E. D., Dalrymple, J. M., and Cole, G. A., 1982, Non-neutralizing monoclonal antibodies can prevent lethal alphavirus encephalitis, *Nature (London)* **297**:70–72.

Schmaljohn, A. L., Kokubun, K. M., and Cole, G. A., 1983, Protective monoclonal antibodies define maturational and pH-dependent antigenic changes in Sindbis virus E1 glycoprotein, *Virology* **130**:144–154.

Seamer, J. H., Boulter, E. A., and Zlotnick, I., 1971, Delayed onset of encephalitis in mice passively immunized against Semliki Forest Virus, *Br. J. Exp. Pathol.* **52**:408–414.

Seay, A. R., and Wolinsky, J. S., 1982, Ross River virus-induced demyelination. I. Pathogenesis and histopathology, *Ann. Neurol.* **12**:380–389.

Seay, A. R., Griffin, D. E., and Johnson, R. T., 1981, Experimental viral polymyositis: Age dependency and immune response to Ross River virus infection in mice, *Neurology* **31**:656–661.

Sheahan, B. J., Barrett, P. N., and Atkins, G. J., 1981, Demyelination in mice resulting from infection with a mutant of Semliki Forest virus, *Acta Neuropathol.* **53**:129–136.

Shinefield, M. R., and Townsend, T. E., 1953, Transplacental transmission of western equine encephalitis, *J. Pediatr.* **43**:21–25.

Shope, R. E., 1980, Medical significance of togaviruses: An overview of diseases caused by togaviruses in man and in domestic and wild vertebrate animals, in: *The Togaviruses: Biology, Structure and Replication* (R. W. Schlesinger, ed.), pp. 47–82, Academic Press, New York.

Shope, R. E., 1985, Alphavirus diseases, in: *Virology* (B. N. Fields, ed.), pp. 931–953, Raven Press, New York.

Shore, H., 1961, O'nyong-yong fever: An epidemic virus disease in East Africa. I. Some clinical and epidemiological observations in the Northern province of Uganda, *Trans. R. Soc. Trop. Med. Hyg.* **55**:361–373.

Sissons, J. G. P., Schreiber, R. D., Cooper, N. R., and Oldstone, M. B. A., 1982, The role of antibody and complement in lysing virus-infected cells, *Med. Microbiol. Immunol.* **170**:221–227.

Smillie, J., Pusztai, R., and Smith, H., 1973, Studies of the influence of host defence mechanisms on infection of mice with an avirulent or virulent strain of Semliki Forest virus, *Br. J. Exp. Pathol.* **54**:260–266.

Smith, C., Wolcott, J. A., Wust, C. J., and Brown, A., 1985, Detection of immunologically cross-reacting capsid protein of alphaviruses on the surfaces of infected L929 cells, *J. Virol.* **53**:198–204.

Spertzel, R. O., Crabbs, C. L., and Vaughn, R. E., 1972, Transplacental transmission of Venezuelan equine encephalomyelitis virus in mice, *Infect. Immun.* **6**:339–343.

Spiegelberg, H. L., 1974, Biological activities of immunoglobulins of different classes and subclasses, *Adv. Immunol.* **19**:259–294.

Staab, E. V., Normann, S. J., and Craig, C. P., 1970, Alterations in reticuloendothelial function by infection with attenuated Venezuelan equine encephalitis (VEE) virus, *J. Reticuloendothel. Soc.* **8**:342–348.

Stanley, J., Cooper, S. J., and Griffin, D. E., 1985, Alphavirus neurovirulence: Monoclonal antibodies discriminating wild-type from neuroadapted Sindbis virus, *J. Virol.* **56**:110–119.

Stanley, J., Cooper, S. J., and Griffin, D. E., 1986, Monoclonal antibodies can protect and promote recovery from fatal Sindbis virus encephalitis *J. Virol.* (in press).

Stollar, V., 1975, Immune lysis of Sindbis virus, *Virology* **66**:620–624.

Stollar, V., Stollar, U. D., Kou, R., Harrap, K. A., and Schlesinger, R. W., 1976, Sialic acid content of Sindbis virus from vertebrate and mosquito cells: Equivalence of biological and immunological viral properties, *Virology* **69**:104.

Suckling, A. J., Webb, H. E., Chew-Lim, M., and Oaten, S. W., 1976, Effect of an inapparent viral encephalitis on the levels of lysosomal glycosidases in mouse brain, *J. Neurol. Sci.* **29**:109–116.

Suckling, A. J., Jagelman, S., and Webb, H. E., 1977, Brain lysosomal glycosidase activity in immunosuppressed mice infected with avirulent Semliki Forest virus, *Infect. Immun.* **15**:386–391.

Suckling, A. J., Jagelman, S., and Webb, H. E., 1978a, A comparison of brain lysosomal enzyme activities in four experimental togavirus encephalitides, *J. Neurol. Sci.* **35**:355–364.

Suckling, A. J., Pathak, S., Jagelman, S., and Webb, H. E., 1978b, Virus-associated demyelination: A model using avirulent Semliki Forest virus infection of mice, *J. Neurol. Sci.* **39**:147–154.

Sudia, W. D., and Newhouse, V. F., 1975, Epidemic Venezuelan equine encephalitis in North America: A summary of virus–vector–host relationships, *Am. J. Epidemiol.* **101**:1–13.

Sudia, W. D., Newhouse, V. F., and Henderson, B. E., 1971, Experimental infections of horses with three strains of Venezuelan equine encephalomyelitis virus. II. Experimental vector studies, *Am. J. Epidemiol.* **93**:206–211.

Sussman, O., Cohen, D., Gerende, J. E., and Kissling, R. E., 1958, Equine encephalitis vaccine studies in pheasants under epizootic and pre-epizootic conditions, *Ann. N. Y. Acad. Sci.* **70**:328–341.

Symington, J., McCann, A. K., and Schlesinger, M. J., 1977, Infectious virus–antibody complexes of Sindbis virus, *Infect. Immun.* **15**:720–725.

Tasker, J. B., Miesse, M. L., and Berge, T. O., 1962, Studies on the virus of Venezuelan equine encephalomyelitis. III. Distribution in tissues of experimentally infected mice, *Am. J. Trop. Med. Hyg.* **11**:844–850.

Taylor, R. M., Hurlbut, H. S., Work, T. H., Kingsbury, J. R., and Frothingham, T. E., 1955, Sindbis virus: A newly recognized arthropod-transmitted virus, *Am. J. Trop. Med. Hyg.* **4**:844–846.

Tazulakhova, E. B., Novakhatsky, A. S., and Yershov, F. I., 1973, Interferon induction by, and antiviral effect of poly (rI)-poly (rC) in experimental viral infection, *Acta Virol.* **17**:487–492.

Tesh, R. B., McLean, R. G., Shroyer, D. A., Calisher, C. H., and Rosen, L., 1981, Ross River virus (Togaviridae: *Alphavirus*) infection (epidemic polyarthritis) in American Samoa, *Trans. R. Soc. Trop. Med. Hyg.* **75**:426–431.

Thind, I. S., and Price, W. H., 1969, The effect of cyclophosphamide treatment on experimental arbovirus infections, *Am. J. Epidemiol.* **90**:62–68.

Thiruvengadam, K. V., Kalyanasundaram, V., and Rajgopal, J., 1965, Clinical and pathological studies on Chikungunya fever in Madras City, *Ind. J. Med. Res.* **53**:729–744.

Tyzzer, E. E., and Sellards, A. W., 1941, Pathology of equine encephalomyelitis in young chickens, *Am. J. Hyg. Sect. B* **33**:69–81.

Victor, J., Smith, D. G., and Pollack, A. D., 1956, The comparative pathology of Venezuelan equine encephalomyelitis, *J. Infect. Dis.* **98**:55–66.

Vilcek, J., 1964, Production of interferon by newborn and adult mice infected with Sindbis virus, *Virology* **22**:651–652.

Walder, R., and Bradish, C. J., 1975, Venezuelan equine encephalomyelitis virus (VEEV): Strain differentiation and specification of virulence markers, *J. Gen. Virol.* **26**:265–275.

Walker, D. H., Harrison, A., Murphy, K., Flemister, M., and Murphy, F. A., 1976, Lymphoreticular and myeloid pathogenesis of Venezuelan equine encephalitis in hamsters, *Am. J. Pathol.* **84**:351–370.

Weinbren, M. P., Kokernut, R. H., and Smithburn, K. C., 1956, Strains of Sindbis like virus isolated from culicene mosquitoes in the Union of South Africa. I. Isolation and properties, *S. Afr. Med. J.* **30:**631–636.

Wenger, F., 1977, Venezuelan equine encephalitis, *Teratology* **16:**359–362.

Willems, W. R., Kaluza, G., Boschek, C. B., Barrier, H., Hager, H., Schutz, H. J., and Feistner, H., 1979, Semliki Forest virus: Cause of a fatal case of human encephalitis, *Science* **203:**1127–1129.

Williams, M. C., and Woodall, J. P., 1961, O'nyong-nyong fever: An epidemic virus disease in East Africa. II. Isolation and some properties of the virus, *Trans. R. Soc. Trop. Med. Hyg.* **55:**135–141.

Wolcott, J. A., Gates, D. W., Wust, C. J., and Brown, A., 1982a, Cross reactive, cell associated antigen on L929 cells infected with temperature sensitive mutants of Sindbis virus, *Infect. Immun.* **36:**704–709.

Wolcott, J. A., Wust, C. J., and Brown, A., 1982b, Immunization with one alphavirus cross primes cellular and humoral immune responses to a second alphavirus, *J. Immunol.* **129:**1267–1271.

Wolcott, J. A., Wust, C. J., and Brown, A., 1984, Identification of immunologically cross-reactive proteins of Sindbis virus: Evidence for unique conformation for E1 glycoprotein from infected cells, *J. Virol.* **49:**379–385.

Woodman, D. R., McManus, A. T., and Eddy, G. A., 1975, Extension of mean time to death of mice with a lethal infection of Venezuelan equine encephalomyelitis virus by antihymocyte serum treatment, *Infect. Immun.* **12:**1006–1011.

Worthington, M., and Baron, S., 1971, Late therapy with an interferon stimulator in an arbovirus encephalitis in mice, *Proc. Soc. Exp. Biol. Med.* **136:**323–327.

Wyckoff, R. W. G., and Tesar, W. C., 1939, Equine encephalomyelitis in monkeys, *J. Immunol.* **37:**329–343.

Zichis, J., and Shaughnessy, H. J., 1940, Experimental Western equine encephalomyelitis: Successful treatment with hyperimmune rabbit serum, *J. Am. Med. Assoc.* **115:**1071–1078.

Zichis, J., and Shaughnessy, H. J., 1945, Successful treatment of experimental Western equine encephalomyelitis with hyperimmune rabbit serum, *Am. J. Public Health.* **35:**815–823.

Zlotnick, I., and Harris, W. J., 1970, The changes in cell organelles of neurons in the brains of adult mice and hamsters during Semliki Forest virus and Louping ill encephalitis, *Br. J. Exp. Pathol.* **51:**37–42.

Zlotnick, I., Grant, D. P., and Batter-Hatton, D., 1972a, Encephalopathy in mice following inapparent Semliki Forest virus (SFV) infection, *Br. J. Exp. Pathol.* **53:**125–129.

Zlotnick, I., Peacock, S., Grant, D. P., and Batter-Hatton, D., 1972b, The pathogenesis of Western equine encephalitis virus (WEE) in adult hamsters with special reference to the long and short term effects on the CNS of the attenuated clone 15 variant, *Br. J. Exp. Pathol.* **53:**59–77.

CHAPTER 9

The Use of Monoclonal Antibodies in Studies of the Structural Proteins of Togaviruses and Flaviviruses

JOHN T. ROEHRIG

I. INTRODUCTION

Since the initial reports of the isolation and characterization of monoclonal antibodies (MAb's) specific for alphaviruses (Roehrig *et al.*, 1980)
and flaviviruses (Dittmar *et al.*, 1980), the arbovirus community has experienced a virtual explosion in the research applications of MAb's to the
antigenic analysis of these important human and veterinary pathogens.
This research, reviewed in the following pages, has revolutionized our
understanding of both the molecular and the clinical immunology of togaviruses and flaviviruses.

The application of MAb's to the antigenic analysis of togaviruses and
flaviviruses has followed what has become a classic virological approach,
which was first introduced in studies with influenza virus (Webster and
Laver, 1980) and murine ecotropic leukemia virus (Lostrom *et al.*, 1979).
After a unique battery of MAb's are identified, the antibodies are used to
map antigenic determinants by combining results obtained from antibody
cross-reactivity analysis, using a large number of naturally occurring or

JOHN T. ROEHRIG • Division of Vector-Borne Viral Diseases, Center for Infectious Diseases, Centers for Disease Control, Public Health Service, U.S. Department of Health and
Human Services, Fort Collins, Colorado 80522.

artifically selected antigenic variants, with results from antigen-dependent biological assays such as virus neutralization (N) and hemagglutination inhibition (HI). This combined analysis generates groupings of antibodies that define distinct antigenic determinants or epitopes.

The spatial arrangement of these epitopes with respect to one another is defined in a competitive antibody-binding assay (CBA), in which the binding of one MAb to antigen is blocked by either previous or simultaneous binding of another MAb. It is assumed that antibodies that mutually block each other's binding (providing they are of similar binding avidity) define epitopes that are closely associated spatially. Investigations with various viruses (Heinz et al., 1984a) have demonstrated that binding of one MAb can actually enhance the binding of another MAb to a spatially unrelated epitope.

Once the CBA has been completed, the basic epitope framework can be designed and used as the foundation for investigations of other epitope characteristics such as conformational stability or reactivity changes during virus maturation. While this experimental approach may have some drawbacks, it is the most efficient method to quickly gain insight into the antigenic structure of a virus agent.

II. ALPHAVIRUSES

A. Serology and Classification of Alphaviruses

Serological classification of alphaviruses has been defined using the HI test, the plaque-reduction N test (PRNT), and the complement-fixation test (Casals, 1967; Dalrymple et al., 1973). This classification has been extensively reviewed elsewhere (Casals, 1957; Karabatsos, 1975; Chanas et al., 1976; Porterfield et al., 1978; Calisher et al., 1980; Porterfield, 1980). Alphaviruses can be divided into six complexes, the prototypes of which are Western equine (WEE), Venezuelan equine (VEE), and Eastern equine encephalitis (EEE) viruses and Middleburg, Ndumu, and Semliki Forest (SFV) viruses. Several of these complexes are composed of multiple subtypes, and some of these subtypes can be divided into multiple variants. These variants can often be distinguished only by PRNT or specialized modifications of the HI test (kinetic HI). For the VEE virus complex, it has been demonstrated that polyclonal rabbit antisera specific for the E2 of the VEE virus subtypes and variants could differentiate these viruses in standard HI or PRNT tests (Trent et al., 1979; Kinney et al., 1983). All alphaviruses have antigenically similar nucleocapsid proteins. Recent studies with MAb's specific for the envelope glycoproteins have been instrumental in defining these complex alphavirus antigenic cross-reactivities at the molecular level.

TABLE I. Protocols for Production of Antialphavirus Monoclonal Antibodies

Virus	Antigen	Days between final boost and fusion	Fusing agent[a]	Myeloma partner	Reference
Sindbis	Pure virus	3–4	PEG 1500	Sp2/0-Ag-14	Roehrig et al. (1980)
Sindbis	Infected mouse brain	3	PEG 1000	P3X63-Ag8	Schmaljohn et al. (1983)
Sindbis	Pure virus	4	PEG 1000	P3X63-Ag8U1	Righi et al. (1983)
Sindbis	Infected mouse brain	3	PEG 4000	P3X63-Ag8.653	Chanas et al. (1982)
WEE	Pure virus	3	PEG 1500	Sp2/0-Ag-14	Hunt and Roehrig (1985)
VEE	Pure virus	4	PEG 1500	Sp2/0-Ag-14	Roehrig et al. (1982a)
EEE	Pure virus	4	PEG 1500	Sp2/0-Ag-14	Hunt (unpublished data)
SF	Pure virus	4	PEG 4000	P3X63-Ag8.653	Boere et al. (1984)

[a] (PEG) Polyethylene glycol.

B. Antigenic Mapping of Alphavirus Structural Proteins Using Monoclonal Antibodies

1. Preparation of Hybridomas

A wide variety of immunization and cell-fusion protocols have been used in deriving antialphavirus MAb's (Table I). It is apparent from Table I that there is no one standard method for production of MAb's. In general, either purified virus or virus from infected mouse brain has been used as the primary immunogen. In our hands, better results are obtained by using purified virus immunogens. One common characteristic for all reported fusions is that hybridization usually takes place 3–4 days following the final immunogen boost. A wide variety of myeloma fusion partners have been used, each giving what appears to be reasonable success, although most authors fail to report their hybridoma isolation efficiency. In our laboratory, it is not unusual to observe 90% of the original fusion wells to be positive for antiviral antibody, following a good immunization. It should be noted that high antivirus antibody titers in the serum of immunized animals do not necessarily predict successful fusions.

Success in isolation of virus-specific MAb's is related to the efficiency and speed of the antibody-screening procedure. Screening is usually performed by radioimmunoassay (RIA) or enzyme-linked immunosorbent assay (ELISA). The use of ELISA allows for the rapid identification of antivirus-antibody-producing clones, without the need for radioisotopes

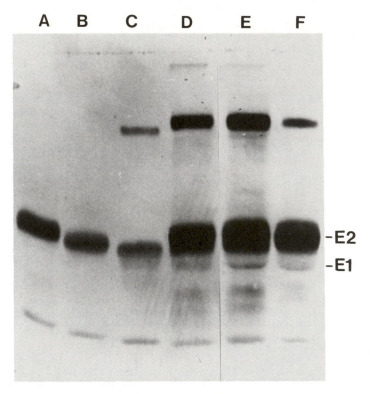

FIGURE 1. Immunoblot specificity testing with VEE virus anti-E2^c MAb. SDS-PAGE show-ing reactions with subtype II (A), subtype ID (B), subtype IC (C), subtype IB (D), subtype IA, TRD (E), and subtype IA, TC-83 (F) purified viruses. The antibody reacts primarily with the E2 of all tested viruses. From Roehrig and Mathews (1985).

(Voller *et al.*, 1976; Roehrig *et al.*, 1980). After the hybridoma has been cloned, antibody characterization, which consists of antigen specificity testing, isotyping, analysis by sodium dodecyl sulfate–polyacrylamide gel electrophoresis (SDS-PAGE) (Laemmli, 1970), isoelectric focusing (Ni-colatti *et al.*, 1979), N, and HI testing is performed. Specificity testing is usually done by radioimmune precipitation (RIP) with solubilized virus antigen or immunoblot (Fig. 1) using modifications of the procedures of Goding (1967) or Towbin *et al.* (1979). Care must be taken in performing the specificity testing. It has been demonstrated by Clegg *et al.* (1983) and Hunt and Roehrig (1985) that the reactivity of monoclonal antibodies can be dependent on detergent treatment of antigen. These investigators demonstrated that anti-E1 specificity could be confused with E1–E2 cross-reactivity because the E1–E2 heterodimer often failed to dissociate in the presence of nonionic detergents. Addition of low amounts of SDS (0.1%) was necessary to determine specificity. A related caveat was that inclusion of a higher concentration of SDS (0.5%) in the RIP could unfold the protein and destroy antigenicity. Once characterization of the MAb's has been completed, high-titer antibody preparations can be prepared by injecting hybridoma cells into the peritoneal cavity of syngeneic mice and harvesting the resultant ascitic fluids. Antibodies can be quantitated after purification by salt precipitation of the ascitic fluids and chromatography on Protein A–Sepharose (Ey et al., 1978). It is these purified antibodies that should be used for virus antigenic analysis, because it is easier to compare reactivities of different MAb isolates if exact concentrations of antibodies are known.

2. Alphavirus E2-Glycoprotein Epitopes

Historically, the alphavirus E2 glycoprotein has been identified as the antigen that elicits antiviral neutralizing antibody following immunization with either infectious virus or purified glycoprotein (Pedersen and Eddy, 1974; Dalrymple et al., 1976; Trent et al., 1979; Kinney et al., 1983). Isolation of anti-E2 MAb's has been reported for Sindbis virus (Roehrig et al., 1980, 1982b; Schmaljohhn et al., 1982, 1983; Righi et al., 1983; Olmstead et al., 1984), VEE viruses (Roehrig et al., 1982a; Roehrig and Mathews, 1985), and SFV (Boere et al., 1983, 1984). The earliest study with Sindbis virus identified two anti-E2 MAb's, one of which neutralized virus infectivity and the other had no biological activity (Roehrig et al., 1980, 1982b). Another early study with Sindbis virus identified two more anti-E2 MAb's with N activity (Schmaljohn et al., 1982). Neither of these studies, however, made attempts to identify and characterize the E2 epitopes.

A more complete analysis of the E2 glycoprotein antigenic structure using MAb's specific for various subtypes of VEE viruses identified eight E2 epitopes (Table II) (Roehrig et al., 1982a; Roehrig and Mathews, 1985). One of the E2 epitopes (E2^c) was capable of eliciting high-titer neutralizing and hemagglutination-inhibiting MAb's (Table II). As little as 5 ng anti-E2^c MAb neutralized 100% of virus infectivity in the PRNT. CBA clustered these eight epitopes into four spatially related sites (Fig. 2). The results from this study indicated that antibodies that bound virion E2 had variable N or HI activity, depending on which virus was used as antigen in the serological assay. These observations correlated with the CBA results, which predicted that the ability of an epitope to elicit neutralizing or hemagglutination-inhibiting MAb appeared to be a function of the spatial proximity of this epitope to the E2^c epitope. For example, the MAb that defined the E2^e epitope failed to neutralize TC-83 virus infectivity or compete with anti-E2^c MAb in the CBA using TC-83 virus as antigen. With the P676 VEE virus, however, anti-E2^e MAb neutralized virus infectivity and competed with anti-E2^c MAb in CBA. The conclusion from this study indicated that the E2^c was the hub of the N site on the VEE virus and any epitope spatially proximal to the E2^c could elicit neutralizing antibody. This hypothesis was confirmed by measuring MAb reactivities with protease-derived peptides of VEE virus. MAb's that failed

TABLE II. Epitopes on the E2 Glycoprotein of Venezuelan Equine Encephalitis Virus Strain TC-83[a]

Virus	Representative MAb	Isotype of MAb	Epitope	N[b]	HI[b]	Protection (%)[c]	ELISA cross-reactivity with VEE complex viruses[d]							
							IA TC-83	IA TRD	IB	IC	ID	IE	IF	2
TC-83	5B4D-6	IgG2A	E2[a]	–	–	0	+	–	–	–	–	–	–	–
TC-83	2A4B-12	IgG2B	E2[b]	–	–	ND	+	+	+	–	–	–	–	–
TC-83	3B4C-4	IgG1	E2[c]	++++	++++	100	+	+	+	+	+	–	+	+
EVE	1A6C-3	IgG2A	E2[d]	–	–	ND	+	+	+	+	–	–	+	+
P676	1A3A-5	IgG2A	E2[e]	–	–	ND	+	+	+	+	+	–	–	–
TRD	1A4D-1	IgG2A	E2[f]	+	–	ND	+	+	+	+	+	+	–	–
PTF	1A3A-9	IgG2A	E2[g]	+	–	100	+	+	+	+	+	+	+	–
PTF	1A3B-7	IgG2A	E2[h]	+	–	100	+	+	+	+	+	+	+	+

[a] From Roehrig and Mathews (1985).

[b] N and HI titers of purified MAb's were converted to a grading scale (++++ ≥8192 – ≤4).

[c] Percentage protection of mice from lethal virus challenge following passive immunization with 20 μg purified MAb (N = 10). (ND) Not determined.

[d] Ratios of absorbance on heterologous/homologous purified virus antigens were calculated. Crosss-reactivities of 50% or more were considered + (N = 5). Anti-E2[h] MAb's reacted with VEE virus subtypes 3, 4, and 5. None of these MAb's reacted with WEE virus or EEE virus. Prototype viruses used were: subtype IA, Trinidad donkey (TRD); subtype IB, PTF; IC, P676; ID, 3880; IE, Mena II; IF, 78V-3531; II, Everglades, Fe3-7c.

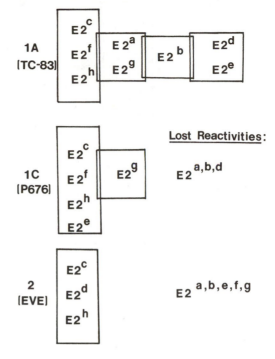

FIGURE 2. CBA maps of E2 for subtypes IA (TC-83), IC (P676), and II (Everglades) VEE viruses as determined with the antibodies listed in Table II. Lost reactivities are those epitopes that cannot be identified on the respective virus. Overlapping boxes indicate competition. From Roehrig and Mathews (1985).

to neutralize virus and compete with anti-E2^c MAb in the CBA identified epitopes that segregated to different peptides than the E2^c epitope (Roehrig and Mathews, 1985).

Similar studies with SF virus identified five epitopes (E2^a–E2^e) on the E2 glycoprotein (Boere *et al.*, 1984). Two of these epitopes, which were spatially unrelated in the CBA, elicited neutralizing MAb's. Neutralization of 100% of virus infectivity required 10 µg anti-E2^c MAb. Anti-E2^d MAb was more efficient in neutralizing virus, requiring only 0.1 µg to neutralize 100% of virus infectivity.

Thus far, all MAb studies have identified E2 epitopes that do not participate in virus infectivity. These results indicate that not all the E2 protein functions in the adsorption of virions to susceptible cells. The observation that VEE virus anti-E2 MAb's block hemagglutination (HA) mediated by VEE virions corroborated the earlier report of HI activity associated with polyclonal anti-VEE E2 antisera (Trent *et al.*, 1979; Kinney *et al.*, 1983).

3. Alphavirus E1-Glycoprotein Epitopes

The alphavirus E1 glycoprotein has been associated with HA of red blood cells (Dalrymple *et al.*, 1976). Isolation of E1 MAb's has been reported for Sindbis virus (Roehrig *et al.*, 1980, 1982b; Schmaljohn *et al.*, 1982, 1983; Chanas *et al.*, 1982; Clegg *et al.*, 1983), VEE virus (Roehrig *et al.*, 1982a), WEE virus (Hunt and Roehrig, 1985), and SFV (Boere *et al.*,

TABLE III. Epitopes on the E1 Glycoprotein of Western Equine Encephalitis Virus (McMillan)[a]

Representative MAb	Isotope of MAb	Epitope[b]	N[b]	HI[b]	Protection (%)[c]	ELISA cross-reactivity with alphaviruses[d]								
						MCM	HJ	FM	SIN	KYZ	WHA	VEE	EEE	SFV
2A6C-7	IgG2B	E1^{a1}	–	±	50	+	+	+	+	+	+	–	–	–
1B5B-6	IgG1	E1^{a2}	–	+++	10	+	+	+	+	+	+	–	–	–
2A3D-5	IgG2A	E1^{b1}	–	±	60	+	+	+/–	+	+	+/–	–	–	–
2A2A-2	IgG1	E1^{b2}	–	–	ND	+	+	+/–	+	+	+/–	–	–	–
1B1D-7	IgG2A	E1^{c}	–	–	20	+	+	+	–	–	–	–	–	–
2B6B-2	IgG2A	E1^{d}	–	++++	100	+	–	–	+	+	+	+	+	+
2B1C-6	IgG2B	E1^{e1}	–	+++	0	+	–	–	–	–	–	–	–	–
2B3B-5	IgG2A	E1^{e2}	–	–	10	+	–	–	–	–	–	–	–	–

[a] From Hunt and Roehrig (1985).
[b] Micrograms of purified MAb required for N and HI were converted to a grading scale (+ + + + = 0.005 to – >25).
[c] Percentage protection of mice from WEE virus challenge following passive immunization with 100 μg purified MAb (N = 10). (ND) Not determined.
[d] Ratios of absorbance on heterologous/homologous purified virus antigens were calculated. Cross-reactivities greater than 50% were considered + (N = 5–14). WEE virus strains used were: McMillan (MCM), Highlands J (HJ), Ft. Morgan (FM), Sindbis (SIN), Kyzylagach (KYZ), and Whatoroa (WHA). The New Jersey Original (NJO) strain of EEE virus, original strain of SFV, and TC-83 strain of VEE virus were also used.

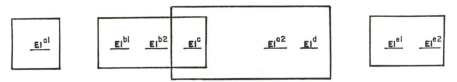

FIGURE 3. CBA map of the E1 of WEE virus (McMillan), as determined with the antibodies listed in Table III. Overlapping boxes indicate competition. From Hunt and Roehrig (1985).

1984). Cross-reactivity and CBA analysis of VEE virus E1 glycoprotein identified three overlapping epitopes (E1^b, E1^c, and E1^d) and one spatially distinct epitope (E1^a) (Roehrig *et al.*, 1982a). MAb's that defined the E1^d epitope demonstrated alphavirus-group-reactive HI activity with virus antigen from infected mouse brain. A surprising finding was that the MAb that defined the E1^b epitope had low-level N activity, requiring 50 µg to neutralize 100% of virus infectivity. This N activity was attributed to the proximal spatial relationship that this epitope had to the E2^c in CBA.

A similar analysis of Sindbis virus and SFV identified five and six epitopes, respectively, on the E1 of these viruses (Schmaljohn *et al.*, 1983; Boere *et al.*, 1984). Both these studies identified neutralizing anti-E1 MAb's, but the spatial relationship of these epitopes to the E2 N site, as measured by the CBA, was not reported. CBA analysis of Sindbis virus defined three overlapping epitopes and two spatially unrelated epitopes (Schmaljohn *et al.*, 1983). CBA analysis of SFV identified six spatially distinct E1 epitopes (Boere *et al.*, 1984). The CBA pattern with SFV differs from those identified with other alphaviruses. In addition to demonstrating N activity with anti-Sindbis virus E1 MAb's, Chanas *et al.* (1982) observed that virus-specific hemolysis of red blood cells was associated with an E1 epitope.

An analysis of the alphavirus E1 glycoprotein of WEE virus divided this antigen into eight epitopes (Hunt and Roehrig, 1985) (Table III). CBA analysis of these epitopes revealed two spatially overlapping sites, similar to those identified on the E1 glycoproteins of VEE and Sindbis viruses, and two spatially distinct epitopes (Fig. 3). These MAb's identify an alphavirus-group-reactive HA site (E1^d), a WEE-virus-specific HA site (E1^{e1})and a WEE-complex-reactive HA site (E1^{a2}). CBA analysis in our laboratory using group-reactive anti-E1 MAb's isolated from WEE-, VEE-, and EEE-virus-immunized mice indicated that these viruses share the same cross-reactive epitope.

4. Alphavirus Glycoprotein Spike Structure

The analysis of E1 and E2 glycoprotein epitopes discussed above must be defined in the context of the overall alphavirus glycoprotein spike structure. This structure most probably consists of heterodimers of E1 and E2, except in the case of SFV, in which E3 is also included (Ziemieki and Garoff, 1978). Epitopes could therefore be derived from protein in-

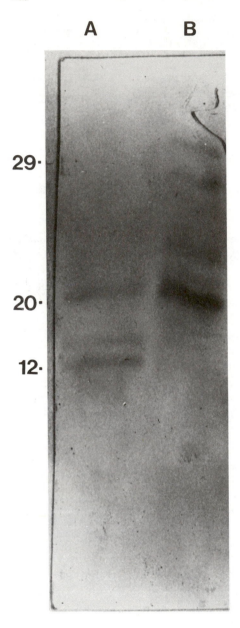

FIGURE 4. Immunoblot reactivity of protein fragments with anti-VEE virus E2^c MAb. (A) Reactivity with fragments generated by codigestion of purified TC-83 virus with *S. aureus* V8 protease and α-chymotrypsin; (B) reactivity with fragments generated by digestion with α-chymotrypsin alone. Markers: carbonic anhydrase (29 kd), soybean trypsin inhibitor (20 kd), and cytochrome C (12 kd).

teractions at the secondary, tertiary, or quaternary levels of organization. Initial studies with Sindbis virus showed that at least one E1 epitope could be destroyed by treating protein with SDS and β-mercaptoethanol (Roehrig *et al.*, 1982b). These results were corroborated by Hunt and Roehrig (1985), who demonstrated that one half of the E1 epitopes on WEE virus were sensitive to SDS treatment. Thus, epitopes sensitive to detergent or reducing reagent must be discontinuous; i.e., they require native conformation for expression of antigenicity.

Virus binding studies with Sindbis virus anti-E1 MAb's demonstrated that many of the E1 epitopes were temporally dependent on virus expression and occurred only on the surface of infected cells (Schmaljohn *et al.* 1982, 1983). Antibody virus binding occurred only at low pH, and antiglobulin added at neutral pH failed to enhance virus N mediated by these E1 MAb's. Hunt and Roehrig (1985) confirmed these results by observing that one half of the WEE virus E1 epitopes could not be recognized by anti-E1 MAb–enzyme conjugates when virus–conjugate mixtures prepared at neutral pH were analyzed by isopyknic centrifugation.

Epitopes on the E2 glycoprotein of VEE viruses appear to be more continuous. Only the E2^d epitope could be denatured by SDS and β-mercaptoethanol treatment (Roehrig and Mathews, 1985). All the other E2 epitopes were stable to this treatment, and the E2^c was stable to proteolytic cleavage of all reactive VEE viruses. Codigestion of TC-83 virus with both *Staphylococcus aureus* V8 protease and chymotrypsin resulted in the production of a 12,000-dalton peptide that retained reactivity with anti-E2^c MAb in immunoblots (Fig. 4).

Righi *et al.* (1983) identified an epitope on Sindbis virus that appeared to be discontinuous, requiring both E1 and E2 for antigenicity. This conclusion was based on the observation that their MAb's precipitated both E1 and E2 in the presence of 1% Triton X-100, but failed to recognize either E1 or E2 after they were separated by chromatography on DEAE–Sepharose A50 columns. To fully understand these results, further studies delineating this antigen–MAb interaction and its sensitivity to various detergents must be performed.

5. Expression of Glycoprotein Epitopes on Infected Cell Surfaces and in the Insect Vector

Very little information is available on the expression of E1 or E2 epitopes on the surface of the infected cell. Schmaljohn *et al.* (1982) demonstrated that E1 epitopes could be detected on the surface of infected cells by complement-mediated cell cytolysis measured by ^{51}Cr release. VEE virus glycoprotein epitopes also may be differentially expressed on the surface of the infected cell (Roehrig *et al.*, unpublished observation). This means that caution should be used when employing virus-infected cells as antigen source. Because not all epitopes are present on the cell surface at the same time postinfection, an element of confusion could be introduced when using virus-infected cells to compare epitope reactivities. To date, MAb's have not been applied to investigating the expression of epitopes in alphavirus-infected mosquito vectors.

C. Alphavirus Immunity

1. Role of Humoral Immunity in Protection from Alphavirus Disease

Because the E2 glycoprotein elicits neutralizing antibody, it has been assumed that this glycoprotein is of primary importance in immunity to alphavirus infection. The first detailed study examining the role of both E1 and E2 in protective immunity was published by Mathews and Roehrig (1982), who followed the progress of a lethal VEE virus infection in 3-week-old mice that had been passively immunized 24 hr previously with known quantities of purified MAb's. Several MAb's were able to protect animals from virus infection. As little as 5 μg anti-E2^c MAb protected 70% of challenged animals. This antibody severely limited virus replication and abrogated disease, provided that it was administered prior to virus entry into the neural tissue. Lower amounts of antibody did not preempt a nascent host immune response, while they still protected animals. Subsequent studies with more broadly cross-reactive, but neutralizing, anti-E2 MAb's demonstrated similar levels of protection (see Table II). Some nonneutralizing anti-E2 MAb's failed to protect animals.

Anti-E1 MAb's were, in general, less protective. Neutralizing anti-E1^b MAb protected animals, but 10-fold more antibody was required compared to the protective anti-E2^c MAb. Nonneutralizing anti-E1^d alphavirus-group-reactive MAb's were also capable of protecting animals from VEE and WEE virus infection, but 100-fold more antibody than anti-E2^c MAb was required (Table III).

Similar studies with SFV confirmed the results obtained with VEE virus (Boere *et al.*, 1983, 1984, 1985). Anti-SFV E2^d MAb's were most efficient in protecting animals from SFV challenge, requiring only 0.1 μg MAb to protect 100% of challenged mice. Nonneutralizing anti-E2^a MAb required 100-fold more antibody than E2^d MAb, and low-activity neutralizing anti-E1^a MAb required 10-fold more antibody than E2^d MAb to protect animals from peripheral virus challenge.

Investigations with Sindbis virus identified the importance of nonneutralizing anti-E1 MAb's in protection from an intracerebral challenge with a neural-adapted strain of virus (Schmaljohn *et al.*, 1982). These results were difficult to interpret quantitatively because antibodies were administered in the form of crude ascitic fluids. In general, however, nonneutralizing anti-E1 MAb's were able to protect animals from Sindbis virus challenge. Subsequent investigation revealed that the sensitivity of this protection was similar to that observed with VEE virus and SFV. These authors hypothesized that protection demonstrated by nonneutralizing MAb's that did not bind to intact virus was probably mediated by complement (C') lysis of virus-infected cells. Similar studies using WEE-virus-specific MAb's corroborated the potential of nonneutralizing anti-E1 MAb's in protection from virus infection (Hunt and Roehrig, 1985).

2. Role of Complement in Protection from Alphavirus Infection

The hypothesis that antialphavirus MAb's protected animals from virus challenge by C'-mediated lysis of virus-infected cells led to an in-

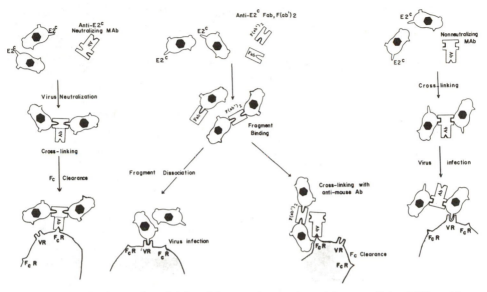

FIGURE 5. Hypothesized model for alphavirus interaction with neutralizing MAb; with neutralizing MAb and Fab and F(ab')₂ fragments, in the presence and absence of antiglobulin; and with nonneutralizing MAb. (VR) Virus receptor; (F_cR) F_c receptor.

vestigation of the role of C' in protection from virus infection by neutralizing MAb (Mathews *et al.*, 1985). The ability of F(ab')₂ and Fab fragments of anti-VEE virus MAb's to protect animals from VEE virus challenge and the ability of anti-E2 neutralizing MAb's to protect C'5 genetically deficient mice or mice depleted of C'3 by cobra venom factor was investigated. Monovalent Fab failed to efficiently neutralize virus infectivity in the PRNT, except in the presence of rabbit anti-mouse IgG antiserum. This fragment could bind virus in the ELISA, but it failed to protect animals from virus challenge. F(ab')₂ antibody fragment could neutralize virus in the PRNT; however, it too was unable to protect animals from virus infection. These results indicate that cross-linking was probably necessary for PRNT and that Fc was important for *in vivo* protection. While neutralizing antibody could lyse virus-infected cells in the presence of C', the ability of these MAb's to protect animals from virus challenge was not diminished following loss of *in vivo* C' activity by either C'3 or C'5 depletion. From these results, it appears that while C' is not necessary for *in vivo* protection, antibody bivalency is important for virus N and the Fc domain functions in clearance of virus, possibly via Fc-receptor-bearing cells. No similar analyses of the role of C' in protection mediated by nonneutralizing MAb's have been reported. A model for the mechanism of protection by neutralizing MAb is shown in Fig. 5.

TABLE IV. Serologically Important Alphavirus Monoclonal Antibodies

Virus	MAb	Isotype	E1/E2	HI	N	Reactivity
WEE	2A2C-3	IgG2A	E1	+	−	Alphavirus group[a]
WEE	2B1C-6	IgG2A	E1	+	−	WEE virus[a]
WEE	1B5B-6	IgG2A	E1	+	−	WEE complex[a]
HJ	2A6B-5	IgG1	E2	−	−	HJ virus[b]
SIN	49	IgG2A	E2	−	+	SIN virus[c]
SF	UM5.1	IgG2A	E2	−	+	SF virus[d]
VEE (IB)	1A3A-9	IgG2A	E2	+	+	VEE complex[e]
VEE (IA)	5B4D-6	IgG2A	E2	−	−	TC-83 virus[f]
VEE (IC)	1A3A-5	IgG2A	E2	+	+	Epizootic VEE[e]
VEE (IE)	1A1B-9	IgG2A	E2	−	−	Enzootic VEE[b]
VEE (II)	1A6C-3	IgG2A	E2	+	+	VEE subtype II[e]
VEE (III)	1B4B-1	IgG2A	E2	−	+	VEE subtype III[b]
VEE (IV)	1A3A-9	IgG1	E2	−	−	VEE subtype IV[b]
VEE (V)	1A4D-3	IgM	E2	+	+	VEE subtype V[b]
EEE (NJO)	1B4A-6	IgG1	E1	+	−	EEE complex[b]
EEE (NJO)	1B5C-3	IgG1	E1	+	−	N. American EEE virus[b]
EEE (BeAn 5122)	1A2C-1	IgG1	E2	ND[g]	ND[g]	S. American EEE virus[b]

[a] Hunt and Roehrig (1985).
[b] CDC prepared reagent (as yet unpublished data).
[c] Schmaljohn et al. (1983a).
[d] Boere et al. (1984).
[e] Roehrig and Mathews (1985).
[f] Roehrig et al. (1982a).
[g] Not determined as yet.

D. Monoclonal Antibodies in Alphavirus Serology and Epidemiology

A current list of those alphavirus MAb's that are possibly useful in serology is given in Table IV. Antigenic cross-reactivity between different alphavirus complexes is limited in standard serological tests with hyperimmune polyclonal reagents. MAb reagents provide an unlimited source of homogeneous, well-characterized reagents for serological use. Many of these reagents can be used to differentiate virus subtypes or variants.

Perhaps the most promising use of MAb reagents is in the identification of etiological agents by antigen-capture immunoassays. In this test, polyclonal or monoclonal antiviral antibody is coated to an insoluble matrix such as a 96-well microtiter plate. Samples such as ground mosquito pools or human sera are than added. If virus antigen is present, it will be captured by the antiviral antibody. Captured antigen can then be directly detected using a MAb-enzyme conjugate. It is important to note that depending on the biochemical characteristics of the MAb, certain antibodies may not function well in capturing antigen from complex solutions.

Another significant application of antiviral MAb's in viral identification is the use of virus-specific antibody in the indirect immunofluorescence assay (IFA) (see below).

E. Vaccine Strategy

Much of the reason for studying the antibody requirements for protection from alphavirus infection centers on the development of new and more effective virus vaccines. As the elements that lead to antivirus immunity become more clearly defined, more concerted efforts can be made at interfacing molecular immunology with recombinant DNA techniques in designing biosynthetic or synthetic vaccines. Current efforts with VEE virus center on the characterization of the 12,000-dalton proteolytic fragment of E2 that contains the E2^c epitope. A similar analysis is currently under way with SFV (W. Boere, personal communication). Fragments such as this could serve as a model for a synthetic peptide vaccine. Because of their chemical instability, the broadly cross-reactive E1 epitopes will probably not be useful as models for synthetic vaccines.

Applications of antiidiotypic (anti-id) antibodies in vaccine strategy and virus receptor analysis for alphaviruses are in their infancy. Anti-id antibodies have been prepared for neutralizing VEE virus anti-E2^c MAb (Roehrig *et al.*, 1985. This rabbit anti-id appears to contain the internal image of the virus antigen because it will cross-react with anti-VEE virus MAb's, and the id–anti-id reaction can be inhibited with virus. The ability of this anti-id to elicit virus-neutralizing antibody needs to be investigated. Application of anti-id reagents to alphavirus serology or vaccine strategy will be practical only if the anti-id is a MAb.

III. FLAVIVIRUSES

A. Serology and Classification of Flaviviruses

The classification of flaviviruses has been extensively reviewed elsewhere (DeMadrid and Porterfield, 1974; Porterfield, 1980) and will not be reviewed here. Flaviviruses can be divided into multiple serocomplexes by the PRNT. Viruses that compose a complex are considered to be serotypes. Serotypic determinants appear to reside on the envelope glycoprotein [(E) 50,000–56,000 daltons] as determined by CBA with polyclonal antisera (Trent, 1977). Flavivirus-group-reactive and complex-reactive determinants can also be identified on the E glycoprotein. The capsid protein is composed of flavivirus-group-reactive determinants only. The E glycoprotein binds to red blood cells and is probably the protein involved in HA. This protein appears to be the only structural protein that can elicit neutralizing antibody.

TABLE V. Protocols for Production of Antiflavivirus Monoclonal Antibodies

Virus[a]	Antigen	Days between final boost and fusion	Fusing agent[b]	Myeloma partner	Reference
DEN	Infected mouse brain/pure virus	3	PEG 1000	P3X63-Ag8	Gentry et al. (1982)
JE	Pure virus	3	PEG 2000	NS-1	Kimura-Kuroda and Yasui (1983)
JE	Infected mouse brain	3	PEG 4000	P3X63-Ag8.653	Kobayashi et al. (1984)
SLE	Pure virus	3	PEG 1500	Sp2/0-Ag14	Roehrig et al. (1983)
TBE	Pure glycoprotein	3	PEG 4000	P3X63-Ag8.653	Heinz et al. (1983)
WN	Infected mouse brain	3	PEG 1000	NS-1	Peiris et al. (1982)
YF	Pure virus	3	PEG 1000	P3X63-Ag8.653	Schlesinger et al. (1983)

[a] (DEN) Dengue; (JE) Japanese encephalitis; (SLE) St. Louis encephalitis; (TBE) tick-borne encephalitis; (WN) West Nile; (YF) yellow fever.
[b] (PEG) Polyethylene glycol.

B. Antigenic Mapping of Flavivirus Structural Proteins Using Monoclonal Antibodies

1. Preparation of Hybridomas

As discussed previously for alphaviruses, a wide variety of immunization techniques have been used to elicit flavivirus-specific hybridomas. A summary of these conditions is shown in Table V. Most immunization protocols use either virus-infected mouse brain or purified virus as antigen. Both techniques have yielded satisfactory results. In our hands, however, isolation of alphavirus-specific hybridomas is much more efficient than isolation of flavivirus-specific hybridomas because fewer alphavirus fusions are necessary to produce large numbers of hybridoma cultures. All the published accounts of anti flavivirus hybridoma isolation have relied on RIA or ELISA for virus-specificity testing. To date, only the antigenic structure of the E glycoprotein has been studied in detail.

2. Flavivirus E-Glycoprotein Epitopes

The first isolation of antiflavivirus MAb's was reported using dengue (DEN) 3 virus as antigen (Dittmar et al., 1980). While this investigation

TABLE VI. Passive Transfer with Purified St. Louis Encephalitis Anti-Ė Glycoprotein Epitope Monoclonal Antibodies[a]

Representative MAb	Specificity[b]	Epitope[c]	Biological activity[d]		Antibody inoculated i.v. (μg)[e]	
			N	HI	5–40	50–200
3B4C-7	Type	E-1[a]	<50	20	20[f]	40
1B2C-5	Type	E-1[b]	<50	8,000	17	40
6B5A-2	Type	E-1[c]	64,000	32,000	100	100
4A4C-4	Type	E-1[d]	800	20	13	20
1B5D-1	Subcomplex	E-2	<50	8,000	20	47
2B5B-3	Subgroup	E-3	<50	8,000	4	38
2B6B-2	Group	E-4[a]	<50	<10	8	40
6B6C-1	Group	E-4[b]	200	3,200	12	36

[a] All purified antibodies were adjusted to 1 mg/ml. End-point ELISA values ranged between 20,480 and 81,920.
[b] Type: SLE; Subcomplex: SLE, WN, Murray Valley, JE, YF; Supercomplex: SLE, JE; Group: all flaviviruses.
[c] Epitopes were determined by cross-reactivity analysis and differing biological activities.
[d] Seventy percent plaque reduction and HI end-point titers.
[e] Percentage survivors of total mice inoculated intraperitoneally with 100 IPLD$_{50}$ of SLE (MSI-7) virus 24 hr after intravenous inoculation of monoclonal antibody ($N = 10$–35 mice from two or more independent observations). Results for all passive transfer experiments were considered to be valid only if 90–100% of the SLE virus controls (virus + diluent) died.
[f] Percent survivors.

was preliminary in nature, it identified DEN-3-type-specific and flavivirus-group-reactive determinants using antigen-absorption techniques. Although antigen specificities were not determined, these antibodies demonstrated low-level HI activity and were probably E-glycoprotein-specific.

Other preliminary observations were subsequently published using West Nile (WN) virus (Peiris *et al.*, 1982) and DEN viruses (Henchal *et al.*, 1982; Gentry *et al.*, 1982). Although of limited scope and lacking important antibody characterizations, both studies identified similar epitopes that ranged in cross-reactivity from type-specific to group-reactive. Anti-WN virus MAb's defined a type-specific epitope that participated in the PRNT (Peiris *et al.*, 1982). Anti-DEN virus MAb's defined a type-specific epitope involved in the PRNT and a separate epitope involved in HI (Gentry *et al.*, 1982).

The most complete analyses of the flavivirus glycoproteins have been performed with tick-borne encephalitis (TBE) virus and St. Louis encephalitis (SLE) virus; however, many similarities exist among all serologically distinct flaviviruses. The antigenic analysis of TBE virus identified eight epitopes. The antigenic cross-reactivity of these epitopes ranged from flavivirus-group-reactive to TBE-type-specific (Heinz *et al.*, 1983a; Stephenson *et al.*, 1984; Gresikova and Sekeyova, 1984). Similarly, the E glycoprotein of SLE virus could be divided into eight epitopes using antigen cross-reactivity assay, IFA, PRNT, and HI assay (Table VI). MAb's that defined type-specific epitopes demonstrated a range of biological reactivities. One of the type-specific epitopes (E-1^c) elicited MAb's that had extremely high PRNT and HI titers. This epitope was postulated to be located in a domain on the glycoprotein that was particularly critical for virus adsorption. Antigenic variation of the type-specific epitopes in a large number of SLE virus strains was investigated using the PRNT and HI assay. With a few exceptions, these epitopes demonstrated very little antigenic drift. The E-1^c epitope appeared to be antigenically stable. A subsequent analysis of the ability of these antibodies to protect animals from virus challenge will be discussed later.

Results similar to those reported with TBE virus and SLE virus have been published for Japanese encephalitis (JE) virus (Kimura-Kuroda and Yasui, 1983; Kobayashi et al., 1984) and yellow fever (YF) virus (Schlesinger et al., 1983; Gould et al., 1985a; Monath et al., 1984a). In each case, type-specific, complex-reactive, subgroup-reactive, and group-reactive epitopes were defined on the E glycoprotein. Investigations with YF-virus-specific MAb's identified antigenic variation among various YF vaccine strains as well as variation between the 17D vaccine strain and the virulent YF wild-type parent virus strain (Schlesinger et al., 1983; Monath et al., 1983).

3. Flavivirus Glycoprotein Spike Structure

An antigenic model for the E-glycoprotein spike of TBE virus is shown in Fig. 6. This model was deduced using virus cross-reactivity assays, CBA's, PRNT, and HI assays (Heinz et al., 1983a). The eight E-glycoprotein epitopes clustered into three domains in CBA's. Domain A contained cross-reactive epitopes; domain B contained epitopes that were shared among only the members of the TBE serocomplex. Domain C consisted of one subtype-specific epitope. Subsequent CBA analysis generally demonstrated that antibodies reactive with one of these domains enhanced the binding of antibodies reactive with the other two domains (Fig. 6) (Heinz et al., 1984a). Binding-constant analysis indicated that this enhanced binding was apparently caused by a conformational change in the E-glycoprotein surface structure. Each of these domains also demonstrated differing antigenic stability characteristics following denaturation by reduction and alkylation or cleavage with CNBr or site-specific proteases (Heinz et al., 1983b). Domain A completely lost its antigenicity following denaturation. Antigenicity of domain B was conserved following denaturation, and reactivity of some of the MAb's defining epitopes in domain B could react with a 9000-dalton peptide fragment of the E glycoprotein (Heinz et al., 1983b). This reactivity could be abolished if the fragment was alkylated prior to reacting with MAb. CBA analysis indicated that the SLE epitopes were arranged in a continuum of six overlapping sites. While these authors did not investigate the enhancement phenomenon with SLE virus, subsequent studies with JE

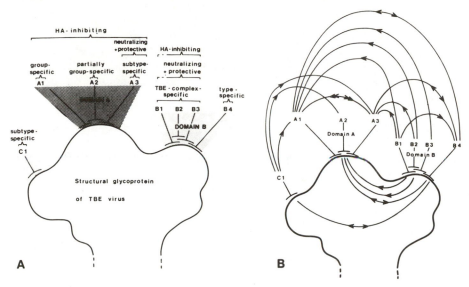

FIGURE 6. Model for the TBE virus E glycoprotein. (A) Biological reactivities and serological cross-reactivities of the epitopes; (B) enhancement of binding (indicated by arrows) between and among domains. From Heinz *et al.* (1983a,b, 1984a).

virus, YF virus, and DEN viruses corroborate the observation of enhanced antibody binding (K. Yasui, J. Schlesinger, and D. Burke, personal communications). Unpublished observations from our laboratory indicate that the type-specific SLE virus epitopes are also stable to reduction denaturation.

4. Expression of Epitopes on the Surface of Infected Cells and in the Insect Vector

To date, there are no published investigations of the kinetics of appearance of E-glycoprotein epitopes on the surface of virus-infected cells. Unpublished observations from our laboratory indicate that not all epitopes are expressed on the cell surface at the same time (Fig. 7). The reasons for this differential expression are unknown at this time. Although the expression of antigenic domains in infected mosquito vectors is a topic of investigation in numerous laboratories, no conclusive information is currently available.

C. Flavivirus Immunity

1. Role of Humoral Immunity in Protection from Flavivirus Infection

Most of our knowledge about the role of humoral immunity in flavivirus disease prevention comes from investigations using SLE virus

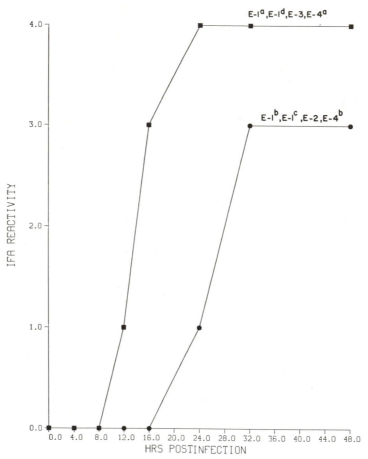

FIGURE 7. Expression of SLE E-glycoprotein epitopes on the surface of unfixed, virus-infected SW-13 human adenocarcinoma cells as a function of time postinfection. IFA reactivity is measured in a relative 1+ to 4+ scale.

(Mathews and Roehrig, 1984) and TBE virus (Heinz *et al.*, 1983a). These investigators used passive antibody transfer to determine the efficiency that each antiepitope MAb had in protecting animals from a lethal challenge with a virulent flavivirus. When administered in large doses (50–200 μg/animal), most of the anti-SLE MAb's were marginally protective (Table VI). Anti-E-1^c MAb, however, was extremely efficient in protecting animals from virus challenge. As little as 50 ng antibody was needed to protect 100% of challenged animals (Mathews and Roehrig, 1982). This antibody could also protect previously exposed animals from death if it were administered prior to viral neural invasion. If MAb's that defined the cross-reactive SLE E-glycoprotein epitopes were combined prior to injection, a synergistic protective effect was observed. Subsequent analysis of cross-protection from challenge with various flaviviruses (SLE virus, JE virus, and Murray Valley encephalitis virus) indicated that non-neutralizing, cross-reactive MAb's could protect animals from challenge with these viruses (L. Staudinger, unpublished observations). The efficiency of this protection appeared to be related to the virulence of the challenge virus and the positioning of the various epitopes on the glycoprotein of these viruses.

Protection studies with TBE-virus-challenged animals yielded results similar to those for SLE virus (Heinz *et al.*, 1983a). Protection mediated by TBE-complex-specific neutralizing antibody was most efficient. These results have led investigators to hypothesize the presence of a "critical" site on the flavivirus E glycoprotein. MAb's that react with epitopes that are located within this site are extremely efficient in protecting animals from virus challenge and neutralizing virus infectivity. Recent observations with JE virus corroborate this hypothesis (K. Yasui, personal communication); however, current investigations with YF virus indicate that N efficiency may also be related to cell type for some MAb's (J. Schlesinger, personal communication).

2. Antibody-Mediated Enhancement of Flavivirus Replication

Since the initial observation that nonneutralizing antibody, complexed to virus, could enhance the replication of that virus in Fc-receptor-bearing cells, the role that this mechanism has in naturally occurring disease has been hotly debated (Halstead and O'Rourke, 1977; Peiris and Porterfield, 1979; Halstead *et al.*, 1980). Epidemiological evidence appeared to indicate that the severity of DEN virus infections correlated to the immune status of the individual. Patients with a previous exposure to one DEN virus serotype exhibited severe disease symptoms when subsequently infected with DEN virus of another serotype (Halstead *et al.*, 1973). The availability of flavivirus-type-specific and cross-reactive MAb's presented an opportunity to experimentally test the hypothesis of antibody-enhanced replication.

The first experiments studying this phenomenon centered on using panels of virus-specific MAb's to determine the virus cross-reactivities of those MAb's that demonstrated the ability to enhance virus replication in macrophage like cell lines. The hypothesis of these studies was that the more cross-reactive MAb's should be more efficient at mediating enhancement of virus replication. The results of a limited study with WN virus indicated that all MAb's, regardless of their specificity, could enhance virus replication in the P388D1 mouse macrophage cell line (Peiris *et al.*, 1982).

A subsequent more detailed study using a larger panel of DEN-specific MAb's gave results that were unlike those found with WN virus (Brandt *et al.*, 1982). These investigators monitored the ability of anti-DEN MAb's to enhance replication of DEN 2 virus in the U-937 human monocyte cell line. Their results indicated that neither type-specific neutralizing MAb, diluted beyond its end point, nor weakly neutralizing HI-positive DEN-serotype-specific MAb could enhance virus infection. On the other hand, MAb's that cross-reacted with DEN and JE viruses demonstrated infection-enhancement titers much greater than their PRNT or HI titers. These investigators hypothesized that the flavivirus conserved epitopes were most probably located on the tips of the glycoprotein spike structure, which allowed interaction of virus antibody complexes with Fc receptors.

A more recent study with YF virus yielded results more similar to those reported with WN virus (Schlesinger and Brandriss, 1983). These studies demonstrated that 13 MAb's specific for the E glycoprotein of YF virus could mediate enhancement of virus replication in the P388D1 cell line regardless of their cross-reactivity.

The inconsistencies of these reports, as yet unresolved, are probably an indication of the complex nature of the true immune-enhancement phenomenon. Whether these differences are caused by antigenic variation among viruses or by intrinsic differences in the macrophagelike cell lines used in the studies is unknown. It is apparent, in this case at least, that extrapolation of *in vitro* results to *in vivo* disease syndromes is difficult.

D. Monoclonal Antibodies in Flavivirus Serology and Epidemiology

Perhaps the contribution of hybridoma technology that has been most significant is in the area of rapid flavivirus identification. Due to the close antigenic relatedness of this virus family, specific MAb reagents are particularly valuable. Conventional diagnostic techniques usually required 1–2 weeks to identify a suspected flavivirus agent. Serologically, PRNT was usually the test of choice in making such identifications. Type-specific MAb reagents have drastically shortened the time required for such identifications by allowing application of other, more rapid techniques, such as IFA or ELISA. Enzyme-conjugates of the more broadly cross-reactive antiflavivirus MAb's are currently being used as standard reagents in various types of antibody- or antigen-capture binding assays, which allows for more rapid diagnosis of the antibody status or of possible flavivirus infection (Monath and Nystrom, 1984; Monath *et al.*, 1984b, 1986; Gould *et al.*, 1985a, Kuno *et al.*, 1986). A list of potentially useful flavivirus MAb's is given in Table VII.

These same specific MAb reagents are being applied to identification of virus agents in mosquito vector populations. A sensitive and rapid identification of virus agents and infected mosquito populations has been of great assistance to arbovirus epidemiologists in predicting and tracking potential or active disease epidemics.

E. Vaccine Strategy

As with alphaviruses, much of the reason for studying the immune requirements for protection from flavivirus infection centers on the de-

TABLE VII. Serologically Useful Flavivirus E Glycoprotein Monoclonal Antibodies

Virus	MAb	Isotype	HI	N	Reactivity
SLE	6B5A-2	IgG2a	+	+	SLE virus[a]
SLE	6B6C-1	IgG2a	+	−	Flavivirus group[a]
JE	6B4A-10	IgG2a	−	−	SLE complex[b]
JE	NARMA 13	Ig3	+	ND[i]	JE virus[c]
WN	F7/101	IgG1	+/−	+/−	WN subgroup[d]
TBE	4D9	IgG1	+	+	TBE complex[e]
DEN 1	1F1-3	ND	−	−	DEN 1 viruses[f]
DEN 2	3H5-1-15	IgG1	+	−	DEN 2 viruses[g]
DEN 3	8A1-12	IgG2a	−	+	DEN 3 viruses[f]
DEN 4	1H10-6-7	IgG2a	+	−	DEN 4 viruses[g]
DEN 4	5C9-1	IgG2a	−	−	DEN complex[f]
YF	2D12	IgG2a	+	+/−	YF virus[h]

[a] Roehrig et al. (1983).
[b] CDC prepared (as yet unpublished data).
[c] Kobayashi et al. (1984); PRNT was not done.
[d] Peiris et al. (1982).
[e] Heinz et al. (1983a).
[f] Henchal et al. (1982).
[g] WRAIR prepared (as yet unpublished data).
[h] Schlesinger et al. (1983).
[i] (ND) Not determined.

velopment of vaccines for these agents. To date, only the live attenuated 17D YF virus vaccine is licensed for use in the United States. An inactivated TBE virus vaccine is available in Europe. A live attenuated JE virus vaccine is currently in use in Japan, and efforts are being made to have it licensed for limited human use in the United States. Attempts to isolate live attenuated vaccine candidates for other medically important flaviviruses, such as DEN or SLE virus, have been wholly unsuccessful. Concerted efforts are being made to interface immunologists with recombinant DNA technologists in attempts to devise artificial immunogens. The observation that antigenicity can be maintained on a small peptide fragment, as has been shown with TBE virus, is promising (Heinz et al., 1984b). It may be possible to produce synthetic vaccine candidates using the amino acid sequences of these reactive fragments as blueprints. The possibility of deriving artificial vaccine candidates composed of synthetic peptides, vaccinia-vectored genes, or antiidiotypic antibodies is currently the research focus of a number of laboratories.

IV. OTHER VIRUSES

While most of the MAb research has centered on alphaviruses and flaviviruses, there are a number of laboratories working with other togaviruses. Rubella virus and lactic dehydrogenase virus are currently being investigated. A study with rubella virus identified six spatially distinct epitopes on the E1 glycoprotein, four of which appeared to be involved in virus hemagglutination (Waxham and Wolinsky, 1985).

V. ASSESSMENT OF FUTURE RESEARCH GOALS

Hybridoma technology has made a significant contribution to our understanding of the complex immunology of alphaviruses and flaviviruses. One area for further investigation is determining the contribution of various epitopes to the interaction with T-lymphocytes. These types of studies are important for answering two questions: (1) Are CTL responses necessary for protection or recovery from disease? (2) What are the epitopes involved in helper T-lymphocyte recognition that lead to an anamnestic immune response?

Application of these and other monoclonal reagents to the analysis of the virus-coded nonstructural proteins should enable us to acquire some answers as to their nature and function. It has recently been shown that MAb's specific for a YF virus nonstructural protein (gp48) can protect animals from a lethal encephalitis (J. Schlesinger, personal communication). Homogeneous antibody reagents that are available in large quantity can be used in affinity purification procedures as well as for immunological markers of identity. Use of antiidiotypic antibodies specific for neutralizing MAb's should aid in identification of receptor complexes on the surface of susceptible cells (Kauffman et al., 1983).

A broader application of these reagents to the study of antigenic drift, with an eye toward vaccine production and identification of virulence factors, will allow for a more complete understanding of the epidemiology and diversity of these viral agents. Recent observations with Sindbis virus indicate that changes in the E2 glycoprotein as detected with MAb's lead to a loss of virulence (Olmstead et al., 1984). These observations are consistent with the antigenic changes that correlate with virulence changes in VEE viruses (Kinney et al., 1983; Roehrig and Mathews, 1985).

A better understanding of what leads to the development of successful antivirus immunity should allow us to develop effective vaccines for this virus group, which has been especially reluctant in yielding to vaccine isolations by current production technology. With further more concerted effort, the mystery of antibody-enhanced virus replication will probably be unraveled. Finally, it is to be hoped that development of successful vaccine prophylaxis will reduce the impact that these virus agents have on the human and animal populations of the world.

ACKNOWLEDGMENTS. The author would like to thank Drs. T. A. Brawner, H. G. Riggs, Jr., M. J. Schlesinger, D. W. Trent, and T. P. Monath for their support and help in much of the research mentioned here. Thanks also go to J. H. Mathews, A. R. Hunt, R. M. Kinney, and S. C. Ure for technical and secretarial assistance. A final thanks goes to the arbovirus hybridoma community for their kind support in the development of this manuscript.

REFERENCES

Boere, W. A. M., Benaissa-Trouw, B. J., Harmsen, M., Kraaijeveld, C. A., and Snippe, H., 1983, Neutralizing and non-neutralizing monoclonal antibodies to the E1 glycoprotein of Semliki Forest virus can protect mice from lethal encephalitis, *J. Gen. Virol.* **64:**1405–1408.

Boere, W. A. M., Harmsen, M., Vinje, J., Benaissa-Trouw, B. J., Kraaijeveld, C. A., and Snippe, H., 1984, Identification of distinct determinants on Semliki forest virus by using monoclonal antibodies with different antiviral activities, *J. Virol.* **52:**575–582.

Boere, W. A. M., Benaissa-Trouw, B. J., Harmsen, T., Erich, T., Kraaijeveld, C. A., and Snippe, H., 1985, Mechanisms of monoclonal antibody-mediated protection against virulent Semliki Forest virus, *J. Virol.* **54:**546–551.

Brandt, W. E., McCown, J. M., Gentry, M. K., and Russell, P. K., 1982, Infection enhancement of Dengue type 2 virus in the U-937 human monocyte cell line by antibodies to flavivirus cross-reactive determinants, *Infect. Immun.* **36:**1036–1041.

Calisher, C. H., Shope, R. E., Brandt, W. E., Casals, J., Karabatsos, N., Murphy, F. M., Tesh, R. B., and Wiebe, M. E., 1980, Proposed antigenic classification of registered arboviruses. I. Togaviridae, alphaviruses, *Intervirology* **14:**229–232.

Casals, J., 1957, The arthropod-borne group of animal viruses, *Trans. N. Y. Acad. Sci.* **19:**219–235.

Casals, J., 1967, Immunological techniques for animal viruses, in: *Methods in Virology* (H. Maramarosch and H. Koprowski, eds.), pp. 113–198, Academic Press, New York.

Chanas, A. C., Johnson, B. K., and Simpson, D. I. H., 1976, Antigenic relationships of alphaviruses by a simple microculture cross-neutralization method, *J. Gen. Virol.* **32:**295–300.

Chanas, A. C., Gould, E. A., Clegg, J. C. S., and Varma, M. G. R., 1982, Monoclonal antibodies to Sindbis virus glycoprotein E1 can neutralize, enhance infectivity, and independently inhibit haemagglutination or haemolysis, *J. Gen. Virol.* **58:**37–46.

Clegg, J. C. S., Chanas, A. C., and Gould, E. A., 1983, Conformational changes in Sindbis virus E1 glycoprotein induced by monoclonal antibody binding, *J. Gen. Virol.* **64:**1121–1126.

Dalrymple, J. M., Vogel, S. N., and Teramoto, A. Y., 1973, Antigenic components of group A arbovirus virions, *J. Virol.* **12:**1034–1042.

Dalrymple, J. M., Schlesinger, S., and Russell, P. K., 1976, Antigenic characterization of two Sindbis envelope glycoproteins separated by isoelectric focusing, *Virology* **69:**93–103.

DeMadrid, A. T., and Porterfield, J. S., 1974, The flaviviruses (group B arboviruses): A cross-neutralization study, *J. Gen. Virol.* **23:**91–96.

Dittmar, D., Haines, H. G., and Castro, A., 1980, Monoclonal antibodies specific for dengue virus type 3, *J. Clin. Microbiol.* **12:**74–78.

Ey, P. L., Prowse, S. J., and Jenkin, C. R., 1978, Isolation of pure IgG1, IgG2a, and IgG2b immunoglobulin from mouse serum using protein A-Sepharaose, *Immunochemistry* **15:**429–436.

Gentry, M. K., Henchal, E. A., McCown, J. M., Brandt, W. E., and Dalrymple, J. M., 1982, Identification of distinct antigenic determinants on dengue-2 virus using monoclonal antibodies, *Am. J. Trop. Med. Hyg.* **31:**548–555.

Goding, J. W., 1967, Use of staphylococcal A as an immunological reagent, *J. Immunol. Methods* **20:**241–253.

Gould, E. A., Buckley, A., and Cammack, N., 1985a, Use of biotin streptavidin interaction to improve flavivirus detection by immunofluorescence and ELISA tests, *J. of Virol. Methods* **11:**41–48.

Gould, E. A., Buckley, A., Cammack, N., Barrett, A. D. T., Clegg, J. C. S., Ishak, R., and Varma, M. G. R., 1985b, Examination of the immunological relationships between fla-viviruses using yellow fever virus monoclonal antibodies, *J. Gen. Virol.* **66:**1369–1382.

Gresikova, M., and Sekeyova, M., 1984, Antigenic relationships among viruses of the tick-borne encephalitis complex as studied by monoclonal antibodies, *Acta. Virol.* **28:**64–68.

Halstead, S. B., and O'Rourke, E. J., 1977, Dengue viruses and mononuclear phagocytes. I. Infection enhancement by nonneutralizing antibody, *J. Exp. Med.* **146:**201–217.

Halstead, S. B., Chow, J. S., and Marchette, N. J., 1973, Immunological enhancement of dengue virus replication, *Nature (London) New Biol.* **243:**24–26.

Halstead, S. B., Porterfield, J. S., and O'Rourke, E. J., 1980, Enhancement of dengue virus infection in monocytes by flavivirus antisera, *Am. J. Trop. Med. Hyg.* **29:**638–642.

Heinz, F. X., Berger, R., Tuma, W., and Kunz, C., 1983a, A topological and functional model of epitopes on the structural glycoprotein of tick-borne encephalitis virus defined by monoclonal antibodies, *Virology* **126:**525–537.

Heinz, F. X., Berger, R., Tuma, W., and Kunz, C., 1983b, Location of immunodominant antigenic determinants on fragments of the tick-borne encephalitis virus glycoprotein: Evidence for two different mechanisms by which antibodies mediate neutralization and hemagglutination inhibition, *Virology* **130:**485–501.

Heinz, F. X., Mandl, C., Berger, R., Tuma, W., and Kunz, C., 1984a, Antibody-induced con-formational changes result in enhanced avidity of antibodies to different antigenic sites on the tick-borne encephalitis virus glycoprotein, *Virology* **133:**25–34.

Heinz, F. X., Tuma, W., Guirakhoo, F., Berger, R., and Kunz, C., 1984b, Immunogenicity of tick-borne encephalitis virus glycoprotein fragments: Epitope-specific analysis of the antibody response, *J. Gen. Virol.* **65:**1921–1929.

Henchal, E. A., Gentry, M. K., McCown, J. M., and Brandt, W. E., 1982, Dengue virus-specific and flavivirus group determinants identified with monoclonal antibodies by indirect immunofluorescence, *Am. J. Trop. Med. Hyg.* **31:**830–836.

Hunt, A. R., and Roehrig, J. T., 1985, Biochemical and biological characteristics of epitopes on the E1 glycoprotein of western equine encephalitis virus, *Virology* **142:**334–346.

Karabatsos, N., 1975, Antigenic relationships of group A arboviruses by plaque reduction neutralization testing, *Am. J. Trop. Med. Hyg.* **24:**527–532.

Kauffman, R. S., Noseworthy, J. H. Nepom, J. T., Finberg, R., Fields, B. N., and Greene, M. I., 1983, Cell receptors for the mammalian reovirus. II. Monoclonal anti-idiotypic an-tibody blocks viral binding to cells, *J. Immunol.* **131:**2539–2541.

Kimura-Kuroda, J., and Yasui, K., 1983, Topographical analysis of antigenic determinants on envelope glycoprotein V3 (E) of Japanese encephalitis virus, using monoclonal an-tibodies, *J. Virol.* **45:**124–132.

Kinney, R. M., Trent, D. W., and France, J. K., 1983, Comparative immunological and bio-chemical analyses of viruses in the Venezuelan equine encephalitis complex, *J. Gen. Virol.* **64:**135–147.

Kobayashi, Y., Hasegawa, H., Oyama, T., Tamai, T., and Kusaba, T., 1984, Antigenic analysis of Japanese encephalitis virus by using monoclonal antibodies, *Infect. Immun.* **44:**117–123.

Kuno, G., Gubler, D. J., and Santiago de Weil, N., 1986, Antigen capture ELISA for the identification of Dengue viruses, *J. Virol. Methods*, in press.

Laemmli, U. K., 1970, Cleavage of structural proteins during the assembly of the head of bacteriophage T4, *Nature (London)* **277:**680–685.

Lostrom, M. E., Stone, M. R., Tam, M., Burnette, W. M., Pinter, A., and Nowinski, R. C., 1979, Monoclonal antibodies against murine leukemia viruses: Identification of six antigenic determinants on the p15(E) and gp70 envelope proteins, *Virology* **98:**336–350.

Mathews, J. M., and Roehrig, J. T., 1982, Determination of the protective epitopes on the glycoproteins of Venezuelan equine encephalomyelitis virus by passive transfer of mon-oclonal antibodies, *J. Immunol.* **129:**2763–2767.

Mathews, J. M., and Roehrig, J. T., 1984, Elucidation of the topography and determination of the protective epitopes on the E glycoprotein of Saint Louis encephalitis virus by passive transfer with monoclonal antibodies, *J. Immunol.* **132:**1533–1537.

Mathews, J. H., Roehrig, J. T., and Trent, D. W., 1985, Role of complement and Fc portion if immunoglubulin G in immunity to Venezuelan equine encephalomyelitis virus infection with glycoprotein-specific monoclonal antibodies, *J. Virol.* **55:**594–600.

Monath, T. P., and Nystrom, R. R., 1984, Detection of yellow fever virus in serum by enzyme immunoassay, *Am. J. Trop. Med. Hyg.* **33:**151–157.

Monath, T. P., Kinney, R. M., Schlesinger, J. J., Brandriss, M. W., and Bres, P., 1983, Ontogeny of yellow fever 17D vaccine: RNA oligonucleotide fingerprint and monoclonal antibody analyses of vaccine produced world-wide, *J. Gen. Virol.* **64:**627–637.

Monath, T. P., Schlesinger, J. J., Brandriss, M. W., Cropp, C. B., and Prange, W. B. 1984a, Yellow fever monoclonal antibodies: Type-specific and cross-reactive determinants identified by immunofluorescence, *Am. J. Trop. Med. Hyg.* **33:**695–698.

Monath, T. P., Nystrom, R. R., Bailey, R. E., Calisher, C. H., and Muth, D. J., 1984b, Immunoglubulin M antibody capture enzyme-linked immunosorbent assay for diagnosis of St. Louis encephalitis, *J. Clin. Microbiol.* **20:**784–790.

Monath, T. P., Hill, L. J., Brown, N. V., Cropp, C. B., Schlesinger, J. J., Saluzzo, J. F., and Wands, J. R., 1986, Sensitive and specific monoclonal immunoassay for detecting yellow fever virus in laboratory and clinical specimens, *J. Clin. Microbiol.*, in press.

Nicolatti, R. A., Briles, D. E., Schroer, J., and Davie, J. M., 1979, Isoelectric focusing of immunoglobulins: Improved methodology, *J. Immunol. Methods* **33:**101–115.

Olmstead, R. A., Baric, R. S., Sawyer, B. A., and Johnston, R. E., 1984, Sindbis virus mutants selected for rapid growth in cell culture display attenuated virulence in animals, *Science* **225:**424–426.

Pedersen, C. E., and Eddy, E. A., 1974, Separation, isolation, and immunochemical studies of the structural proteins of Venezuelan equine encephalomyelitis virus, *J. Virol.* **14:**740–744.

Peiris, J. S. M., and Porterfield, J. S., 1979, Antibody-mediated enhancement of flavivirus replication in macrophage cell lines, *Nature (London)* **282:**509–511.

Peiris, J. S. M., Porterfield, J. S., and Roehrig, J. T., 1982, Monoclonal antibodies against the flavivirus West Nile, *J. Gen. Virol.* **58:**283–289.

Porterfield, j. S., 1980, Antigenic characteristics and classification of togaviridae, in: *The Togaviruses: Biology, Structure, and Replication* (R. W. Schlesinger, ed.), pp. 13–46, Academic Press, New York.

Porterfield, J. S., Casals, J., Chumakov, M. P., Gaidamovich, S. Y., Hannoun, J., Holmes, I. H., Horzinek, M. C., Mussgay, Oker-Blom, N., Russell, P. K., and Trent, D. W., 1978, Togaviridae, *Intervirology* **9:**129–148.

Righi, M., Radaelli, A., Ricciardi, P., Liboi, E., and de Giuli Morghen, C., 1983, Identification by monoclonal antibodies of a new epitope in the glycoprotein complex of Sindbis virus, *J. Virol. Methods* **6:**203–214.

Roehrig, J. T., and Mathews, J. H., 1985, The neutralization site on the E2 of Venezuelan equine encephalomyelitis virus is composed of multiple conformationally stable epitopes, *Virology* **142:**347–356.

Roehrig, J. T., Corser, J. A., and Schlesinger, M. J., 1980, Isolation and characterization of hydridoma cell lines producing monoclonal antibodies against the structural proteins of Sindbis virus, *Virology* **101:**41–49.

Roehrig, J. T., Day, J. W., and Kinney, R. M., 1982a, Antigenic analysis of the surface glycoproteins of a Venezuelan equine encephalomyelitis virus (TC-83) using monoclonal antibodies, *Virology* **118:**269–278.

Roehrig, J. T., Gorski, D., and Schlesinger, M. J., 1982b, Properties of monoclonal antibodies directed against the glycoproteins of Sindbis virus, *J. Gen. Virol.* **59:**421–425.

Roehrig, J. T., Mathews, J. H., and Trent, D. W., 1983, Identification of epitopes on the E glycoprotein of Saint Louis encephalitis virus using monoclonal antibodies, *Virology* **128:**118–126.

Roehrig, J. T., Hunt, A. R., and Mathews, J. H., 1985, Identification of anti-idiotype antibodies that mimic the neutralization site of Venezuelan equine encephalomyelitis virus, *In High-Technology Route to Virus Vaccines*, (G. R. Dreesman, J. G. Bronson, and R. C. Kennedy, eds.), American Society for Microbiology, Washington, D. C. pp. 142–153.

Schlesinger, J. J., and Brandriss, M. W., 1983, 17D yellow fever virus infection of P388D1 cells mediated by monoclonal antibodies: Properties of the macrophage Fc receptor, *J. Gen. Virol.* **64:**1255–1262.

Schlesinger, J. J., Brandriss, M. W., and Monath, T. P., 1983, Monoclonal antibodies distinguish between wild and vaccine strains of yellow fever virus by neutralization, hemagglutination inhibition, and immune precipitation of the virus envelope protein, *Virology* **125:**8–17.

Schmaljohn, A. L., Johnson, E. D., Dalrymple, J. M., and Cole, G. A., 1983a, Non-neutralizing monoclonal antibodies can prevent lethal alphavirus encephalitis, *Nature (London)* **297:**70–72.

Schmaljohn, A. L., Kokubun, J. M., and Cole, G. A., 1983b, Protective monoclonal antibodies define maturational and pH-dependent antigenic changes in Sindbis virus E1 glycoprotein, *Virology* **130:**144–154.

Stephenson, J. R., Lee, J. M., and Wilton-Smith, P. D., 1984, Antigenic variation among members of the tick-borne encephalitis complex, *J. Gen. Virol.* **65:**81–89.

Towbin, H., Staehelin, T., and Gordon, J., 1979, Electrophoretic transfer of proteins from polyacrylamide gels to nitrocellulose sheets: Procedures and some applications, *Proc. Natl. Acad. Sci. U.S.A.* **76:**4350–4354.

Trent, D. W., 1977, Antigenic characterization of flavivirus structural proteins separated by isoelectric focusing, *J. Virol.* **22:**608–618.

Trent, D. W., Clewley, J. P., France, J. K., and Bishop, D. H. L., 1979, Immunochemical and oligonucleotide fingerprint analyses of Venezuelan equine encephalomyelitis complex viruses, *J. Gen. Virol.* **43:**365–381.

Voller, A., Bidwell, D., and Bartlett, A., 1976, Microplate enzyme immuno-assay for the immunodiagnosis of virus infections, in: *Handbook of Clinical Microbiology* (N. R. Rose and H. Friedman, eds.), pp. 456–462, American Society for Microbiology, Washington, D. C.

Waxham, M. N., and Wolinsky, J. S., 1985, Detailed immunologic analysis of the structural polypeptides of rubella virus using monoclonal antibodies, *Virology* **143:**153–165.

Webster, R. G., and Laver, W. G., 1980, Determination of the number of non-overlapping antigenic areas on Hong Kong (H3N2) influenza virus hemagglutinin with monoclonal antibodies and the selection of variants with potential epidemiological significance, *Virology* **104:**139–148.

Ziemieki, A., and Garoff, H., 1978, Subunit composition of the membrane glycoprotein complex of Semliki Forest virus, *J. Mol. Biol.* **122:**259–269.

CHAPTER 10

Structure of the Flavivirus Genome

CHARLES M. RICE, ELLEN G. STRAUSS, AND
JAMES H. STRAUSS

I. INTRODUCTION

The flaviviruses were formerly classified as a genus in the family Togaviridae. They have now been elevated to family status, family Flaviviridae, in part because of differences in replication and assembly (Westaway, 1980) (Chapter 11) and in part because their genome structure is
quite different from that of the alphaviruses (Rice *et al.*, 1985) (compare
with Chapter 3). With the determination of the complete nucleotide sequence of the yellow fever virus genome and a large portion of the Murray
Valley encephalitis virus genome, it has become clear that these viruses
represent a distinct group among the plus-stranded RNA viruses. This
chapter will focus on the implications of these and other recent sequence
data on flavivirus gene expression, replication, and evolution.

II. PHYSICAL STRUCTURE OF THE GENOME

The flavivirus genome consists of a single-stranded RNA nearly 11
kilobases in length (Westaway, 1980; Deubel *et al.*, 1983). This RNA is
plus-stranded and infectious, and thus the incoming RNA must be translated to produce any enzymes required for RNA replication. No subgenomic messenger RNAs (mRNAs) have been identified in flavivirus-infected cells, and it is generally believed that the virion RNA (or RNA
molecules virtually identical to this RNA) is the only flavivirus-specific

CHARLES M. RICE, ELLEN G. STRAUSS, AND JAMES H. STRAUSS • Division of Biology, California Institute of Technology, Pasadena, California 91125.

mRNA (Boulton and Westaway, 1977; Naeve and Trent, 1978; Wengler *et al.*, 1978; Westaway, 1980). The genomic RNA has a type 1 cap at the 5' end of the form $m^7GpppAmp$ (Wengler *et al.*, 1978; Cleaves and Dubin, 1979) and lacks cap-associated and internal base methylated adenine residues (Cleaves and Dubin, 1979). It is of interest that this cap structure differs from that of alphaviruses, which have a type 0 cap (see Chapter 3), and from most vertebrate mRNAs, which have a base methylated adenine in the penultimate position. Because of these differences and since RNA replication occurs in the cytoplasm, the cap may be added by one or more viral enzymes. Several studies have shown that flavivirus RNAs do not contain a 3'-terminal polyadenylate [poly(A)] tract (Wengler *et al.*, 1978; Wengler and Wengler, 1981; Vezza *et al.*, 1980; Deubel *et al.*, 1983), but instead terminate with a nucleotide sequence that shows a fairly high degree of conservation among viruses examined to date. This conserved sequence, which may be part of a 3'-terminal secondary structure, is probably important in RNA replication and perhaps encapsidation and is discussed in more detail below.

In the case of yellow fever virus RNA (17D vaccine strain), the nucleotide sequence of the genome has been determined in its entirety from complementary DNA (cDNA) clones and is 10,862 nucleotides in length (Rice *et al.*, 1985). The base composition is 27.3% A, 23.0% U, 28.4% G, and 21.3% C, and the molecular weight of this RNA, calculated as the Na^+ form, is 3.75×10^6. An examination of this sequence for open reading frames is shown in Fig. 1. The RNA contains a single long open reading frame of 10,233 nucleotides beginning from the first AUG in frame 2, followed by three closely spaced termination codons. This open reading frame spans virtually the entire length of the RNA, excluding only 118 nucleotides of 5' untranslated and 511 nucleotides of 3' untranslated sequence. No other long open reading frames are found within the RNA, either in the genome sense or in the complementary-strand sense, as illustrated in Fig. 1, and it is unlikely that any products are translated from the yellow fever genome other than those translated from this very long open reading frame. Support for this view comes from partial sequence analysis of cDNA clones of Murray Valley encephalitis virus (Dalgarno *et al.*, 1986) and dengue virus type 2 (V. Vakharia, T. Yaegashi, R. Feighny, S. Kohlekar, and R. Padmanabhan, unpublished data). These sequences contain single long open reading frames encoding polypeptides homologous to those of yellow fever virus (see below). Thus far, no conservation of alternative open reading frames in either the genome or complementary strands has been found among these viruses. However, before the possibility that minor flavivirus-specific products could be encoded in alternative reading frames is dismissed, more extensive sequence comparisons are needed. The length of the 5' untranslated region of Murray Valley encephalitis virus has been determined by primer extension and is approximately 97 nucleotides, or about 20 nucleotides shorter than that of yellow fever virus (Dalgarno *et al.*, 1986).

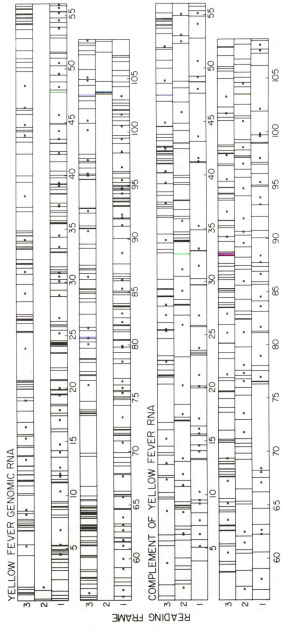

FIGURE 1. Open reading frame analysis. Graphic representations of the distribution of termination codons (vertical lines) in all three possible reading frames are shown for the complete yellow fever virus RNA sequence and its complement (Rice *et al.*, 1986a). (*) First methionine codon in each open reading frame.

III. GENOME ORGANIZATION

A. Gene Order

As mentioned above, the flavivirus genomic RNA is believed to be the only virus-specific mRNA species, and it is translated into the three structural proteins (C, M, and E) as well as two large nonstructural proteins (NV4 and NV5) and a complex set of smaller nonstructural proteins (reviewed in Westaway, 1980). Profiles of yellow-fever-virus-specific polypeptides (17D strain) synthesized in various cell types are shown in Fig. 2. Of these virus-specified polypeptides, three are usually glycoproteins: E, the virion envelope protein, which for yellow fever virus (Schlesinger *et al.*, 1983) (Fig. 2b) and Kunjin virus (Wright *et al.*, 1981; Wright, 1982) is found in both glycosylated and nonglycosylated forms, and intracellular proteins NV2 and NV3 [referred to as prM and NS1, respectively (see Table I and below)]. In the case of yellow fever virus, E, NS1 (NV3), and prM (NV2) show considerable heterogeneity (Schlesinger *et al.*, 1983) (Fig. 2): Multiple forms of cell-associated E are present, at least some of which are also found in purified virus; NS1 appears at least as a doublet; and prM is also made up of at least two forms the relative prevalence of which varies among host cells. Similar observations have been made for dengue 2 virus prM (Smith and Wright, 1985) and Kunjin virus E, in which the mature nonglycosylated virion protein (Wright, 1982) may be produced by processing of intracellular glycosylated precursors [both by trimming of carbohydrate chains and by proteolytic cleavage (Wright *et al.*, 1981)]. The pattern of other small nonstructural proteins is even more variable when compared among different host cells, with the exception of a species of about 29 kilodaltons [the intense band migrating slightly slower than prM, possibly NVX (see below)] and the smallest nonstructural protein (possibly NV1) (Fig. 2a). Definitive comparisons of the small nonstructural proteins among different flaviviruses are even more difficult; for this reason we propose an alternative nomenclature (Table I) that will be discussed later in this section. The mobilities of the largest nonstructural proteins, NV4 and NV5 [referred to as NS3 and NS5, respectively, in Fig. 2a (see Table I)], are constant when the proteins are isolated from different cell types, but the relative yields can differ markedly.

The complete translated nucleotide sequence of the yellow fever virus genome is shown in Fig. 3, and the positions of some of the virus-specific proteins have been mapped in the RNA sequence by the expedient of isolating these proteins and obtaining N-terminal amino acid sequence for them (Fig. 4) (Bell *et al.*, 1985; Rice *et al.*, 1986b). The start points of these proteins (indicated in Fig. 3) are found in the translated sequence of the long open reading frame and include the three structural proteins C, M, and E, as well as several nonstructural proteins (discussed below). From these results, it is clear that the flavivirus genome is organized into

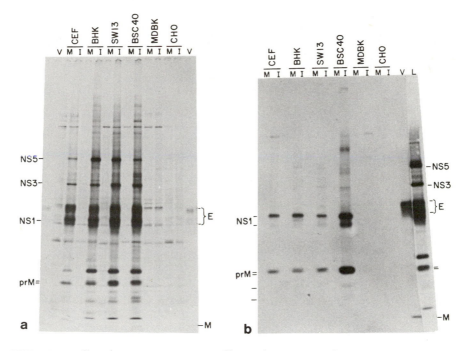

FIGURE 2. Yellow fever virus proteins. Cell monolayers were infected (I) with 20 PFU/cell of yellow fever virus (17D strain) or mock-infected (M) and labeled from 14 to 36 hr after infection with either [^{35}S]methionine (a) or [^{3}H]mannose (b). Postnuclear supernatants were prepared after lysis of the cells with Triton X-100, and equivalent portions of each lysate were immunoprecipitated using yellow-fever-virus-specific hyperimmune ascitic fluid. The samples were reduced, denatured, and separated on 8–20% linear gradient acrylamide slab gels in a discontinuous buffer system. Purified [^{35}S]methionine-labeled virus markers are shown (V); however, the intact capsid protein does not appear in this virus preparation and is often lost when purified virus samples are frozen. In (b), a portion of the infected SW13 immunoprecipitate labelled with [^{35}S] methionine was run as a marker in (L). Chicken embryo fibroblasts (CEF), baby hamster kidney (BHK-21), SW13, and BSC-40 (a variant cell line resistant to high temperature derived from BSC-1) cells are permissive for yellow fever virus growth; (MDBK) and Chinese hamster ovary (CHO) are nonpermissive. The positions of the structural (E and M) and the large nonstructural proteins (NS1, NS3, NS5) are indicated using the nomenclature proposed in Table I. prM, the putative glycosylated precursor to M, appears as a doublet, as does the E protein and perhaps NS1. The positions of two small minor glycoproteins are indicated in (b). (From Rice *et al.* 1986b)

two regions: The 5'-terminal one fourth encodes the structural proteins of the virus, and the nonstructural proteins are encoded in the 3'-terminal three fourths of the genome.

The open reading frame begins with the nucleocapsid protein, C (formerly called V2). The first methionine in this reading frame is found immediately adjacent to the N-terminal serine of the mature protein. Thus, in agreement with data from *in vitro* translation of the genomic RNAs of tick-borne encephalitis virus (Svitkin *et al.*, 1978, 1981, 1984), West Nile virus (Wengler *et al.*, 1979), and Kunjin virus (Monckton and

TABLE I. Flavivirus Polypeptides

Protein[a]		M_r[b]	M_pred.[c]	Glycosylated?	Comments
Proposed nomenclature	Old nomenclature				
Structural region					
C	V2	13,000–16,000	13,432	No	Nucleocapsid protein
prM	(NV1½) (NV2) (NV2½)	19,000–23,000	18,813	Yes	Precursor to M
M	V1	8,000–8,500	8,526	No	Virion envelope protein
E	V3	51,000–60,000	53,712	Both forms[d]	Major virion envelope protein
Nonstructural region					
NS1	NV3	44,000–49,000	45,869	Yes	Soluble complement-fixing antigen
ns2a	(NV2½) (NV2)	16,000–21,000	18,086	No	Hydrophobic; function unknown
ns2b	(NV1½)	12,000–15,000	13,823	No	Hydrophobic; function unknown
NS3	NV4	67,000–76,000	69,319	No	Replicase component?
ns4a	(NVX) (NV2½)	24,000–32,000	31,196	No	Hydrophobic; function unknown

| ns4b | (NV1) | 10,000–11,000 | 12,159 | No | Hydrophobic; function unknown |
| NS5 | NV5 | 91,000–98,000 | 104,079 | No | Replicase component? |

[a] Flavivirus proteins ordered according to location in the yellow fever genome. An alternative nomenclature for flavivirus nonstructural proteins is proposed that is based on the yellow fever virus gene order determined by nucleic acid and protein sequence analysis (see the text and Fig. 3). The structural proteins, C, M, and E have been definitively mapped and are encoded in the order shown. M is believed to be derived from a precursor prM located between C and E. The large nonstructural proteins (formerly NV3, NV4, and NV5) have also been definitively mapped and are numbered in order of appearance in the genome (5' → 3') with an upper-case NS designation. The remaining coding sequences in the nonstructural region probably encode several small intracellular polypeptides that are designated by a lower-case ns. Numerous small proteins (some of which may actually represent mixtures of more than one polypeptide species) have been identified in flavivirus-infected cells and were formerly called NV1, NV1½, NV2, NV2½, and NVX (Westaway et al., 1980) (see Chapter 11). Their tentative assignments to ns polypeptides or to prM are shown in parentheses. For alternative nomenclature, see the text and Westaway et al., (1980). Minor virus-specific protein species that have been detected include two small glycoproteins $M_r \approx 13,000$ and 17,000 (Schlesinger et al., 1983; Smith and Wright, 1985); see also Fig. 2b) and NV4½ (apparently related to NV4 (Svitkin et al., 1981; Smith and Wright, 1985), perhaps equivalent to ns2b + NS3).

[b] Range of flavivirus protein sizes estimated from acrylamide gel electrophoresis. Some of these proteins have not yet been identified for all flaviviruses thus far examined (for comparative analyses, see Westaway, 1973; Westaway et al., 1977; Heinz and Kunz, 1982). In particular, definitive comparisons between NV3, NV2, NV2½, NVX, and NV1½ are difficult because of the complexity of flavivirus protein patterns in the $8000 < M_r < 45,000$ size range and are thus far tenuously based only on relative size. Alternative pathways of posttranslational cleavage may be used by different viruses or the same virus in different host cells for the production of the smaller nonstructural and larger nonstructural polypeptides (see, for instance, Fig. 2a). However, given the relatively consistent pattern of structural and larger nonstructural proteins, it seems likely that many of the small nonstructural proteins are also conserved, but their apparent migration on acrylamide gels and their labeling efficiency are influenced by differences in amino acid composition.

[c] Polypeptide molecular weights calculated according to the cleavage sites shown in Fig. 3.

[d] Both glycosylated and nonglycosylated forms of E have been identified for yellow fever virus (Schlesinger et al., 1983) and Kunjin virus (Wright et al., 1981; Wright, 1982).

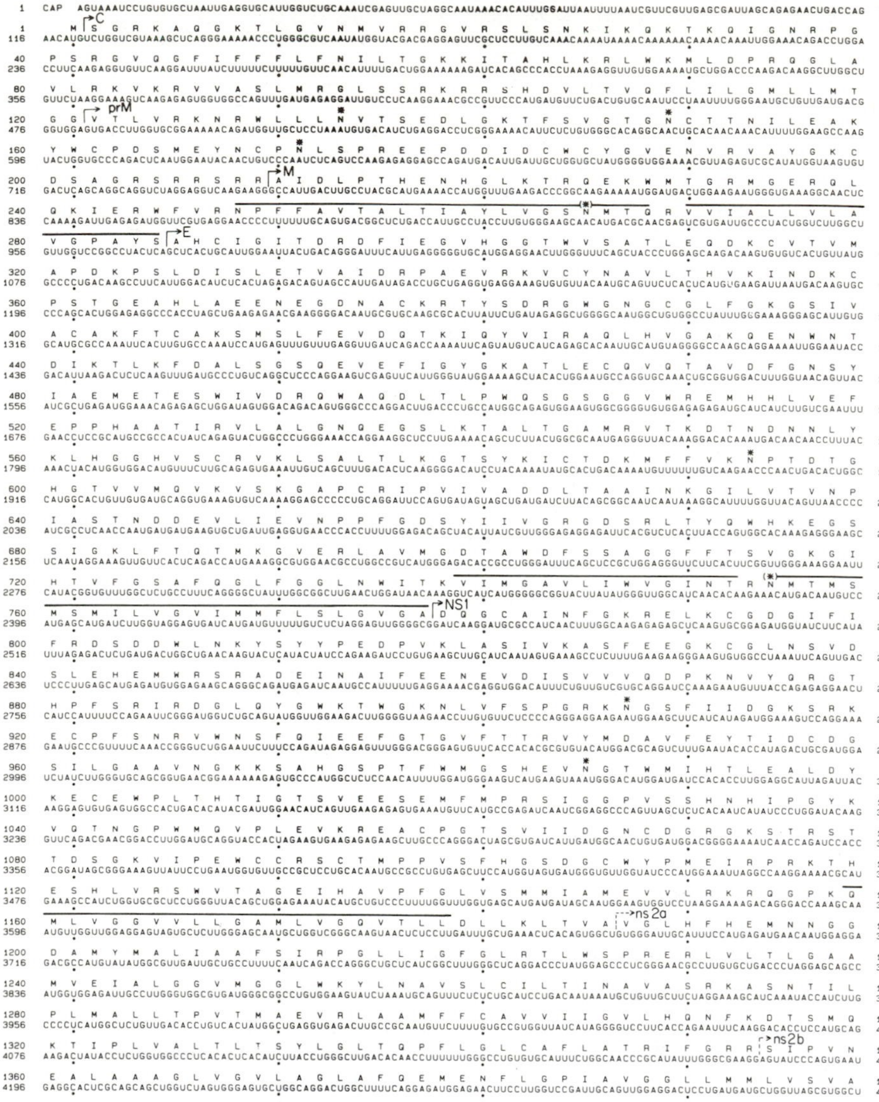

FIGURE 3. Entire sequence of the RNA of yellow fever virus, 17D strain. Nucleotides are numbered from the 5' terminus. Amino acids are numbered from the first methionine in the polyprotein sequence. The beginning of each protein is labeled (see Table I and the text for nomenclature); tentative assignments are indicated by dashed arrows. Putative hydrophobic membrane-associated segments in the structural region and in NS1 are overlined. Potential N-linked glycosylation sites (in prM, E, and NS1) are denoted by asterisks. The

```
1800  T I L M T A T P P G T S D E F P H S N G E I E D V Q T D I P S E P W N T G H D W  1839
5516  ACAAUCUUGAUGACAGCCACCCGCCUGGGACUAGUGAUGAUGAAAUGCAAAUGGAAUGAAGAGAUUGAUCAGGUGUAAACGGACAUGACUGGAUGACAUGACUGG  5635

1840  I L A D K R P T A W F L P S I R A A N V M A A S L R K A G K S V V V L N R K T F  1879
5636  AUCCUAGCUGACAAAAGGCCCACGGCAUUGAGUGUUCUUCCAUCCAUCAAGGUCAAAUGGCUGAAAGGCUGGAAAGGGUUCCUGAAACAGGAAAACCUUU  5755

1880  E R E Y P T I K Q K K P D F I L A T D I A E M G A N L C V E R V L D C R T A F K  1919
5756  GAGAGAGAAAUACCCCACGAUAAAGCAGAAGAAACCUGACUGAAUGCCACUGACAUAGCUGAAAUGGGAGCGAGGUCGUGUGAGCGAGUGCUGGAUGACGGCUUAAG  5875

1920  P V L V D E G R K V A I K G P L R I S A S S A A Q R R G R I G R N P N R D G D S  1959
5876  CCUGUGCUUGUGGAUGAAGGGAGGAAGGUGGCAAUAAAAGGGCCACUUGUAUCCUCGCAUCCUCGCUGCUCAAAGGAGGGGGCGCAUUGGGAGAAACCCAACAGAGGACGACAUCA  5995

1960  Y Y Y S E P T S E N N A H H V C W L E A S M L L D N M E V R G G M V A P L Y G V  1999
5996  UACUACUAUUCUGAGCCUACAAGUGAAAAUAAUGCCCACCACGUCUGCUGGUUGGAGGCCUCAAUGCUCUUGGACAACAUGGAGGUGAGGGGUGGAAUGGUGCGCCCCACUCUAUGGCGGU  6115

2000  E G T K T P V S P G E M R L A D D Q R K V F R E L V R N C D L P V W L S W Q V A  2039
6116  GAAGGAACUAAAAACACCAGUUCCCCUGGUGAAAUGACUGAGGUGAUCCAGAGGAAAGUCAGAGAACUAGUGAGGAAUUGUGACCUUCCCGUUUGGCUUUCUUGGCAAGUGGCC  6235

2040  K A G L K T N D R K W C F E G P E E H E I L N D S G E T V K C R A P G A A K K P  2079
6236  AAGGCUGGUUUGAAGACUAAUGACAGGAAAUGGUGUUUAGAGGGUCCUGAGGAACAUGAGAUCUUGAAUGACAGCGGUGAAACUGUGAAGUGCCGGGCCCCUGGAGGAGCAAAGAAGCCU  6355

2080  L R P R W C D E R V S S D Q S A L S E F I K F A E G R R I G A A E V L V V L S E L  2119    →ns4a
6356  CUGCGCCCAAGGUGGUGUGACGAAAGGGGUGUCGUCGGAUCAAUCUGCCUUGUCGGAAUUUAUAAAGUUCGCAGAGGGACGGCGAAUAGGUGCAGCAGAAGUGCUUGUCGUACUCAGUGAACUC  6475

2120  P D F L A K K G G E A M D T I S V F L H S E E G S R A Y R N A L S M M P E A M T  2159
6476  CCUGAUUUCCUAGCCAAAAAGGGUGGAGAGGCAAUGGAUACCAUCAGUGUGUUCCUUCACUCUGAGGAAGGUUCCCGAACAGCUUACCGCAAUGCACUAUCAAUGAUGCCUGAGGCAAUGACA  6595

2160  I V M L F I L A G L L T S G M V I F F M S P K G I S R M S A M A G T M A G C G Y  2199
6596  AUAGUCAUGCUGUUUAUACUGGCAGGACUGCUGACUUCGGGAAUGGUCAUCUUCUUCAUGUCCCCAAAGGGCAUCAGUAGAAUGUCAGCUAUGGCAGGCACUAUGGCUGGAUGCGGCUAUUU  6715

2200  L M F L G G V K P T H I S Y V M L I F F V L M V V V I P E P G Q Q R S I Q D N Q  2239
6716  CUCAUGUUCCUUGGAGGCGUCAAACCCACUCACAUCUCUUAUGUCAUGCUCAUAUUCUUUGUCCUGAUGGUGGUUGUGAUCCCGGAGCCAGGGCAACAAAGGUCCAUCCAAGACAACCAA  6835

2240  V A Y L I I G I L T L V G A V A A N E L G M I E K T K E D L F Q K K N L I P E S  2279
6836  GUGGCAUACCUCAUUAUUGGCAUCUUGACGCUGGUUGGCGCGGUGGCUGCCAACGAGCUAGGCAUGUGGAGAAAACCAAAGACCAAGAGGACCUCUUUGGGGAAGAAGAACCUCAUCCCUGAAUCAUAGU  6955

2280  A S P W S W P D L D L K P G A A W T V Y V G I V T M L S P M L H H W I K V E Y G  2319
6956  GCUUCACCCUGGAGCUGGCCGGAUCUUGACCUGAAGCCUGGAGCAGCGUGGACAGUGUACGUGGGCAUUGUUACAAUGCUGUCUCCAAUGUUGCACCACUGGAUCAAAGUCGAAUAUGGC  7075

2320  N L S L S G I A Q S A S V L S F M D K G I P F M K M N I S V I M L L V S G W N S  2359
7076  AACCUGUCCUUGUCUGGAAUAGCCCAGUCAGCCUCAGUCCUGUCAUUCAUGGACAAGGGGAUACCAUUCAUGAAGAUGAACAUUUCGGUCAUCAUGUUGCUGGUCAGUGGCUGGAAUUCA  7195

2360  I T V M P L L C G I G C A M L H W S L I L P G I K A Q Q S K L A Q R R I V H G V  2399    →ns4b
7196  AUAACAGUGAUGCCUCUUCUUUGUGGCAUAGGAUGCGCAAUGCUGCACUGGUCAUUGAUCCUUCCUGGAAUCAAAGCCCAGCAAAGCAAAUUGGCUCAGCGCAGAAUAGUUCAUGGCGUU  7315

2400  A E N P V V D G N P T V D I E E A P E M P A L Y E K K L A L Y L L L A L S L A S  2439
7316  GCCGAGAACCCAGUGGUUGAUGGAAAUCCAACAGUAGACAUCGAAGAAGCUCCUGAAAUGCCUGCACUAUACGAGAAAAAGCUGGCUCUAUAUCUCCUGCUGGCCUUAAGCCUCGCUUCU  7435

2440  V A M C R T P F S L A E G I V L A S A A L G P L I E G N T S L L W N G P M A V S  2479
7436  GUUGCCAUGUGCAGAACGCCCUUUUCAUUGGCUGAAGGCAUUGUCCUAGCAUCAGCUGCCUUAGGGCCGCUCAUAGAGGGAAACACUAGUUUGUUGAAUGGACCUAUGGCUGUCUCC  7555

2480  M T G V M R G N H Y A F V G V M Y N L W K M K T G R R G S A N G K T L G E V W K  2519    →NS5
7556  AUGACAGGAGUGAUGAGAGGAAUCAUUAUGCUGUGGGGAGUCAUGUACAAUCUAUGGAAGAUGAAAACUGGACGCCGGGGGAGUGCUAACGGGAAAACUUUGGGUGAAGUGUGGAAG  7675

2520  R E L N L L D K R Q F E L Y K R T D I V E V D R D T A R R H L A E G K V D T G V  2559
7676  AGGGAACUGAAUCUGUUGGACAAGCGACAAUUCGAGUUGUAUAAAAGGACUGACAUUGUGGAGGUGGAUCGGGACACGGCACGCAGGCAUUUGGCCGAAGGGAAGGUAGACACCGGGGUG  7795

2560  A V S R G T A K L R W F H E R G Y V K L E G R V I D L G C G R G G W C Y Y A A A  2599
7796  GCGGUCUCCAGGGGGACCGCAAAGUUAAGGUGGUUCCAUGAGCGUGGUUAUGUCAAGCUGGAAGGGAGAGUUAUUGAUCUGGGCUGUGGCCGUGGAGGCUGGUGCUAUCGCCAUGCGCGC  7915

2600  Q K E V S G V K G F T L G R D G H E K P M N V Q S L G W N I I T F K D K T D I H  2639
7916  CAAAAGGAAGUAUCUGGGGUCAAAGGAUUCACUCUUGGGAGAGACGGCCAUGAAAAACCCAUGAAUGUGCAAAGUCUGGGAUGGAACAUCAUCACCUUCAAAGACAAAACUGACAUUCAC  8035

2640  A L E P V K C D T L L C D I G E S S G G G V T E G E R T V R V L D T V E K W L A  2679
8036  CGCCUAGAACCAGUGAAAUGCGAUACCCUUUUGUGUGAUAUCGGAGAGUCAUCUGGAGGUGGGGUUACUGAGGGAGAGCGCACUGUAAGAGUUCUGGACACUGUUGAGAAAUGGCUGGCG  8155

2680  C G V D N F C V K V L A P Y M P D V L E K L E L L Q R R F G G T V I R N P L S R  2719
8156  UGUGGGGUAGACAACUUCUGUGUGAAGGUGUUAGCCCCCUACAUGCCAGAUGUCCUCGAGAAACUGGAAUUGCUCCAAAGGAGGUUUGGGGGAACAGUGAUCAGGAACCCUCUCUCCAGG  8275

2720  N S T H E M Y Y V S G A R S N V T F T V N Q T S R L L M R R M R R P T G K V T L  2759
8276  AAUUCCACUCACGAAAUGUACUACGUGUCGGAGCCGCAGCAAUGCACAUUUACUGUGAAUCAAACAUCCCGCCUCCUGAUGAGGAGAAUGAGGCGUCCAACUGGGAAAGUGACCCUGG  8395

2760  E A D V I L P I G T R S V E T D K G P L D K E A I E E R V E R I K S E Y M T S W  2799
8396  GAGGCUGACGUCAUCCUCCCAAUUGGGACACGCAGUGUUGAGACAGAUAAGGGACCCCUGGACAAAGAGGCCAUAGAAGAAAGGGUUGAGAGGAUUAAGUCAGAGUACAUGACAUCGUGG  8515

2800  F Y D N D N P Y R T W H Y C G S Y V T K T S G S A A S M V N G V V K I L T Y P W  2839
8516  UUUUAUGACAAUGACAACCCCUACAGGACCUGGCACUACUGUGGGUCUUAUGUCACAAAAACCUCAGGAAGUGCGGCUAGCAUGGUUAAUGGGGUGGUGAAAAUUCUGACAUAUCCAUGG  8635

2840  D R I E E V T R M A M T D T T P F G Q Q R V F K E K V D T R A K D P P A G T R K  2879
8636  GACAGGAUAGAGGAGGUUACAAGAAUGGCAAUGACUGACACCACCCCAUUUGGGCAGCAAAGAGUGUUUAAAGAGAAAGUUGACACGCGAGCAAAGGAUCCUCCAGCGGGAACUAGGAAG  8755

2880  I M K V V N R W L F R H L A R E K N P R L C T K E E F I A K V R S H A A I G A Y  2919
8756  AUCAUGAAAGUUGUCAACAGGUGGCUGUUCCGCCACCUGGCCAGAGAAAAGAACCCCAGACUGUGCACAAAGGAAGAGUUCAUAGCAAAGGUCCGAAGUCAUGCAGCUAUUGGAGCUUAC  8875

2920  L E E Q E Q W K T A N E A V Q D P K F W E L V D E E R K L H Q Q G R C R T C V Y  2959
8876  CUGGAAGAACAAGAGCAGUGGAAGACUGCCAAUGAAGCCGUCCAAGACCCAAAGUUCUGGGAACUGGUGGAUGAAGAGAGGAAGCUGCACCAACAAGGGAGGUGUCGCACGUGCGUUUAC  8995

2960  N M M G K R E K K L S E F G K A K G S R A I W Y M W L G A R Y L E F E A L G F L  2999
8996  AACAUGAUGGGGAAAAGAGAAAAGAAACUUUCAGAGUUUGGCAAAGCCAAGGGAAGCCGUGCCAUAUGGUAUAUGUGGCUGGGAGCGCGCUAUCUUGAGUUCGAGGCCCUUGGAUUCCUG  9115

3000  N E D H W A S R E N S G G G V E G I G L Q Y L G Y V I R D L A A M D G G G F Y A  3039
9116  AAUGAAGACCAUUGGGCUUCCAGGGAAAACUCAGGAGGAGGUGUAGAAGGCAUUGGCUUACAAUACCUAGGAUAUGUGAUUAGGGACCUAGCAGCCAUGGAUGGUGGAGGGUUUUACGCA  9235

3040  D D T A G W D T R I T E A D L D D E Q E I L N Y M S P H H K K L A Q A V M E M T  3079
9236  GAUGACACCGCUGGAUGGGACACGCGCAUCACAGAGGCAGACCUUGAUGAUGAACAGGAGAUCUUGAACUACAUGAGCCCACAUCACAAAAAACUGGCACAAGCAGUGAUGGAAAUGACA  9355

3080  Y K N K V V K V L R P A P G G K A Y M D V I S R R D Q R G S G Q V V T Y A L N T  3119
9356  UACAAGAACAAAGUGGUGAAAGUGCUGAGACCCGCCGGAGGGAAAGCCUACAUGGAUGUCAUAAGUCGGAGAGAUCAGCGAGGGAGUGGUCAAGUUGUUACUUAUGCUCUGAACACC  9475

3120  I T N L K V Q L I R M A E A E M V I H H Q H V Q D C D E S V L T R L E A W L T E  3159
9476  AUCACUAAUCUUGAAAGUGGAUCAAUUGGAGAAAUGGCGAGAGCAGAGAUGGUGAUACACCAUCAGCAUGUGCAGGACUGUGAUGAGUCGGUGCUGACCAGGCUGGAGGCCUGGCUCACUGAGG  9595

3160  H G C D R L K R M A V S G D D C V V R P I D D R F G L A L S H L N A M S K V R K  3199
9596  CACGGAUGUGACAGACUGAAGAGGAUGGCGGUGAGUGGAGACGACUGCGUUGUCCGGCCCAUUGAUGACAGAUUUGGCCUGGCCCUGUCCCAUCUCAACGCCAUGUCCAAGGUUAGAAAG  9715

3200  D I S E W Q P S K G W N D W E N V P F C S H H F H E L Q L K D G R R I V V P C R  3239
9716  GACAUAUCUGAAUGGCAGCCAUCAAAAGGGUGGAAUGAUUGGGAGAAUGUUCCCUUCUGUUCCCACCACUUCCAUGAGCUGCAGCUGAAGGAUGGCAGGAGGAUUGUGGUGCCUUGCCGA  9835

3240  E Q D E L I G R G R V S P G N G W M I K E T A C L S K A Y A N M W S L M Y F H K  3279
9836  GAACAGGAUGAACUUAUUGGGAGAGGGAGGGUCUCACCUGGGAAUGGAUGGAUGAUCAAGGAAACCGCUUGCCUCAGCAAAGCCUACGCCAACAUGUGGUCACUCAUGUACUUCCACAAA  9955

3280  R D M R L L S L A V S S A V P T S W V P Q G R T W S I H G K G E W M T T E D M  3319
9956  AGGGACAUGAGACUCUUGUCUUUGGCAGUUUCCUCUGCUGUGCCUACGUCCUGGGUUCCACAAGGGAGGACAUGGACGCAGGAGAAGGGGGAGUGGAUGACCACAGAAGACAUG  10075

3320  L E V W N R V W I T N N P H M Q D K T M V K K W R D V P Y L T K R Q D K L C G S  3359
10076  CUUGAGGUGUGGAACAGAGUAUGGAUAACCAACAACCCACACAUGCAGGACAAGACAAUGGUGAAAAAAUGGAGAGAUGUCCCUUAUCUAACCAAAAGACAAGACAAGCUGUGCGGAUCA  10195

3360  L I G M T N R A T W A S H I M L V I H R I R T L G Q E K Y T D Y L T V M D R Y  3399
10196  CUGAUUGGAAUGACCAAUAGGGCCACCUGGGCUCCCCACAUGCAUUUAGUCAUCCAUCGUAUCCGAACGCUGGGACAGGAGAAAUACACUGACUAUCUAACAGUCAUGGACAGGUAU  10315

3400  S V D A D L Q G E L I  3411
10316  UCUGUGGAUGCUGACCUACAACUGGGAUGCUUACUGGGUGAACGGAAUCAUCAUGGGGAUCUAAAACAGGAUUAAACCGGGGAUACAAACCACGGGUGGAAACCGGACUCCCCACAACCUGAAACCGGGGAUAAAA  10435

10436  CCACGGCUGGAGAACCGGGCUCCGCACUUAAAAUGGAAACAGAAACCGGGAUAAAAACUACGGAUGGAGAACCGGACUCCACAUUGGACAGAAGAAGUUGCAGCCCAGAACCCCACA  10555

10556  CGAGUUUUGCCACUGGCUAAGCUGUGGAGGCAGUGCAGGGUCAGCCGAUCCUCAGGUUGCGAAAAUGUGGUUCUGGACCUCCCACCCCAGAGUAAAAGAACGGAGCCUCCGCUA  10675

10676  CCACCCUCCCACGGGGUAGAAAGGACGGGGUCUAGAGGUUAGAGGAGACCCUCCAGGGAACAAUGGGACCAUAUUGACGCGGCAGGGAAAGACCGGAGUGGUUUCUCUGCUUUUUCCU  10795

10796  CCAGAGGACGUCUGGGAGCACAGUUUGCUCAAGAAUAAAGCAGACCUUUUGGUGACAAAACCACACU  10862
```

region of NS5 homologous to other RNA viruses (see Section VI and Fig. 15) is enclosed by brackets, and the conserved Gly-Asp-Asp sequence is boxed. Repeated nucleotide sequences are underlined (see also Fig. 14). Closely spaced in-phase stop codons that terminate the long open reading frames are boxed. The single-letter amino acid code is used: (A) Ala; (C) Cys; (D) Asp; (E) Glu; (F) Phe; (G) Gly; (H) His; (I) Ile; (K) Lys; (L) Leu; (M) Met; (N) Asn; (P) Pro; (Q) Gln; (R) Arg; (S) Ser; (T) Thr; (V) Val; (W) Trp; (Y) Tyr.

Westaway, 1982), the translation of the yellow fever genome initiates with the capsid protein, and the amino-terminal methionine is removed during maturation of the protein (Bell *et al.*, 1985). Nucleotide sequences adjacent to these AUG initiation codons are quite different for the two flaviviruses for which sequence data exist in this region: G-A-A-C-A-A-U-G-U for yellow fever virus (Rice *et al.*, 1985) and U-U-C-A-A-A-U-G-U for Murray Valley encephalitis virus (Dalgarno *et al.*, 1986), as compared to the Kozak consensus sequence for eukaryotic translation initiation sites, C-C-$\frac{A}{G}$-C-C-A-U-G-(G) (Kozak, 1984a). The translational efficiency of flavivirus mRNAs relative to host or other viral mRNAs is at present unknown. We note that alternative in-frame AUG codons occur 14 (yellow fever virus) and 15 (Murray Valley encephalitis virus) codons downstream of the methionine initiating translation of C; these AUG codons have surrounding sequences more similar to the consensus sequence mentioned above (U-C-A-A-U-A-U-G-G for yellow fever virus) and could be used to produce alternative intracellular capsid-related polypeptides.

Following the capsid protein is protein prM (NV2), a putative nonstructural glycoprotein of molecular weight 19,000–23,000. This assignment is based on the alignment of the deduced protein sequences of yel-

```
NUCLEOCAPSID  (C)
                   10        20
              |    |    |    |    |
YF      SGRKAQ-GKTLGVNMVRRGVRSLSN
MVE     .KKPGGP..PRV...LK..IPRVFP
SLE     ?KKPGKP.RNRV...LK..SSRVNP
TBE     VKKAILK..GR.PPRRV

GLYCOPROTEIN  (E)
                 10        20        30        40        50
            |    |    |    |    |    |    |    |    |    |
YF      AHCIGITDRDFIEGVHGGTWVSATLEQDKCVTVMAPDKPSLDISLETVAIDRPAE
MVE     FN.L.MSS......AS.A...DLV..G.S.I.I..A...T...RMMNIEATNL.L
SLE     FN.L.TSN...V..AS.A..IDLV..GGS?......E..?..FKVMKMEASE?.S
DEN     ???I..SN...V...S..S..DIV..HGR?.PT..KNT
```

FIGURE 4. Comparison of flavivirus structural protein N-terminal sequences. Protein sequence data for yellow fever virus (YF), St. Louis encephalitis virus (SLE), and dengue 2 virus (DEN) are from Bell *et al.* (1985); data for tick-borne encephalitis virus (TBE) are from Boege *et al.* (1983). The overlines in YF (Rice *et al.*, 1985) and Murray Valley encephalitis virus (MVE) (Dalgarno *et al.*, 1986) indicate residues deduced solely from nucleic acid sequence. Underlining indicates some uncertainty; ? indicates that no assignment was made; a dot indicates that the residue is the same as in YF; regions of highest conservation are shaded. Modified from Bell *et al.* (1985) are reprinted with the permission of Academic Press.

low fever virus (Rice *et al.*, 1985) and Murray Valley encephalitis virus (Dalgarno *et al.*, 1986) in that region with N-terminal sequence data from West Nile virus prM (Castle *et al.*, 1985). Contained within the sequence of prM, at the C-terminal end, is the sequence of the mature M protein, suggesting that prM is a precursor to M. M has not been found within the infected cell, and these sequence data are consistent with the hypothesis that during maturation of the virion, prM is cleaved to produce M (Shapiro *et al.*, 1972; Russell *et al.*, 1980) and a glycoprotein of polypeptide molecular weight about 12,000. For this reason, we refer to this polypeptide as prM, denoting its probable precursor relationship to M (see below and Table I).

The sequence of M is followed by that of the E protein, the major envelope protein that contains both the virus hemagglutinin and the neutralization epitopes. Thus, the structural proteins (or their precursors) are found adjacent to one another and occupy the 5'-terminal 2.5 kilobases of the genome.

The nonstructural proteins are encoded in the remainder of the genome. In addition to prM, at least 4 and as many as 12 nonstructural proteins have been described in flavivirus-infected cells (Westaway, 1973, 1980; Westaway *et al.*, 1977; Heinz and Kunz, 1982; Smith and Wright, 1985) (see also Fig. 2). Some of these proteins must be active in the replication of the viral RNA. The start points of the three largest nonstructural proteins (NV3, NV4, and NV5) (see Fig. 3) have been located by N-terminal amino acid sequence analysis (Fig. 3) (Rice *et al.*, 1986b). As previously suggested by peptide mapping of the corresponding nonstructural proteins from other flaviviruses (Wright *et al.*, 1977; Wengler *et al.*, 1979; Svitkin *et al.*, 1981; Heinz and Kunz, 1982), the sequence data show that these proteins map to nonoverlapping segments in the yellow fever virus nonstructural region (see Fig. 3 and below). The gene order of yellow fever virus that has thus far been established is 5'-C-prM(M)-E-NV3-?-NV4-?-NV5-3'. Definitive mapping of the smaller nonstructural proteins (NVX, NV2½, NV2, NV1½, NV1), which as mentioned above exhibit considerable heterogeneity both in comparisons between flaviviruses (Westaway, 1977, 1980; Russell *et al.*, 1980; Heinz *et al.*, 1982) and when the same virus is grown in different host cells [see, for instance, Fig 2 and Smith and Wright (1985)], awaits further N-terminal sequence analysis. However, the assignments of the larger nonstructural proteins and their predicted sizes leave two regions in the polyprotein for which polypeptide products have not yet been identified. These probably encode at least some of the smaller proteins.

In an attempt to simplify the description of flavivirus-encoded nonstructural polypeptides, in particular the smaller proteins, we suggest a modified nomenclature (described in Table I) based on the linear order of these proteins in the yellow fever virus genome as an alternative to designations based on apparent molecular weights (Westaway *et al.*, 1980). This approach assumes that members of Flaviviridae will have similar genome organization, expressing similar proteins from comparable regions of their genomes. This assumption has been partially verified by an extensive sequence comparison of Murray Valley encephalitis virus and yellow fever virus (see Sections III.C and IV).

The yellow fever gene order in this alternative nomenclature would be 5'-C-prM(M)-E-NS1-ns2a-ns2b-NS3-ns4a-ns4b-NS5-3' (NS1, NS3, and NS5 correspond to NV3, NV4, and NV5, respectively). The regions between each of the large nonstructural proteins have been subdivided on the basis of the presence of putative cleavage sites that are conserved among the flaviviruses thus far examined and homologous to those used for production of NS3 and NS5 (see Section III.C).

B. Translation Strategy

The existence of a single long open reading frame in yellow fever RNA suggests that translation initiates at the AUG codon at position 119–121 and could proceed down virtually the entire RNA. The individual protein products would be released from the precursor polyprotein by proteolytic cleavages. It seems intuitively unlikely that such a long open reading frame would be maintained if a major portion of any gene product were to be made by independent initiation of translation (see below). Furthermore, a logical and consistent set of proteolytic cleavages can be postulated to generate the final protein products (see Section III.C). In addition, translation of the flavivirus genome in vitro produces polypeptides related to the structural proteins (Svitkin et al., 1978, 1981; Wengler et al., 1979; Monckton and Westaway, 1982), which in the presence of appropriate membrane fractions can be processed efficiently to yield C and E (Svitkin et al., 1984). Peptide mapping of in vitro translation products (Wengler et al., 1979) as well as selective incorporation of N-formyl-methionine (Svitkin et al., 1981) suggest that the only initiation event that can be detected in vitro corresponds to the methionine residue adjacent to the amino terminus of the mature capsid protein.

Westaway (Westaway, 1977; Westaway et al., 1984) has presented evidence, however, that the translation strategy of flaviviruses involves the use of multiple independent initiation sites for translation of proteins from the virion-length mRNA. Somewhat heretical in concept, this hypothesis envisages internal binding sites for ribosome attachment that can lead to initiation of protein synthesis. Such a mechanism would be appealing for translation of the nonstructural proteins early in the infectious cycle when enzymes necessary for RNA replication are needed, rather than the structural proteins located at the 5' end of the genome. Although commonly used by bacterial RNA viruses, such internal initiation is apparently rare in animal RNA viruses and nonviral eukaryotic mRNAs (Kozak, 1983). A few examples of internal initiation are noted here. In the case of infectious pancreatic necrosis virus, internal initiation appears to occur, at least *in vitro*, because three polypeptides unrelated by peptide mapping and labeled with *N*-formyl-methionine are produced from what is believed to be the same 3.6-kilobase mRNA (Mertens and Dobos, 1982). These polypeptides, of 29, 31, and 62 kilodaltons, are indistinguishable from those produced *in vivo*. The primary structure of the mRNA has not been determined as yet, and it is unknown whether these polypeptides are encoded in alternative overlapping reading frames near the 5′ terminus of the RNA or in separate cistrons, possibly punctuated by termination codons. There are numerous examples in eukaryotes of initiation *in vivo* at alternative AUG codons near the 5′ end of an mRNA (see, for instance, Lomedico and McAndrew, 1982; Kozak, 1983) as well as reinitiation of translation of eukaryotic ribosomes at downstream AUG codons after encountering termination triplets (Kozak, 1984b; Hughes *et al.*, 1984; Dixon and Hohn, 1984; Liu *et al.*, 1984). None of these examples is relevant (with the possible exception of infectious pancreatic necrosis virus), however, to translation of the flavivirus nonstructural proteins, which are located between 2.4 and 7.6 kilobases from the 5′ end of their mRNA and found in the same open reading frame as the 5′-terminal structural proteins. Translation of these proteins by internal initiation would not involve a modification of the scanning model (Kozak, 1983), but would entail a novel mechanism that has thus far been observed in eukaryotes only *in vitro* (see, for instance, Dorner *et al.*, 1984).

In the original version Westaway's hypothesis stated that at least seven flavivirus proteins initiated independently (Westaway, 1977), but this was later redefined to postulate only two or three independent ribosome-binding sites accompanied by limited proteolytic processing of polypeptide precursors (Westaway *et al.*, 1984). The evidence for internal initiation comes from essentially four sets of experiments, which will be discussed in some detail. As will be noted, now that the yellow fever sequence is known, one of these lines of evidence actually supports a polyprotein translation model.

The first line of evidence for independent initiation was the failure to observe large precursor polypeptides in flavivirus-infected cells during short pulse-labeling periods, even when amino acid analogues or protease inhibitors were included (Westaway, 1973; Westaway and Shew, 1977). Although such polyprotein precursors can be seen in brief pulses in cells infected with poliovirus (a picornavirus) or alphaviruses in which post-translational processing is known to occur (see Chapter 3), the failure to observe them in flavivirus-infected cells could also be explained by a very rapid processing of nascent polyprotein precursors. It is perhaps of note that inclusion of amino acid analogues or protease inhibitors was not found to inhibit the production of flavivirus proteins such as E that clearly mature by posttranslational cleavage; the success of these approaches depends greatly on the properties of the proteolytic cleavage sites and the corresponding protease(s). More recent experiments, in fact, with dengue 2 virus (G. Cleaves, personal communication) and with Japanese en-

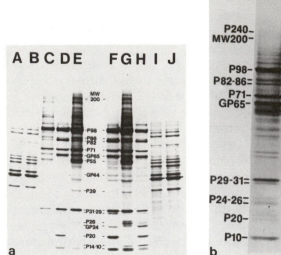

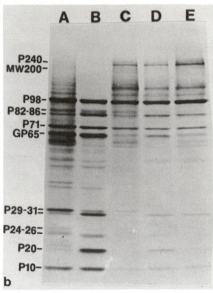

FIGURE 5. High-molecular-weight virus-specific polypeptides in dengue 2 virus (DEN2)-infected baby hamster kidney (BHK) cells. (a) Infected (lanes C–H) or mock-infected (lanes A, B, I, and J) cultures at 25 hr postinfection were pulse-labeled with [^{35}S]methionine for 7 min (lanes A–D), 11 min (lanes E, F, I, and J), or 15 min (lanes G and H) and either harvested immediately (lanes A, C, E, G, and I) or chased for 30 min in the presence of excess unlabeled methionine (lanes B, D, F, H, and J). The cultures had been treated with actinomycin D for 7–24 hr postinfection to inhibit host macromolecular synthesis. The denatured samples were resolved on a discontinuous sodium dodecyl sulfate–polyacrylamide gel system containing a gradient from 7 to 15% acrylamide in the resolving gel. (b) BHK cell cultures were infected with DEN2 and treated with actinomycin D as described above. At 25.5 hr post-infection, the cultures were pulse-labeled with [^{35}S]methionine for 15 min in the absence (lanes A and B) or the presence (lanes C–E) of L-1-tosylamide-2-phenylethylchloromethyl ketone (TPCK) (10 μg/ml) and 1 mM ZnCl$_2$. After the pulse-labeling period, excess unlabeled methionine was added to all cultures, and cultures A and C were harvested. The remaining cultures were incubated for an addiitional 60 min in the presence (lane E) or absence (lanes B and D) of TPCK and ZnCl$_2$. The samples were analyzed as described above. From Dr. Graham R. Cleaves (unpublished data).

cephalitis virus (P. Eastman, J. Mecham, and C. Blair, personal communication), using pulse-label, pulse–chase experiments, and protease inhibitor experiments, have found evidence for precursor polyproteins. In these experiments, proteins larger than 200,000 daltons could be found after brief pulses. In the case of dengue 2 virus, these large polypeptides were found to chase into smaller products (Fig. 5a), and processing could be inhibited by a combination of protease inhibitors, zinc ions, and L-1-tosylamide-2-phenylethylchloromethyl ketone (Fig. 5b). In the case of Japanese encephalitis virus, the large polypeptides are immunoprecipitable with virus-specific antiserum and were found by peptide mapping to contain the sequences of smaller viral proteins.

The second line of evidence involved pactamycin inhibition of translation initiation followed by runoff labeling of Kunjin virus proteins (Westaway, 1977). Because labeling of proteins was correlated in general with increasing molecular weight, Westaway argued that each protein initiated independently. However, reexamination of his data in light of the known gene order found in yellow fever virus supports instead the hypothesis that translation initiates at only a single site adjacent to the capsid protein and proceeds sequentially through the entire genome, as is suggested by the presence of the long open reading frame. By a perhaps unfortunate circumstance, the major flavivirus proteins are found in the genome in order of increasing molecular weight, thus giving rise to the observed labeling order. However, NV1 (possibly ns4b), a small, prominent, nonstructural polypeptide, is more resistant to pactamycin treatment than NV4 (NS3) (Westaway, 1977), consistent with its position in the genome (see Fig. 3) rather than its size. Also of note are results of *in vivo* translation studies of both Japanese encephalitis virus (Shapiro *et al.*, 1973) and Kunjin virus (Westaway and Shew, 1977) in the presence of puromycin. Since puromycin causes premature termination of translation, the inhibitory effect of the drug should increase with the size of the translation product. In the case of a polyprotein, the production of polypeptides immediately following the site of translation initiation should be least inhibited, with distal products showing the greatest inhibition. In both studies, the incorporation of label into the nonstructural proteins (with the exception of prM) was markedly inhibited in the presence of the drug relative to C, prM, and E. These are the expected results given the established order of flavivirus gene products (see Section III.A) if translation *in vivo* begins only with the capsid protein

The third line of evidence involved the kinetics of labeling of Kunjin proteins after synchronous initiation of translation with high-salt treatment followed by a shift to isotonic medium. Because most Kunjin proteins are quickly labeled, Westaway concluded that each initiated translation independently. However, these results are not compelling because the high-salt treatment was only 20 min long, and it is unclear whether all ribosomes have run off in this time period. In fact, Westaway states that a longer treatment with high salt leads to such low incorporation of radiolabel that the products cannot be seen clearly on gels. An alternative explanation for the results is that RNA secondary structures or the presence of codons for which transfer RNA (tRNA) is in limiting concentrations (Grantham *et al.*, 1981) or both can lead to long ribosome transit times and that the labeling observed was due to preinitiated translation products. Resumption of protein synthesis after complete runoff may be slow because of inefficient reinitiation at the 5' end of the RNA due to a less than ideal context for initiation (Kozak, 1984a) surrounding the 5'AUG codon (Section III.A).

Finally, UV mapping data were found to be consistent with two ribosome-binding sites, one of which is internal (Westaway *et al.*, 1984). However, the gene order deduced from these experiments (for Kunjin virus) is different from the gene order found for yellow fever virus and for Murray Valley encephalitis virus, with regard to both structural proteins and nonstructural proteins, and it seems very likely that these experiments are giving the wrong answer. There are two possible explanations for the anomalous UV mapping results: (1) If the flavivirus genome is translated as a polyprotein and processed by virus-encoded proteases to obtain the final gene products, then the location of these protease genes relative to other products could affect the mapping results. (2) If translation of the flavivirus genome is nonuniform as suggested above because of RNA secondary structure or regulation by use of minor codons, continued translation of downstream gene products could make the translation of these genes appear quite resistant to UV inactivation.

In conclusion, the presence of a long open reading frame in yellow fever virus and Murray Valley encephalitis virus RNAs, the fact that the final proteins found do not initiate with methionine but appear to arise from a consistent set of proteolytic cleavages (discussed in more detail below), the results of the pactamycin runoff experiments of Westaway (1977), the *in vitro* translation data, and the recent evidence for polyprotein precursors all support the view that translation of the flavivirus genome *in vivo* initiates with the capsid protein near the 5' end of the genome and proceeds sequentially through the genome to produce one precursor polyprotein. Cleavage of this precursor is rapid and occurs during translation, so that the precursor is not seen in its entirety. Although internal translation initiation cannot be formally excluded, production of the downstream products could also be regulated by premature termination and/or by nonuniform translation of the genome such that translation of the nonstructural proteins is slow; the amounts of the translated products present could be regulated as well by differential stability of the final products.

C. Proteolytic Cleavages in Processing of Flavivirus Proteins

Proteolytic processing of a polyprotein precursor appears to be required to produce the final protein products, and these cleavages fall into three classes: (1) cleavage after the initiating methionine; (2) cleavages after double basic residues, probably catalyzed by virus-encoded enzymes or in one case perhaps by a cellular enzyme found in the Golgi apparatus; (3) cleavages after serine, alanine, or other short side-chain amino acid residues probably catalyzed by the cellular protease(s) called signalase. A summary of these proposed cleavage sites is presented in Table II.

Amino-terminal protein sequence data of the capsid proteins of yellow fever virus (Bell *et al.*, 1985) and tick-borne encephalitis virus (Boege *et al.*, 1983) reveal the first amino acid in mature capsid protein is Ser or Val, respectively, indicating that the initiating methionine had been removed, presumably by a cellular protease.

TABLE II. Flavivirus N-Terminal Cleavage Sites[a]

Protein	Virus	Cleavage site		Protease
C	YF	M	S GRKAQ	Host?
	MVE	M	SKKPGG	
	SLE	?	? KKP GK	
	TBE	?	VKKAI L	
prM	YF	I LG MLL MTGG	VTLVRKNRWLL L N	Signalase?
	MVE	I FML I GF AAA	LKLS TFQGK I MMT	
	WN	LLGL I AC AGA	VTLSNFQGK VMMT	
M	YF	SRRSRR	AI DL P THE NHGLKTR	Golgi?
	MVE	SKRSRR	S I TVQTHGE S TLVNK	
	SLE		? I S VQHHGD? LAP KN	
	WN	SRRSRR	SLTVQTHGE S TLANK	
E	YF	VGPAYS	AHCI GI TDRDFI	Signalase?
	MVE	VAPAYS	FNCLGMS S RDFI	
	SLE		FNCLGT SNRDFV	
	DEN2		? ? ? I GI SNRDFV	
NS1	YF	LS LGVGA	DQGCAINFGKREL	Signalase?
	MVE	LATNVHA	DTGCAIDI TRREL	
	SLE		A L	
(ns2a)			?	Signalase?
(ns2b)	YF	TRI F GRR	S I PVNE	Viral?
	MVE	CNPNK KR	GWPATE	
NS3	YF	VRGARR	S GDVLWDI PTPKI	Viral?
	MVE	LKYTKR	GG VFWDTPS PKV	
	SLE		AL V KV	
(ns4a)	YF	FI KFAEGRR	GAAE VLVV	Viral?
	MVE	FKDFA? GKR	S AI GFFEV	

(continued)

TABLE II. (*Continued*)

Protein	Virus	Cleavage site		Protease
(ns4b)	YF	KL AQRR	VF HGVA	Viral?
	MVE	RAAQKR	TAAGI M	
	DEN2	RE AQKR	AAAGI M	
NS5 (1)	YF	MKT GRR	GSAN GKTLGEVWKRE	Viral?
	MVE	K PA F KR	GRAG GRTLGEQWKEK	
	DEN2	NTTNTRR	GTGNIGE TLGEK WKS R	
	SLE		**K ATL T K**	
(2)	YF	DRDTARR	HL AEGK VDTGVAVS	
	MVE	DRTE ARR	ARREGNKVVGHPVS	
	DEN2	DRT LAKE GIKR	GE TDH HAVS	

a Putative N-terminal flavivirus protein cleavage sites (see the text for discussion). Amino acid residues determined by N-terminal protein sequence analysis are shown in boldface type. Protein sequence data for the structural proteins (C, M, E, and also prM) are from Bell *et al.*, (1985) or in the case of West Nile virus (WN) from Castle *et al.*, (1985). The yellow fever virus (YF) nonstructural N-terminal protein sequences and the St. Louis encephalitis (SLE) nonstructural protein sequences are from Rice *et al.*, (1986b). In the case of SLE, only the positions of five labeled amino acids were established. The data for tick-borne encephalitis virus (TBE) are from Boege *et al.* (1983). The remaining sequences were deduced solely from the YF [Rice *et al.*, 1985], Murray Valley encephalitis virus [MVE] [Dalgarno *et al.*, 1986], and dengue 2 (DEN2) [V. Vakharia, T. Yaegashi, R. Feighny, S. Kohlekar, and R. Padmanabhan, unpublished data] nucleotide sequence data (see also Figs. 3, 6, and 9). Tentative cleavage sites, indicated by parentheses, are based on homology with confirmed cleavage sites and the sizes of YF-specific polypeptides observed in infected cells (Schlesinger *et al.*, 1983; J. Pata and C. M. Rice, unpublished data). Less homologous alternative cleavage sites in the nonstructural region occur after residue 1946 (Gln-Arg-Arg ↓ Gly), residue 2548 [Ala-Arg-Arg ↓ His; shown above as NS5 (2)]], residue 2707 (Gln-Arg-Arg ↓ Phe), and residue 3104 (Ser-Arg-Arg ↓ Asp).

The site of the cleavage to release the capsid protein from the po-lyprotein precursor has not been positioned precisely. The start of the following protein, prM, has now been positioned (Table II) and it im-mediately follows a stretch of uncharged residues (see Fig. 5) that might function as an internal signal sequence for the insertion of prM across the membrane (see below). This cleavage could be catalyzed by a host enzyme such as signalase. The observation that during *in vitro* translation of tick-borne encephalitis virus RNA, C is produced more efficiently when membrane fractions are included would seem to support this (Svit-kin *et al.*, 1984). It is not known if C extends to the start of prM, in which case it might be an integral membrane protein anchored by a C-terminal membrane spanning domain, or whether a second cleavage, either during translation or during virion maturation, leads to a shortened version of C. One possibility in the latter case is a cleavage in the sequence Ser-Arg-Lys-Arg-Arg ↓ Ser at the position noted by the arrow. Such a cleavage would remove the membrane spanning domain. During translation, such a cleavage might allow the hydrophobic domain to serve as an N-terminal signal sequence, or such a cleavage during virus assembly could release C from membrane association. It is obviously of importance to establish the C-terminus of C, as well as of other viral proteins.

The difficulty in relating published patterns of C protein production *in vivo* or *in vitro* to the sequence data and interpreting their significance for possible cleavage sites is that the intracellular form of C or the form of C produced by *in vitro* translation differs from the form of C found in the virion (Wengler *et al.*, 1979; Svitkin *et al.*, 1981, 1984; Westaway *et al.*, 1977). The virion C migrates *more slowly* than the other forms and differs in one or more tryptic peptides. The differences involved are un-clear. Removal of the N-terminal Met could conceivably be the only modification involved, and the intracellular form could be a precursor to the virion form, although why this should cause the electrophoretic mo-bility to decrease is obscure. Alternatively, a second initiation codon could be used (Section III.A), and the intracellular or *in vitro* forms might not be precursors to the virion form. Thirdly, as discussed above, the presence or absence of a C-terminal anchor could be involved. Other possible modifications are also conceivable, and the identity of the var-ious forms needs to be rigorously established.

A string of uncharged amino acids is found beginning at position 105 in the yellow fever virus polyprotein (Figs. 6 and 7) that could function as a signal sequence (Perlman and Halvorson, 1983) and lead to the in-sertion of prM into the endoplasmic reticulum (Walter and Blobel, 1982). The N terminus of prM has been established from data for West Nile virus (Castle *et al.*, 1985) and it appears that this uncharged sequence functions as an internal signal sequence and forms the C terminus of C. Although the orientation of prM has not been proven, the conserved gly-cosylation sites near the N terminus and the C-terminal hydrophobic domain (Figs. 6 and 7) suggest a transmembrane protein with an external

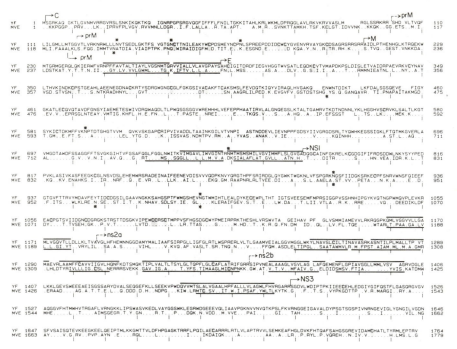

FIGURE 6. Alignment of Murray Valley encephalitis virus (MVE) and yellow fever virus (YF) protein sequences. The YF protein sequence shown in Fig. 3 was aligned with the corresponding sequence of MVE (Dalgarno *et al.*, 1986) by introducing gaps to maximize homology. Residues are numbered from the beginning of the polyprotein. The start points of the proteins for which N-terminal sequence data exist are indicated (solid arrows) (Bell *et al.*, 1985; Rice *et al.*, 1986b). Less certain assignments (see the text) are indicated by dashed arrows. Contiguous stretches of uncharged amino acids are overlined (YF) or underlined (MVE), and potential N-linked glycosylation sites are marked by asterisks. Cysteine residues are shaded.

glycosylated portion and a C terminus associated with the cytoplasmic face of the lipid bilayer. The cleavage of prM to produce M occurs after the sequence Arg-Ser-Arg-Arg ↓ Ala and is a late event, probably occurring during virus maturation and release. No M· has been detected in the infected cell, but Shapiro *et al.* (1972) have shown that prM (NV2) can be incorporated, in place of M, into viruslike particles of low specific infectivity in the presence of Tris buffer, and on this basis postulated that prM was a precursor of M (reviewed in Russell *et al.*, 1980). The sequence data are consistent with this hypothesis, and it appears that during maturation of virus particles, a glycopeptide of 11.4 kilodaltons (not including carbohydrate) is removed from the precursor prM, leaving the nonglycosylated M protein embedded in the virion membrane. Trace quantities of small virus-specific glycoproteins have been detected in cytoplasmic extracts (Schlesinger *et al.*, 1983; Smith and Wright, 1985) (see Fig. 2b), but whether this glycopeptide fragment remains cell-associated and is rapidly degraded or is released into the extracellular medium is at present un-

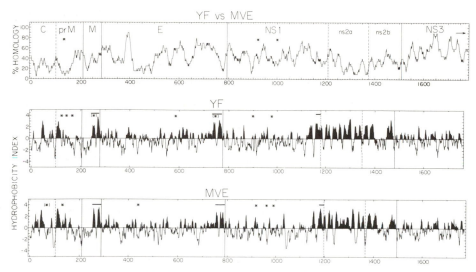

FIGURE 7. Homology and hydrophobicity analyses of yellow fever virus (YF) and Murray
Valley encephalitis virus (MVE) proteins. The top panel shows a moving average percentage
homology plot of the protein sequences as aligned in Fig. 6 using a window of 20 amino
acids. The middle and bottom panels are hydrophobicity analyses of YF and MVE proteins,
respectively, using the program of Kyte and Doolittle (1982) and a search length of seven
residues. Hydrophobic regions above the midline have been shaded. Putative hydrophobic,
membrane-associated domains of the structural proteins and NS1 are indicated by solid
bars; potential N-linked glycosylation sites are denoted by asterisks; glycosylation sites
conserved between YF and MVE are indicated on the homology profile. The nomenclature
for the proteins is described in Table I and the text. The arrow after NS3 indicates that not
all this protein sequence is available for comparison. Cleavage sites are indicated by solid
(confirmed sites) or dashed (tentative assignments) vertical lines.

known. This cleavage site has the same canonical sequence as that shown
by a number of virus glycoproteins that are cleaved late in virus maturation, including alphavirus PE2, fowl plague virus hemagglutinin, and
retrovirus glycoproteins, namely, Arg-X-$\genfrac{}{}{0pt}{}{\text{Arg}}{\text{Lys}}$-Arg, where X can vary (Dalgarno *et al.*, 1983; reviewed in Strauss and Strauss, 1985). Thus, similar
host enzymes are probably responsible for all these cleavages and are
postulated to be one or more proteases of the Golgi apparatus that cleave
after double basic residues and may be similar to the cathepsins (Docherty
et al., 1982; Takio *et al.*, 1983) or related enzymes that process proinsulin,
proalbumin, and several polypeptide hormones. Alternatively, prM could
possess a proteolytic activity that is activated late in maturation such
that prM acts as an autoprotease.

The fact that the timing and subcellular localization of the prM cleavage appear to be different from those of the other apparently rapid cotranslational cleavage events involved in flavivirus nonstructural protein
maturation (see below) also suggests that a different enzyme is responsible for processing prM, although double basic amino acid residues ap-

pear to form at least part of the recognition sequence for both types of cleavage.

The orientation of E has also not been established, but again appears to be N terminus outside and C terminus spanning the bilayer (Fig. 7) (see also Section IV). Because the C terminus of prM has not been precisely localized, the nature of the signal sequence responsible for inserting E is not clearly defined, but it appears reasonable that the C terminus of prM contains a stop transfer signal consisting of a 21-residue uncharged stretch (presumably a membrane-spanning segment) flanked by basic residues, followed by a second hydrophobic segment of 15 amino acids (Figs. 3 and 7) acting as an internal signal sequence that leads to a looping back through the bilayer of the C terminus of prM and the N terminus of E. A similar mechanism has been proposed for the insertion of E1 of the alphaviruses (Rice and Strauss, 1981a) (Chapter 3). The cleavage to release the N terminus of E follows a Ser residue and thus could be catalyzed by signalase on the lumenal side of the rough endoplasmic reticulum.

Protein NS1 (GP48 or NV3) follows E, and the amino acid sequence suggests that the same mechanisms are involved in inserting the N terminus of NS1 through the bilayer and cleavage (after Ala) by signalase to separate NS1 (see Section IV). The C terminus of E has not been positioned, but long hydrophobic stretches are found just before the start of NS1 (Fig. 7) that could serve as stop-transfer signals (and thus membrane-spanning segments) as well as an internal signal sequence. Whether portions of the protein are removed [as in the "6 K protein" of alphaviruses located between PE2 and E1 (Welch and Sefton, 1979; Garoff *et al.*, 1980a; Rice and Strauss, 1981a)] is not yet known. It is possible that such processing events could account for some of the apparent precursor–product relationships between larger intracellular forms of E-related polypeptides and the mature E of Kunjin virus and West Nile virus (Wright *et al.*, 1981; Wright and Warr, 1985).

The C terminus of NS1 has also not been positioned. Molecular-weight estimates suggest that it is near residue 1187. There is a possible stop-transfer signal just upstream (see Figs. 3, 6, and 7) that could anchor the protein in the bilayer, and cleavage to separate it from the following nonstructural proteins could also be catalyzed by signalase. In any event, it is of interest that the three glycoproteins of flaviviruses, prM, E, and NS1, follow one another and are probably sequentially inserted into the endoplasmic reticulum during translation, utilizing one or more host enzymes for cotranslational processing. The sequence data support the hypothesis that each has the usual membrane protein topology of N terminus outside and a C-terminal hydrophobic anchor. However, additional experiments are required to rigorously establish their orientation with respect to the lipid bilayer and their exact C termini.

Downstream from NS1, the start points of NS3 (NV4) and NS5 (NV5) have been positioned from N-terminal protein sequence analysis (see Table II). In the case of yellow fever virus, each appears to arise by cleavage following two Arg residues: NS3 in the sequence Gly-Ala-Arg-Arg ↓ Ser-Gly and NS5 in the sequence Thr-Gly-Arg-Arg ↓ Gly-Ser. Note that in both cases, the double arginine residues are surrounded by short-chain amino acids, often glycine. NS5 appears to be the C-terminal protein in the putative polyprotein precursor, since molecular-weight considerations predict it to extend to the end of the open reading frame.

As mentioned above, these assignments leave two regions in the polyprotein for which polypeptide products have not yet been identified. Assuming that other nonstructural proteins will be produced from these regions by the same protease responsible for N-terminal cleavage of NS3 and NS5, we have scanned the remaining sequences for additional cleavage sites (Table II). Molecular-weight estimates (Schlesinger *et al.*, 1983) have positioned the C terminus of NS1 near residue 1187, as noted above. The next potential cleavage sequence, Gly-Arg-Arg ↓ Ser, at residue 1355 would produce two small nonstructural polypeptides of approximately 18 kilodaltons (ns2a) and 14 kilodaltons (ns2b) located between NS1 and NS3 (see Fig. 3 and Tables I and II). The C terminus of NS3 may be produced by cleavage at the sequence Glu-Gly-Arg-Arg ↓ Gly (after residue 2107) and would produce a polypeptide the calculated molecular weight of which agrees well with the observed size of NS3 on polyacrylamide gels (Schlesinger *et al.*, 1983; J. Pata and C. M. Rice, unpublished data). Between this site and the N terminus of NS5, a single potential cleavage site (Ala-Gln-Arg-Arg ↓ Val) is found preceding residue 2395. Cleavage here would result in two methionine-rich, hydrophobic polypeptides of 31 kilodaltons (ns4a) and 12 kilodaltons (ns4b) (Fig. 3 and Table II). Polypeptides of these approximate sizes (10, 14, 18, and 30 kilodaltons) do exist in yellow-fever-infected cells, but definitive mapping of these polypeptides as well as other minor species, awaits additional N-terminal sequence data. Similarly, in the absence of C-terminal sequence data, we cannot be sure of the exact termini of any of the flavivirus proteins. Some heterogeneity in flavivirus polypeptides may result from variable exopeptidase digestion of the C-terminal residues or alternative internal cleavages (see Fig. 2 and the Table I and II footnotes).

We hypothesize, therefore, that flaviviruses encode one or more proteases that cleave after two basic residues surrounded by short-side-chain amino acids, often Gly, to produce the nonstructural proteins, with the probable exception of NS1, from the polyprotein precursor. The location of this proteolytic activity has yet to be established. Several possibilities exist for the protease involved in producing the capsid protein. If a cleavage occurs after two basic residues, the same virus-specific protease or one similar to that proposed above for the production of the nonstructural proteins could be responsible. This proteolytic activity would be present in the structural region, since the capsid cleavage occurs in expression systems in which the nonstructural region of the yellow fever genome is not present and therefore translated (C. M. Rice and J. H. Strauss, unpublished data). In this case, the capsid protease may reside within C

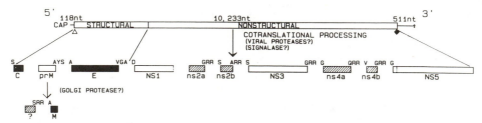

FIGURE 8. Organization and processing of proteins encoded by the yellow fever virus genome. Untranslated regions of the genome are shown as single lines and the translated region
as an open box. (△) Initiation codon (AUG); (◆) termination codon (UGA). The protein
nomenclature is described in Table I. The single-letter amino acid code is used for sequences
flanking assigned cleavage sites (solid lines). Two other potential cleavage sites are shown
as dotted lines. Structural proteins, identified nonstructural proteins, and hypothesized nonstructural proteins (see the text) are indicated by solid, open, and hatched boxes, respectively.
Other potential cleavage sites have been found and are described in Table II. Reproduced
from Rice *et al.* (1985) with permission.

itself or within prM, and in this event, whether the same protease also
cleaves the nonstructural proteins or whether a second nonstructural
protease is found in that region of the genome is unknown. Alternatively,
the cleavage that occurs after the subsequent uncharged stretch of amino
acids could be catalyzed by the host enzyme signalase. Determination of
the C terminus of the capsid protein and the N-terminal sequence of prM
from yellow fever virus or Murray Valley encephalitis virus should be
useful for sorting out these various possibilities. We postulate that the
remaining cleavage events involving proteolytic processing of the glycoproteins from the precursor involve cellular organelle-bound enzymes,
possibly signalase to separate the glycoproteins prM, E, and NS1 from
one another and either an autocatalytic proteolytic activity or a Golgi
protease to cleave prM to produce M. A schematic diagram of the processing events for yellow fever virus is shown in Fig. 8. This figure also
contains a schematic of the genome organization, location of the virus
genes, and the amino acid sequences around the putative cleavage sites.

Alignment of deduced protein sequences from Murray Valley encephalitis virus (Dalgarno *et al.*, 1986) (see Fig. 6) and dengue virus type
2 (V. Vakharia, T. Yaegashi, R. Feighny, S. Kohlekar, and R. Padmanabhan, personal communication) (Fig. 9) homologous to regions of the yellow fever polyprotein sequence reveals potential cleavage sites of specificity similar to yellow fever (Table II). For the nonstructural proteins
(following NS1), it is of interest that in the case of Murray Valley encephalitis virus, for both of the potential cleavage sites sequenced, the
double basic residues are Lys-Arg, whereas in the case of dengue 2 virus,
for the two potential cleavage sites that can thus far be compared, one
is Lys-Arg (ns4b) and the other is Arg-Arg (NS5).

FIGURE 9. Alignment of yellow fever virus (YF) and dengue 2 virus (DEN2) nonstructural protein sequences. Data for YF are from Rice *et al.* (1985) and for the DEN2 sequence from V. Vakharia, T. Yaegashi, R. Feighny, S. Kohlekar, and R. Padmanabhan (unpublished data). Gaps have been introduced to maximize homology, and the putative start points of the nonstructural proteins are indicated as described in the Fig. 3 caption. Dots indicate homology with the YF sequence, cysteine residues are shaded, and contiguous stretches of uncharged amino acids are overlined (YF) or underlined (DEN2).

IV. CHARACTERISTICS OF FLAVIVIRUS PROTEINS FROM SEQUENCE DATA

The amino acid compositions of the ten predicted yellow fever proteins as well as partial data for Murray Valley encephalitis, tick-borne encephalitis, and dengue type 2 viruses are shown in Table III. Also included are molecular weights based on the sequence data and proposed cleavage sites. Alignments of homologous flavivirus proteins have been presented in Figs. 6 and 9, and in Figs. 7 and 10 are shown graphic representations of the homologies of these proteins at the amino acid level and a comparison of their hydrophobicity profiles.

The capsid protein is quite basic in character, with 23–25% of the approximately 120 amino acids being Arg or Lys (Table III). The function of these positive charges is presumably to partially neutralize the negative charges in the RNA encapsidated by this protein. Unlike the situation with alphavirus capsid proteins, in which the basic residues are clustered in the N-terminal half (Garoff *et al.*, 1980b; Rice and Strauss, 1981a; Boege *et al.*, 1981; Dalgarno *et al.*, 1983), these positive charges are distributed over the entire length of the flavivirus capsid protein, and no long colinear regions of pronounced homology exist between the capsid proteins of yellow fever and Murray Valley encephalitis viruses (see Figs. 6 and 7). The most highly conserved region in the capsid proteins is be-

TABLE III. Putative Sizes and Amino Acid Compositions of Flavivirus Proteins[a]

	C			prM		M			E		
	YF	MVE	TBE	YF	MVE	YF	MVE	TBE	YF	MVE	TBE
Position in polyprotein[b]:	2–121	2–125	—	122–285	126–292	211–285	218–292	—	286–778	293–793	—
Calculated mol. wt.[c]	13,432	13,756	—	18,813	18,745	8,526	8,281	—	53,712	53,961	—
Composition[d]											
Ala	3.4	7.3	10.4	6.1	7.2	9.4	8.0	8.9	7.4	9.0	7.7
Arg	11.7	8.9	11.2	9.2	5.4	8.0	2.7	4.4	3.3	3.6	3.3
Asn	3.4	3.3	—	6.1	4.8	4.1	5.4	—	4.1	4.6	—
Asp	1.7	1.4	—	4.3	6.6	1.4	2.7	—	5.5	4.2	—
Asx	—	—	3.9	—	—	—	—	5.6	—	—	8.9
Cys	0	0	0	3.7	3.6	0	0	n.d	2.5	2.4	1.6
Gln	5.1	0.9	—	2.5	3.0	5.4	2.7	—	3.3	2.0	—
Glu	0	1.4	—	6.1	3.0	5.4	2.7	—	4.9	4.8	—
Glx	11.7	9.7	6.8	—	—	—	—	11.4	—	—	9.9
Gly	1.7	0.9	10.8	6.8	6.0	6.7	6.7	12.9	10.2	10.4	12.8
His	5.1	4.1	1.3	1.3	1.2	2.7	1.4	3.0	2.7	3.0	3.5
Ile	13.4	11.3	2.6	3.7	7.8	5.4	5.4	2.7	6.3	4.6	3.3
Leu	10.9	13.0	9.0	9.2	7.8	10.7	13.4	17.2	7.1	8.4	9.8
Lys	4.2	4.9	12.1	4.3	6.6	4.1	6.7	4.4	6.1	5.0	6.7
Met	5.1	6.5	4.8	2.5	2.4	4.1	1.4	1.3	3.3	3.4	2.3
Phe	2.6	5.7	1.2	2.5	1.8	4.1	1.4	1.4	3.9	4.2	3.6
Pro	6.7	4.1	7.0	4.3	4.2	2.7	2.7	3.4	2.7	3.4	5.7
Ser	5.1	5.7	3.7	5.5	5.4	8.0	6.7	7.2	6.3	8.2	5.4
Thr	0.9	0.9	6.7	7.4	9.6	2.7	12.0	7.2	8.0	8.4	8.8
Trp	0	0	n.d	3.1	3.0	2.7	4.1	n.d	2.3	1.8	n.d.
Tyr	8.4	10.5	0	3.7	3.6	9.4	4.1	2.6	1.9	2.4	2.7
Val	1.7	3.3	8.6	8.6	7.2	6.7	10.7	6.7	9.2	7.4	6.9
a	24.2	22.6	—	10.4	9.6	14.7	5.4	—	10.4	9.0	—
b	47.6	51.7	24.6	14.7	13.2	54.7	10.7	11.8	12.0	11.6	13.5
n	26.7	22.6	—	42.1	44.4	24.0	49.4	—	45.1	43.2	—
p	—	—	—	33.0	33.0	—	34.7	—	32.7	36.4	—
Homology (%)[e]		27			33		36			45	

	NS1		(ns2a)		(ns2b)		NS3	(ns4a)	(ns4b)		NS5
	YF	MVE	YF	MVE	YF	MVE	YF	YF	YF	DEN2	YF
Position in polyprotein[b]:	779–1187	794–1207	1188–1354	1208–1372	1355–1484	1373–1503	1485–2107	2108–2394	2395–2506	—	2507–3411
Calculated mol. wt.[c]	45,869	46,746	18,086	17,861	13,823	14,255	69,319	31,196	12,159	12,310	104,079
Ala	4.0	6.3	11.4	14.0	13.1	11.5	8.6	7.7	11.7	9.0	6.0
Arg	5.2	6.3	5.4	5.5	3.9	3.1	7.4	2.1	3.6	5.4	8.2
Asn	3.7	3.4	3.6	4.9	1.6	0.8	3.7	2.5	5.4	5.4	3.5
Asp	5.2	5.4	1.2	1.9	2.4	7.7	5.0	2.5	1.8	2.7	6.3
Asx	—	—	—	—	—	—	—	—	—	—	—
Cys	3.0	2.9	1.8	1.3	0	0.8	1.7	1.1	0.9	1.8	1.9
Gln	2.5	2.0	1.8	1.9	2.4	0.8	2.5	2.8	0	2.7	3.2
Glu	6.9	8.0	2.4	3.1	10.0	3.9	7.9	3.9	6.3	2.7	7.1
Glx	—	—	—	—	—	—	—	—	—	—	—
Gly	9.8	7.8	9.0	6.1	10.8	84.	8.7	8.8	9.0	8.1	7.5
His	3.0	2.2	1.8	1.3	1.6	0.8	2.5	1.8	1.8	0	2.9
Ile	5.7	5.4	6.6	6.1	2.4	6.2	5.0	8.1	2.7	7.2	4.4
Leu	5.9	10.0	16.8	15.2	17.0	12.3	7.6	12.6	13.4	11.7	8.2
Lys	5.7	5.6	2.4	3.7	3.1	3.1	5.5	4.9	3.6	3.6	6.7
Met	3.5	1.5	5.4	6.1	3.1	3.9	2.3	8.1	6.3	5.4	3.6
Phe	4.2	3.2	5.4	4.3	3.9	3.1	3.4	3.9	2.7	3.6	2.4
Pro	5.2	3.9	3.6	4.3	3.1	4.6	5.5	4.9	6.3	5.4	3.0
Ser	8.1	5.8	4.2	7.3	7.0	6.2	6.1	9.5	5.4	3.6	5.1
Thr	5.4	8.0	7.2	5.5	0.8	7.7	5.3	3.5	4.5	11.7	6.5
Trp	3.0	3.4	1.2	1.3	2.4	5.4	2.5	2.1	1.8	2.7	3.2
Tyr	2.5	2.2	1.8	1.3	0.8	2.3	1.7	2.1	3.6	1.8	3.4
Val	8.6	7.5	7.2	6.1	11.6	8.4	8.1	8.1	9.9	6.3	7.9
a	12.0	13.3	3.6	4.9	12.4	11.5	12.9	6.3	8.1	5.4	13.3
b	13.7	14.1	9.6	10.4	8.5	6.9	15.3	8.8	9.0	9.0	17.8
n	42.1	42.8	59.3	58.8	58.5	55.8	45.0	57.9	54.5	53.6	41.5
p	32.3	30.0	27.6	26.1	20.8	26.0	27.0	27.2	28.6	32.2	27.6
Homology (%)[e]	44		27		35				35		

[a] Murray Valley encephalitis virus (MVE) and yellow fever virus (YF) data are taken from Figs. 3 and 6 (Dalgarno et al., 1986; Rice et al., 1985); amino acid compositions from purified tick-borne encephalitis (TBE) proteins are from Boege et al. (1983) and dengue 2 virus (DEN2) from V. Vakharia, T. Yaegashi, R. Feighny, S. Kohlekar, and R. Padmanabhan (unpublished data)/ All the C-terminal amino acid positions are tentative (see the text). In addition, the N termini of prM, ns2a, ns2b, ns4a, and ns4b have not been established by N-terminal sequence analysis (see the text).

[b] Tentative locations of the proteins in the polyprotein sequences of YF (Fig. 3) and MVE (Fig. 6).

[c] Calculated polypeptide molecular weights.

[d] Expressed as mole percent. Functional groups of amino acids are defined as acidic [a (Asp, Glu)]; basic [b (His, Lys, Arg)]; neutral [n (Ala, Phe, Ile, Leu, Met, Pro, Gln, Val, Trp)]; or polar [p (Cys, Gly, Asn, Ser, Thr, Tyr)].

[e] Percentage homology with alignment shown in Fig. 6 for YF and MVE; YF vs. DEN2 alignment for ns4b shown in Fig. 9. Gaps were counted as mismatches.

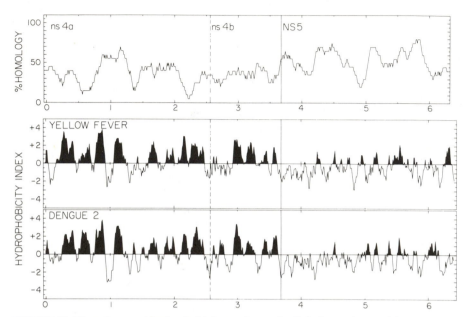

FIGURE 10. Homology and hydrophobicity analyses of yellow fever virus and dengue 2 virus nonstructural proteins. Analyses were carried out as described in the Fig. 7 caption. The dengue 2 virus sequence is from V. Vakharia, T. Yaegashi, R. Feighny, S. Kohlekar, and R. Padmanabhan (unpublished data), and the yellow fever virus sequence is from Rice *et al.* (1985); the alignment is that shown in Fig. 9. Potential cleavage sites are indicated by vertical lines; the protein nomenclature is described in Table I. Amino acid residues are numbered in hundreds.

tween residues 38 and 55 (numbered from the N terminus of yellow fever virus C) and is relatively rich in uncharged amino acids (see Figs. 6 and 7). This domain may be involved in protein–protein or specific protein–RNA interactions or both during nucleocapsid assembly and acquisition of the lipoprotein envelope.

Yellow fever virus prM has several potential glycosylation sites of the form Asn-X-Ser/Thr. Three of these are found within the region not shared with M and are closely spaced, namely, Asn-13, Asn-29, and Asn-51 (the numbering assumes that prM begins just after the double glycine residues shown in Fig. 3 and Table II, thus consisting of 164 amino acids, and may not be precise as discussed in Section III.C). The first of these potential glycosylation sites is conserved in Murray Valley encephalitis virus (Figs. 6 and 7). There is in addition a potential site at Asn-145, in the region shared with M, but this is found in the hydrophobic belt predicted to span the bilayer and thus would be inaccessible for glycosylation. This interpretation is consistent with the fact that M is not glycosylated. The N-terminal domain of prM is relatively hydrophobic and shows little sequence conservation (Figs. 6 and 7), consistent with its putative function as a signal sequence. It is unclear whether this domain is removed by signalase during maturation of prM. In contrast, the domain of prM preceding the N terminus of the mature M protein (~50 amino acids) is highly conserved (Figs. 6 and 7), containing five conserved cysteine residues. In this region is also found a conserved sequence Cys-Trp (see Fig. 6), a characteristic sequence that often comprises part of the active site of thiol proteases (Takio *et al.*, 1983). The question of whether prM functions as a protease for flavivirus protein processing will require further investigation, but the capsid protein cleavage or the cleavage of prM itself, or both, could conceivably be catalyzed by prM.

Since only 75 amino acids separate the beginning of M from the N terminus of E, M must be quite short, and its estimated molecular weight from acrylamide gels is approximately 8 kilodaltons. M is hydrophobic in nature (Figs. 6 and 7), especially in the C-terminal 37 residues, which contain only a single charged residue (an arginine in yellow fever virus matched with a lysine in Murray Valley encephalitis virus). The possible functions of this region as a membrane-spanning domain and internal signal sequence have been discussed earlier. However, these uncharged domains of M show rather higher sequence conservation (Figs. 6 and 7) than is usual for membrane anchors or signal sequences, suggesting that the C terminus of M may interact specifically with the E protein or the nucleocapsid or both during virus assembly.

Yellow fever virus E, the major external envelope protein on the surface of mature virions, which has been shown to contain a number of flavivirus antigenic determinants, is 493 residues in length, if it extends to the N terminus of NS1, and has two potential carbohydrate attachment sites of the type Asn-X-Ser/Thr, at Asn-309 and Asn-470. The second of these is found in the hydrophobic C-terminal domain (Figs. 6 and 7) and is probably not available for glycosylation. Therefore, Asn-309 is probably the sole site of glycosylation and, since nonglycosylated forms of E are found (Schlesinger *et al.*, 1983) (see Fig. 2), is not always glycosylated. The reason for the failure to glycosylate this site at times is not clear, but may have to do with the fact that Pro is the bridging amino acid (the site is Asn-Pro-Thr). Neither one of these potential glycosylation sites is conserved with Murray Valley encephalitis virus E (Fig. 6), which contains only a single potential N-linked glycosylation site at Asn-154, considerably closer to the putative N terminus of E than Asn-309 of yellow fever virus. Adjacent to the predicted C terminus of E, there are two hydrophobic domains punctuated by basic amino acids (as with M and prM) that presumably function to anchor E in the lipid bilayer as well as to promote the translocation of NS1 across the membrane. In contrast to M, the C-terminal hydrophobic domain of E shows very little sequence conservation between yellow fever and Murray Valley virus (Fig. 6), consistent with its function as a transmembrane segment or signal sequence. Other regions of the E protein show both remarkable conservation and divergence, depending on the region (Figs. 6 and 7). For instance, residues 92–123 of the yellow fever virus E are 79% homologous with the corresponding region of Murray Valley encephalitis virus E, whereas residues

124–174 are only 14% homologous (counting gaps as mismatches). Future experiments using monoclonal or peptide-specific antibodies should yield interesting correlates between these data and the group-, type-, and strain-specific flavivirus epitopes with important biological functions that are associated with the E protein (Heinz *et al.*, 1982, 1983a; Gentry *et al.*, 1982; Henchal *et al.*, 1982; Peiris *et al.*, 1982; Schlesinger *et al.*, 1983, 1984; Kimura-Kuroda and Yasui, 1983; Roehrig *et al.*, 1983; Stephenson *et al.*, 1984) (see also Chapter 11). In addition, the eventual determination of the three-dimensional structure of a flavivirus E protein coupled with experiments using the expanding battery of monoclonal antibodies with interesting biological properties should prove to be a very fruitful area of investigation. Initial experiments of this type using proteolytic fragments of the tick-borne encephalitis virus E protein have already been reported (Heinz *et al.*, 1983b, 1984).

The features of the structural proteins mentioned above predict that the majority of the E protein and approximately half the M protein should be exposed on the mature virion surface and therefore be sensitive to digestion by appropriate proteases. Protease digestion of purified tick-borne encephalitis virus (Heinz and Kunz, 1979; Franz Heinz, personal communication) and also of yellow fever virus (J. Pata and C. M. Rice, unpublished data) support this hypothesis. The properties of M (or prM) and E that allow their accumulation in intracellular membranes at sites of virus assembly and maturation (similar to the location of bunyavirus assembly) and that distinguish them from other virus envelope proteins such as those of alphaviruses that undergo vectorial transport to the plasma membrane (reviewed in Strauss and Strauss, 1985) are unknown.

The protein NS1 has several noteworthy features. Unlike the E protein, NS1 contains two potential N-linked glycosylation sites the positions of which are conserved between yellow fever virus and Murray Valley encephalitis virus (Figs. 6 and 7), at positions 130 and 208 from the yellow fever virus NS1 N terminus. Murray Valley encephalitis virus contains a third potential site at position 175. As previously mentioned, NS1 is in fact glycosylated (Westaway, 1975; Schlesinger *et al.*, 1983; Smith and Wright, 1985), and monoclonal antibodies against NS1 are capable of mediating complement-dependent lysis of yellow-fever-infected cells, suggesting its presence at the plasma membrane (J. Schlesinger, personal communication). Additional evidence for cell-surface localization of NS1 has been obtained in immunocytochemical studies of dengue-2-infected cells (Cardiff and Lund, 1976). Overall, NS1 is hydrophilic, but the probable C terminus of NS1 from molecular-weight estimates could contain a hydrophobic sequence for anchoring the protein in the membrane (Fig. 7). The function of NS1 is unknown, but it could be involved in virus assembly rather than RNA replication. In this regard, it is of interest that NS1 has been shown to be the soluble complement-fixing antigen for dengue 2 (Smith and Wright, 1985), and suggestive evidence exists that it is also the soluble complement-fixing antigen for yellow fever virus (reviewed in Russell *et al.*, 1980; Schlesinger *et al.*, 1983). These membrane-associated and "soluble" forms of NS1 could conceivably differ by the presence or absence of the putative C-terminal hydrophobic segment mentioned above; however, no direct evidence exists to support this hypothesis. Differences among flaviviruses in the steady-state levels of these two forms may account for the apparently low and variable quantities of NS1 detected in Kunjin-virus-infected cells (Westaway, 1973; Wright *et al.*, 1977) as opposed to the high levels of NS1 (relative to the other large nonstructural proteins) in cells infected with dengue virus type 2 (Smith and Wright, 1985) or yellow fever virus (Schlesinger *et al.*, 1983).

In the N-terminal region of the flavivirus polyprotein including NS1, the positions of cysteine residues are highly conserved (see Fig. 6) in contrast to the other regions encoding nonstructural polypeptides that have thus far been examined (Fig. 6 and also Fig. 7). This may reflect a need for additional conformational stability in the virion structural proteins and in NS1. These polypeptides may require very specific three-dimensional configurations to interact with one another during virion maturation as well as in their interactions with the extracellular environment. Similar structural constraints are seen in the alphavirus structural proteins (see Chapter 3).

The two largest nonstructural proteins, NS3 and NS5, are both fairly hydrophilic, containing no long stretches of hydrophobic amino acids (Figs. 3, 7, and 10), and carry net positive charges (15 in the case of NS3, 34 in the case of NS5). One or both is probably involved in RNA replication (see Section VI), and the net positive charge and hydrophilic character may reflect this. From the regions for which data are available (see Figs. 6 and 7 for comparison of yellow fever and Murray Valley encephalitis viruses and Figs. 9 and 10 for comparison of yellow fever and dengue 2 viruses), NS3 and NS5 would appear to be the most highly conserved of the flavivirus proteins, which is also consistent with their possible enzymatic role in RNA replication.

The four unassigned regions of the polyprotein sequence that may encode polypeptides ns2a, ns2b, ns4a, and ns4b, as discussed in Section III.C, are all rather rich in hydrophobic amino acids (Figs. 3, 7, and 10), possessing long strings of uncharged residues. From the limited amount of comparative sequence data available (Figs. 6 and 9 and Table II), these polypeptides (or corresponding regions of the polyprotein) appear to be among the least conserved of the flavivirus proteins in terms of linear sequence homology. However, as can be seen in Figs. 7 and 10, the distribution of hydrophilic and hydrophobic amino acid residues is quite highly conserved, as shown by the fact that the hydrophobicity profiles are virtually superimposable. It should again be emphasized that given the heterogeneity in polyacrylamide gel patterns of the small flavivirus nonstructural proteins [as well as heterogeneity in E and NS1 (see Wright *et al.*, 1981; Wright, 1982; Schlesinger *et al.*, 1983)], both among flavi-

FIGURE 11. Nucleotide homology between yellow fever virus (YF) and West Nile virus (WN) [data are from Wengler and Wengler (1981) and Rice *et al.* (1985)] at the 3' termini of the genomic (+) strand and complementary (−) strand RNAs. Nucleotide identities in the 3'-terminal sequences of (+) and (−) strands are circled; those that are homologous between yellow fever and West Nile RNAs are underlined [(−) strand] or overlined [(+) strand]. Reproduced from Rice *et al.*, (1985) with permission.

viruses and for the same flavivirus grown in different cell types [see, for example, Fig 2 and Smith and Wright (1985)], it is likely that alternative cleavage sites exist, giving rise to other as yet unidentified virus-specified polypeptides. The scheme depicted in Fig. 8 and described above is a working hypothesis and is presented in the hope that this may stimulate the rigorous identification of these or other virus-specific proteins in infected cells and the characterization of their functions in the virus life cycle.

V. STRUCTURES IN THE FLAVIVIRUS GENOME

Several sequence elements and potential secondary structures have been found in the flavivirus genome that could be important for RNA replication, translation, or encapsidation. The first of these is illustrated in Fig. 11, in which the 5' and 3' ends of West Nile virus RNA (Wengler and Wengler, 1981) and of yellow fever virus RNA (Rice *et al.*, 1985) are shown. There are nucleotide sequences in common between the 3' ends of the plus and the minus strand of each virus (Wengler and Wengler, 1981) (circled nucleotides), and these sequences are largely conserved between West Nile and yellow fever viruses (underlined nucleotides). The nucleotide conservation between two viruses representing two different serotypes within the mosquito-borne flaviviruses suggests that these nucleotides form part of a recognition sequence for the viral replicase. The 3'-terminal sequence of West Nile virus (strain E101) genomic RNA reported by Brinton *et al.* (1986) differs from that of Wengler and Wengler (1981) by the insertion of an additional A residue three nucleotides from the 3'-terminus. This sequence exhibits even greater homology with the 3' terminus of West Nile minus-strand RNA and the yellow fever virus 3'-terminal sequences. The fact that at least part of the hypothetical recognition sequence is shared by both plus and minus strands suggests that a similar enzyme complex is involved in the initiation of both plus- and

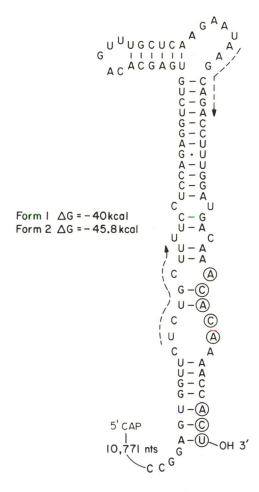

FIGURE 12. Possible secondary structures at the 3' terminus of yellow fever virus genomic RNA. Circled nucleotides are shared with the 3'-terminus of the yellow fever (−) strand (Rice *et al.*, 1985) (see Fig. 11). ΔG values were calculated according to Tinoco *et al.* (1973). A more stable conformation than the one shown (form 1) can be formed if the two overlined sequences are base-paired to one another (form 2). Reproduced from Rice *et al.* (1985) with permission.

minus-strand synthesis. This result contrasts with that found for alphaviruses (Chapter 3) and is reminiscent instead of the situation for negative-strand viruses (reviewed in Strauss and Strauss, 1983).

A second feature of the 3' end of flavivirus genomic RNAs may be the presence of a stable hairpin loop involving the 3' end. It has been reported that West Nile RNA is difficult to label at the 3' end (Wengler and Wengler, 1981), indicating that it may be involved in secondary structure, and similar results have been obtained with yellow fever RNA (Deubel *et al.*, 1983; Rice *et al.*, 1986a). A possible secondary structure for the 3' end of yellow fever genomic RNA is shown in Fig. 12. This structure has a calculated stability of 40–46 kcal/mole, using the formulas of Tinoco *et al.* (1973), and thus is likely to exist in solution. Two points should be noted: First, the 3'-terminal U is hydrogen-bonded in this structure and thus is not readily available as a substrate for enzymatic modification, as has been found experimentally. Second, the conserved pentanucleotide -A-C-A-C-A forms a loop in which only two of the nucleotides could be

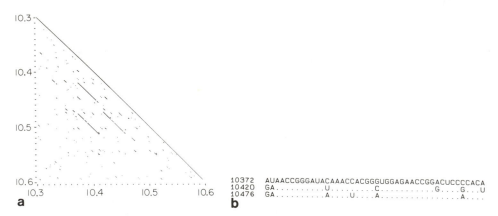

FIGURE 13. Repeated sequences in the 3′ noncoding region of yellow fever virus (YF). (a) A dot matrix homology analysis of the YF noncoding region from 10.3 to 10.6 kilobases from the 5′ end of the genomic RNA. An identity of five out of six nucleotides was scored as a match. (b) Sequences thus identified aligned and numbered relative to the 5′ terminus of YF RNA (Fig. 3). Dots indicate nucleotides that are identical with the top sequence.

hydrogen-bonded, and this sequence could be recognized and bound by a viral replicase even when a secondary structure was present at the 3′ end. A similar structure based on sequence data as well as mapping of RNase-sensitive sites has been proposed for the 3′ terminus of West Nile virus genomic RNA (Brinton *et al.*, 1986) (see Chapter 11). The function of this structure is unclear. An attractive hypothesis is that it is involved in RNA encapsidation and that binding of capsid protein to this structure initiates encapsidation while preventing replication of the RNA. A second possibility is that it is required for RNA replication *per se*.

A third sequence element of note in yellow fever genomic RNA is a set of closely spaced repeated sequences that are found between nucleotides 10,374 and 10,520 (underlined in Fig. 3). These sequences, each about 40 nucleotides long, contain four to six changes between them in pairwise comparisons (Fig. 13). The significance of the repeated sequences in flavivirus replication is at present unknown; they may be important in replication, or they may reflect relaxed evolutionary constraints on this region of the genome. Repeated sequences have also been found in the 3′untranslated region of alphavirus RNAs, and these repeated sequences tend to be conserved among alphaviruses (Ou *et al.*, 1982).

One of the cDNA clones from Murray Valley encephalitis genomic RNA contains sequences representing a portion of the 3′untranslated region of this RNA (Rice *et al.*, 1986a), and this sequence is compared to the corresponding region of yellow fever virus in the dot matrix shown in Fig. 14. Although most of the portion of the Murray Valley encephalitis virus 3′ untranslated region thus far sequenced is not homologous to yellow fever virus, there are two highly conserved domains between these two viruses, the second of which terminates immediately 5′ to the pro-

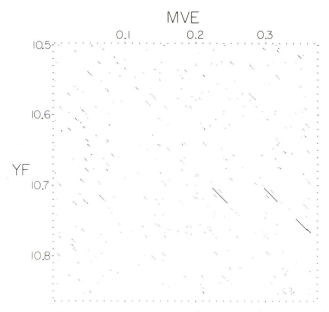

FIGURE 14. Conserved nucleotide sequences in the 3' noncoding region of yellow fever virus (YF) and Murray Valley encephalitis virus (MVE) RNAs. A dot matrix comparison is shown for the YF RNA sequence between 10.5 kilobases from the 5' end and the 3' terminus with an MVE clone that contains a portion of the 3' noncoding sequences of MVE. An identity of six out of eight nucleotides was scored as a match.

posed 3'-terminal secondary structure of yellow fever virus. At least the second of these conserved regions is also found in West Nile virus genomic RNA at a similar distance from the 3'-terminus (Brinton *et al.*, 1986). The Murray Valley encephalitis virus cDNA clone begins at this position and thus may have arisen by a self-priming event during first-strand cDNA synthesis due to the presence of a homologous 3' terminal secondary structure in Murray Valley encephalitis virus genomic RNA similar to the one shown for yellow fever virus in Fig. 12. In addition, the first of these highly conserved sequences is repeated at least once in Murray Valley encephalitis virus RNA, but not in yellow fever virus RNA (Fig. 14). The conservation of these sequences leads to the hypothesis that they are important for RNA replication, and it should be noted that one of these highly conserved sequences could be juxtaposed close to the conserved pentanucleotide -A-C-A-C-A by virtue of the 3' terminal secondary structure, and thus both sequences could be bound by a replicase during initiation of RNA replication.

Finally, a potential secondary structure can be predicted in yellow fever virus RNA in the junction region between the structural and nonstructural proteins (not shown). Whether this or other RNA structures are involved in regulation of protein synthesis is unclear, and comparative studies with other flavivirus RNAs will be required to ascertain whether this feature is common to other flaviviruses.

VI. EVOLUTION OF FLAVIVIRUSES

Although the flaviviruses were once classified with the Togaviridae, it is becoming clear that they deserve their recent reclassification as a separate family. The mature virions are morphologically similar to alphaviruses, in that they have a single-stranded RNA (+) sense genome encapsidated in an icosahedral nucleocapsid and surrounded by a lipid bilayer containing virus-specified polypeptides, but they differ markedly in genome organization and replication strategy. The location of the genes encoding the structural proteins at the 5' end of the genome, the single long reading frame, and the lack of the subgenomic message are all characteristics shared with picornaviruses rather than alphaviruses (reviewed in Strauss and Strauss, 1983).

Members of the Flaviviridae have been grouped according to their arthropod vectors: the mosquito-borne flaviviruses, the tick-borne flaviviruses, and those of no known vector (Chamberlain, 1980). Their relationships have been extensively studied using polyclonal antisera (reviewed in Porterfield, 1980), but relatively few comparative sequence data exist. All these viruses share common antigenic determinants as evidenced by broad cross-reactions in hemagglutination-inhibition tests. Relationships deduced from cross-neutralization tests, which are more specific, clearly separate the mosquito-borne and tick-borne viruses into distinct groups (Porterfield, 1980). The mosquito-borne flaviviruses have been further divided into several subgroups, although the issue is complex because different antisera give somewhat different results and because the relationships among the flaviviruses fall more on a continuum than into discrete units. Porterfield (1980) recognizes five subgroups of mosquito-borne flaviviruses: the dengue subgroup, the Ugandan S subgroup, the Ntaya subgroup, the Spondweni subgroup, and the Japanese encephalitis subgroup, which also includes St. Louis encephalitis, Murray Valley encephalitis, Kunjin, and West Nile viruses (Porterfield, 1980). Furthermore, yellow fever virus usually gives monospecific or only weak reactions in cross-neutralization tests (Porterfield, 1980; C. Calisher, personal communication) and together with its distinct epidemiology is best classified as a separate subgroup (Chamberlain, 1980). Thus, for the purposes of this review, the sequence data obtained for yellow fever, Murray Valley encephalitis, and dengue 2 RNAs, and the protein sequence data for yellow fever, St. Louis encephalitis, and dengue 2 E proteins, represent data for three different subgroups of the mosquito-borne flaviviruses.

Recent experiments using monoclonal antibodies have demonstrated both group cross-reactive and type-specific epitopes associated with the flavivirus E protein (Chapter 9) (Heinz *et al.*, 1982, 1983a; Gentry *et al.*, 1982; Henchal *et al.*, 1982; Peiris *et al.*, 1982; Schlesinger *et al.*, 1983, 1984; Kimura-Karoda and Yasui, 1983; Roehrig *et al.*, 1983; Stephenson *et al.*, 1984; Blok *et al.*, 1984). In most cases, the biological activities of these monoclonal antibodies (hemagglutination inhibition or neutralization or both) follow the trend found for polyclonal antisera; i.e., neutralizing monoclonal antibodies (which can also inhibit hemagglutination) tend to be more specific than antibodies that inhibit hemagglutination. There are, however, a growing number of exceptions in which neutralizing epitopes are subgroup-cross-reactive in neutralization tests. This suggests that during infection of a vertebrate host by a single flavivirus, the predominant neutralizing antibodies produced are those that react with divergent epitopes of E important for neutralization. While these observations are undoubtedly relevant to flavivirus pathology, epidemiology, host range, and divergent evolution, they also point out the need to examine flaviviruses at the level of protein and nucleic acid sequence homology to determine their overall relatedness. From limited comparative sequence data for the E proteins of four flaviviruses representing three serological subgroups (Fig. 4), it appears that the N-terminal 40 amino acids are 52–60% conserved between viruses belonging to different serogroups and 77% conserved in the two viruses from the same subgroup (St. Louis encephalitis and Murray Valley encephalitis). Over its entire length, the E protein of yellow fever virus and Murray Valley encephalitis virus is 45% conserved.

It is likely that the broad cross-reactions among flaviviruses commonly observed in hemagglutination-inhibition tests and occasionally seen in neutralization tests are due to immune recognition of conserved envelope protein domains that mediate important biological functions in the virus life cycle such as specific binding to host cells or virus fusion with intracellular membranes during penetration. As described in Section IV, the yellow fever and Murray Valley encephalitis E proteins have domains that are highly conserved as well as domains that are quite divergent. Thus far, evolutionary relationships as deduced from the serological and sequence comparisons (Figs. 4, 6, 7, 9–12, and 14) are in general agreement, and it appears clear that all flaviviruses descended from a common ancestor. The amino acid sequence conservation found in the flavivirus E proteins is not dissimilar to that previously found among alphavirus surface glycoproteins (Bell et al., 1984; Rice and Strauss, 1981a; Dalgarno et al., 1983).

An alternative method to estimate relative nucleotide sequence homology that has been used for several different flaviviruses involves the measurement of S1 nuclease resistance of cDNA–RNA hybrids (Blok et al., 1984). These data clearly demonstrate sequence conservation among the four dengue serotypes (as well as with other subgroups), but also indicate a close relationship between dengue type 2 virus and Edge Hill virus that might not have been expected from either serological tests using polyclonal antisera (Porterfield, 1980) or from epidemiological considerations, since these viruses are distinct in their vertebrate host range, mosquito vectors, and geographic distributions. Future sequence comparisons should be of great use in elucidating the evolutionary relationships among the diverse members of Flaviviridae.

To investigate the possibility of more distant evolutionary relationships of flaviviruses to other RNA viruses, we have searched for homologies within the putative polymerase genes of various plant and animal viruses. Significant homologies have been found between alphaviruses and plant viruses (Haseloff *et al.*, 1984; Ahlquist *et al.*, 1985) and less extensive homologies between picornaviruses and alphaviruses (Kamer and Argos, 1984). Kamer and Argos (1984) have aligned the polymerase gene of poliovirus with those of several viruses including alfalfa mosaic virus, bromegrass mosaic virus, tobacco mosaic virus, Sindbis virus, foot-and-mouth disease virus, encephalomyocarditis virus, and cowpea mosaic virus. The deduced amino acid sequence of yellow fever virus NS5 between residues 3037 and 3181 has also been aligned with this collection of diverse RNA virus proteins in Fig. 15, as well as the comparable regions of Middelburg virus and of human rhinovirus. In this figure, the location of the aligned sequences relative to the open reading frame has been given. For the four genera of picornaviruses, these sequences lie within the region of the polyprotein that encodes the viral polymerase; the comparable region of the cowpea mosaic virus genome is present on the B RNA (Franssen *et al.*, 1984). The bromegrass mosaic virus and alfalfa mosaic sequences are from RNA 2, which encodes a portion of the viral replicase. The regions of tobacco mosaic virus, Sindbis virus, and Middelburg virus shown are all downstream of their respective termination codons (and therefore produced solely by readthrough) (Pelham, 1978; Strauss *et al.*, 1983) in the polypeptide called 183.3 K for tobacco mosaic virus or nsP4 for Sindbis and Middelburg viruses (see Fig. 18 of Chapter 3). These homologous regions are quite short and separated from one another by intervening sequences of varying length, depending on the virus, and probably represent conserved functional domains for particular RNA-dependent RNA replicase functions.

It is also instructive to compare the doublet frequencies found in yellow fever and Murray Valley encephalitis RNAs (Table IV) with those found for other eukaryotic RNA viruses. Yellow fever and Murray Valley RNAs demonstrate a pronounced deficiency in the CG doublet, a characteristic shared with many other viruses of birds and mammals, as well as with vertebrate DNA (Russell *et al.*, 1976; Bird, 1980). The flavivirus RNAs are deficient in the UA doublet as well. In contrast, insect viruses, plant viruses, and alphaviruses contain the CG frequency predicted by their base compositions (see Chapter 3). The significance of the low CG frequency in flaviviruses is unclear. RNA viruses lack the C-methylation found in DNA that is invoked as the rationale for a reduced CG content (Salser, 1977; Bird, 1980). They also mutate rapidly, so that the evolutionary origin of the virus would not be expected to contribute to dinucleotide frequencies. The most satisfactory rationale for a nonrandom dinucleotide frequency is that it might in some way be involved in ad-

```
                                                                    AMINO ACID #
YF      ..FYADDTAGWDTRIT EADLDDEQEILNYMS...      3037-3065
SIN     ..VLETDIASFDKS QDDAMALTGLMILEDLG...      2269-2297
MID     ..VLETDIASFDKS QDDSLAYTGLMLLEDLG...
TMV     ..VLELDISKYDKS QNEFHCAVEYEIWRRLG...       1381-1409
BMV     ..FLEADLSKFDKS QGELHLEFQREILLALG...       461-489     (RNA2)
A1MV    ..FKEIDFSKFDKS QNELHHLIQERFLKYLG...       526-554     (RNA2)
CoMV    ..VLCCDYSSFDGLLSKQVMDVIASMINELCG...       1431-1460   (RNAB)
POLIO   ..FA FDYTGYDASLS PAW  FEAL  KMVL...       1976-1999
FMD     ..VWDVDYSAFDANHCSDAM   NIMFEEVFR...       2099-2125
EMC     ..VYDVDYSNFDSTHS VAM  FRLLAEEFFT...       2061-2087
RHINO   ..MA FDYSNFDASLS PVW  FVCL EK VL...       1949-1972
          *+   *++ **        +*    +  + ++
             *        *


YF      ..MDVISRRDQRGSGQVVTYALNTITNLKVQL...       3098-3127
SIN     ..GTRFKFGAMMKSGMFLTLFVNTVLNVVIAS...       2321-2350
MID     ..GTRFKFGAMMKSGMFLTLFVNTMLNMTIAS...
TMV     ..GIKTCIWYQRKSGDVTTFIGNTVIIAACLA...       1433-1462
BMV     ..KVGMSVSFQRRTGDAFTYFGNTLVTMAMIA...       513-542     (RNA2)
A1MV    ..GVFFNVDFQRRTGDALTYLGNTIVTLACLC...       578-607     (RNA2)
CoMV    ..NTVWRVECGIPSGFPMTVIVNSIFNEILIR...       1485-1514   (RNAB)
POLIO   ..NKTYCVKGGMPSGCSGTSIFNSMINNLIIR...       2023-2052
FMD     ..NKRITVEGGMPSGCSATSIINTILNNIYVL...       2150-2179
EMC     ..EKRFLITGGLPSGCAATSMLNTIMNNIIIR...        2112-2141
RHINO   ..DEIYVVEGGMPSGCSGTSIFNSMINNIIIR...        1995-2024
          + +        **  +* * ****
            *        *     *


YF      ..GCDRLKRMAVS GDD CVVRPID...               3162-3181
SIN     ..RLKTSRCAAFI GDD NIIGVV ...              2356-2376
MID     .. LTNSKCAAFI GDD NIVHGVK..
TMV     ..PMEKIIKGAFC GDD SLLYFPK...               1466-1486
BMV     .. CD   CAIFS GDD SLIISKV...              550-566     (RNA2)
A1MV    ..       FVVAS GDD SLIGTVE...             619-633     (RNA2)
CoMV    ..SFDKLIGLVTY GDD NLISVNA...              1533-1553   (RNAB)
POLIO   .. LDHLK MIAY GDD VIASYPH...              2064-2082
FMD     .. LDTYT MISY GDD IVVASDY...              2191-2209
EMC     .. FDDVK VLSY GDD LLVATNY...              2153-2171
RHINO   .. LDKLK ILAY GDD LIVSYPY...              2036-2054
           +      +*  +***  **
                       ***
```

FIGURE 15. Homologous domains of polymerase genes of a number of RNA viruses. Three noncontiguous stretches of homologous amino acid sequence are shown. The amino acids are numbered from the beginning of the open reading frame in each case. Plus signs, asterisks, and double asterisks indicate increasing degrees of conservation. The canonical Gly-Asp-Asp (GDD) sequence thought to be essential for RNA-dependent polymerases is boxed. The sequences shown were reported in the following references: (YF) yellow fever virus (Rice *et al.*, 1985); (SIN) Sindbis virus (Strauss *et al.*, 1984); (MID) Middelburg virus (Strauss *et al.*, 1983); (TMV) tobacco mosaic virus (Goelet *et al.*, 1982); (BMV) bromegrass mosaic virus RNA2 (Ahlquist *et al.*, 1984); (AlMV) alfalfa mosaic virus RNA2 (Cornelissen *et al.*, 1983); (CoMV) cowpea mosaic virus RNA B (Lomonossoff and Shanks, 1983); (POLIO) poliovirus (Kitamura *et al.*, 1981); (FMD) foot-and-mouth disease virus (Carroll *et al.*, 1984; (EMC) encephalomyocarditis virus (Palmenberg *et al.*, 1984); (RHINO) human rhinovirus 14 (Stanway *et al.*, 1984). These three regions are indicated by horizontal bars in Fig. 21 of Chapter 3.

TABLE IV. Dinucleotide Frequencies[a]

First nucleotide	Second nucleotide			
	A	G	U	C
A	7.94 (8.05)	7.70 (7.51)	6.04 (6.30)	5.58 (6.22)
G	8.95 (8.21)	8.76 (7.60)	5.36 (5.25)	5.33 (5.76)
U	2.81 (3.70)	9.62 (8.95)	5.54 (6.68)	5.00 (4.49)
C	7.55 (8.10)	2.32 (2.76)	6.04 (5.60)	5.39 (4.79)

[a] Doublet frequency (%) for yellow fever virus 17D strain [Rice *et al.*, 1985 (10,862 nucleotides; 27.3% A, 28.4% G, 23.0% U, 21.3% C)] and Murray Valley encephalitis virus [Dalgarno *et al.*, 1986 (shown in parentheses; nucleotides 5–5436; 26.4% A, 28.7% G, 24.4% U, 20.7% C)].

aptation to growth in the host. As one possibility, if the host has a low CG frequency, the distribution of tRNAs might reflect this, and a rapid and efficient translation of mRNA might require matching the codon preference (see below) of the host. But this line of argument seems to break down when alphaviruses and flaviviruses are compared. Alphaviruses replicate quite efficiently in both insects and higher vertebrates and have evolved to alternate in insect and vertebrate hosts; they have a normal (random) CG frequency (Rice and Strauss, 1981b). Flaviviruses also replicate quite efficiently in both insects and higher vertebrates and have evolved to alternate in insect and vertebrate hosts, yet they have a low CG frequency. In this sense, yellow fever seems to be adapted to life in mammals even though it possesses the ability to replicate in its mosquito vector, whereas alphaviruses show a CG frequency characteristic of insect viruses. Although its functional significance is unclear, this difference again suggests that the evolutionary history of the two groups of viruses may be quite distinct.

The codon usage in yellow fever RNA is shown in Table V. Codon usage is nonrandom. As an example, note that the Leu codon CUG is used more than 10 times as often as UUA and 3 times as often as CUA. Similarly, the GUG codon for Val is 8 times more common than GUA and 2 times as common as either GUU or GUC. In addition, codons containing the CG doublet are infrequently used. The four CGN codons of Arg represent only 60 of 210 Arg codons, whereas AGA and AGG represent the remainder. Similarly, note the infrequent use of UCG for Ser, CCG for Pro, ACG for Thr, and GCG for Ala. This is consistent with the low CG doublet frequency in the RNA and different from the result found for alphaviruses (Chapter 3).

For most coding regions of alphavirus genomes for which comparative sequence data are available (Strauss *et al.*, 1983; Dalgarno *et al.*, 1983), even where amino acids are conserved, nucleotide divergence has been extensive enough to virtually randomize codon usage when multiple codons are available. A comparable analysis was conducted on highly conserved regions of the E protein (amino acids 377–406) and NS3 protein

TABLE V. Yellow Fever Virus (17D Strain) Codon Usage

Phe	UUU	64	Ser	UCU	44	Tyr	UAU	35	Cys	UGU	32
	UUC	52		UCC	42		UAC	44		UGC	32
Leu	UUA	8		UCA	56	Ochre	UAA	1	Opal	UGA	0
	UUG	72		UCG	8	Amber	UAG	0	Trp	UGG	85
Leu	CUU	46	Pro	CCU	40	His	CAU	50	Arg	CGU	12
	CUC	49		CCC	28		CAC	32		CGC	25
	CUA	37		CCA	56	Gln	CAA	42		CGA	12
	CUG	100		CCG	12		CAG	51		CGG	11
Ile	AUU	63	Thr	ACU	53	Asn	AAU	56	Ser	AGU	34
	AUC	69		ACC	51		AAC	68		AGC	31
	AUA	44		ACA	73	Lys	AAA	92	Arg	AGA	67
Met	AUG	129		ACG	21		AAG	101		AGG	83
Val	GUU	69	Ala	GCU	83	Asp	GAU	70	Gly	GGU	36
	GUC	69		GCC	80		GAC	88		GGC	68
	GUA	16		GCA	58	Glu	GAA	108		GGA	124
	GUG	132		GCG	23		GAG	102		GGG	73

[a] Codon usage for the yellow fever virus, 17D strain, long open reading frame of 10,233 nucleotides (3411 codons excluding the termination codon). Data from Rice *et al.* (1985).

(amino acids 1558–1696) of yellow fever and Murray Valley encephalitis viruses. In these regions, there were 113 conserved amino acids (67% overall homology). Normalizing the data for overall codon usage frequencies, we would predict that for 43.5 of these, the same codon would be used on the basis of random chance alone, while the actual number of matches is 48. Thus, during the evolutionary divergence of Murray Valley encephalitis virus from yellow fever virus, codon usage has been very nearly randomized for conserved amino acids. On the other hand, the nonrandom codon preferences mentioned above are in general similar for these two viruses. Since RNA genomes acquire silent mutations in codons for conserved amino acids, sequence comparisons and homology searches in coding regions are best done at the level of the deduced amino acid sequence rather than the nucleotide sequence.

VII. CONCLUDING REMARKS

The subject of the evolution of RNA viruses is of considerable interest, but difficult to approach because of the lack of a fossil record. Present-day viruses can be examined for sequence and structural homologies or similarities in genome organization in an attempt to deduce their evolutionary relationships. The work of Kamer and Argos (1984) mentioned in this chapter raises the possibility that all RNA viruses may have evolved from a common ancestral RNA virus that arose at some unknown time in the past. However, even if these weak homologies found do reflect an evolutionary relationship among the replicase enzymes, it is still possible that RNA viruses arose more than once from a cellular gene or genes that had in turn descended from a common ancestor. In this regard, it is of interest that the conserved block containing the Gly-Asp-Asp sequence (Fig. 15) also shows limited homology not only with the β subunit of bacteriophage MS2 RNA polymerase, but also with several RNA-dependent DNA polymerases (Kamer and Argos, 1984). It is possible that, during RNA virus evolution, new functions were acquired by capturing host genes in some way. The various protease activities possessed by RNA viruses seem logical candidates for such capture, especially in view of limited homologies shown by the alphavirus capsid protease and animal serine proteases (Boege *et al.*, 1981; Hahn *et al.*, 1985). However, such virus-encoded proteolytic activities may also have arisen independently during virus evolution. As more becomes known about virus and host proteases, it will be of interest to see whether some virus proteases show homologies to other classes of animal proteases.

If all RNA viruses have evolved from a common ancestor, the distinct genome organization and replication strategy of flaviviruses, the scant sequence homology with other RNA viruses, and the lack of a poly(A) tract suggest that these viruses represent a separate branch of the RNA virus evolutionary tree from other plus-stranded viruses. In addition, the presence of a conserved sequence, albeit short, at the 3' ends of both plus and minus strands is unlike the situation for other plus-stranded viruses sequenced to date and is reminiscent instead of negative-stranded viruses (reviewed in Strauss and Strauss, 1983), suggesting that the flaviviruses might fall on the evolutionary branch that led to negative-stranded viruses.

Given the very high mutation rate of RNA viruses (Reanney, 1982; Holland *et al.*, 1982) and the fact that these viral populations contain genomic RNAs with a particular average sequence but in which a significant fraction of molecules differ in one or more nucleotides from this average sequence (Domingo *et al.*, 1978), the stability of flaviviruses observed in nature is surprising. Relatively few oligonucleotide differences (probably accounting for less than 1% difference in the RNA sequences) were found when the yellow fever virus 17D strain and the parent Asibi strain were compared (Monath *et al.*, 1983), even though these strains are separated by more than 200 passages. Some strains of dengue virus isolated 20 years apart have been found to have very few changes (Trent *et al.*, 1983; D. Trent, personal communication). This constancy may reflect an as yet uncharacterized adaptive significance (perhaps relating to RNA secondary and tertiary structures) for even silent nucleotide changes in the protein-coding regions.

During the next few years, many more flavivirus nucleic acid and protein sequences will become available for comparison, leading, no doubt, to considerable refinement in our understanding of flavivirus genome structure and organization. This information immediately allows new approaches to the study of flavivirus immunobiology, and future directions would seem to include the development of *in vitro* replication systems capable of utilizing exogenous templates and the manipulation of the viral genome and its encoded products via recombinant DNA technology. These advances should greatly facilitate the design of direct experiments to test hypotheses formulated from primary sequence data and comparative analyses and should help to elucidate aspects of flavivirus replication that are at present only poorly understood.

ACKNOWLEDGMENTS. We are grateful to the following colleagues for making their data available to us prior to publication: V. Vakharia, T. Yaegashi, R. Feighny, S. Kohlekar, and R. Padmanabhan (dengue 2 sequence); G. Cleaves (dengue 2 polyprotein precursors); P. Eastman, J. Mecham, and C. Blair (Japanese encephalitis polyproteins); and also to J. Schlesinger and F. Heinz for allowing the quotation of their unpublished results. We are also grateful to T. Hunkapiller for computer programs used in producing many of the figures presented here and to L. Hood for the use of the computer facilities. The work of the authors is supported by Grants AI 10793 and AI 20612 from the National Institutes of Health and by Grant DMB 8316856 from the National Science Foundation.

REFERENCES

Ahlquist, P., Dasgupta, R., and Kaesberg, P., 1984, Nucleotide sequence of the brome mosaic virus genome and implications for viral replication, *J. Mol. Biol.* **172**:369–383.

Ahlquist, P., Strauss, E. G., Rice, C. M., Strauss, J. H., Haseloff, J., and Zimmern, D., 1985, Sindbis virus proteins nsP1 and nsP2 contain homology to nonstructural proteins from several RNA plant viruses, *J. Virol.* **53**:536–542.

Bell, J. R., Kinney, R. M., Trent, D. W., Strauss, E. G., and Strauss, J. H., 1984, An evolutionary tree relating eight alphaviruses, based on amino-terminal sequences of their glycoproteins, *Proc. Natl. Acad. Sci. U.S.A.* **81**:4702–4706.

Bell, J. R., Kinney, R. M., Trent, D. W., Lenches, E. M., Dalgarno, L., and Strauss, J. H., 1985, N-terminal amino acid sequences of structural proteins of three flaviviruses, *Virology* **143**:224–229.

Bird, A. P., 1980, DNA methylation and the frequency of CpG in animal DNA, *Nucleic Acids Res.* **8**:1499–1504.

Blok, J., Henchal, E. A., and Gorman, B. M., 1984, Comparison of dengue viruses and some other flaviviruses by cDNA–RNA hybridization analysis and detection of a close relationship between dengue virus serotype 2 and Edge Hill virus, *J. Gen. Virol.* **65**:2173–2181.

Boege, U., Wengler, G., Wengler, G., and Wittmann-Liebold, B., 1981, Primary structure of the core proteins of the alphaviruses Semliki Forest virus and Sindbis virus, *Virology* **113**:293–303.

Boege, U., Heinz, F. X., Wengler, G., and Kunz, C., 1983, Amino acid compositions and amino-terminal sequences of the structural proteins of a flavivirus, European tick-borne encephalitis virus, *Virology* **126**:651–657.

Boulton, R. W., and Westaway, E. G., 1977, Togavirus RNA: Reversible effect of urea on genomes and absence of subgenomic viral RNA in Kunjin virus-infected cells, *Arch. Virol.* **55**:201–208.

Brinton, M. A., Fernandez, A. V., and Amato, J., 1986, Sequence analysis of the 3' terminus of West Nile virus, strain E101, genome RNA,

Cardiff, R. D., and Lund, J. K., 1976, Distribution of dengue-2 antigens by electron immunocytochemistry, *Infect. Immun.* **13:**1699–1709.

Carroll, A. R., Rowlands, D. J., and Clarke, B. E., 1984, The complete nucleotide sequence of the RNA coding for the primary translation product of foot-and-mouth disease virus, *Nucleic Acids Res.* **12:**2461–2472.

Castle, E., Nowak, T., Leidner, U., Wengler, G., and Wengler, G., 1985, Sequence analysis of the viral core protein and the membrane-associated proteins V1 and NV2 of the flavivirus West Nile virus and of the genome sequence for these proteins, *Virology* **145:**227–236.

Chamberlain, R. W., 1980, Epidemiology of arthropod-borne Togaviruses: The role of arthropods as hosts and vectors and of vertebrate hosts in natural transmission cycles, in: *The Togaviruses* (R. W. Schlesinger, ed.), pp. 175–228, Academic Press, New York.

Cleaves, G. R., and Dubin, D. T., 1979, Methylation status of intracellular dengue type 2 40 S RNA, *Virology* **96:**159–165.

Cornelissen, B. J. C., Brederode, F. T., Veeneman, G. H., van Boom, J. H., and Bol, J. F., 1983, Complete nucleotide sequence of alfalfa mosaic virus RNA 2, *Nucleic Acids Res.* **11:**3019–3025.

Dalgarno, L., Rice, C. M., and Strauss, J. H., 1983, Ross River virus 26S RNA: Complete nucleotide sequence and deduced sequences of the encoded structural proteins, *Virology* **129:**170–187.

Dalgarno, L., Strauss, J. H., and Rice, C. M., 1986, Partial nucleotide sequence of Murray Valley encephalitis virus: Comparison of the encoded polypeptides with yellow fever virus structural and nonstructural proteins, *J. Mol. Biol.*, in press.

Deubel, V., Crouset, J., Bénichou, D., Digoutte, J.-P., Bouloy, M., and Girard, M., 1983, Preliminary characterization of the ribonucleic acid of yellow fever virus, *Ann. Virol.* **134E:**581–588.

Dixon, L. K., and Hohn, T., 1984, Initiation of translation of the cauliflower mosaic virus genome from a polycistronic mRNA: Evidence from deletion mutagenesis, *EMBO J.* **3:**2731–2736.

Docherty, K., Carroll, R. J., and Steiner, D. F., 1982, Conversion of proinsulin to insulin: Involvement of a 31,500 molecular weight thiol protease, *Proc. Natl. Acad. Sci. U.S.A.* **79:**4613–4617.

Domingo, E., Sabo, D. Taniguchi, T., and Weissman, C., 1978, Nucleotide sequence heterogeneity of an RNA phage population, *Cell* **13:**735–744.

Dorner, A., J. Semler, B. L., Jackson, R. J., Hanecak, R., Duprey, E., and Wimmer, E., 1984, *In vitro* translation of poliovirus RNA: Utilization of internal initiation sites in reticulocyte lysate, *J. Virol.* **50:**507–514.

Franssen, H., Leunissen, J., Goldbach, R., Lomonossoff, G., and Zimmern, D. 1984, Homologous sequences in nonstructural proteins from cowpea mosaic virus and picornaviruses, *EMBO J.* **3:**855–861.

Garoff, H., Frischauf, A.-M., Simons, K., Lehrach, H., and Delius, H., 1980a, Nucleotide sequence of cDNA coding for Semliki Forest virus membrane glycoproteins, *Nature (London)* **288:**236–241.

Garoff, H., Frischauf, A.-M., Simons, K., Lehrach, H., and Delius, H., 1980b, The capsid protein of Semliki Forest virus has clusters of basic amino acids and prolines in its amino-terminal region, *Proc. Natl. Acad. Sci. U.S.A.* **77:**6376–6380.

Gentry, M. K., Henchal, E. A., McCown, J. M., Brandt, W. E., and Dalrymple, J. M., 1982, Identification of distinct antigenic determinants on dengue-2 virus using monoclonal antibodies, *Am. J. Trop. Med. Hyg.* **31:**548–555.

Goelet, P., Lomonossoff, G. P., Butler, P. J. G., Akam, M. E., Gait, M. J., and Karn, J., 1982, Nucleotide sequence of tobacco mosaic virus RNA, *Proc. Natl. Acad. Sci. U.S.A.* **79:**5818–5822.

Grantham, R., Gautier, C., Guoy, M., Jacobzone, M., and Mercier, R., 1981, Codon catalog usage is a genome strategy modulated for gene expressivity, *Nucleic Acids Res.* **9**:r43–r74.

Hahn, C. S., Strauss, E. G., and Strauss, J. H., 1985, Sequence analysis of three Sindbis virus mutants temperature-sensitive in the capsid protein autoprotease, *Proc. Natl. Acad. Sci. U.S.A.* **82**:4648–4652.

Haseloff, J., Goelet, P., Zimmern, D., Ahlquist, P., Dasgupta, R., and Kaesberg, P., 1984, Striking similarities in amino acid sequence among nonstructural proteins encoded by RNA viruses that have dissimilar genomic organization, *Proc. Natl. Acad. Sci. U.S.A.* **81**:4358–4362.

Heinz, F. X., and Kunz, C., 1979, Protease treatment and chemical crosslinking of a flavivirus: Tick-borne encephalitis virus, *Arch. Virol.* **60**:207–216.

Heinz, F. X., and Kunz, C., 1982, Molecular epidemiology of tick-borne encephalitis virus: Peptide mapping of large non-structural proteins of European isolates and comparison with other flaviviruses, *J. Gen. Virol.* **62**:271–285.

Heinz, F. X., Berger, R., Majdic, O., Knapp, W., and Kunz, C., 1982, Monoclonal antibodies to the structural glycoprotein of tick-borne encephalitis virus, *Infect. Immun.* **37**:869–874.

Heinz, F. X., Berger, R., Tuma, W., and Kunz, C., 1983a, A topological and functional model of epitopes on the structural glycoprotein of tick-borne encephalitis virus defined by monoclonal antibodies, *Virology* **126**:525–537.

Heinz, F. X., Berger, R., Tuma, W., and Kunz, C., 1983b, Location of immunodominant antigenic determinants on fragments of the tick-borne encephalitis virus glycoproteins: Evidence for two different mechanisms by which antibodies mediate neutralization and hemagglutination inhibition, *Virology* **130**:485–501.

Heinz, F. X., Tuma, W., Guirakhoo, F., Berger, R., and Kunz, C., 1984, Immunogenicity of tick-borne encephalitis virus glycoprotein fragments: Epitope-specific analysis of the antibody response, *J. Gen. Virol.* **65**:1921–1929.

Henchal, E. A., Gentry, M. K., McCown, J. M., and Brandt, W. E., 1982, Dengue virus-specific and flavivirus group determinants identified with monoclonal antibodies by indirect immunofluorescence, *Am. J. Trop. Med. Hyg.* **31**:830–836.

Holland, J., Spindler, K., Horodyski, F., Grabau, E., Nichol, S., and VandePol, S., 1982, Rapid evolution of RNA genomes, *Science* **215**:1577–1585.

Hughes, S., Mellstrom, K., Kosik, E., Tamanoi, F., and Brugge, J., 1984, Mutation of a termination codon affects *src* initiation, *Mol. Cell. Biol.* **4**:1738–1746.

Kamer, G., and Argos, P., 1984, Primary structural comparison of RNA-dependent polymerases from plant, animal and bacterial viruses, *Nucleic Acids Res.* **12**:7269–7282.

Kimura-Kuroda, J., and Yasui, K., 1983, Topographical analysis of antigenic determinants on envelope glycoprotein V3 (E) of Japanese encephalitis virus, using monoclonal antibodies, *J. Virol.* **45**:124–132.

Kitamura, N., Semler, B. L., Rothberg, P. G., Larsen, G. R., Adler, C. J., Dorner, A. J., Emini, E. A., Hanecak, R., Lee, J. J., van der Werf, S., Anderson, C. W., and Wimmer, E., 1981, Primary structure, gene organization and polypeptide expression of poliovirus RNA, *Nature (London)* **291**:547–553.

Kozak, M., 1983, Comparison of initiation of protein synthesis in procaryotes, eucaryotes, and organelles, *Microbiol. Rev.* **47**:1–45.

Kozak, M., 1984a, Compilation and analysis of sequences upstream from the translational start site in eukaryotic mRNAs, *Nucleic Acids Res.* **12**:857–872.

Kozak, M., 1984b, Selection of initiation sites by eucaryotic ribosomes: Effect of inserting AUG triplets upstream from the coding sequence for preproinsulin, *Nucleic Acids Res.* **12**:3873–3893.

Kyte, J., and Doolittle, R. F., 1982, A simple method for displaying the hydropathic character of a protein, *J. Mol. Biol.* **157**:105–132.

Liu, C.-C., Simonsen, C. C., and Levinson, A. D., 1984, Initiation of translation at internal AUG codons in mammalian cells, *Nature (London)* **309**:82–85.

Lomedico, P. T., and McAndrew, S. J., 1982, Eukaryotic ribosomes can recognize preproin-

324 CHARLES M. RICE *et al.*

sulin initiation codons irrespective of their position relative to the 5' end of mRNA, *Nature (London)* **299:**221–226.

Lomonossoff, G. B., and Shanks, M., 1983, The nucleotide sequence of cowpea mosaic virus B RNA, *EMBO J.* **2:**2253–2258.

Mertens, P. P. C., and Dobos, P., 1982, Messenger RNA of infectious pancreatic necrosis virus is polycistronic, *Nature (London)* **297:**243–246.

Monath, T. P., Kinney, R. M., Schlesinger, J. J., Brandriss, M. W., and P. Brès, 1983, Ontogeny of yellow fever 17D vaccine: RNA oligonucleotide fingerprint and monoclonal antibody analyses of vaccines produced world-wide, *J. Gen. Virol.* **64:**627–637.

Monckton, R. P., and Westaway, E. G., 1982, Restricted translation of the genome of the flavivirus Kunjin *in Vitro, J. Gen. virol.* **63:**227–232.

Naeve, C. W., and Trent, D. W., 1978, Identification of Saint Louis encephalitis virus mRNA, *J. Virol.* **25:**535–545.

Ou, J.-H., Trent, D. W., and Strauss, J. H., 1982, The 3' noncoding regions of alphavirus RNAs contain repeating sequences, *J. Mol. Biol.* **156:**719–730.

Palmenberg, A. C., Kirby, E. M., Janda, M. R., Drake, N. L., Duke, G. M., Potratz, K. F., and Collett, M. S., 1984, The nucleotide and deduced amino acid sequence of the encephalomyocarditis viral polyprotein coding region, *Nucleic Acids Res.* **12:**2969–2985.

Peiris, J. S. M., Porterfield, J. S., and Roehrig, J. T., 1982, Monoclonal antibodies against the flavivirus West Nile, *J. Gen. Virol.* **58:**283–289.

Pelham, H. R. B., 1978, Leaky UAG termination codon in tobacco mosaic virus RNA, *Nature (London)* **272:**469–471.

Perlman, D., and Halvorson, H. O. 1983, A putative signal peptidase recognition site and sequence in eukaryotic and prokaryotic signal peptides, *J. Mol. Biol.* **167:**391–409.

Porterfield, J. S., 1980, Antigenic characteristics and classification of Togaviridae, in: *The Togaviruses* (R. W. Schlesinger, ed.), pp. 13–46, Academic Press, New York.

Reanney, D. C., 1982, The evolution of RNA viruses, *Annu. Rev. Microbiol.* **36:**47–73.

Rice, C. M., and Strauss, J. H., 1981a, Nucleotide sequence of the 26S mRNA of Sindbis virus and deduced sequence of the encoded virus structural proteins, *Proc. Natl. Acad. Sci. U.S.A.* **78:**2062–2066.

Rice, C. M., and Strauss, J. H., 1981b, Synthesis, cleavage, and sequence analysis of cDNA complementary to the 26S mRNA of Sindbis virus, *J. Mol. Biol.* **150:**315–340.

Rice, C. M., Lenches, E. M., Eddy, S. R., Shin, S. J., Sheets, R. L., and Strauss, J. H., 1985, Nucleotide sequence of yellow fever virus: Implications for flavivirus gene expression and evolution, *Science* **229:**726–733.

Rice, C. M., Dalgarno, L., Strauss, E. G., and Strauss, J. H., 1986a, cDNA cloning of flavivirus genomes for comparative analysis and expression (submitted).

Rice, C. M., Aebersold, R., Teplow, D. B., Pata, J., Bell, J. R., Vorndam, A. V., Trent, D. W., Brandriss, M. W., Schlesinger, J. J., and Strauss, J. H., 1986b, Partial N-terminal amino acid sequences of three nonstructural proteins of two flaviviruses, submitted.

Roehrig, J. T., Mathews, J. H., and Trent, D. W., 1983, Identification of epitopes on the E glycoprotein of St. Louis encephalitis virus using monoclonal antibodies, *Virology* **128:**118–126.

Russell, G. J., Walker, P. M. B., Elton, R. A., and Subak-Sharpe, J. H., 1976, Doublet frequency analysis of fractionated vertebrate nuclear DNA, *J. Mol. Biol.* **108:**1–23.

Russell, P. K., Brandt, W. E., and Dalrymple, J. M., 1980, Chemical and antigenic structure of flaviviruses, in: *The Togaviruses* (R. W. Schlesinger, ed.), pp. 503–529, Academic Press, New York.

Salser, W., 1977, Globin mRNA sequences: Analysis of base-pairing and evolutionary implications, *Cold Spring Harbor Symp. Quant. Biol.* **42:**985–1002.

Schlesinger, J. J., Brandriss, M. W., and Monath, T. P., 1983, Monoclonal antibodies distinguish between wild and vaccine strains of yellow fever virus by neutralization, hemagglutination inhibition, and immune precipitation of the virus envelope protein, *Virology* **125:**8–17.

Schlesinger, J. J., Walsh, E. E., and Brandriss, M. W., 1984, Analysis of 17D yellow fever virus envelope protein epitopes using monoclonal antibodies, *J. Gen. Virol.* **65**:1637–1644.

Shapiro, D., Brandt, W. E., and Russell, P. K., 1972, Change involving a viral membrane glycoprotein during morphogenesis of group B arboviruses, *Virology* **50**:906–911.

Shapiro, D., Kos, K. A., and Russell, P. K., 1973, Protein synthesis in Japanese encephalitis virus-infected cells, *Virology* **56**:95–109.

Smith, G. W., and Wright, P. J., 1985, Synthesis of proteins and glycoproteins in dengue type 2 virus-infected Vero and *Aedes albopictus* cells, *J. Gen. Virol.* **66**:559–571.

Stanway, G., Hughes, P. J., Mountford, R. C., Minor, P. D., and Almond, J. W., 1984, The complete nucleotide sequence of a common cold virus: Human rhinovirus 14, *Nucleic Acids Res.* **12**:7859–7875.

Stephenson, J. R., Lee, J. M., and Wilton-Smith, P. D., 1984, Antigenic variation among members of the tick-borne encephalitis complex, *J. Gen. Virol.* **65**:81–89.

Strauss, E. G., and Strauss, J. H., 1983, Replication strategies of the single stranded RNA viruses of eukaryotes, *Curr. Top. Microbiol. Immunol.* **105**:1–98.

Strauss, E. G., and Strauss, J. H., 1985, Assembly of enveloped animal viruses, in: *Virus Structure and Assembly* (S. J. Casjens, ed.), pp. 205–234, Jones and Bartlett, Portola Valley, California.

Strauss, E. G., Rice, C. M., and Strauss, J. H., 1983, Sequence coding for the alphavirus nonstructural proteins is interrupted by an opal termination codon, *Proc. Natl. Acad. Sci. U.S.A.* **80**:5271–5275.

Strauss, E. G., Rice, C. M., and Strauss, J. H., 1984, Complete nucleotide sequence of the genomic RNA of Sindbis virus, *Virology* **133**:92–110.

Svitkin, Y. V., Lyapustin, V. N., Lashkevich, V. A., and Agol, V. I., 1978. A comparative study on translation of flavivirus and picornavirus RNAs *in vitro*: Apparently different modes of protein synthesis, *FEBS Lett.* **96**:211–215.

Svitkin, Y. V., Ugarova, T. Y., Chernovskaya, T. V., Lyapustin, V. N., Lashkevich, V. A., and Agol, V. I., 1981, Translation of tick-borne encephalitis virus (flavivirus) genome *in vitro*: Synthesis of two structural polypetides, *Virology* **110**:26–34.

Svitkin, Y. V., Lyapustin, V. N., Lashkevich, V. A., and Agol, V. I., 1984, Differences between translation products of tick-borne encephalitis virus RNA in cell-free systems from Krebs-2 cells and rabbit reticulocytes: Involvement of membranes in the processing of nascent precursors of flavivirus structural proteins, *Virology* **135**:536–541.

Takio, K., Towatari, T., Katunuma, N., Teller, D. C., and Titani, K., 1983, Homology of amino acid sequences of rat liver cathepsins B and H with that of papain, *Proc. Natl. Acad. Sci. U.S.A.* **80**:3666–3670.

Tinoco, I., Borer, P. N., Dengler, B., Levine, M. D., Uhlenbeck, O. C., Crothers, D. M., and Gralla, J., 1973, Improved estimation of secondary structure in ribonucleic acids, *Nature (London) New Biol.* **246**:40–41.

Trent, D. W., Grant, J. A., Rosen, L., and Monath, T. P., 1983, Genetic variation among dengue 2 viruses of different geographic origin, *Virology* **128**:271–284.

Vezza, A. C., Rosen, L., Repik, P., Dalrymple, J., and Bishop, D. H. L., 1980, Characterization of the viral RNA species of prototype dengue viruses, *Am. J. Trop. Med. Hyg.* **29**:643–652.

Walter, P., and Blobel, G., 1982, Mechanism of protein translocation across the endoplasmic reticulum, *Biochem. Soc. Symp.* **47**:183–191.

Welch, W. J., and Sefton, B. M., 1979, Two small virus-specific polypeptides are produced during infection with Sindbis virus, *J. Virol.* **29**:1186–1195.

Wengler, G., and Wengler, G., 1981, Terminal sequences of the genome and replicative-form RNA of the flavivirus West Nile virus: Absence of poly(A) and possible role in RNA replication, *Virology* **113**:544–555.

Wengler, G., Wengler, G., and Gross, H. J., 1978, Studies on virus-specific nucleic acids synthesized in vertebrate and mosquito cells infected with flaviviruses, *Virology* **89**:423–437.

Wengler, G., Beato, M., and Wengler, G., 1979, *In vitro* translation of 42S virus-specific RNA from cells infected with the flavivirus West Nile virus, *Virology* **96**:516–529.

Westaway, E. G., 1973, Proteins specified by group B togaviruses in mammalian cells during productive infections, *Virology* **51**:454–465.

Westaway, E. G., 1975, The proteins of Murray Valley encephalitis virus, *J. Gen. Virol.* **27**:283–292.

Westaway, E. G., 1977, Strategy of the flavivirus genome: Evidence for multiple internal initiation of translation of proteins specified by Kunjin virus in mammalian cells, *Virology* **80**:320–335.

Westaway, E. G., 1980, Replication of flaviviruses, in: *The Togaviruses* (R. W. Schlesinger, ed.), pp. 531–581, Academic Press, New York.

Westaway, E. G., and Shew, M., 1977, Proteins and glycoproteins specified by the flavivirus Kunjin, *Virology* **80**:309–319.

Westaway, E. G., McKimm, J. L., and McLeod, L. G., 1977, Heterogeneity among flavivirus proteins separated in slab gels, *Arch. Virol.* **53**:305–312.

Westaway, E. G., Schlesinger, R. W., Dalrymple, J. M., and Trent, D. W., 1980, Nomenclature of flavivirus-specified proteins, *Intervirology* **14**:114–117.

Westaway, E. G., Speight, G., and Endo, L., 1984, Gene order of translation of the flavivirus Kunjin: Further evidence of internal initiation *in vivo*, *Virus Res.* **1**:333–350.

Wright, P. J., 1982, Envelope protein of the flavivirus Kunjin is apparently not glycosylated, *J. Gen. Virol.* **59**:29–38.

Wright, P. J., and Warr, H. M., 1985, Peptide mapping of envelope-related glycoproteins specified by the flaviviruses Kunjin and West Nile, *J. Gen. Virol.* **66**:597–601.

Wright, P. J., Bowden, D. S., and Westaway, E. G., 1977, Unique peptide maps of the three largest proteins specified by the flavivirus Kunjin, *J. Virol.* **24**:651–661.

Wright, P. J., Warr, H. M., and Westaway, E. G., 1981, Synthesis of glycoproteins in cells infected by the flavivirus Kunjin, *Virology* **109**:418–427.

CHAPTER 11

Replication of Flaviviruses

Margo A. Brinton

I. INTRODUCTION

Flaviviruses were classified as members of the togavirus family until 1984, when the International Committee for the Nomenclature of Viruses voted to make Flaviviridae a separate family (Westaway *et al.*, 1986). The togavirus family had originally been defined using morphological criteria. The change in classification was the result of recent research that clearly demonstrated that flaviviruses, although generally similar to alphaviruses in their morphology, differ markedly from the alpha togaviruses in their virion structure, strategy of replication, and morphogenesis.

There are currently 64 identified flaviviruses, of which yellow fever (YF) virus is the prototype. The family name is derived from the Latin word *flavus*, which means "yellow." All flaviviruses share a group-specific antigen and the flaviviruses have been subdivided into subgroups and complexes on the basis of their serological cross-reactivity and the type of arthropod vector by which each virus is transmitted during its natural cycle (Table I) (Chamberlain, 1980; Calisher *et al.*, 1986). Virus abbreviations used in the text are given in Table I.

A number of the flaviviruses are human pathogens that regularly cause significant human morbidity and mortality throughout the world. Dengue virus infections in Asia and also, recently, in the Caribbean are of special public health concern because dengue virus infections can lead to an often fatal hemorrhagic fever termed dengue shock syndrome (Shope, 1980; Halstead, 1981). Japanese (JE), Murray Valley (MVE), St. Louis encephalitis (SLE), West Nile (WN), and Rocio encephalitis viruses are periodically responsible for epidemics or scattered cases of human central nervous system (CNS) disease and fever often distributed over

MARGO A. BRINTON • The Wistar Institute of Anatomy and Biology, Philadelphia, Pennsylvania 19104.

TABLE I. Antigenic and Vector Classification of Flaviviruses

Mosquito-borne	Anti-genic group[a]	Tick-borne	Anti-genic group[a]	Vector unknown	Anti-genic group[a]
Alfuy	III	Kadam	I	Carey Island	I
Bussuquara	III	Karshi	I	Negishi	I
Japanese encephalitis (JE)	III	Kyasanur Forest disease	I	Phnom Penh bat	I
		Langat	I	Cowbone Ridge	II
Kunjin	III	Louping ill	I	Jutiapa	II
Murray Valley encephalitis (MVE)	III	Omsk hemorrhagic fever	I	Moc	II
				Sal Vieja	II
		Powassan	I	Erlita	II
Saint Louis encephalitis (SLE)	III	Royal Farm	I	Koutango	III
		Saumarez Reef	I		
Stratford	III	Tick-borne encephalitis (TBE)	I	Israel turkey meningoencephalitis	VII
Usutu	III			Apoi	VII
West Nile (WN)	III	Russian spring-summer	I	Bukalasae bat	VII
				Dakar bat	VII
Bagaza	IV	Tyuleniy	U	Entebbe bat	VII
Kokobera	IV			Rio Bravo	VII
Ntaya	IV			Saboya	VII
Tembusu	IV				
Yokose	IV			Aroa	U
				Cacipacore	U
Banzi	V			Gadgets Gully	U
Bouboui	V			Montana myotis leukoencephalitis	U
Edge hill	V				
Uganda S	V			Rocio	U
				Sokuluk	U
Dengue-1	VI				
Dengue-2	VI				
Dengue-3	VI				
Dengue-4	VI				
Ilheus	U				
Jugra	U				
Naranjal	U				
Sepik	U				
Sponweni	U				
Yellow fever (YF)	U				
Wesselsbron	U				
Zika	U				

[a] Antigenic groups are designated I, II, III, IV, V, VI, VII, and U (unknown).

extensive geographic areas. Infections with many of the other flaviviruses cause morbidity and mortality preferentially in the very young and the elderly, but humans of all ages can be afflicted (Shope, 1980). The tick-borne flaviviruses, such as Powassan, louping ill, Kyasanur Forest disease, tick-borne encephalitis (TBE), Russian spring summer, and Omsk hemorrhagic fever viruses, induce human meningitis or encephalitis with a case fatality rate ranging from 1 to 30% (Shope, 1980). Disease can result from replication of flaviviruses in peripheral organs or in cells in the CNS (Harrison et al., 1982; Grimley, 1983). A recent study by Monath et al. (1983) indicates that during the course of a natural infection initiated by the bite of an arthropod vector, virus may gain access to the CNS, not by passive diffusion through brain capillaries, but by a neural olfactory pathway. Their results obtained with intraperitoneal SLE virus infections of Syrian hamsters indicated that a low-level viremia led to the infection of highly susceptible Bowman gland cells in the olfactory neuroepithelium. This was followed by centripetal axon transport of virus to the olfactory bulb and subsequent spread of virus throughout the CNS. Infection of cells in the nasal mucosa and the presence of virus in nasal mucus could result in non-arthropod-borne virus transmission.

For more extensive coverage of work published on flaviviruses prior to 1980, the reader is referred to several previous reviews (R. W. Schlesinger, 1977; J. H. Strauss and Strauss, 1977; Monath, 1980; Westaway, 1980; Russell et al., 1980; Halstead, 1980).

II. VIRION MORPHOLOGY AND COMPOSITION

A. Morphology

Flaviviruses contain a central nucleocapsid surrounded by a lipid bilayer. Virions are spherical and have a diameter of about 40–50 nm. Their cores are 25–30 nm in diameter. Along the outer surface of the virion envelope are projections that are 5–10 nm long with terminal knobs 2 nm in diameter (Murphy, 1980).

B. Physicochemical Properties and Chemical Composition

The buoyant density of flavivirions is 1.22–1.24 g/cm^3 in CsCl and 1.18–1.20 g/cm^3 in sucrose. The sedimentation coefficient for virions in sucrose has been reported to be 175–218 S. The infectivity of virions is most stable in the range of pH 7–9. Virions are rapidly inactivated at 50°C, with infectivity decreasing by 50% in 10 min at this temperature (Porterfield et al., 1978). Virion infectivity is also sensitive to inactivation by ultraviolet light, ionic and nonionic detergents, and trypsin digestion. The chemical composition of a representative flavivirus, SLE virus, is

TABLE II. Composition of Flaviviruses[a]

Component	Percentage of total dry weight
Protein	66
RNA	6
GC content = 48%	
Carbohydrate	12
Neutral carbohydrate	3
Hexosamine	6
N-Acetyl neuraminic acid	3
Lipids	17
Molar ratio of cholesterol to phopholipid = 0.29	
Phospholipid	12
Neutral lipid	5

[a] Values given were obtained for SLE virus by Trent and Naeve (1980).

shown in Table II. The majority of the carbohydrate in virions is present in glycoprotein, but a small amount of glycolipid is also present. Various types of tissue-culture cells that differ in their phospholipid composition have been found to produce progeny flavivirions with a phospholipid composition similar to that of the host cell (Trent and Naeve, 1980). For instance, Vero cells and SLE grown in these cells contain less sphingomyelin and phosphatidylserine than do baby hamster kidney (BHK) cells or virus grown in BHK cells (Trent and Naeve, 1980). These results indicate that the virion phospholipids are derived from preexisting host-cell lipids. No matter what cell type was used, however, the percentage of the total phospholipid represented by sphingomyelin and phosphatidylserine was always higher in the virions than in the host cells, whereas the percentage of phosphatidylcholine was similar in virions and host cells. The lipid composition of the virion appears to be determined by the intracellular membrane site of virus maturation. This aspect of viral morphogenesis is discussed in more detail in Section IV.

C. Structural Proteins

Flavivirions contain three types of polypeptides (Fig. 1). The glycosylated envelope (E) protein ($M_r \approx 51–59 \times 10^3$) is located in the virion envelope and forms the observed spike projections on the outer surface of the virions. It is currently not known whether these spikes represent monomers, dimers, or trimers of the E protein. The carbohydrate moieties of the E protein have been reported to be primarily mannose and complex glycans. The membrane (M) protein ($M_r \approx 7–9 \times 10^3$) is also associated with the virion envelope, but is not glycosylated. Both the E and M proteins are solubilized after treatment of virions with a nonionic detergent

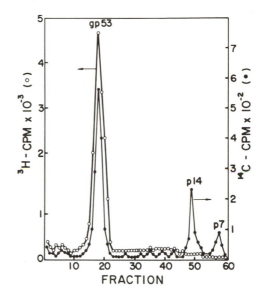

FIGURE 1. Virion proteins of SLE. Coelectrophoresis of [^{14}C]amino-acid-labeled (●) and [^{3}H]glucosamine-labeled (○) SLE structural proteins on a 13% sodium dodecyl sulfate–polyacrylamide gel. From Trent and Naeve (1980).

(Westaway and Reedman, 1969). The genome RNA is contained within a central core structure composed exclusively of the capsid (C) protein ($M_r \approx 13–16 \times 10^3$). The symmetry of the nucleocapsid structure has not been determined rigorously, but is thought to be icosahedral.

Tryptic peptide maps of the three structural proteins of several of the flaviviruses indicate that each is a unique protein (Wright and Westaway, 1977). The isoelectric points of detergent-solubilized flavivirus E and C proteins have been determined to be 7.6–7.8 and 10.2–10.4, respectively (Trent, 1977).

The primary immune response to flaviviruses during infection is directed toward the surface epitopes of the E protein. The hemagglutinating activity of flavivirions resides in the E protein. E protein from JE virions that had been purified by means of sodium dodecyl sulfate (SDS)–polyacrylamide gels and subsequently renatured by ion-exchange chromatography was found to agglutinate red blood cells and to bind to hemagglutination-inhibiting antibodies in anti-JE virus sera. The purified E protein was also shown to be able to elicit neutralizing antibody when injected into mice (Takegami *et al.*, 1982). Purified C protein elicited no neutralizing antibody, whereas neutralizing activity was elicited by purified M protein. The C protein has been found to have group-reactive antigenic determinants. The role of the M protein in flavivirus replication and in the assembly of virions is currently unknown. Lee and Schloemer (1981) reported that if uninfected host cells were pretreated with purified Banzi virus M protein, interference was observed with a subsequent Banzi virus infection. These data suggest that a part of the M protein may be exposed on the surface of flavivirions and that the M protein may interact with the cell surface. The E protein also presumably contains a site that interacts with the cell surface during virion attachment.

The 64 different viruses that constitute the flavivirus family demonstrate a variety of complex antigenic interrelationships with each
other. Recently, panels of monoclonal antibodies have been generated
against several flaviviruses, such as TBE virus (Stephenson *et al.*, 1984;
Heinz *et al.*, 1983), YF virus (J. J. Schlesinger *et al.*, 1983), WN virus (Peiris
et al., 1982), SLE virus (Roehrig *et al.*, 1983), and the four serotypes of
dengue virus (Henchal *et al.*, 1982). With the use of these monoclonal
antibodies, attempts have been made to identify the epitopes on the viral
E glycoprotein that are involved in virus neutralization, hemagglutination, immune enhancement, and type-, subcomplex-, super-complex-,
and group-specific antigenic cross-reactivity. Heinz *et al.* (1983), using a
panel of TBE virus monoclones, J. J. Schlesinger *et al.* (1983), using a panel
of YF virus monoclones, and Roehrig *et al.* (1983), using a panel of SLE-
monoclones, each defined at least eight distinct epitopes on the E glycoprotein (see Chapter 9).

Flavivirions have sometimes been found to contain a nonstructural
glycoprotein incorporated into virion envelopes in addition to or in lieu
of the M protein (Shapiro *et al.*, 1972a–c; Westaway and Shew, 1977;
Wright, 1982). Such virions have been observed extracellularly when virus
is produced in the presence of Tris-buffered medium and intracellularly
when virus is grown with normal tissue-culture media. Recent nucleotide
sequence data obtained with YF virus indicate that the M protein is indeed
cleaved from a glycosylated precursor protein ($M_r \approx 19–23 \times 10^3$) (Rice
et al., 1985) (see Chapter 10).

Differences in the extent of glycosylation of the flavivirion E protein
have been reported. G. W. Smith and Wright (1985) found that the E
protein from dengue 2 virions grown in Vero cells migrated more slowly
in SDS-containing polyacrylamide gels than did the E protein from dengue
2 virions grown in C6/36 mosquito cells. The contribution of carbohydrate to the molecular weight of the E glycoprotein is approximately 4000
daltons. An increased electrophoretic mobility of the envelope glycoproteins of two alphaviruses, Sindbis virus and Semliki Forest virus, grown
in mosquito cells was previously shown to result from the absence of
sialic acid residues (Stollar *et al.*, 1976; Luukkonen *et al.*, 1977). Virus-
specified glycoproteins of virus grown in cultured mosquito cells presumably lack sialic acid because mosquito cells do not possess a sialyl
transferase (Stollar *et al.*, 1976). However, in the case of flaviviruses, the
absence of sialic acid could not account entirely for the E-protein migration differences observed. Comparison of glycopeptides indicated that the
E proteins are modified at a late stage of their maturation by proteolysis
or glycan processing or both (G. W. Smith and Wright, 1985). Kunjin virus
grown in Vero cells has been found to contain E protein that is not glycosylated (Wright, 1982). The E protein of extracellular Kunjin virions
could not be labeled with radioactive galactose, mannose, or glucosamine,
had a density in CsCl consistent with that of a protein lacking carbohydrate, and did not bind to concanavalin A–agarose. Interestingly, two

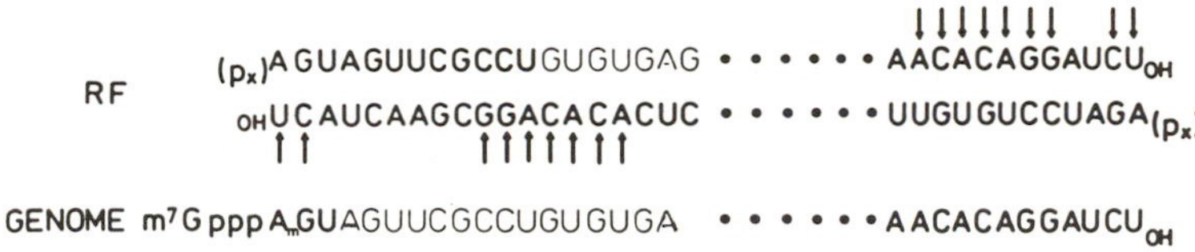

```
                                                      |||||||  ||
                (px)A GUAGUUCGCCUGUGUGAG • • • • • • AACACAGGAUCUOH
        RF
                OHUC AUCAAGCGGACA CACUC • • • • • • UUGU GUCCUAGA(px)
                   ↑↑          ↑↑↑↑↑↑↑

   GENOME  m7G pppA GUAGUUCGCCUGUGUGA • • • • • • AACACAGGAUCUOH
                     m
```

FIGURE 2. Terminal sequences of WN-virus-specific RNAs. Identical sequences are present in the 3'-terminal regions of both the genome and replicative form (RF) RNA as indicated by arrows. The nature of the 5'-terminal structures of the RF is not known, but may consist of one or more phosphates. From Wengler and Wengler (1981).

precursors of E, gp66 and gp59, and the intracellular form of the E protein in Kunjin-infected cells have been shown to be glycosylated (Wright and Warr, 1985). The E proteins obtained from extracellular virions produced during all other flavivirus infections studied so far definitely do contain oligosaccharides. However, at least two forms of the E protein with different molecular weights have been found in YF strain 17D virions (J. J. Schlesinger et al., 1983) and TBE virions (Heinz and Kunz, 1982). SLE-virus-specified intracellular E protein was separable into several fractions by DEAE–cellulose chromatography. The E proteins present in these fractions differed markedly in the ratio of glucosamine to leucine (Qureshi and Trent, 1973a–c).

D. Genome RNA

Flavivirions contain a single-stranded RNA (ssRNA) with a molecular weight of about 4×10^6 and a sedimentation coefficient in sucrose gradients of 40–44 S. Examination of YF virus RNA under the electron microscope in the presence of 4 M urea and 80% formamide revealed a linear molecule of about 11,000 bases (Deubel et al., 1983). The purified genome RNA is infectious and is designated as "plus"-stranded. The RNA contains a type 1 5'-terminal $M^7GpppAmp N_1$ cap structure (Cleaves and Dubin, 1979; Wengler et al., 1978).

The 3' terminus of the flavivirus genome is not polyadenylated. Instead, the genome has been shown to terminate in CU_{OH} (Wengler and Wengler, 1981; Rice et al., 1985; Brinton et al., 1986). The same heptanucleotide, (3') GGACACA (5'), was found to be present near the 3' termini of both the plus-and minus-strand RNAs of WN and YF and near the 3' termini of JE and SLE genomic RNAs (Fig. 2) (Wengler and Wengler, 1981; Rice et al., 1985; Dalrymple and Trent, personal communications). Although the first two 3'-terminal nucleotides and the heptanucleotide sequence appear to be highly conserved, the intervening nucleotides appear not to be conserved. Even two strains of WN virus were found to differ by one nucleotide in this region (Wengler and Wengler, 1981; Brinton et al., 1986)(see Figs. 2 and 3). It has been suggested that the conserved

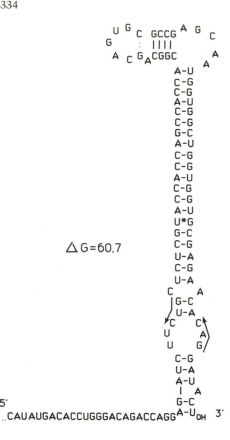

FIGURE 3. 3′-Terminal sequence of WN virus, strain E101. The first 84 nucleotides form a stable stem and loop structure. From Brinton *et al.* (1986).

sequences may be important for polymerase recognition or binding or both.

The entire genome RNA of the 17D strain of YF virus has recently been sequenced from complementary DNA (cDNA) (Rice *et al.*, 1985), and direct RNA sequence data have been obtained for the 3′ end of the WN virus, strain E101, genome (Brinton *et al.*, 1986). A fairly stable, complex stem and loop structure (Fig. 3) appears to be located at the 3′ end of the genome RNA (Brinton *et al.*, 1986). The existence of this structure was demonstrated by RNase digestion patterns of 3′-labeled WN virion RNA. Nucleotides within the stem structure were inefficiently digested, while nucleotides within the loops located at the top of the stem were cut efficiently (Brinton *et al.*, 1986). A DNA primer complementary to the 5th through the 21st 3′ nucleotides of WN virion RNA was unable to initiate reading within the hairpin stem, even when primer and template were heated together and then slow cooled and/or the dideoxy sequencing reactions were incubated at 50°C. Although a consistent sequence was obtained with this primer, this sequence was not found in the first 150 nucleotides from the 3′ termini of the RNA. This result indicates that this primer binds to a second site farther downstream in the RNA (Brinton *et al.*, 1986). A 3′-terminal stem and loop structure has also been predicted from the cDNA sequence generated by Rice *et al.* (1985) (see Chapter 10). Although the 3'-terminal stem and loop structures of WN virus, strain E101, and YF virus, strain 17D, differ significantly in sequence, they are very similar in length and form. A second heptanucleotide sequence located on the descending arm of the stem and aligning, but not base-pairing, with the conserved heptanucleotide sequence identified by Wengler and Wengler (1981) also appears to be highly conserved. The 3' stem and loop structure may be important in regulating the initiation of minus-strand RNA synthesis, which is less efficient than plus-strand synthesis. This structure could also be involved in genome RNA interactions with capsid proteins. Further information about the structure and arrangement of the genes can be found in the flavivirus genome in Chapter 10.

The arrangement of genes within the flavivirus RNA is quite different from the gene arrangement of the alpha togaviruses and more closely resembles that of the picornaviruses; the virion structural proteins are encoded at the 5' end of the genome. The first gene encoded at the 5' end of the flavivirus genome is the C-protein gene, followed by the M-protein gene and then the E-protein gene (5'-C-M-E-3') (Rice *et al.*, 1985).

III. INFECTION OF CULTURED CELLS

Flaviviruses can replicate in a wide variety of cultured cells of vertebrate and arthropod origin. In mammalian and avian cells, flaviviruses replicate at a significantly slower rate than the alphaviruses. The initial latent phase of the flavivirus replication cycle lasts about 12 hr, after which progeny virus begins to be released from infected cells. Maximal titers of virus do not begin to be produced until 24 hr after infection (Trent and Naeve, 1980).

In many types of vertebrate- and arthropod-cell cultures, flavivirus infection is not cytopathic. However, replication of flaviviruses in cultures of primary duck and chicken embryo cells, two lines of monkey kidney cells (Vero and MK_2), a line of porcine kidney cells (PS), and baby hamster kidney cells results in the development of cytopathogenic changes of sufficient magnitude for use of these cells in plaque assays (Pfefferkorn and Shapiro, 1974; Trent and Naeve, 1980). Flavivirus infection of arthropod cells can be either cytocidal or noncytocidal (Mussgay *et al.*, 1975; Trent and Naeve, 1980). Some flaviviruses induce syncytium formation in mosquito-cell lines (Igarashi, 1978, 1979; Ng and Westaway, 1980; Stollar, 1980). Flavivirus progeny produced by some types of arthropod cells have been found to differ phenotypically from the parental virus used to infect the culture. Dengue (type 2) virions produced by *A. albopictus* cells were found to lack hemagglutinating activity (Sinarachatanant and Olson, 1973). Ng and Westaway (1983) found that Kunjin virions, but not dengue (type 2) or JE virus, produced by *A. albopictus* cells lacked hemagglutinating activity. The *A. albopictus*-grown Kunjin virions were found to be unstable during routine purification procedures and to contain E protein that migrated more rapidly in polyacrylamide gels. These virions, which lacked hemagglutinating activity, were nevertheless able to bind to erythrocytes and block hemagglutination by the parental virus. The altered virion phenotype reverted to that of the parental virus after a single cycle of growth in Vero cells.

It has been shown that togaviruses enter their host cells by adsorptive endocytosis and are then delivered to intracellular vacuoles and lysosomes (see Chapter 4). The viral genome is subsequently released into the cytoplasm by a low-pH-induced fusion of the virion and lysosomal membranes (White *et al.*, 1980). Under normal infection conditions, fusion between the cellular plasma membrane and the viral envelope does not occur. However, such a fusion event can be artificially triggered by addition of acidic medium (White *et al.*, 1980). pH-dependent fusion has also been demonstrated for representative viruses of the rhabdovirus, myxovirus, and paramyxovirus families (White *et al.*, 1981).

Extensive attempts to demonstrate a similar fusion phenomenon with WN virus in BHK cells have not succeeded (Brinton, Amato, and Gonda, unpublished data). Even though a number of different experimental conditions were tried, including attempts to fuse infected cells with each other or extracellular virus with uninfected cells, the precise requirements for flavivirus fusion were not achieved. Electron-microscopic data obtained by Gollins and Porterfield (1985) indicate that free flavivirions and virus–antibody complexes do enter the cell by an adsorptive endocytotic process which delivers them to intracellular vacuoles and lysosomes. Also, pH-dependent flavivirus lysis has been demonstrated with extracellular virus and chick red blood cells with a pH optimum of 5.4 (Commack and Gould, 1985).

A. Effect on Host-Cell Macromolecular Synthesis

Host-cell macromolecular synthesis is not dramatically decreased by flavivirus infection. In one report, it was estimated that infection of BHK cells with SLE virus inhibited host protein synthesis 30% by 2 hr after infection (Trent *et al.*, 1969). The synthesis of high levels of cell proteins continues throughout the flavivirus replication cycle. Takehara (1971, 1972) reported that during JE virus or dengue virus infections in Vero cells, host-cell DNA synthesis and DNA-dependent RNA polymerase activity were reduced by 50–60% and 22–35%, respectively, by 8 hr after infection. However, Trent *et al.* (1969) reported that cellular DNA synthesis in BHK cells infected with SLE virus declined only at 18 hr after infection, when virus-induced cytopathic effects became apparent.

B. Immune Enhancement

Antibody-dependent enhancement of flavivirus infection has been demonstrated (Peiris and Porterfield, 1979; Halstead et al., 1980). This opsonic phenomenon occurs when cells bearing Fc receptors on their surfaces are infected with a flavivirus inoculum that has been mixed with antiviral antibody at concentrations insufficient to cause viral neutralization. Immune enhancement has been observed during infections of the murine macrophage cell lines, P388D1 (Peiris and Porterfield, 1979), and MK1 and Mm1 (Hotta et al., 1984), the human macrophage cell line U-937 (Brandt et al., 1982), human peripheral-blood leukocytes (Halstead and O'Rourke, 1977), and human adherent monocytes (Brandt et al., 1979) with WN virus E101, YF virus 17D, and dengue 2 virus. Virus yields are enhanced by 20-to 1000-fold, depending on the strain of virus, cell type, and antibody used. Gollins and Porterfield (1984) have shown that the proportion of WN virus that bound to the surface of P388D1 cells and subsequently entered the cells was higher in the presence of subneutralizing concentrations of antiviral antibody than in their absence. The increased yields of virus were found to be due to an increased number of cells becoming productively infected in the presence of antiviral antibody (Peiris et al., 1981b).

It is clear that antibody enhancement is dependent on the presence of Fc receptors on the target-cell surface. Antibody enhancement could be successfully blocked by a monoclonal anti-Fc receptor antibody (Peiris et al., 1981a,b), and neither F(ab')$_2$ fragments nor antiviral immunoglobulin M (IgM) were found to mediate antibody enhancement (Halstead and O'Rourke, 1977; Peiris and Porterfield, 1982). In the absence of antibody, flavivirus particles presumably bind to a cell-surface receptor other than the Fc receptor (Daughaday et al., 1981). It is not known whether the uncoating processes differ during antibody-mediated and normal infections. However, data obtained by Gollins and Porterfield (1985) indicate that antibody-complexed WN virus enters the cell in the same manner as free WN virus. Antibody-mediated enhancement of YF virus 17D infection in P388D1 cells was not inhibited by concentrations of cytochalasin B that inhibited cellular phagocytosis and chloroquine also did not inhibit antibody enhancement of virus infection (Brandriss and Schlesinger, 1984).

Initially, antibody-enhancement experiments were carried out with hyperimmune sera. Species homology between the macrophage and the immunoglobulin was found not to be essential (Peiris and Porterfield, 1981). Antisera produced in rabbits, mice, humans, and sheep all enhanced flavivirus replication in murine P388D1 cells. However, in chick embryo fibroblast cultures, avian, but not mammalian flavivirus antisera, induced enhancement (Kliks and Halstead, 1980). Study of the antigenic specificity of antibody enhancement indicated that with flaviviruses, the phenomenon was group-reactive. Infection with a particular flavivirus could be enhanced with antisera specific for both closely related and distantly related flaviviruses (Peiris and Porterfield, 1982). In contrast, enhancement of alphaviruses occurs only with antisera specific for closely related viruses. Studies have subsequently been carried out with panels of flavivirus monoclonal antibodies using P388D1 and YF virus 17D. J. J. Schlesinger and Brandriss (1983) tested 13 monoclonal antibodies for their ability to induce antibody enhancement. All the monoclonal antibodies, whether they were YF virus 17D type-specific or flavivirus cross-reactive, were found to enhance infection. The degree to which a particular monoclonal antibody enhanced a YF virus 17D infection did not correlate with its titer of neutralizing or hemagglutination-inhibition activity. Of 12 dengue 2 monoclonal antibodies tested during dengue 2 infections in U-937 cells, only the group-specific antibodies were found to mediate enhancement (Brandt et al., 1982). U-937 cells could not be infected by dengue 2 virus in the absence of antibody. Interestingly, these cells were susceptible to YF virus 17D replication in the absence of antibody. TBE virus infection in P388D1 cells was enhanced by 13 TBE-virus-complex-specific monoclonal antibodies, but not by a group-specific monoclonal antibody (Phillpotts et al., 1985). Using 5 monoclonal antibodies, Halstead et al. (1984) found that the same strain of dengue 2 virus (NGC) used by Brandt et al. (1982) was enhanced by one of two type-specific and two of three group-specific monoclonal antibodies in both U-937 and P388D1 cells. Seven different dengue 2 strains were also tested with the same 5 monoclonal antibodies (Halstead et al., 1984). Although all seven dengue 2 virus strains were neutralized by each of the 5 monoclonal antibodies in assays carried out in LLC-MK$_2$ cells, the viruses varied in their ability to be enhanced by the monoclonal antibodies during infections of P388D1 and U-937 cells. Two of the dengue 2 strains were enhanced by only 3 of the monoclonal antibodies, two by 4 of the antibodies, and three by all 5 of the antibodies. Experiments with monoclonal antibodies have not yet led to a clear definition of the characteristics of an immune-enhancing antibody.

Dengue viruses replicate in mononuclear phagocytes during human infection. In some instances, dengue virus infection induces a severe hemorrhagic disease termed "dengue shock syndrome" (Halstead, 1980, 1981, 1982; Sangkawibha et al., 1984). The factors that lead to the occurrence of dengue shock syndrome are not completely understood. However, it has been hypothesized that enhancing antibodies produced in an individual during a secondary infection with a dengue virus of a different serotype than the one responsible for the initial infection enhances viral replication in monocytes. These infected monocytes then become the targets of immune elimination, which is possibly T-cell-mediated, and release chemical mediators of shock and hemorrhage characteristic of dengue shock syndrome (Pang, 1983). Secondary infections with dengue 2 virus following primary infections with dengue 1, dengue 3, or dengue

4 virus (in descending order of risk) appear to be most often associated with dengue shock syndrome in Asia (Sangkawibha *et al.*, 1984). The extent of such an enhancement phenomenon might be expected to vary with the infecting virus strain, the specificity of the immune response, and the previous exposure history of the individual host (Halstead *et al.*, 1983, 1984). That immune enhancement may occur during *in vivo* infections has been indicated by the observation that cultures of peripheral-blood leukocytes from dengue-immune donors yield higher titers of virus than do peripheral-blood leukocytes from normal donors after *in vitro* infection (Halstead *et al.*, 1976). However, YF virus was found to grow equally well in leukocytes from immune and nonimmune donors (Liprandi and Walder, 1983).

Antibody-dependent enhancement of flavivirus infection has also been found to be mediated by a complement receptor (Cardosa *et al.*, 1983). In the presence of fresh normal serum, anti-WN virus IgM was shown to enhance WN virus infection of P388D1 cells. Further, this enhancement could be blocked by a monoclonal antibody, M1/70, that is known to inhibit the binding of the C3 complement component. The two types of antibody enhancement (Fc-and complement-receptor-mediated) may function in a synergistic manner. Some subclasses of IgG also fix complement, and many types of cells (such as erythrocytes, lymphocytes, neutrophils, monocytes, macrophages, and kidney epithelial cells) can express complement receptors on their surfaces.

In other studies, treatment of mouse peritoneal cultures with bacterial lipopolysaccharide, phytohemagglutinin, bacterial cell walls, peptidoglycans, or a water-soluble polymer of peptidoglycan subunits for 3 days prior to dengue 2 virus infection resulted in an increased production of virus (Hotta *et al.*, 1983). The increased yield appeared to correlate with an increase in the number of virus-infected cells. The treated macrophages were found to phagocytize latex particles more efficiently than untreated cells. These authors suggest that dengue 2 virus infection can be established by virus taken up by phagocytosis.

IV. VIRAL MORPHOGENESIS

Progeny virions are readily observed in thin sections of flavivirus-infected cells by 8–12 hr after infection within the cisternae of endoplasmic reticulum vesicles. These vesicles are located in the perinuclear area of the cytoplasm (Murphy, 1980; Westaway, 1980). Hypertrophy of the rough and smooth endoplasmic reticulum and in some areas the Golgi membranes is a characteristic feature of flavivirus infections (Fig. 4A and B). Dense accumulations of convoluted microtubules develop near the nucleus (Fig. 4C). These mesh-like structures are often located near intracellular progeny virions, and it has been suggested that they may be somehow involved in virus maturation (Westaway, 1980). Mature virions

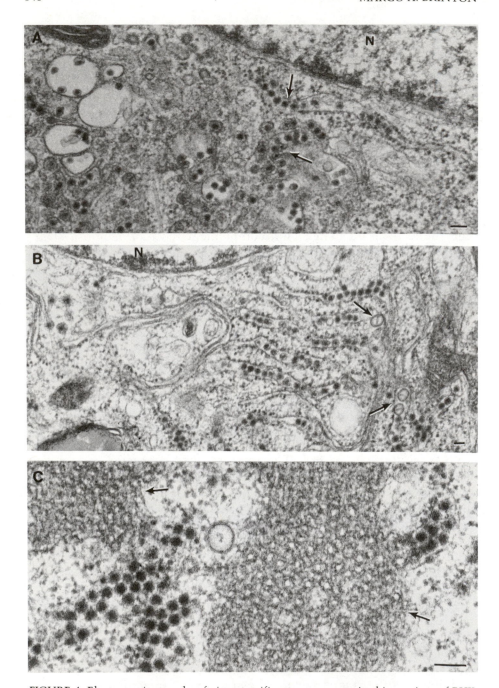

FIGURE 4. Electron micrographs of virus-specific structures seen in thin sections of BHK-21/WI2 cells 18 hr after infection with WN virus E101 at a multiplicity of infection of 5. (A) Virions are seen within cisternae of the endoplasmic reticulum in the perinuclear area of the cell. ×43,000. (B) Round lucent, double-membrane vesicles are often seen in areas of virus maturation. These vesicles have internal reticular webs and are located within

accumulate in vacuoles with smooth membranes, within the Golgi complex, and in vesicles with electron-dense particles, presumably ribosomes, along their outer surfaces. Cytoplasmic accumulations of nonenveloped nucleocapsids are not observed in flavivirus-infected cells, and nucleocapsids cannot be isolated from infected cells. Intracellular virions have been reported to be composed of E, C, and a glycosylated nonstructural protein instead of the M protein (Shapiro et al., 1972a–c, 1973). In JE-virus-infected cells, this nonstructural protein was designated gp17–19. Recent nucleotide sequence data indicate that in YF virus, the nonstructural protein of this molecular weight is not glycosylated, but that a 19–23 $\times$ 10^3 dalton protein is the glycosylated precursor of the M protein that is found in extracellular virions (see Chapter 10). No direct evidence for envelopment of flaviviruses by an intracellular budding process has been obtained by either thin-section or freeze–fracture studies of infected cells (Murphy et al., 1968; Demsey et al., 1974; Westaway, 1980; Murphy, 1980). Leary and Blair (1980) have suggested that virion assembly may occur within lucent vesicles budded from cytopathic vacuoles. Intracellular virions have been observed to exit from infected cells by at least two routes. The lamellae of smooth membranes in Kunjin-infected cells have been found to open directly to the exterior (Westaway, 1980). Virus-containing vesicles have been observed to move to the periphery of the cytoplasm and fuse with the plasma membrane (Filshie and Rehacek, 1968; Dalton, 1972). The final processing of the M protein apparently occurs during the release of virions from cells. The efficiency of flavivirus release from infected cells can be enhanced by addition of 40–80 mM MgCl$_2$ to the culture medium at 10 hr after infection (Matsumura et al., 1972). In mosquito salivary gland cells, SLE virions have been reported to exit from the apical end of the cells by means of membrane fusion or a localized breakdown of the plasma membrane (Whitfield et al., 1973). Virions released by cells lysed during cytocidal infections often remain enclosed within membrane vesicles (Murphy et al., 1968; Sriurairatna et al., 1973; Calberg-Bacq et al., 1975).

A. Intracellular Sites of Viral Macromolecular Synthesis

In flavivirus-infected cells, essentially all the virus-specified protein present appears to be associated with detergent-labile membranes (Shapiro et al., 1973). No preferred cellular membrane site was observed for the incorporation of either the E protein or the C protein, since all the flavivirus proteins were present in all the cellular membrane fractions separated on discontinuous sucrose gradients by the method of Caliguiri and Tamm (1970) (Shapiro et al., 1972 a–c; Kos et al., 1975; Stohlman

---

cisternae of endoplasmic reticulum. $\times$ 27,500. (C) Cytoplasmic membranes often proliferate in infected cells to form meshlike structures. $\times$ 98,500. (N) Nucleus. Scale bars: 200 nm.

et al., 1975; Boulton and Westaway, 1976). The incorporation of the viral proteins into the plasma membranes was unexpected, since there is no evidence that the plasma membrane plays any role in flavivirus replication or maturation. Recently, however, J. J. Schlesinger *et al.* (1986) were unable to detect the E protein of YF virus on the surfaces of infected neuroblastoma cells using 14 anti-E monoclonal antibodies. The nonstructural protein gp44–48 (NV3) was readily detected on the surfaces of the infected neuroblastoma cells. Immunofluorescent studies indicate that within the infected cell, the viral E protein first accumulates around the nucleus in a narrow band that progressively broadens into the cytoplasm with time after infection.

Studies utilizing various cell-fractionation methods to attempt to identify the intracellular membrane sites of viral RNA and protein synthesis have provided equivocal results. Westaway and Ng (1980) found that viral RNA labeled by a 10-min pulse of [^{3}H]uridine was located primarily in the light membrane fraction obtained from the cytoplasm of infected cells. The incorporated [^{3}H]uridine in these membranes was present in an RNase-resistant form. RNase-sensitive viral RNA was found to accumulate on heavy membrane fractions obtained from the nuclear pellet by shearing and nonionic detergent treatment after a 3-hr pulse of [^{3}H]uridine. In previous studies, dengue 2 virus RNA and protein synthesis were both found to occur on the same dense membrane fraction (Stohlman *et al.*, 1975). In other studies, when cells were disrupted with nonionic detergents such as NP40, no viral RNA was found to be associated with outer nuclear membranes (Naeve and Trent, 1978; Westaway, 1977). The use of detergent to lyse the plasma membrane could also result in the release of viral replication complexes and polysomes from membranes as well as the release of viral cores from intracellular virions (Westaway and Ng, 1980).

Although it is clear that flavivirus protein and RNA synthesis as well as virion assembly occur in the perinuclear region of the infected cell, conflicting results have been obtained as to whether the host-cell nucleus is required during any phase of the virus replication cycle. JE-virus-like particles were reported to have been observed within the nuclei of infected cells (Yazuzumi *et al.*, 1964), and JE virus replication was reported not to occur in cells enucleated by cytochalasin B prior to infection (Kos *et al.*, 1975). However, virus replication did continue, but at a reduced level, in cells enucleated 4 hr or later after infection. Cordycepin, an inhibitor of polyadenylation, decreased both JE virus and SLE virus replication by 50–60% when added to cells up to 9 hr after infection (Kos *et al.*, 1976; Brawner *et al.*, 1979). The apparent effect of cordycepin on flavivirus replication could be a secondary effect of decreased cell viability or might reflect the involvement of host proteins in flavivirus replication. Mitomycin C, an inhibitor of DNA synthesis, had no effect on JE virus replication (Leary and Blair, 1983). Disruption of microtubules by vinblastine treatment of cells reduced Kunjin virus replication by at least

10-fold (Ng *et al.*, 1983). Conflicting results have been obtained with actinomycin D (Trent and Neave, 1980). Some studies indicated that flavivirus replication could be reduced by addition of 1 μg/ml or more of actinomycin D during the first 9 hr after infection (Brawner *et al.*, 1979). However, when actinomycin D was added for a few hours prior to infection, after the latent phase, or at low doses, flavivirus replication was unaffected. More recent studies by Leary and Blair (1983) indicate that the apparent inhibition of flavivirus replication by actinomycin D may have been a secondary result of a progressive loss of cellular viability due to drug toxicity. The much longer growth cycle of flaviviruses requires that cells remain viable for 20–30 hr to obtain normal maximal viral yields. Addition of 5 μg/ml of actinomycin D to the culture medium for 2-hr periods from 1 to 6 hr after infection had no effect on the yield of JE virus harvested at 24 hr after infection (Leary and Blair, 1983). No immunofluorescence has been detected in the nuclei of infected cells with either anti-double-stranded RNA (dsRNA) antibody or antiviral antibody. To date, there is no evidence of a specific nuclear involvement in flavivirus replication. However, the outer nuclear membranes do provide sites for flavivirus replication and maturation.

Since the two largest nonstructural proteins (p96 and p67), which presumably function as viral RNA-dependent RNA polymerases (Westaway, 1977), are found associated with all the cellular membrane fractions, little is known about the intracellular location of the active viral replication complexes. However, flavivirus polymerase activity has consistently been found to be higher in cell fractions enriched for outer nuclear membranes (Zebovitz *et al.*, 1974; Chu and Westaway, 1984; Grun and Brinton, 1986a). In Kunjin-virus-infected cells, the rough endoplasmic reticulum in the perinuclear region of the cell has been observed to change from a fine network to a coarse one during the first 4–12 hr of infection (Ng *et al.*, 1983). Using an antibody specific for dsRNA, Ng *et al.* (1983) found that immunofluorescence in Kunjin-virus-infected cells was distributed in a fine network in the perinuclear region. The network contained many foci. Ribosome like particles are often observed to stud the cytoplasmic side of the vesicles within which mature flavivirions accumulate in infected cells (Matsumura *et al.*, 1971). Autoradiographs of actinomycin-D-treated, JE-virus-infected cells labeled with [^{3}H]uridine for 10–30 min showed an increasing accumulation of radioactivity on rough endoplasmic reticulum (Takeda *et al.*, 1978; Lubiniecki and Henry, 1974). These results indicate that the flavivirus proteins may be synthesized on membranes at the site of viral RNA synthesis and virion maturation.

V. VIRAL RNA SYNTHESIS

After uncoating, the flavivirus genome RNA must first be translated to provide the viral replicase and transcriptase activities needed for RNA synthesis. The genome 40 S RNA is the initial template, from which a minus-strand complementary RNA is synthesized. This minus strand then becomes the template for the synthesis of progeny plus-strand RNAs. The newly synthesized plus-strand 40 S RNAs are utilized as templates for the production of minus-strand RNAs, as messenger RNA (mRNA) for the translation of viral proteins, and as molecules for encapsidation into progeny virions. It is interesting to note that among the plus-strand RNA animal viruses, only picornaviruses and flaviviruses produce no subgenomic mRNAs and have the polymerase gene(s) at the 3' end of the genome coding region.

Intracellular viral RNA synthesis has been detected as early as 3 hr after infection (Takeda *et al.*, 1978). Trent *et al.* (1969) found that flavivirus RNA synthesis was biphasic, with a small peak of synthesis occurring at 6 hr after infection and then a steady increase in RNA synthesis beginning by 13 hr after infection. The majority of the viral RNA produced during infection is plus-strand 40 S genome RNA (Stollar *et al.*, 1967). It has been estimated that the ratio of plus- to minus-strand RNAs produced is 10:1 (Cleaves *et al.*, 1981).

Several types of flavivirus-specific RNAs have been identified in infected cells. In addition to single-stranded 40 S genome RNA, RNase-resistant RNA (20–22 S) and RNA that is partially (50–70%) resistant to RNase (20–28 S) have been observed in cells infected with a number of different flaviviruses (Stollar *et al.*, 1967; Trent *et al.*, 1969; Zebovitz *et al.*, 1972; Wengler *et al.*, 1978; Cleaves *et al.*, 1981; Chu and Westaway, 1985). Pulse–chase studies showed that [³H]uridine appeared first in the 20–28 S RNA and subsequently in the 40 S RNA (Cleaves *et al.*, 1981). These studies demonstrated that the 20–28 S RNA represented replicative-intermediate (RI) RNA. It was further suggested that the RNase-resistant 20–22 S RNA was a replicative-form (RF) RNA (Cleaves *et al.*, 1981). The RF RNA was soluble in 2 M LiCl and, when rebanded on a sucrose gradient after LiCl fractionation, migrated as a single sharp band. The RI RNA was not soluble in LiCl and migrated heterogeneously in gradients. Although these studies clearly suggested that there was a precursor–product relationship between the heterogeneous 20–28 S RI RNA and the 40 S virion RNA, the precise role of the RF RNA in the replication process has not been delineated. The RF molecules were labeled during the experimental pulse periods. Complete denaturation of the 20–28 S heterogeneous RNA resulted in a molecular-weight shift, indicating that the RI RNA was composed of genome-length RNA and a continuum of subgenomic length RNAs. In contrast, the RF RNA contained only genome-length RNA (Cleaves *et al.*, 1981). The 30–50% RNase susceptibility of the radioactivity incorporated into the RI RNA during pulse–chase experiments carried out at 42 hr after infection was attributed to incomplete base-pairing due to an excess of plus-strand products over minus-strand templates (Cleaves *et al.*, 1981). Calculation of the number of nascent growing strands per dengue 2 RI structure indicated that there were 5 per RI RNA, and it was estimated that 12–15 min is needed for the completion of a nascent chain (Cleaves *et al.*, 1981).

Somewhat different results have been reported recently by Chu and Westaway (1985), using Kunjin-infected cells. Isotope was incorporated only into RF and RI RNAs within a 10-min pulse period, but after a 10- to 20-min chase, the majority of the radioactivity was found in progeny genome RNA. In contrast to the findings of Cleaves *et al.* (1981), Chu and Westaway (1985) observed a relatively high RNase resistance of the flavivirus RI RNA, rapid labeling of the RF RNA, and slow appearance of label in 40 S RNA. Chu and Westaway (1985) have proposed a model for flavivirus plus-strand RNA synthesis that hypothesizes that the RF RNA is a double-stranded molecule which acts as a recycling template for semiconservative and asymmetric replication. According to their model, plus-strand RNA synthesis is initiated on the minus strand template of an RF molecule in a semiconservative manner. The synthesis of the new plus strand proceeds rapidly, converting the RF to an RI RNA. A short delay occurs in the release of the displaced strand, so that the majority of RI RNAs at any one time consist of the minus-strand template, one displaced completed plus strand, and one base-paired almost completed nascent plus strand. When release of the displaced plus strand does occur, the incomplete one finishes, but remains attached along its entire length to the minus-strand template and an RF molecule is regenerated. According to this model, during a pulse–chase experiment, labeled 40 S plus-strand RNA would first be released only after the second round of synthesis on any particular minus-strand template had occurred. This model does not explain how the initial minus strand is initiated from the genome of the infecting virion, nor does it explain how selective amplification of plus-strand synthesis is accomplished.

Minus-strand synthesis may be carried out by a different mechanism and by a different enzyme complex than plus-strand synthesis. It is currently not known which of the viral nonstructural proteins provide the replicase functions or whether the initiation and elongation activities reside in the same or different proteins. Part of the evidence provided by Chu and Westaway (1985) in support of their replication model is the observation that only labeled 40 S RNA was detected when RI RNAs were totally denatured. However, Cleaves *et al.* (1981) did observe the production of labeled, heterogeneous small nascent RNAs under similar experimental conditions. In addition, RI RNAs synthesized *in vitro* by a nuclear extract of Kunjin-virus-infected cells contained heterogeneous ssRNA with molecular weights ranging from 0.5 to 4.2×10^6 (Chu and Westaway, 1985). RI RNAs produced *in vitro* by WN-infected cell extracts also contain heterogeneous subgenomic-sized RNAs as well as 40 S RNA (Grun and Brinton, unpublished data). These data indicate that the rate of synthesis of nascent plus strands is not so rapid as to preclude detection of partially completed nascent strands after denaturation of flavivirus RI RNAs, as was suggested by Chu and Westaway (1985).

Flavivirus *in vitro* polymerase activity has been detected in crude infected cellular extracts by several laboratories (Zebovitz *et al.*, 1974; Takeda *et al.*, 1978; Brawner *et al.*, 1979; Cardiff *et al.*, 1973; Qureshi and Trent, 1972; Chu and Westaway, 1985; Grun and Brinton, 1986a and b). Polymerase activity was found to be reduced by treatment with nonionic detergents (Cardiff *et al.*, 1973), indicating an association of the active replication complexes with membranes. Polymerase activity is associated with cytoplasmic membranes, the mitochondrial fraction, and nuclear-associated membranes. However, activity was consistently found to be highest in nuclear-associated membranes (Zebovitz *et al.*, 1974; Takeda *et al.*, 1978; Chu and Westaway, 1985; Grun and Brinton, 1986a). This finding is not unexpected, since the majority of viral RNA synthesis occurs in the perinuclear region of infected cells (Westaway, 1980). Qureshi and Trent (1972) isolated a 250 S "replication complex" from SLE-infected cells by treatment of cell extracts with EDTA. These complexes contained pulse-labeled (15 min) genome RNA and 20–26 S partially RNase-resistant RNA as well as polymerase activity.

Although a number of laboratories have successfully detected *in vitro* flavivirus polymerase activity in infected cell extracts, limited information has been reported on the optimal reaction conditions. Optimal conditions for WN virus polymerase activity in cell-free BHK21/WI2 extracts have recently been determined (Grun and Brinton, 1986b). Further experiments are necessary to determine whether these conditions are also optimal for other flaviviruses and cell extracts. By analogy with other RNA virus polymerases, it seems likely that the two largest nonstructural proteins, p96 and p67, do have polymerase function, and antibody to p67 has been found to inhibit *in vitro* polymerase activity (Brinton and Grun, unpublished data).

RNA synthesis must be initiated on the 3' ends of both the plus-strand and the minus-strand RNAs. Since the extent of synthesis of these two strands is regulated quite differently during the course of infection, it is not unreasonable to postulate that different proteins or protein complexes may be involved in initiation of synthesis on the plus- and minus-strand RNAs. Once RNA synthesis has been initiated on a template, an elongation activity is required to complete the nascent strand. Further work is needed to determine which viral proteins, individually or in complex, provide these various functions. Investigation of the involvement of host-cell proteins in flavivirus RNA synthesis is also of interest, since host factors have been reported to be involved in viral RNA synthetic functions during infections with picornaviruses and several RNA bacteriophages (Zinder, 1975; Dasgupta *et al.*, 1980).

The RNA products synthesized in *in vitro* polymerase reactions sediment heterogeneously between 20 and 28 S on sucrose gradients (Zebovitz *et al.*, 1974; Cardiff *et al.*, 1973). Zebovitz *et al.* (1974) found this RNA to be resistant to RNase treatment, while Cardiff *et al.* (1973) found it to be sensitive. More recently, Chu and Westaway (1985) reported that both RI and RF RNA were labeled during an *in vitro* polymerase reaction, only trace amounts of labeled free ssRNA was detected. We have found that RI RNA is more extensively labeled than RF RNA and that significant amounts of 40 S RNA are labeled and released during *in vitro* reactions with WN-infected BHK cell extracts (Grun and Brinton, 1986a and b). Although cell extracts made from WN-infected cells 16h after infection primarily synthesize plus-strand genome RNA from minus-strand templates, addition of 1 to 2 μg of exogenous genome RNA to the *in vitro* reactions stimulated polymerase activity by two-fold (Grun and Brinton, 1986b). This observation indicates that RNA synthesis can be initiated *in vitro* from the added genomic RNA templates.

Small intracellular (8–15 S) RNAs have also been observed in actinomycin-D-treated flavivirus-infected cells (Zebovitz *et al.*, 1972, 1974; Takeda *et al.*, 1978; Naeve and Trent, 1978). Wengler *et al.* (1978) reported that three sizes of virus-specific low-molecular-weight flavivirus RNAs could be detected. A 5×10^4 dalton species was synthesized in both *A. albopictus* and BHK-21 cells infected with Uganda S virus. Two low-molecular-weight RNAs (6.5×10^4 and 4.2×10^4) were observed in WN-infected BHK cells, while only the 6.5×10^4 dalton RNA was synthesized in the WN-infected mosquito cells. Further study of the 6.5×10^4 dalton RNA revealed that it was of plus-strand polarity, not "capped" at its 5' end, and did not contain a 3' poly(A) sequence. The mode of synthesis of these small virus-specific RNAs is not known. These RNAs may represent prematurely terminated transcripts or defective interfering RNAs (see Section VIII). If they are not defective interfering RNAs, it is possible that these small RNAs may play an as yet undetermined role in flavivirus replication.

VI. VIRAL PROTEIN SYNTHESIS

At all times after infection, the only viral RNA found in polysomes purified from flavivirus-infected cells is the full-sized genome RNA (Naeve and Trent, 1977; Cleaves and Schlesinger, 1977; Trent and Naeve, 1980). No poly(A) sequence has been found at the 3' end of progeny plus-strand RNAs isolated from polysomes. All the available evidence indicates that the 40 S RNA that serves as the only viral messenger RNA is identical to the genome RNA that is encapsidated into virions. *In vitro* translation of virion RNA indicated that translation begins at a single initiation site (Wengler *et al.*, 1979; Svitkin *et al.*, 1981; Monckton and Westaway, 1982). However, only viral structural protein polypeptides were produced by *in vitro* translation, suggesting that the structural proteins were encoded at the 5' end of the genome. This gene arrangement is similar to that of the picornaviruses, rather than that of the alphaviruses (E. G. Strauss and J. H. Strauss, 1983). The nucleotide sequence of the genome RNA of YF virus demonstrates that the structural genes are arranged from the 5' end of the RNA in the order 5'-C-M-E (Rice *et al.*, 1985). In addition, the YF virus nucleotide sequence was found to contain a single, very long open reading frame. N-terminal amino acid sequences have been obtained for the E, C, M, and gp19–23 (precursor of M) proteins of several flaviviruses (Bell *et al.*, 1985). The demonstration that these viral proteins were not blocked and did not have a methionine at their N termini suggested that they were produced from a polyprotein precursor by posttranslational cleavage. Although there has been some evidence that a large (200,000- to 300,000-dalton) viral specific precursor is produced in flavivirus-infected cells, a rigorous demonstration of such a precursor polyprotein has not yet been reported. The posttranslational cleavage of the flavivirus polyprotein precursor in infected cells must occur very rapidly and with relatively few large intermediate precursors, since pulse–chase experiments have failed to demonstrate transient precursor polypeptides (Westaway, 1977, 1980; Westaway *et al.*, 1984). The viral-specific proteins detected in infected cells each appear to be unique by peptide mapping and together do not exceed the coding capacity of the flavivirus genome. Pactamycin mapping studies indicated that viral proteins were completed in the order of their size (Westaway, 1977). Rapid labeling of all viral proteins was observed when protein synthesis was reinitiated after release of a high-salt block (Westaway, 1977). On the basis of these data, Westaway (1977, 1980) proposed that the translation of each of the flavivirus proteins is separately initiated and terminated on the genome mRNA. Data from UV mapping experiments suggested that the flavivirus genome RNA contains two separate translational units (Westaway *et al.*, 1984). A further discussion of these observations and the flavivirus translational strategy can be found in Chapter 10.

Since flavivirus infection does not appreciably inhibit host-cell protein synthesis, it is necessary to treat infected cells with actinomycin D or cycloheximide (Trent and Qureshi, 1971; Shapiro *et al.*, 1971), to synchronize translation reinitiation by high-salt treatment (Westaway, 1977), or to use a double-label subtraction method (Westaway, 1973) to detect the intracellular synthesis of viral-specific proteins. Nine virus-specified proteins have been identified in [^{35}S]methionine-labeled extracts from flavivirus-infected cells (Fig. 5; Rice *et al.*, 1985). Each of these proteins appears to be unique by peptide mapping. One of these proteins is the structural E glycoprotein, gp51–59 (V3), and one a nonstructural glycoprotein gp19–23 (NV2 or NV2$\frac{1}{2}$) that is the precursor of the M protein. The M protein is not found in infected cells, but only in extracellular virions (Shapiro *et al.*, 1972a–c; Westaway, 1980). The form of the C protein [p13–16 (NV1$\frac{1}{2}$)] present intracellularly migrates slightly faster in polyacrylamide gels than the C protein (V2) isolated from virions (Wright and Westaway, 1977). One additional nonstructural protein, gp44–48 (NV3), is also glycosylated. Five further nonglycosylated, nonstructural proteins usually detected in flavivirus-infected cells are p91–98 (NV5), p67–75 (NV4), p24–32 (NVX), p19–21 (NV2 or NV2$\frac{1}{2}$), and p10 (NV1). The

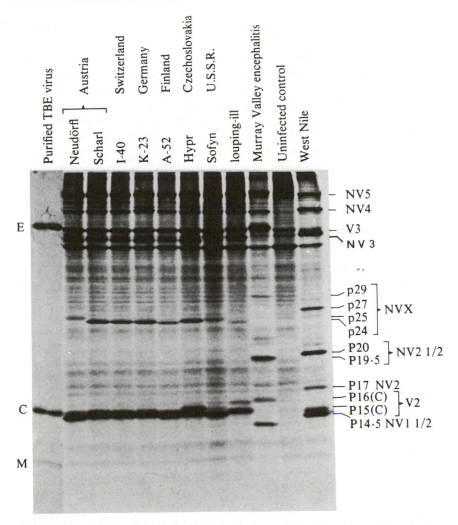

FIGURE 5. Intracellular flavivirus-specific proteins. [³⁵S]Methionine-labeled cell extracts from primary chick embryo cells infected with various flaviviruses were analyzed on a 17% SDS–polyacrylamide gel. Proteins were labeled from 24 to 30 hr after infection. Purified TBE virions and an uninfected cell extract were used as controls. From Heinz and Kunz (1982).

former designation of each protein is indicated in parentheses (Westaway *et al.*, 1980). Some confusion has resulted because all the small intracellular proteins are not always detected in radioactively labeled infected cell extracts. Also, the migration of some of these proteins varies with the host cell and flavivirus used. Identification of NV2 and NV2½ has been especially difficult. On the basis of new information obtained from the sequencing and mapping of the YF virus, 17D, genome RNA, a new nomenclature for the flavivirus nonstructural proteins has been proposed (see Chapter 10).

The functions of the majority of the nonstructural proteins are not currently known. It is presumed that the two largest proteins function in polymerase complexes. The gp44–48 (NV3) protein has recently been identified as the soluble complement-fixing antigen, which is released by flavivirus-infected cells (G. W. Smith and Wright, 1985). Immunization of mice with gp44–48 (NV3) obtained from YF-virus-infected cells was found to provide protection against virus challenge in the absence of neutralizing antibody (J. J. Schlesinger *et al.*, 1986). Complement-fixing monoclonal antibodies to gp44–48 were also protective, apparently through complement-mediated cytolysis of infected cells expressing gp44–48 on their plasma membranes (J. J. Schlesinger *et al.*, 1986).

Comparisons of the migration patterns in polyacrylamide gels of proteins produced by various flaviviruses indicate that there is significant variation in the apparent molecular weights of both structural and nonstructural proteins (Shapiro *et al.*, 1971; Trent and Qureshi, 1971; Westaway, 1973; Westaway and Shew, 1977; Heinz and Kunz, 1982; Wright *et al.*, 1983) (Fig. 5). It is interesting that serologically distinct flaviviruses show heterogeneity in the molecular weights and peptide maps of their two largest nonstructural proteins, p91–98 and p67–75, as well as in their virion E glycoprotein, gp51–59 (Heinz and Kunz, 1982). In the case of togaviruses and certain plant viruses, viral polymerase protein sequences have been found to be highly conserved even during divergent virus evolution (Ahlquist *et al.*, 1985). Comparison of the sequences of the two largest nonstructural proteins of various flaviviruses will indicate the extent of conservation (see Chapter 3).

VII. GENETICALLY CONTROLLED RESISTANCE TO FLAVIVIRUSES

A gene coding for resistance to flavivirus-induced disease has been found to segregate within several mouse populations. Heritable susceptibility to YF virus was first reported by Sawyer and Lloyd (1931) in Rockefeller Institute mice. From the same outbred mouse colony, Webster (1933) developed a resistant line [(BRVR) bacteria-resistant, virus-resistant] by breeding mice that survived infections with louping ill or SLE virus. An inbred susceptible mouse strain (BSVS) was also developed by Webster. Sabin (1952a,b) subsequently demonstrated that Princeton Rockefeller Institute (PRI) mice were resistant to YF virus, but susceptible to other types of viruses. Using PRI mice as the source of the resistance gene, a third inbred resistant strain, C3H/RV, was developed that is congenic to susceptible C3H/He. A study of wild mice obtained from California and Maryland demonstrated the presence of the flavivirus resistance gene among wild mouse populations (Darnell *et al.*, 1974).

In all instances, flavivirus resistance is inherited as an autosomal dominant allele. Since virus infections often induce permanent impairment or death in their hosts, it is not unreasonable to expect that they could exert a selective pressure for the maintenance of host alleles that fortuitously confer a reduced susceptibility to viral diseases in host populations (Brinton and Nathanson, 1981; Brinton et al., 1984). To date, murine genes that can specifically influence the outcome of infections with members of three families of DNA viruses and five families of RNA viruses have been identified (Brinton et al., 1984). Each of these resistance genes influences infections with only one family of viruses or, in some cases, one strain of virus. This virus-specific feature distinguishes genetically controlled resistance from other types of host defense mechanisms and implies that the product of each resistance gene affects a step in the infection cycle characteristic of only one type of virus. Such an interaction could occur at any step during the infectious process.

Flaviviruses can replicate in mice that possess the resistance allele, but in resistant mice, virus titers in tissues are lower (by 1000- to 10,000-fold) than in susceptible mice, and the spread of infection is slower and usually self-limiting. Factors such as the age of the host, the degree of virulence of the infecting flavivirus, and the route of infection have been found to influence the extent of the phenotypic expression of resistance. Although resistant mice can be killed by large doses of virulent flaviviruses administered by the intracerebral route or by infection after experimental immunosuppression, a reduced production of infectious virus is always observed in the brains of C3H/RV mice as compared to the brains of similarly treated susceptible C3H/He mice. The titer of WN virus per gram of brain is shown in Fig. 6A for adult mice sacrificed on successive days after an intracerebral injection with $10^{5.5}$ plaque-forming units (PFU) of WN virus E101. Under these conditions of infection, all the C3H/He mice died on day 4, while about 50% of the C3H/RV mice died on day 7. The titer of virus in C3H/RV brains rose somewhat on day 7 in mice that were moribund. Even though resistant mice produce less virus than susceptible mice, a functioning immune system is required for the survival of infected resistant animals. The levels of neutralizing antibody and interferon that are produced in response to flavivirus infection are higher in susceptible animals than in resistant ones, correlating with the higher titers of virus synthesized in susceptible animals. Interferon is not specifically involved in mediating the expression of flavivirus resistance, as is the case with resistance to myxoviruses (Brinton et al., 1982). Other types of viruses replicate equally well in resistant and susceptible cell cultures and animals.

Cultured cells obtained from resistant mice produce lower yields of flaviviruses than do comparable cultures obtained from susceptible animals, even though the cells in both cultures are equally infectable (Fig. 6B) (Webster and Johnson, 1941; Darnell and Koprowski, 1974). This difference in virus yield is observed with embryo fibroblasts, kidney cells, brain cells, and macrophages. Also, the differential ability to produce flaviviruses was maintained by established lines of SV40-transformed em-

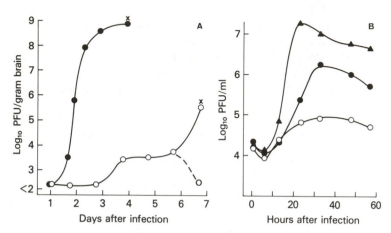

FIGURE 6. Growth of West Nile virus, strain E101. (A) In the brains of adult susceptible C3H/He (●) and resistant C3H/RV (○) mice. Animals were injected with $10^{5.5}$ plaque-forming units (PFU) of virus by the intracerebral route. With this dose of virus, 100% of the C3H/He animals died by day 5 and 50% of the C3H/RV mice by day 8. Virus titers increased on day 7 in C3H/RV mice that were moribund. (×) Moribund animals. (B) In cell cultures of resistant C3H/RV (○) primary mouse embryo fibroblasts, susceptible C3H/He (●) primary mouse embryofibroblasts, and BHK21/WI2 cells (▲). Cultures were infected with a multiplicity of infection of 10. Adapted from Brinton (1981).

bryo fibroblasts prepared from C3H/RV and C3H/He mice (Darnell and Koprowski, 1974). The rate of progeny genome RNA synthesis in resistant cells is reduced as compared to the rate in susceptible cells (Brinton, 1981). In addition, the production of a greater proportion of defective interfering virus particles by resistant cells seems to be a specific manifestation of the expression of the resistance-gene product (Brinton, 1983).

One plausible hypothesis for the observed effects of the resistance-gene product is that it functions at the level of the viral replication complex. If a host protein does interact with the flavivirus replication complex affecting the recognition, binding, or copying of the template RNA by the polymerase, then an alteration in this host protein could affect the efficiency or faithfulness, or both, of viral RNA synthesis.

VIII. DEFECTIVE INTERFERING PARTICLES

The production of and the interference by defective interfering (DI) particles has been difficult to detect during acute flavivirus infections. Because virus yields are relatively low compared to other types of viruses, flaviviruses have been routinely grown in the most permissive host cells available to facilitate experimental study. Further, if a flavivirus infection induces a rapid cytopathic effect in these permissive cells, the possibility of multiple cycles of viral replication within the same host cell is eliminated. Small virus-specific RNAs (8–15 S) have often been observed in flavivirus-infected cells (Zebovitz *et al.*, 1972, 1974; Takeda *et al.*, 1978; Naeve and Trent, 1978). Wengler *et al.* (1978) reported that WN virus produced two low-molecular-weight RNAs (6.5×10^4 and 4.2×10^4) in BHK cells, but only the 6.5×10 dalton species in *A. albopictus* cells. In the same study, Uganda S virus produced a 5×10^4 dalton RNA in both types of cells. Further, these RNAs, although of plus-strand polarity, were not capped or polyadenylated and were inactive in cell-free translation systems (Wengler *et al.*, 1978). These results indicated that the small RNAs did not function as subgenomic mRNAs. However, the size and type of small RNAs associated with these flavivirus infections were characteristic of both the infecting virus and the host cell and apparently not the result of random degradation of 40 S RNA. It is not known whether or not these small intracellular RNAs are able to interfere with the replication of 40 S flavivirus RNA.

Rabbit kidney cells (MA-111) and Vero cells persistently infected with JE virus continue to produce low levels of infectious virus (Schmaljohn and Blair, 1977). Determination of ratios of extracellular physical particles to extracellular infectious particles indicated that defective particles were produced by these cultures. Also, particles from the persistently infected cultures interfered with the replication of wild-type JE virus. In these experiments, the degree of reduction in virus yield was proportional to the amount of virus from the persistently infected culture added. Interference with wild-type WN virus was demonstrated with virus produced by murine embryo fibroblast cultures persistently infected with WN virus (Brinton, 1982). The WN virus produced by long-term persistently infected cultures was no longer able to plaque on BHK cells. However, flaviviruslike particles of normal size were observed by electron microscopy in fluids harvested from these persistently infected cultures. The virions produced by persistently infected cells contained heterogeneous RNA that was primarily smaller than 40 S genome RNA. Three distinct size classes were observed when the RNA was analyzed under denaturing conditions on formaldehyde–agarose gels. RNA of each size class hybridized with ssDNA complementary to 40 S RNA, indicating that these RNAs were virus-specific and of plus-strand polarity. Although the WN virus populations produced by persistently infected cells had relatively low amounts of virions containing full-size genome RNA and a high proportion of particles containing smaller RNAs, the extent of interference with standard WN virus replication in BHK cells during a single growth cycle was relatively low. A maximum reduction in the yield of infectious virus of about 1 log was consistently observed. However, the amount of 40 S RNA synthesized during a 1.5-hr labeling period 24 hr after infection in BHK cells coinfected with persistent and standard virus was reduced by 50% (Brinton, 1982). These results imply that only a portion of the virus particles in the fluids harvested from the persistently infected cultures contained RNA that could interfere substantially with wild-type WN virus replication in BHK cells.

Evidence for the production of DI particles during serial undiluted passage of WN virus in murine embryo fibroblasts was reported by Darnell and Koprowski (1974). Passage of WN virus in either resistant (C3H/RV) or susceptible (C3H/He) embryo fibroblasts resulted in a cyclic rise and fall in the extracellular virus titer characteristic of a Von Magnus effect (Von Magnus, 1954). The titer of virus produced by the C3H/He cells continued to cycle between 10^4 and $10^{6.5}$ PFU/ml during 11 passages. However, passage of WN virus in C3H/RV cells resulted in the elimination of infectious virus by passage 7 (Darnell and Koprowski, 1974; Brinton, 1983). WN virus obtained after three serial undiluted passages in C3H/RV cells was able to interfere with standard WN virus replication, whereas interference was not observed with virus obtained after three passages in C3H/He cells. A. L. Smith (1981) found that Banzi virus populations obtained from the brains of C3H/RV mice that had been inoculated intraperitoneally contained demonstrable interfering activity. No interfering activity could be detected with virus obtained from C3H/RV animals that had been inoculated intracerebrally or with virus obtained from C3H/He animals inoculated by either route.

Sucrose-density-gradient analysis of WN virus E101 virus particles produced during acute infections of C3H/RV and C3H/He cells in culture indicated that the majority of the DI particles were denser than the standard virions (Fig. 7C and H) (Brinton, 1983). This could be due to the packaging of more than one deleted RNA molecule per capsid or to a slight reduction in the overall size of the DI particles compared to the standard virions. The host cell was found to influence significantly the composition of the viral progeny populations produced. WN virus replication in C3H/RV cells yielded progeny with a high proportion of DI particles, whereas progeny virus produced by congenic C3H/He cells consisted primarily of standard virions (Fig. 7D, E, I, and J). The ratio of DI particles to standard virions could be shifted by passaging virus from one cell type to the other (Fig. 8). WN virus pools were prepared in resistant C3H/RV or susceptible C3H/He cells by infecting cells with a multiplicity of infection of 10, harvesting the culture fluid 48 hr after infection, and then concentrating the virus 10-fold. These virus pools were used to infect either C3H/He or C3H/RV cell cultures, and the RNA present in progeny particles produced during the first and second cycles of replication was analyzed on 10–30% SDS sucrose gradients (Fig. 8). The C3H/He cells were able to preferentially amplify virus with full-size RNA, while C3H/RV cultures selectively amplified the particles containing the small-size RNAs. Homologous interference could be demonstrated with WN virus produced by C3H/RV cells, but not with virus produced by C3H/He cells. Four size classes of small RNAs with sedimentation coefficients of approximately 8, 15, 26, and 34 S were consistently observed to be associated with extracellular particles. The four species of small RNAs were presumed to have been present in the stock of WN virus E101 used in these experiments. Cloning of the virus stock by six consecutive

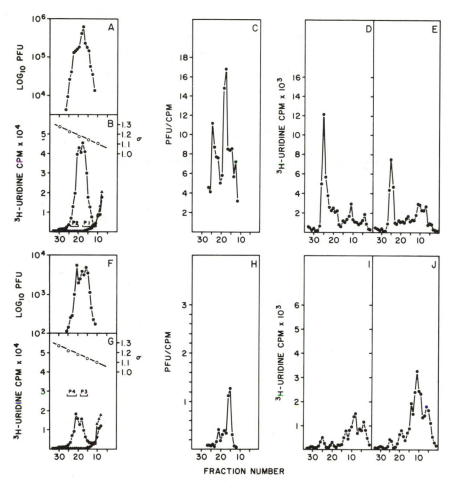

FIGURE 7. Isopycnic density-gradient centrifugation of WN progeny virus produced by susceptible C3H/He and resistant C3H/RV embryofibroblasts. Cultures were labeled with [³H]uridine at 6–32 hr after infection with WN virus at a multiplicity of infection of 10. The infectivity of the infecting virus was determined by plaque assay in BHK cells. Harvested culture fluids were clarified and centrifuged directly on 15–45% isopycnic sucrose density gradients. (A–E) Virus produced by C3H/He cells; (F–J) virus produced by C3H/RV cells. Samples of each gradient fraction were analyzed for acid-insoluble radioactivity (B, G) and infectivity (A, F). The ratio of infectivity to uridine counts was calculated for each gradient fraction (C, H). Fractions were pooled as indicated, and the extracted RNA was sedimented on 15–30% SDS–sucrose gradients (D, E, I, and J for pools 1–4, respectively). (●) Virus-infected culture fluid; (▲) uninfected control culture fluid; (○) σ values. The direction of sedimentation is from right to left. From Brinton (1983).

plaque-purification steps reduced the amount of DI particles present, but did not completely eliminate them. Stocks of other flaviviruses might be expected to differ in the size classes of the DI particle RNAs they contain. Whether the RNAs in each of the four size classes are able to interfere with the replication of wild-type WN virus is not currently

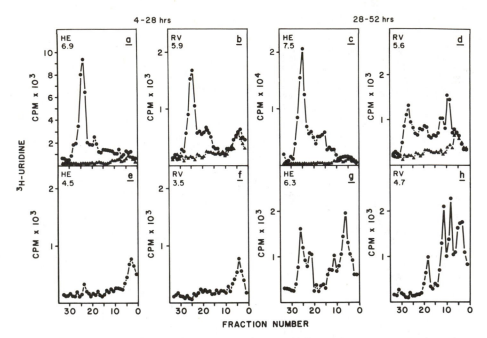

FIGURE 8. Sucrose-gradient sedimentation of extracellular progeny virion RNA produced by resistant C3H/RV or susceptible C3H/He cells infected with pools of WN virus that had been grown in one of the two cell types for 48 hr (see Fig. 7) and concentrated 10-fold. Cultures were incubated with [³H]uridine, and virus was harvested at 28 or 52 hr after infection. The RNA was extracted and sedimented on 15–30% SDS–sucrose gradients. (a–d) Virus produced by cells infected with C3H/He-WN virus (MOI = 1); (e–h) virus produced by cells infected with C3H/RV-WN virus (MOI = 0.1). The host cell is indicated in the upper left-hand corner of each panel, as is the total infectivity ($\log_{10}$) of the harvested progeny virus. (●) Infected; (▲) uninfected control. The direction of sedimentation is from right to left. From Brinton (1983).

known. However, the two smallest species of RNA appeared to be present in the greatest relative abundance in the progeny virus produced by C3H/RV cells, and these virus populations demonstrated the highest interfering activity. It is tempting to speculate that the host proteins specified by the alleles of the murine flavivirus resistance gene may interact specifically at the level of viral RNA synthesis, affecting template–polymerase interactions. If a host protein does interact with the flavivirus replication complex and, further, if the C3H/He and C3H/RV cells do contain different alleles for this protein, then an interaction between the viral polymerase and these host proteins could influence the enrichment of or extent of interference by flavivirus DI particles. Data from both negative-strand RNA (Lazzarini *et al.*, 1981; Schubert *et al.*, 1979) and other positive-strand RNA (Baron and Baltimore, 1982; Dasgupta *et al.*, 1980) virus systems indicate that host proteins can be involved specifically in viral RNA synthesis. It has also been suggested that interactions between viral polymerases and host factors could explain the apparent role of the host cell in DI-particle generation (Holland *et al.*, 1976; Kang and Allen, 1978; Lazzarini *et al.*, 1981; Schubert *et al.*, 1979).

IX. PERSISTENT INFECTIONS

Persistence of flaviviruses in experimental animals has been reported. Webster and Clow (1936) observed that genetically resistant BRVR mice inoculated intranasally with SLE virus developed a subclinical encephalitis that was histologically apparent for 3 months after infection. Virus could be isolated from the brains of these mice for about 1 month after infection. Slavin (1943) reported the persistence of SLE in the brains of mice for over 5 months after infection. Transplacental transmission of JE virus was demonstrated in intraperitoneally infected females during consecutive pregnancies occurring as long as 6 months after infection (Mathur *et al.*, 1982).

Persistence of various tick-borne flaviviruses has been observed more often. Kyasanur Forest virus normally kills mice after inoculation, but occasionally paralyzed infected mice survive for many months (Virus Research Center, Poona, 1963; Price, 1966). When the animals were sacrificed, virus was isolated from the brains and other tissues of these paralyzed mice. Serum neutralizing antibody could often not be detected, even though the paralyzed mice were resistant to subsequent virus challenge. Persistence of louping ill virus in immunosuppressed guinea pigs for more than 50 days after infection was reported by Zlotnik *et al.* (1976). Chronic encephalitis induced in monkeys by TBE virus (Russian spring-summer virus) has been reported by a number of investigators (Ilienko *et al.*, 1974; Zlotnik *et al.*, 1976; Andzhaparidze *et al.*, 1978; Asher, 1979; Malenko *et al.*, 1982; Fokina *et al.*, 1982; Frolova and Pogodina, 1984). Chronic histopathological lesions and infectious virus were detected as late as 789 days after infection. Some indication that flaviviruses may also persist for long periods in humans has been obtained (Edelman *et al.*, 1976; Monath, 1971; Ogawa *et al.*, 1973).

Establishment of persistent infections has been successfully accomplished with a number of different flaviviruses in a variety of mammalian, arthropod, and reptilian cell lines. In some instances, due to the intense cytopathic effect of the particular flavivirus on the host-cell line, persistent infections could be established only when techniques were used to reduce the extent of the virus-induced cell damage. Schmaljohn and Blair (1977) established persistent infections in a line of rabbit kidney cells (MA-111) with JE virus that had previously been subjected to 30–40 serial undiluted passages in MA-111 cells. Infection of some host-cell lines with a particular flavivirus does not result in extensive destruction of the cells. Under these circumstances, it is possible to establish persistent infections simply by routine subculturing of surviving infected cells. This was the case in the establishment of persistent infections with dengue viruses in either HeLa cells (Maguire and Miles, 1965), rhesus monkey testis cells (Hotta and Evans, 1956), green monkey kidney cells (Shimazu *et al.*, 1966), or an *A. albopictus* cell line (Sinarachatanant and Olson, 1973), a line of *Toxorhynchites amboinensis* cells (Kuno, 1982) or human KB cells (Beasley *et al.*, 1960; Schulze and Schlesinger, 1963); WN virus in either murine L-929 cells (Jarman *et al.*, 1968), lines of murine embryo fibroblasts (Brinton, 1982), or monkey LLC-MK$_2$ cells (Katz and Goldblum, 1968); and SLE virus in a turtle heart (TH-1) cell line (Mathews and Vorndam, 1982).

Interferon production, although sometimes demonstrable in persistently infected mammalian-cell cultures, does not appear to be directly involved in the maintenance of flavivirus persistence (Schmaljohn and Blair, 1977; Katz and Goldblum, 1968; Brinton, 1982). DI virus particles were produced by rabbit kidney cells (MA-11) persistently infected with JE virus (Schmaljohn and Blair, 1977) and by murine embryo fibroblast cell lines persistently infected with WN virus (Brinton, 1982). DI virus particles may well play an important role in the establishment and maintenance of flavivirus persistent infections.

Virus produced by persistently infected cultures often progressively changes from the parental virus phenotype. SLE virus from persistently infected turtle heart cells 1 year after infection was less virulent for mice, had an altered cell-culture host range, and had an increased thermal liability as compared to the virus used to initiate the infection (Mathews and Vorndam, 1982). Dengue viruses of all four serotypes obtained 1 year after infection from separate persistently infected nonvector mosquito *T. amboinensis* cultures showed changes in antigenic reactivity, increased temperature sensitivity, and decreased neurovirulence in mice as compared to the original viruses used to initiate the infections (Kuno, 1982). WN virus obtained from persistently infected murine embryo fibroblasts had lost its ability to plaque in BHK-21 cells. Also, temperature-sensitive mutants and a replication-efficient mutant of WN virus were isolated from these persistently infected cell-culture fluids (Brinton, 1981, 1982; Brinton *et al.*, 1985). Analysis of the WN virus progeny populations obtained from persistently infected cultures indicated that these cultures simultaneously replicated several different mutant virus populations.

In some long-term persistently infected cultures, the low levels of infectious virus are produced by only a small proportion of the cells in the culture. Only 4 of 200 cells cloned from rabbit-cell cultures persistently infected with JE virus contained immunofluorescence-detectable viral antigen and released infectious virus (Schmaljohn and Blair, 1979). Sucrose-gradient analysis of [^{3}H]uridine-labeled cell extracts indicated that all the clones produced some intracellular actinomycin-D-resistant RNA. However, culture supernatants were not analyzed for the presence of released virus, which was unable to induce plaques in test-cell monolayers. Superinfection of nonproducer cell clones with JE virus led to the production of infectious virus in the absence of any cytopathic effect

(Schmaljohn and Blair, 1979). JE virus infection normally produces extensive cytopathology in rabbit-cell cultures.

Virus-specific immunofluorescence was observed in 100% of cells in murine-cell cultures persistently infected with WN virus (Brinton, 1982). Some cells (20–70%) showed a bright perinuclear and/or diffuse cytoplasmic fluorescence during the first 15 weekly subcultures. Other cells in the initial subcultures and all cells in long-term persistently infected cultures showed fainter perinuclear and/or focal cytoplasmic viral-specific fluorescence. All the WN-virus-persistently infected cultures continued to release virus particles throughout 2.5 years of cultivation. However, the virus produced by long-term cultures maintained at 37°C could no longer plaque on BHK cells and contained predominantly heterogenous subgenomic viral RNAs (Brinton, 1982). In contrast to what was observed with JE-virus-persistently infected cultures, superinfection of long-term WN-virus-persistently infected cultures with parental virus did not result in the release of progeny virus with the ability to cause plaques. However, superinfection with a replication-efficient mutant (RE-WN virus) did (Brinton and Fernandez, 1983).

The results obtained with persistent flavivirus infections indicate that if a flavivirus causes extensive cytopathology in a particular host cell, a persistent infection can be established only when some method of reducing the viral cytopathic effect is employed during the initial subcultures of the infected cells. Once established, persistently infected cultures produce progeny virus that becomes progressively less cytopathic. Many types of mutant viruses are generated and simultaneously replicated, including deletion mutants with homologous interfering activity and temperature-sensitive mutants.

X. FLAVIVIRUS MUTANTS

A reduction in viral virulence has been observed during serial passage of many of the flaviviruses in cell culture and eggs. Mixed populations of large- and small-plaque variants are commonly observed in passaged virus stocks as well as in field isolates obtained from humans, insects, rodents, and birds (Hammon et al., 1963; Mayer, 1963; Paul, 1966; Price et al., 1963; Theiler, 1951; Eckels et al., 1976). Analysis of selected plaques obtained from such virus populations has shown that in general, plaque size, virulence, and temperature sensitivity are only rarely correlated; however, exceptions have been observed. A small-plaque variant of Central European TBE virus obtained from a persistently infected human amnion-cell culture had reduced virulence for mice (Mayer, 1963). A Kyasanur Forest virus small-plaque mutant obtained after passage in monkey kidney-epithelial-cell cultures was both reduced in virulence and temperature-sensitive (Paul, 1966). A small-plaque mutant of dengue 2 virus, which was isolated after a single passage of a human serum in primary green monkey kidney cells, was temperature-sensitive and showed reduced virulence (Eckels *et al.*, 1976). Persistently infected *A. albopictus* cells yielded a small-plaque dengue 2 virus mutant that was temperature-sensitive as well as of reduced virulence (Igarashi, 1979).

Spontaneously arising temperature-sensitive mutants of both JE virus (Halle and Zebovitz, 1977) and dengue 2 virus (Eckels *et al.*, 1976) have been tested for use as attenuated live virus vaccines. Undiluted passage of the S-1 dengue virus mutant sometimes yielded large-plaque variants that were virulent and not temperature-sensitive (Eckels *et al.*, 1980). Attempts to define the lesions of spontaneous flavivirus temperature-sensitive mutants have rarely been reported. The S-1 temperature-sensitive clone of dengue 2 virus isolated by Eckels *et al.* (1976) was more thermolabile at the nonpermissive temperature of 38.5°C than the parental virus. Viral adsorption studies indicated that the temperature-sensitive protein produced by this mutant was probably a structural protein involved in virus attachment to cells (Eckels *et al.*, 1983).

Flavivirus temperature-sensitive mutants have been isolated after chemical mutagenesis (Table III). Tarr and Lubiniecki (1976) found 7 temperature-sensitive clones of dengue 2 virus among 138 clones isolated from virus grown in primary hamster kidney cells in the presence of 5-azacytidine. Of these dengue mutants, 5 were partially characterized; 3 were RNA-plus mutants, while 2 were RNA-minus mutants. Complementation analysis indicated that these mutants represented four complementation groups. Virus yields in these experiments were assayed by microquantitative complement fixation.

Recently, temperature-sensitive mutants of JE virus (Eastman and Blair, 1985) and SLE virus (Hollingshead *et al.*, 1983) have been efficiently induced by chemical mutagenesis of infected cells also treated with actinomycin D. Both 5-fluorouracil and 5-azacytidine were used to induce SLE mutants. Flavivirus infection does not significantly affect host-cell metabolism, and treatment of host cells with actinomycin D apparently enhances the incorporation of the mutagens into the flavivirus RNA. Nine temperature-sensitive mutants of SLE were obtained; two were RNA-minus, while the remaining seven were RNA-plus mutants. Complementation analysis indicated that these mutants represented four complementation groups, with two of these groups containing more than one of the mutants. Ten mutants of JEV were isolated: five were RNA-plus mutants and five were RNA-minus mutants. Seven complementation groups were defined with these ten mutants. Three of the complementation groups contained RNA-minus mutants. At the nonpermissive temperature, mutants in the three RNA minus complementation groups produced no detectable protein, with the exception of one mutant which produced low levels of protein. The RNA plus mutants produced normal amounts of protein at the non-permissive temperature, again with one exception which produced a low amount of protein. The complementation observed between JEV mutants was sometimes nonreciprocal and

TABLE III. Temperature-Sensitive Mutants of Flaviviruses

Virus	Mutagen[a]	Number of mutants	Phenotype					Reference
			Small-plaque	Reduced virulence	RNA+	RNA−		
Dengue 2	—	1	X	X	—	—		Eckels et al. (1976)
Dengue 2	—	1	X	X	—	—		Igarashi (1979)
Kyasanur Forest	—	1	X	X	—	—		Mayer (1963)
JE	—	1	—	—	—	—		Halle and Zebovitz (1977)
	AzaC	2			0	2		Eastman and Blair (1985)
	5-FU	8			5	3		
WN	—	12+	—	—	2	10		Brinton et al. (1985a)
Dengue 2	AzaC	5	—	—	3	2		Tarr and Lubiniecki (1976)
SLE	AzaC	8	—	—	7	2		Hollingshead et al. (1983)
	5-FU	1						

[a] (AzaC) 5-Azacytidine; (5-FU) 5-fluorouracil.
104

interference between certain pairs of mutants was also observed (Eastman and Blair, 1985).

Temperature-sensitive mutants of WN virus were generated by long-term persistently infected cultures of genetically resistant C3H/RV and susceptible C3H/He murine embryo fibroblasts maintained at 32°C (Brinton *et al.*, 1985). The WN virus used to initiate these infections contained no detectable temperature-sensitive virus. The 40/32°C plaquing ratio of progeny virus obtained from persistently infected cultures maintained at 37°C decreased only slightly during 10–15 subcultures. Thereafter, although these cultures continued to produce WN virus particles, they no longer produced WN virus able to cause plaques on BHK cells. Progeny virus from duplicate cultures shifted to 32°C at subculture 6 was also tested periodically for its ability to plaque at 40 and at 32°C. Viral progeny titers increased after the shift to 32°C, indicating the possible presence of temperature-sensitive mutants. Temperature-sensitive mutants were subsequently isolated from culture fluids obtained from both persistently infected resistant and susceptible cells. However, only in the resistant cell cultures did temperature-sensitive virus become the majority population (Table IV). The resistant cells appear to provide an environment that is advantageous for the amplification of flavivirus temperature-sensitive mutants. It has been previously demonstrated that resistant C3H/RV cells preferentially amplify WN virus defective RNAs (Brinton, 1983). The growth of wild-type virus is reduced in the resistant cells, and this may allow temperature-sensitive mutants to compete more effectively during persistent infections in resistant cells. Preliminary data indicate that temperature-sensitive mutants generated by the resistant cultures are in general less leaky than those obtained from the susceptible cultures. To date, no complementation among the WN virus temperature-sensitive mutants has been demonstrable. The majority of mutants obtained from the C3H/RV cultures have an RNA-minus phenotype. It is possible that the C3H/He and C3H/RV cultures selectively amplify different groups of temperature-sensitive mutants (Brinton *et al.*, 1985).

In addition to the temperature-sensitive mutants, a non-temperature-sensitive mutant of WN virus was isolated from the culture fluid of one of the genetically resistant persistently infected C3H/RV cultures. This mutant produced higher yields of virus and incorporated [3H]uridine into intracellular 40 S RNA more efficiently than the parental virus and grew equally well in resistant and susceptible cell cultures (Brinton, 1981). Analysis of the genome RNA of this replication-efficient mutant, designated RE-WN virus, by oligonucleotide fingerprinting indicated that the mutant RNA differed from the genome RNA of the parent by three unique oligonucleotides. The RE-WN virus was insensitive to interference by a parental WN virus pool enriched in DI particles even during replication in resistant C3H/RV cells (Brinton and Fernandez, 1983). This mutant was also able to superinfect persistently infected murine cell cultures, whereas the parental virus was not.

TABLE IV. Temperature Sensitivity of Plaques Picked from Persistently Infected RV and He Culture Fluids

Culture	Passage	Number of *ts* mutants/ number of plaques picked[a]
Susceptible cell lines		
He 100	9	2/24
	11	1/10
	18	0/20
	33	0/21
	37	1/13
	53	0/8
	57	1/10
He 103	8	0/24
	11	3/12
	17	0/12
	18	2/22
	23	1/27
Resistant cell lines		
RV 100	10	0/40
	12	0/12
	14	1/8
	17	1/17
	21	47/70
	24	20/20
RV 103	13	0/16
	18	0/14
	25	1/10
	27	5/20
	31	19/23
	43	3/7
	57	5/6

[a] Medium from persistently infected cultures was collected at the time of each of the weekly subcultures. Plaques were picked from 32°C titration plates about every 2–4 passages. Virus contained in the picked plaques was then titrated by plaque assay at both 32 and 40°C. Plaque isolates displaying a 40/32°C ratio of 0.01 or less were scored as temperature-sensitive.

Comparison of the 3'-terminal 135 nucleotides of the parental WN virus and the RE-WN virus showed no differences between the two RNAs in this noncoding region of the genome presumed to contain sequences involved in the recognition and binding of the replication complex (Brinton *et al.*, 1986). Future studies will focus on the 3' noncoding region of the minus-strand RNA and the viral polymerase gene sequences. This mutant provides a unique tool for further study of the factors involved in regulating flavivirus RNA synthesis.

XI. SUMMARY

During the past five years, much has been learned about flavivirus infections and replication strategy. By 1984, a sufficient amount of new information had been obtained to institute the reclassification of flaviviruses (Westaway *et al.*, 1986). The most significant recent advance has been provided by the nucleotide sequence of the YF virus genome and the partial map of the viral genes determined by Rice *et al.* (1985) (see Chapter 10). The new sequence data have confirmed the prediction from *in vitro* translation data that the viral structural genes are located at the 5' end of the coding region. The mapping of the flavivirus genes on the nucleotide sequence has provided a precise delineation of the number and sizes of the flavivirus nonstructural proteins and of the made of flavivirus protein translation and processing.

Nine epitopes of the virion glycoprotein have been mapped for three different flaviviruses (Roehrig *et al.*, 1983; J. J. Schlesinger *et al.*, 1983; Heinz *et al.*, 1983) with panels of monoclonal antibodies. Future comparisons of antigenic variants at the level of the glycoprotein nucleotide sequence should greatly enhance our understanding of the complex antigenic relationships known to exist among flaviviruses, the determinants of virulence, and the various biological functions of the E protein such as hemagglutination, attachment, and fusion.

The existence of a very stable stem and loop structure composed of the first 84–87 bases at the 3' terminus of the flavivirus genome has been established (Brinton *et al.*, 1986; Rice *et al.*, 1985). Such a structure could function in polymerase recognition and binding and/or capsid protein binding. Many of the plus-stranded plant viruses also have no poly(A) instead have a transfer-RNA-like structure at their 3' termini (E. R. Strauss and J. H. Strauss, 1983). Future comparison of 5'- and 3'-terminal sequences from a variety of flaviviruses should indicate highly conserved regions that may be important signals for replicase and transcriptase recognition and binding. Comparisons between YF virus and WN virus indicate that a high level of sequence conservation is observed in the first two nucleotides at the 3' termini of both plus and minus strands, a heptanucleotide located near each terminus, and a 27-base sequence beginning at nucleotides 85 or 87 from the 3' termini of the genome RNA (Brinton *et al.*, 1986; Rice *et al.*, 1985).

The generation of flavivirus DI particles has been demonstrated during acute infections with WN virus (Brinton, 1983) and persistent infections with JE virus (Schmaljohn and Blair, 1977) and WN virus (Brinton, 1982). However, the mechanism of generation of the various size classes of deleted viral RNAs observed is not yet known. Also, the molecular basis for the differential ability of various cell cultures to generate and/or amplify these deleted flavivirus RNAs is not understood. Sequence analysis of flavivirus DI RNAs should yield further information about the sequences and secondary structure elements required for transcription and replication of flavivirus RNAs.

The fact that many flaviviruses can replicate in a wide range of vertebrate hosts and in arthropod vectors implies that any functions required from the host cell during the flavivirus replication cycle must be present in a wide range of host species. The existence of the murine flavivirus-specific resistance gene indicates that one of these host proteins can differ in the efficiency with which it can interact with flavivirus macromolecules. Such host factors could affect both the efficiency and the fidelity of flavivirus RNA transcription.

Many questions concerning flavivirus replication remain unanswered. For instance, the functions of the various nonstructural proteins have not been delineated. The mechanism by which flavivirus genome RNAs are capped at their 5' end is not known. Does a host-cell enzyme or a viral enzyme cap the RNA? The mechanism by which plus-strand RNA is made efficiently, while minus-strand RNA is not, is also not understood. The proteolytic enzymes involved in the processing of the flavivirus proteins from precursors have not been identified. Continued study of flaviviruses on a molecular level during the coming years should greatly increase our knowledge of flavivirus replication.

REFERENCES

Ahlquist, P., Strauss, E. G., Rice, C. M., Strauss, J. H., Haseloff, J., and Zimmern, D., 1985, Sindbis virus proteins ns P1 and ns P2 contain homology to nonstructural proteins from several RNA plant viruses, *J. Virol.* **53:**536.

Andzhaparidze, O. G., Rozina, E. E., Bogomolova, N. N., and Boriskin, Yu. S., 1978, Morphological characteristics of the infection of animals with tick-borne encephalitis virus persisting for a long time in cell cultures, *Acta Virol.* **22:**218.

Asher, D. M., 1979, Persistent tick-borne encephalitis infection in man and monkeys: Relation to chronic neurologic disease, in: *Artic and Tropical Arboviruses: Proceedings of the 2nd International Symposium on Artic Arboviruses*, Mont Gabriel, Canada, pp. 179–195, Academic Press, New York.

Baron, M. H., and Baltimore, D., 1982, Purification and properties of a host cell protein required for poliovirus RNA replication *in vitro*, *J. Biol. Chem.* **257:**12,351.

Beasley, A. R., Lichter, W., and Sigel, M. M., 1960, Studies on latent infections of tissue cultures with dengue virus. *Arch. Virusforsch.* **10:**672.

Bell, J. R., Kinney, R. M., Trent, D. W., Lenches, E. M., Dalgarno, L., and Strauss, J. H., 1985, Amino-terminal amino acid sequences of structural proteins of three flaviviruses, *Virology* **143:**224.

Boulton, R. W., and Westaway, E. G., 1976, Replication of the flavivirus Kunjin: Proteins, glycoproteins and maturation associated with cell membranes, *Virology* **69:**416.

Brandriss, M. W., and Schlesinger, J. J., 1984, Antibody-mediated infection of P388D1 cells with 17D yellow fever virus: Effects of chloroquine and cytochalasin B, *J. Gen. Virol.* **65:**791.

Brandt, W. E., McCown, J. M., Top, F. H., Jr., Bancroft, W. H., and Russell, P. K., 1979, Effect of passage history on dengue-2 virus replication in subpopulations of human leukocyte, *Infect. Immun.* **26:**534.

Brandt, W. E., McCown, J. M., Gentry, M. K., and Russell, P. K., 1982, Infection enhancement of dengue type 2 virus in the U-937 human monocyte cell line by antibodies to flavivirus cross-reactive determinants, *Infect. Immun.* **36:**1036.

Brawner, T. A., Trousdale, M. O., and Trent, D. W., 1979, Cellular localization of Saint Louis encephalitis virus RNA replication, *Acta Virol.* **23:**284.

Brinton, M. A., 1981, Isolation of a replication-efficient mutant of West Nile virus from a persistently infected genetically resistant mouse cell culture, *J. Virol.* **39:**413.

Brinton, M. A., 1982, Characterization of West Nile virus persistent infections in genetically resistant and susceptible mouse cells. I. Generation of defective nonplaquing virus particles, *Virology* **116:**84.

Brinton, M. A., 1983, Analysis of extracellular West Nile virus particles produced by cell cultures from genetically resistant and susceptible mice indicates enhanced amplification of defective interfering particles by resistant cultures, *J. Virol.* **46:**860.

Brinton, M. A., and Fernandez, A. V., 1983, A replication-efficient mutant of West Nile virus is insensitive to DI particle interference, *Virology* **129:**107.

Brinton, M. A., and Nathanson, N., 1981, Genetic determinants of virus susceptibility: Epidemiologic implications of murine models, *Epidemiol. Rev.* **3:**115.

Brinton, M. A., Arnheiter, H., and Haller, O., 1982, Interferon independence of genetically controlled resistance to flaviviruses, *Infect. Immun.* **36:**284.

Brinton, M. A., Blank, K. J., and Nathanson, N., 1984, Host genes that influence susceptibility to viral diseases, in: *Concepts in Viral Pathogenesis* (A. L. Notkins and M. B. A. Oldstone, eds.), pp. 71–78, Springer-Verlag, New York.

Brinton, M. A., Davis, J., and Schaefer, D., 1985, Characterization of West Nile virus persistent infections in genetically resistant and susceptible mouse cells. II. Generation of temperature-sensitive mutants, *Virology* **140:**152.

Brinton, M. A., Fernandez, A. V., and Amato, J., 1986, The 3'-nucleotides of flavivirus genomic RNA form a conserved secondary structure, *Virology*, submitted.

Calberg-Bacq, C.-M., Rentier-Delrue, F., Osterrieth, P. M., and Duchesne, P. Y., 1975, Electron microscopy studies on Banzi virus particle and its development in the suckling mouse brain, *J. Ultrastruct. Res.* **53:**193.

Caliguiri, L. A., and Tamm, I., 1970, The role of cytoplasmic membranes in poliovirus biosynthesis, *Virology* **42:**100.

Calisher, C. H., Dalrymple, J. N., Karabatsos, N., Shope, R. E., Bishop, D. H. L., Brandt, W., Casals, J., Porterfield, J. S., Tesh, R. B., and Westaway, E. G., 1986, Antigenic relationships among flaviviruses as determined by cross neutralization: Proposed antigenic classification, submitted.

Cammack, N., and Gould, E. A., 1985, Conditions for haemolysis by flaviviruses and characterization of the haemolysin, *J. Gen. Virol.* **66:**2291.

Cardiff, R. D., Dalrymple, J. M., and Russell, P. K., 1973, RNA polymerase in group B arbovirus (dengue-2) infected cells, *Arch. Gesamte Virusforsch.* **40:**392.

Cardosa, M. J., Porterfield, J. S., and Gordon, S., 1983, Complement receptor mediates enhanced flavivirus replication in macrophages, *J. Exp. Med.* **158:**258.

Chamberlain, R. W., 1980, Epidemiology of arthropod-borne togaviruses: The role of arthropods as hosts and vectors and of vertebrate hosts in natural transmission cycles, in: *The Togaviruses* (R. W. Schlesinger, ed.), pp. 175–227, Academic Press, New York.

Chu, P. W. G., and Westaway, E. G., 1985, Replication strategy of Kunjin virus: Evidence for recycling role of replicative form RNA as template in semiconservative and asymmetric replication, *Virology* **140:**68.

Cleaves, G. R., and Dubin, D. T., 1979, Methylation status of intracellular dengue type 2 40S RNA, *Virology* **96:**159.

Cleaves, G. R., and Schlesinger, R. W., 1977, Characterization of polysomal RNA's from dengue virus infected KB cells, *Abstr. Am. Soc. Microbiol.* **S-50:**287.

Cleaves, G. R., Ryan, T. E., and Schlesinger, R. W., 1981, Identification and characterization of type 2 dengue virus replicative intermediate and replicative form RNAs, *Virology* **111:**73.

Dalton, S., 1972, Infection of neuronal and glial tissue *in vivo* by Langat, a group B arbovirus, *Ann. Inst. Pasteur (Paris)* **123:**489.

Darnell, M. B., and Koprowski, H., 1974, Genetically determined resistance to infection with group B arboviruses. II. Increased production of interfering particles in cell cultures from resistant mice, *J. Infect. Dis.* **129:**248.

Darnell, M. B., Koprowski, H., and Lagerspetz, K., 1974, Genetically determined resistance to infection with group B arboviruses. I. Distribution of the resistance gene among various mouse populations and characteristics of gene expression *in vivo*, *J. Infect. Dis.* **129**:240.

Dasgupta, A., Zabel, P., and Baltimore, D., 1980, Dependence of the activity of the poliovirus replicase on a host cell protein, *Cell* **19**:423.

Daughaday, C. C., Brandt, W. E., McCown, J. M., and Russell, P. K., 1981, Evidence for two mechanisms of dengue virus infection of adherent human monocytes: Trypsin-sensitive virus receptors and trypsin-resistant immune complex receptors, *Infect. Immun.* **32**:469.

Demsey, A., Steere, R. L., Brandt, W. E., and Veltri, B. J., 1974, Morphology and development of dengue-2 virus employing the freeze–fracture and thin-section techniques, *J. Ultruct. Res.* **46**:103.

Deubel, V., Crouset, J., Benichou, D., Digoutte, J.-P., Bouloy, M., and Girard, M., 1983, Preliminary characterization of the ribonucleic acid of yellow fever virus, *Ann. Virol.* **134E**:581.

Eastman, P. S., and Blair, C. D., 1985, Temperature-sensitive mutants of Japanese encephalitis virus, *J. Virol.* **55**:611.

Eckels, K. H., Brandt, W. E., Harrison, V. R., McCown, J. M., and Russell, P. K., 1976, Isolation of a temperature-sensitive dengue-2 virus under conditions suitable for vaccine development, *Infect. Immun.* **14**:1221.

Eckels, K. H., Harrison, V. R., Summers, P. L., and Russell, P. K., 1980, Dengue-2 vaccine: Preparation from a small plaque virus clone, *Infect. Immun.* **27**:175.

Eckels, K. H., Summers, P. L., and Russell, P. K., 1983, Temperature-sensitive events during the replication of the attenuated S-1 clone of dengue type 2 virus, *Infect. Immun.* **39**:750.

Edelman, R., Schneider, R. J., Vejjajiva, A., Pornpibul, R., and Voodhikul, P., 1976, Persistence of virus-specific IgM and clinical recovery after Japanese encephalitis, *Am. J. Trop. Med. Hyg.* **25**:733.

Filshie, B. K., and Rehacek, J., 1968, Studies of the morphology of Murray Valley encephalitis and Japanese encephalitis viruses growing in cultured mosquito cells, *Virology* **34**:435.

Fokina, G. I., Malenko, G. V., Levina, L. S., Koreshkova, G. V., Rzhakhova, O. E., Mamonenko, L. L., Pogodina, V. V., and Frolova, M. P., 1982, Persistence of tick-borne encephalitis virus in monkeys. V. Virus localization after subcutaneous inoculations, *Acta. Virol.* **26**:369.

Frolova, M. P., and Pododina, V. V., 1984, Persistence of tick-borne encephalitis virus in monkeys. VI. Pathomorphology of chronic infection in central nervous system, *Acta. Virol.* **28**:232.

Gollins, S. W., and Porterfield, J. S., 1984, Flavivirus infection enhancement in macrophages: Radioactive and biological studies on the effect of antibody on viral fate, *J. Gen. Virol.* **65**:1261.

Gollins, S. W., and Porterfield, J. S., 1985, Flavivirus infection enhancement in macrophages: an electron microscopic study of viral cellular entry, *J. Gen. Virol.* **66**:1969.

Grimley, P. M., 1983, Arbovirus encephalitis: Which road traveled by makes all the difference?, *Lab. Invest.* **48**:369.

Grun, J. B., and Brinton, M. A., 1986a, Characterization of West Nile viral RNA-dependent RNA polymerase and cellular terminal adenylyl and uridylyl transferase in cell-free extracts, *J. Virol.*, submitted.

Grun, J. B., and Brinton, M. A., 1986b, Enhancement of in vitro West Nile virus RNA-dependent RNA polymerase activity, *J. Virol.*, submitted.

Halle, S., and Zebovitz, E., 1977, A spontaneous temperature sensitive mutant of Japanese encephalitis virus: Preliminary characterization, *Arch Virol.* **54**:165.

Halstead, S. B., 1980, Immunological parameters of togavirus disease syndromes, in: *The Togaviruses* (R. W. Schlesinger, ed.), pp. 107–173; Academic Press, New York.

Halstead, S. B., 1981, The pathogenesis of dengue: Molecular epidemiology in infectious disease: The 1981 Alexander D. Langmuir Lecture, *Am. J. Epidemiol.* **114**:632.

Halstead, S. B., 1982, Immune enhancement of viral infection, in: *Progress in Allergy* (P. Kallos, ed.), pp. 301–364, S. Karger, Basel.

Halstead, S. B., and O'Rourke, E. J., 1977, Dengue viruses and mononuclear phagocytes. I. Infection enhancement by non-neutralizing antibody, *J. Exp. Med.* **146:**201.

Halstead, S. B., Marchette, N. J., Chow, J. S., and Lolekha, S., 1976, Dengue virus replication enhancement in peripheral blood leukocytes from immune human beings, *Proc. Soc. Exp. Biol. Med.* **151:**136.

Halstead, S. B., Porterfield, J. S., and O'Rourke, E. J., 1980, Enhancement of dengue infection in monocytes by flavivirus antisera, *Am. J. Trop. Med. Hyg.* **29:**638.

Halstead, S. B., Rojanasuphot, S., and Sangkawibha, N., 1983, Original antigenic sin in dengue, *Am. J. Trop. Med. Hyg.* **32:**154.

Halstead, S. B., Venkateshan, C. N., Gentry, M. K., and Larsen, L. K., 1984, Heterogeneity of infection enhancement of dengue 2 strains by monoclonal antibodies, *J. Immunol.* **132:**1529.

Hammon, W. McD., Rohitaydhin, S., and Rhim, J. S., 1963, Studies on Japanese B encephalitis virus vaccines from tissue culture. IV. Preparation and characterization of a pool of attenuated Oct-541 line for human vaccine trial, *J. Immunol.* **91:**295.

Harrison, A. K., Murphy, F. A., and Gardner, J. J., 1982, Visceral target organs in systemic St. Louis encephalitis virus infection of hamsters, *Exp. Mol. Pathol.* **37:**292.

Heinz, F. X., and Kunz, C., 1982, Molecular epidemiology of tick-borne encephalitis virus: Peptide mapping of large non-structural proteins of European isolates and comparison with other flaviviruses, *J. Gen. Virol.* **62:**271.

Heinz, F. X., Berger, R., Tuma, W., and Kunz, C., 1983, A topological and functional model of epitopes on the structural glycoprotein of tick-borne encephalitis virus defined by monoclonal antibodies, *Virology* **126:**525.

Henchal, E. A., Gentry, M. K., McCown, J. M., and Brandt, W. E., 1982, Dengue virus-specific and flavivirus group determinants identified with monoclonal antibodies by indirect immunofluorescence, *Am. J. Trop. Med. Hyg.* **31:**830.

Holland, J. J., Villarreal, L. P., and Breindl, M., 1976, Factors involved in the generation and replication of rhabdovirus defective T particles, *J. Virol.* **17:**805.

Hollingshead, P. G., Jr., Brawner, T. A., and Fleming, T. P., 1983, St. Louis encephalitis virus temperature-sensitive mutants. I. Induction, isolation and preliminary characterization, *Arch. Virol.* **75:**171.

Hotta, S., and Evans, C. A., 1956, Cultivation of mouse-adapted dengue virus (type 1) in rhesus monkey tissue culture, *J. Infect. Dis.* **98:**88.

Hotta, H., Hotta, S., Takada, H., Kotani, S., Tanaka, S., and Ohki, M., 1983, Enhancement of dengue virus type 2 replication in mouse macrophage cultures by bacterial cell walls, peptidoglycans, and a polymer of peptidoglycan subunits, *Infect. Immun.* **41:**462.

Hotta, H., Wiharta, A. S., and Hotta, S., 1984, Antibody-mediated enhancement of dengue virus infection in mouse macrophage cell lines, Mk1 and Mm1, *Proc. Soc. Exp. Biol. Med.* **175:**320.

Igarashi, A., 1978, Isolation of Singh's *Aedes albopictus* cell clone sensitive to dengue and chikungunya viruses, *J. Gen Virol.* **40:**531.

Igarashi, A., 1979, Characteristics of *Aedes albopictus* cells persistently infected with dengue viruses, *Nature (London)* **280:**690.

Igarashi, A., Harrap, K. A., Casals, J., and Stoller, V., 1976, Morphological, biochemical and serological studies on a viral agent (CFA) which replicates in and causes fusion of *Aedes albopictus* (Singh) cells, *Virology* **74:**174.

Ilienko, V. I., Platonov, V. G., Komandenko, V. G., Prozorova, I. N., and Panov, A. T., 1974, Pathogenic study on chronic forms of tick-borne encephalitis, *Acta Virol.* **18:**341.

Jarman, R. V., Morgan, P. N., and Duffy, C. E., 1968, Persistence of West Nile virus in L929 mouse fibroblasts, *Proc. Soc. Exp. Biol. Med.* **129:**633.

Kang, C. Y., and Allen, R., 1978, Host function-dependent induction of defective interfering particles of vesicular stomatitis virus, *J. Virol.* **25:**202.

Katz, E., and Goldblum, N., 1968, Establishment, steady state, and cure of a chronic infection of LLC cells with West Nile virus, *Arch. Gesamte Virusforsch.* **25**:69.

Kliks, S. C., and Halstead, S. B., 1980, An explanation for enhanced virus production in chick embryo cells, *Nature (London)* **285**:504.

Kos, K. A., Osborne, B. A., and Goldsby, R. A., 1975, Inhibition of Group B arbovirus antigen production and replication in cells enucleated with cytochalasin B, *J. Virol.* **15**:913.

Kos, K. A., Goldsby, R. A., and Top, F. H., Jr. 1976, Differential inhibition of flavivirus and alphavirus replication by cordycepin, *Abstr. Annu. Meet. Am. Soc. Microbiol.*, p. 244.

Kuno, G., 1982, Persistent infection of a nonvector mosquito cell line (TRA-171) with dengue viruses, *Intervirology* **18**:45.

Lazzarini, R. A., Keene, J. D., and Schubert, M., 1981, The origins of defective interfering particles of the negative-strand RNA viruses, *Cell* **26**:145.

Leary, K. R., and Blair, C. D., 1980, Sequential events in the morphogenesis of Japanese encephalitis virus, *J. Ultrastruct. Res.* **72**:123.

Leary, K. R., and Blair, C. D., 1983, Japanese encephalitis virus replication: Studies on host cell nuclear involvement, *Exp. Mol. Pathol.* **38**:264.

Lee, C., and Schloemer, R. H., 1981, Identification of the antiviral factor in culture medium of mosquito cells persistently infected with Banzi virus, *Virology* **110**:445.

Liprandi, F., and Walder, R., 1983, Replication of virulent and attenuated strains of yellow fever virus in human monocytes and macrophage-like cells (U-937), *Arch. Virol.* **76**:51.

Lubiniecki, A. S., and Henry, C. J., 1974, Autoradiographic localization of RNA synthesis directed by arboviruses in the cytoplasm of infected BHK-21 cells, *Proc. Soc. Exp. Biol. Med.* **145**:1165.

Luukkonen, A., von Bonsdorff, C.-H., and Renkonen, O., 1977, Characterization of Semliki Forest virus grown in mosquito cells: Comparison with the virus from hamster cells, *Virology* **78**:331.

Maguire, T., and Miles, J. A. R., 1965, The arbovirus carrier state in tissue cultures, *Arch. Virusforsch.* **15**:457.

Malenko, G. V., Fokina, G. I., Levina, L. S., Mamonenko, L. E., Rzhakhova, O. E., Pogodina, V. V., and Frolova, M. P., 1982, Persistence of tick-borne encephalitis virus in monkeys. IV. Virus localization after intracerebral inoculation, *Acta Virol.* **26**:362.

Mathews, J. H., and Vorndam, A. V., 1982, Interferon-mediated persistent infection of Saint Louis encephalitis virus in a reptilian cell line, *J. Gen. Virol.* **61**:177.

Mathur, A., Arora, K. L., and Chaturvedi, U. C., 1982, Transplacental Japanese encephalitis virus (JEV) infection in mice during consecutive pregnancies, *J. Gen. Virol.* **59**:213.

Matsumura, T., Stoller, V., and Schlesinger, R. W., 1971, Studies on the nature of dengue viruses. V. Structure and development of dengue virus in Vero cells, *Virology* **46**:344.

Matsumura, T., Stollar, V., and Schlesinger, R. W., 1972, Effect of ionic strength on the release of dengue virus from Vero cells, *J. Gen. Virol.* **17**:343.

Mayer, V., 1963, Two variants of tick-borne encephalitis showing different plaque morphology, *Virology* **20**:372.

Monath, T. P. C., 1971, Neutralizing antibody responses in the major immunoglobulin classes to yellow fever 17D vaccination of humans, *Am. J. Epidemiol.* **93**:122.

Monath, T. (ed.), 1980, *St. Louis Encephalitis*, American Public Health Association, Washington, D. C.

Monath, T. P., Cropp, C. B., and Harrison, B. S., 1983, Mode of entry of a neurotropic arbovirus into the central nervous system, *Lab. Invest.* **48**:399.

Monckton, R. P., and Westaway, E. G., 1982, Restricted translation of the genome of the flavivirus Kunjin *in vitro*, *J. Gen. Virol.* **63**:227.

Murphy, F. A., 1980, Togavirus morphology and morphogenesis, in: *The Togaviruses* (R. W. Schlesinger, ed.), pp. 241–316, Academic Press, New York.

Murphy, F. A., Harrison, A. K., Gary, G. W., Whitfield, S. G., and Forrester, F. T., 1968, St. Louis encephalitis virus infection of mice: Electron microscopic studies of central nervous system, *Lab. Invest.* **19**:652.

Mussgay, M., Enzmann, P.-J., Horzinek, M. C., and Weiland, E., 1975, Growth cycle of arboviruses in vertebrate and arthropod cells, *Prog. Med. Virol.* **19**:257.

Naeve, C. W., and Trent, D. W., 1977, Identification of Saint Louis encephalitis virus mRNA, Abstracts of the Annual Meeting of the American Society of Microbiology, p. 287, Washington, D. C.

Naeve, C. W., and Trent, D. W., 1978, Identification of Saint Louis encephalitis virus mRNA, *J. Virol.* **25**:535.

Ng, M. L., and Westaway, E. G., 1980, Establishment of persistent infections by flavivirus in *Aedes albopictus* cells, in: *Invertebrate Systems in Vitro* (E. Kurstak, K. Maramorosch, and A. Dubendorfer, eds.), pp. 389–402, Elsevier, Amsterdam.

Ng, M. L., and Westaway, E. G., 1983, Phenotypic changes in the flavivirus Kunjin after a single cycle of growth in an *Aedes albopictus* cell line, *J. Gen. Virol.* **64**:1715.

Ng, M. L., Pedersen, J. S., Toh, B. H., and Westaway, E. G., 1983, Immunofluorescent sites in Vero cells infected with the flavivirus Kunjin, *Arch. Virol.* **78**:177.

Ogawa, M., Okubo, H., Tsuji, Y., Yasui, N., and Someda, K., 1973, Chronic progressive encephalitis occurring 13 years after Russian spring–summer-encephalitis, *J. Neurol. Sci.* **19**:363.

Pang, T., 1983, Immunoepidemiology helps to unravel the mysteries of dengue haemorrhagic fever, *Immunol. Today* **4**:334.

Paul, S. D., 1966, Some biological properties of two variants of Kyasanur Forest disease virus, *Indian J. Med. Res.* **54**:419.

Peiris, J. S. M., and Porterfield, J. S., 1979, Antibody-mediated enhancement of flavivirus replication in macrophage cell lines, *Nature (London)* **282**:509.

Peiris, J. S. M., and Porterfield, J. S., 1981, Antibody-dependent enhancement of plaque formation on cell lines of macrophage origin: A sensitive assay for antiviral antibody, *J. Gen. Virol.* **57**:119.

Peiris, J. S. M., and Porterfield, J. S., 1982, Antibody-dependent plaque enhancement: Its antigenic specificity in relation to Togaviridae, *J. Gen. Virol.* **58**:291.

Peiris, J. S. M., Gordon, S., Unkeless, J. C., and Porterfield, J. S., 1981a, Monoclonal anti-Fc receptor IgG blocks antibody enchancement of viral replication in macrophages, *Nature (London)* **289**:189.

Peiris, J. S. M., Gordon, S., Porterfield, J. S., and Unkeless, J. C., 1981b, Antibody mediated enhancement of virus replication in macrophage-like cell lines: Its dependence on the Fc receptor in: *Heterogeneity of the Mononuclear Phagocytes: Proceedings of an International Workshop* (O. Forster and M. Landy, eds.), pp. 469–476, Academic Press, New York.

Peiris, J. S. M., Porterfield, J. S., and Roehrig, J. T., 1982, Monoclonal antibodies against the flavivirus West Nile, *J. Gen. Virol.* **58**:283.

Pfefferkorn, E. R., and Shapiro, D., 1974, Reproduction of small and intermediate RNA viruses, in: *Comprehensive Virology 2* (H. Fraenkel-Conrat and R. R. Wagner, eds.), pp. 171–230, Plenum Press, New York.

Phillpotts, R. J., Stephenson, J. R., and Porterfield, J. S., 1985, Antibody dependent enhancement of tick-borne encephalitis virus infectivity, *J. Gen. Virol.* **66**:1831.

Porterfield, J. S., Casals, J., Chumakov, M. P., Gaidamovich, S. Ya, Hannoun, C., Holmes, I. H., Horzinek, M. C., Mussgay, M., Oker-Blom, N., Russell, P. K., and Trent, D. W., 1978, Togaviruses, *Intervirology* **9**:129.

Price, W. H., 1966, Chronic disease and virus persistence in mice inoculated with Kyasanur Forest disease virus, *Virology* **29**:679.

Price, W. H., Lee, R. W., Gunkel, W. F., and O'Leary, W., 1963, The virulence of West Nile virus and TP21 virus and their application to a group B arbovirus vaccine, *Am. J. Trop. Med. Hyg.* **10**:403.

Qureshi, A. A., and Trent, D. W., 1972, Saint Louis encephalitis viral ribonucleic acid replication complex, *J. Virol.* **9**:565.

Qureshi, A. A., and Trent, D. W., 1973a, Group B arbovirus structural and nonstructural antigens. I. Serological identification of Saint Louis encephalitis virus soluble antigen, *Infect. Immun.* **7**:242.

Qureshi, A. A., and Trent, D. W., 1973b, Group B arbovirus structural and nonstructural antigens. II. Purification of Saint Louis encephalitis virus intracellular antigens, *Infect. Immun.* **8**:985.

Qureshi, A. A., and Trent, D. W., 1973c, Group B arbovirus structural and nonstructural antigens. III. Serological specificity of solubilized intracellular viral proteins, *Infect. Immun.* **8**:933.

Rice, C. H., Lenches, E. M., Dalgarno, L., Eddy, S. R., Shin, S. J., Sheets, R. L., Trent, D. W., and Strauss, J. H., 1985, Nucleotide sequence of yellow fever virus: Implications for flavivirus gene expression and evolution, *Science* **229**:726.

Rochrig, J. T., Mathews, J. H., and Trent, D. W., 1983, Identification of epitopes on the E glycoprotein of Saint Louis encephalitis virus using monoclonal antibodies, *Virology* **128**:118.

Russell, P. K., Brandt, W. E., and Dalrymple, J. M., 1980, Chemical and antigenic structure of flaviviruses, in: *The Togaviruses* (R. W. Schlesinger, ed.), pp. 503–529, Academic Press, New York.

Sabin, A. B., 1952a, Nature of inherited resistance to viruses affecting the nervous system, *Proc. Natl. Acad. Sci. U.S.A.* **38**:540.

Sabin, A. B., 1952b, Genetic, hormonal and age factors in natural resistance to certain viruses, *Ann. N. Y. Acad. Sci.* **54**:936.

Sangkawibha, N., Rojanasuphot, S., Ahandrik, S., Viriyapongse, S., Jatanasen, S., Salitul, V., Phanthumachinda, B., and Halstead, S. B., 1984, Risk factors in dengue shock syndrome: A prospective epidemiologic study in Rayong, Thailand, *Am. J. Epidemiol.* **120**:653.

Sawyer, W. A., and Lloyd, W., 1931, The use of mice in tests of immunity against yellow fever, *J. Exp. Med.* **54**:533.

Schlesinger, J. J., and Brandriss, M. W., 1983, 17D yellow fever virus infection of P388D1 cells mediated by monoclonal antibodies: Properties of the macrophage Fc receptor, *J. Gen. Virol.* **64**:1255.

Schlesinger, J. J., Brandriss, M. W., and Monath, T. P., 1983, Monoclonal antibodies distinguish between wild and vaccine strains of yellow fever virus by neutralization, hemagglutination inhibition, and immune precipitation of the virus envelope protein, *Virology* **125**:8.

Schlesinger, J. J., Brandriss, M. W., and Walsh, E. E., 1985, Protection against 17D yellow fever encephalitis in mice by passive transfer of monoclonal antibodies to the nonstructural glycoprotein gp48 and by active immunization with gp48, *J. Immunol.* **135**:2805.

Schlesinger, R. W., 1977, Dengue viruses in: *Virology Monographs*, No. 16, pp. 1–132, Springer-Verlag, New York.

Schmaljohn, C., and Blair, C. D., 1977, Persistent infection of cultured mammalian cells by Japanese encephalitis virus, *J. Virol.* **24**:580.

Schmaljohn, C. S., and Blair, C. D., 1979, Clonal analysis of mammalian cell cultures persistently infected with Japanese encephalitis virus, *J. Virol.* **31**:816.

Schubert, M., Keene, J. D., and Lazzarini, R. A., 1979, A specific internal RNA polymerase recognition site of VSV RNA is involved in the generation of DI particles, *Cell* **18**:749.

Schulze, I. T., and Schlesinger, R. W., 1963, Plaque assay of dengue and other group B arthropod-borne viruses under methyl cellulose overlay media, *Virology* **19**:40.

Shapiro, D., Brandt, W. E., Cardiff, R. D., and Russell, P. K., 1971, The proteins of Japanese encephalitis virus, *Virology* **44**:108.

Shapiro, D., Brandt, W. E., and Russell, P. K., 1972a, Change involving a viral membrane glycoprotein during morphogenesis of group B arboviruses, *Virology* **50**:906.

Shapiro, D., Kos, K., Brandt, W. E., and Russell, P. K., 1972b, Membrane-bound proteins of Japanese encephalitis virus-infected chick embryo cells, *Virology* **48**:360.

Shapiro, D., Trent, D., Brandt, W. E., and Russell, P. K., 1972c, Comparison of the virion polypeptides of Group B arboviruses, *Infect. Immun.* **6**:206.

Shapiro, D., Kos, K. A., and Russell, P. K., 1973, Japanese encephalitis virus glycoproteins, *Virology* **56**:88.

Shimazu, Y., Aoki, H., and Hotta, S., 1966, Research on dengue in tissue culture. II. Further observations on virus-tissue culture affinity, *Kobe J. Med. Sci.* **12**:189.

Shope, R. E., 1980, Medical significance of togaviruses, in: *The Togaviruses* (R. W. Schlesinger, ed.), pp. 47–82, Academic Press, New York.

Sinarachatanant, P., and Olson, L. C., 1973, Replication of dengue virus type 2 in *Aedes albopictus* cell culture, *J. Virol.* **12**:275.

Slavin, H. B., 1943, Persistence of the virus of St. Louis encephalitis in the central nervous system of mice for over five months, *J. Bacteriol.* **46**:113.

Smith, A. L., 1981, Genetic resistance to lethal flavivirus encephalitis: Effects of host age and immune status and route of inoculation on production of interfering Banzi virus in vivo, *Am. J. Trop. Med. Hyg.* **30**:1319.

Smith, G. W., and Wright, P. J., 1985, Synthesis of proteins and glycoproteins in dengue type 2 virus-infected Vero and *Aedes albopictus* cells, *J. Gen. Virol.* **66**:559.

Sriurairatna, S., Bhamarapravati, N., and Phalavadhtana, O., 1973, Dengue virus infection of mice: Morphology and morphogenesis of dengue type 2 virus in suckling mouse neurons, *Infect. Immun.* **8**:1017.

Stephenson, J. R., Lee, J. M., and Wilton-Smith, P. D., 1984, Antigenic variation among members of the tick-borne encephalitis complex, *J. Gen. Virol.* **65**:81.

Stohlman, S. A., Wisseman, C. L., Jr., Eylar, O. R., and Silverman, D. J., 1975, Dengue virus induced modifications of host cell membranes, *J. Virol.* **16**:1017.

Stollar, V., 1980, Togaviruses in cultured arthropod cells, in: *The Togaviruses* (R. W. Schlesinger, ed.), pp. 583–621, Academic Press, New York.

Stollar, V., Schlesinger, R. W., and Stevens, T. M., 1967, Studies on the nature of dengue viruses. III. RNA synthesis in cells infected with type 2 dengue virus, *Virology* **33**:650.

Stollar, V., Stollar, B. D., Koo, R., Harrap, K. A., and Schlesinger, R. W., 1976, Sialic acid contents of Sindbis virus from vertebrate and mosquito cells, *Virology* **69**:104.

Strauss, E. G., and Strauss, J. H., 1983, Replication strategies of the single stranded RNA viruses of eukaryotes, *Curr. Top. Microbiol. Immunol.* **105**:1.

Strauss, J. H., and Strauss, E. G., 1977, Togaviruses, in: *The Molecular Biology of Animal Viruses* (D. P. Nayak, ed.), pp. 111–166, Marcel Dekker, New York.

Svitkin, Y. V., Ugarova, T. Y., Chernovskaya, T. V., Lyapustin, V. N., Lashkevich, V. A., and Agol, V. I., 1981, Translation of tick-borne encephalitis virus (flavivirus) genome in vitro: Synthesis of two structural polypeptides, *Virology* **110**:26.

Takeda, H., Oya, A., Hashimoto, K., Yasuda, T., and Yamada, M.-A., 1978, Association of virus specific replicative ribonucleic acid with nuclear membrane in chick embryo cells infected with Japanese encephalitis virus, *J. Gen. Virol.* **38**:281.

Takegami, T., Miyamoto, H., Nakamura, H., and Yasui, K., 1982, Biological activities of the structural proteins of Japanese encephalitis virus, *Acta Virol.* **26**:312.

Takehara, M., 1971, Comparative studies on nucleic acid synthesis and virus-induced RNA polymerase activity in mammalian cells infected with certain arboviruses, *Arch. Gesamte. Virusforsch.* **34**:266.

Takehara, M., 1972, Inhibition of nuclear protein synthesis in BHK-21 cells infected with arboviruses, *Arch. Gesamte Virusforsch.* **39**:163.

Tarr, G. C., and Lubiniecki, A. S., 1976, Chemically-induced temperature sensitive mutants of dengue virus type 2. I. Isolation and partial characterization, *Arch. Virol.* **50**:223.

Theiler, M., 1951, The virus, in: *Yellow Fever* (G. K. Strode, ed.), pp. 39–136, McGraw-Hill, New York.

Trent, D. W., 1977, Antigenic characterization of flavivirus structural proteins separated by isoelectric focusing, *J. Virol.* **22**:608.

Trent, D. W., and Naeve, C. W., 1980, Biochemistry and replication, in: *St. Louis Encephalitis* (T. Monath, ed.), pp. 159–199, American Public Health Association, Washington, D. C.

Trent, D. W., and Qureshi, A. A., 1971, Structural and nonstructural proteins of Saint Louis encephalitis virus, *J. Virol.* **7**:379.

Trent, D. W., Swenson, C. C., and Qureshi, A. A., 1969, Synthesis of Saint Louis encephalitis virus ribonucleic acid in BHK-21/13 cells, *J. Virol.* **3**:385.

Virus Research Center, Poona, 1963, Survival of KFD virus in mice surviving paralysis, Annual Report, p. 109.

Von Magnus, P., 1954, Incomplete forms of influenza virus, *Adv. Virus Res.* **2**:59.

Webster, L. T., 1933, Inherited and acquired factors in resistance to infection. I. Development of resistant and susceptible lines of mice through selective breeding, *J. Exp. Med.* **57**:793.

Webster, L. T., and Clow, A. D., 1936, The limited neurotropic character of the encephalitis virus (St. Louis type) in susceptible mice, *J. Exp. Med.* **63**:433.

Webster, L. T., and Johnson, M. S., 1941, Comparative virulence of St. Louis encephalitis virus cultured with brain tissue from innately susceptible and innately resistant mice, *J. Exp. Med.* **74**:489.

Wengler, G., and Wengler, G., 1981, Terminal sequences of the genome and replication from RNA of the flavivirus West Nile virus: Absence of poly(A) and possible role in RNA replication, *Virology* **113**:544.

Wengler, G., Wengler, G., and Gross, H. J., 1978, Studies on virus-specified nucleic acids synthesized in vertebrate and mosquito cells infected with flaviviruses, *Virology* **89**:423.

Wengler, G., Beato, M., and Wengler, G., 1979, *In vitro* translation of 42S virus-specific RNA from cells infected with the flavivirus, West Nile virus, *Virology* **96**:516.

Westaway, E. G., 1973, Proteins specified by group B togaviruses in mammalian cells during productive infections, *Virology* **51**:454.

Westaway, E. G., 1977, Strategy of the flavivirus genome: Evidence for multiple internal initiation of translation of proteins specified by Kunjin virus in mammalian cells, *Virology* **80**:320.

Westaway, E. G., 1980, Replication of flaviviruses, in: *The Togaviruses* (R. W. Schlesinger, ed.), pp. 531–581, Academic Press, New York.

Westaway, E. G., and Ng, N. L., 1980, Replication of flaviviruses: Separation of translation sites of Kunjin virus proteins and of cells proteins, *Virology* **106**:107.

Westaway, E. G., and Reedman, B. M., 1969, Proteins of the group B arbovirus Kunjin, *J. Virol* **4**:688.

Westaway, E. G., and Shew, M., 1977, Proteins and glycoproteins specified by the flavivirus Kunjin, *Virology* **80**:309.

Westaway, E. G., Schlesinger, R. W., Dalrymple, J. M., and Trent, D. W., 1980, Nomenclature of flavivirus specified proteins, *Intervirology* **14**:114.

Westaway, E. G., Speight, G., and Endo, L., 1984, Gene order of translation of the flavivirus Kunjin: Further evidence of internal initiation *in vivo*, *Virus Res.* **1**:333.

Westaway, E. G., Brinton, M. A., Gaidamovich, S. Ya., Horzinek, M. C., Igarashi, A., Kaariainen, L., Lvov, D. K., Porterfield, J. S., Russell, P. K., and Trent, D. W., 1986, Flaviviridae, *Intervirology* **24**:183.

White, J., Kartenbeck, J., and Helenius, A., 1980, Fusion of Semliki Forest virus with the plasma membrane can be induced by low pH, *J. Cell Biol.* **87**:264.

White, J., Matlin, K., and Helenius, A., 1981, Cell fusion by Semliki Forest, influenza, and vesicular stomatitis viruses, *J. Cell Biol.* **89**:674.

Whitfield, S. G., Murphy, F. A., and Sudia, W. D., 1973, St. Louis encephalitis virus: An ultrastructural study of infection in a mosquito vector, *Virology* **56**:70.

Wright, P. J., 1982, Envelope protein of the flavivirus Kunjin is apparently not glycosylated, *J. Gen. Virol.* **59**:29.

Wright, P. J., and Warr, H. M., 1985, Peptide mapping of envelope-related glycoproteins specified by the flaviviruses Kunjin and West Nile, *J. Gen. Virol.* **66**:597.

Wright, P. J., and Westaway, E. G., 1977, Comparisons of the peptide maps of Kunjin virus proteins smaller than the envelope protein, *J. Virol.* **24**:662.

Wright, P. J., Warr, H. M., and Westaway, E. G., 1983, Comparisons by peptide mapping of proteins specified by Kunjin, West Nile and Murray Valley encephalitis viruses, *Aust. J. Exp. Biol. Med. Sci.* **61**:641.

Yazuzumi, G., Tsubo, I., Sugihara, R., and Nakai, Y., 1964, Analysis of the development of Japanese B encephalitis (JBE) virus. I. Electron microscope studies of microglia infected with JBE virus, *J. Ultrastruct. Res.* **11**:213.

Zebovitz, E., Leong, J. K. L., and Doughty, S. C., 1972, Japanese encephalitis virus replication: A procedure for the selective isolation and characterization of viral RNA species, *Arch. Gesamte Virusforsch.* **38**:319.

Zebovitz, E., Leong, J. K. L., and Doughty, S. C., 1974, Involvement of host cell nuclear envelope membranes in the replication of Japanese encephalitis virus, *Infect. Immun.* **10**:214.

Zinder, N. (ed.), 1975, *RNA Phages*, Cold Spring Harbor Press, Cold Spring Harbor, New York.

Zlotnik, I., Grant, D. P., and Carter, J. B., 1976, Experimental infection of monkeys with viruses of the tick-borne encephalitis complex: Degenerative cerebellar lesions following inapparent forms of the disease or recovery from clinical encephalitis, *Br. J. Exp. Pathol.* **57**:200.

CHAPTER 12

Pathobiology of the Flaviviruses

Thomas P. Monath

I. INTRODUCTION

Among arthropod-borne and related viruses, the Flaviviridae are medically the most important group and biologically one of the most intriguing. Elucidation of perplexingly complex virus- and host-specified factors that underlie virulence and pathogenesis has lagged behind other areas of virology, and the available information is largely descriptive. Ultimately, flavivirus biology and pathogenesis will be understood in terms of viral gene expression, virus receptor–host cell membrane interactions, biochemical alterations in host cells, physiological responses, and immune and nonimmune mechanisms that control virus replication and virus spread, subjects to which other chapters in this book are devoted. There will remain, however, a need to understand and synthesize this information with observations relating to infection at the level of the intact organism (virus, vector, and host) and of populations of organisms in nature. It is at these levels that the phenomena that require explanation first present themselves.

Central themes in flavivirus biology are the variation in virulence of virus strains and susceptibility or resistance of vector and host individuals and populations. The remarkable evolutionary diversity and genetic plasticity of the flaviviruses provides both natural and laboratory-derived mutants with differing virulence properties for molecular and biological studies. Natural variation among flavivirus strains in virulence for clinical hosts and infectivity for vectors and reservoir hosts is more

THOMAS P. MONATH • Division of Vector-Borne Viral Diseases, Center for Infectious Diseases, Centers for Disease Control, Public Health Service, U.S. Department of Health and Human Services, Fort Collins, Colorado 80522.

important than is generally appreciated and is discussed further below. The contribution of host genes to the control of flavivirus infections in laboratory models is reviewed in Chapter 11; mention will be made below of evidence for genetic control in clinical and biological hosts.

Some principles that underlie study of viral pathogenesis include: (1) definition of the target cells and tissues for replication and injury; (2) separation of injury due to primary virus-induced cytopathology from secondary immunological or physiological mechanisms; and (3) elucidation of the mechanisms tht control viral spread (R. W. Schlesinger, 1980). These principles will be used as a framework for this review of flavivirus pathobiology, while recognizing, as pointed out by Murphy (1979), that an understanding of the mechanisms involved ". . . is always hindered by an inability to separate the virus–host interactions in each directly and indirectly affected organ and tissue."

This chapter will focus on vertebrate hosts in their role as diseased organisms. The interactions of viruses and arthropod vectors, while interesting and important in flaviviral biology, are beyond the scope of this review.

II. MEDICAL IMPORTANCE OF THE FLAVIVIRUSES

Of the 61 currently recognized viruses, 28 (46%) have been associated with disease in humans (Table I). Three viruses, dengue, yellow fever, and Japanese encephalitis (JE), are prevalent enough to engender global or panregional public health concern. Classic dengue, a self-limited disease characterized by fever, rash, headache, and arthralgia, occurs in epidemic form in the Caribbean basin, Southeast Asia, East Africa, and Oceania, annually affecting hundreds of thousands to millions of persons. Dengue hemorrhagic fever (DHF), first recognized as a nosological entity in 1954, is an immunopathological disease characterized by hemoconcentration and a hemorrhagic diathesis and, in its extreme form [dengue shock syndrome (DSS)], by a protein-losing shock state and a 5–10% case fatality rate. Some 10,000–20,000 cases of DHF/DSS occur annually in Southeast Asia; in 1981, the disease appeared for the first time in epidemic form in the western hemisphere (Cuba; 10,000 cases).

Yellow fever remains an important health problem in tropical America and Africa. In the past 20 years, recurring outbreaks in West Africa have affected nearly 250,000 persons. The disease is characterized by fever, hepatic, renal, and myocardial dysfunction, and hemorrhage and has a case fatality rate of approximately 20%.

JE occurs in endemic and epidemic form over a wide area of Asia. Morbidity and mortality estimates are inaccurate and incomplete, but it is certain that tens of thousands of cases occur annually. In Thailand, for example, the annual incidence in childhood populations is 20–50 cases per 100,000 (C. H. Hoke, S. Jatanesen, and D. S. Burke, unpublished). The

TABLE I. Flaviviruses Associated with Disease Syndromes in Humans, Estimated Total Number of Cases, and Vectors Involved in Transmission

Predominant syndrome	Human disease prevalence (total cases reported)[a]			
	Very high ($>10^5$)	High (10^4–10^5)	Moderate (10^2–10^4)	Low ($<10^2$)
Febrile illness	Dengue (M)	West Nile (M)	—	Banzi (M), Bouboui (M), Bussuquara (M), Ilheus (M), Sepik (M), Spondweni (M), Wesselsbron (M), Zika (M), Rio Bravo (N), Tyuleiny (N)
Encephalitis	Japanese encephalitis (M)	Tick-borne encephalitis (T)	St. Louis encephalitis (M), Rocio (M), Murray Valley encephalitis (M), Kyasanur Forest disease (T)	West Nile, Powassan (T), Omsk hemorrhagic fever, Kunjin (M), louping ill (T), Ilheus, Apoi (N), Negishi (T), dengue, Rio Bravo, ?Modoc (N)
Hemorrhagic fever	Dengue Yellow fever (M)	Kyasanur Forest disese	Omsk hemorrhagic fever (T)	—

^{a} Mode of transmission: (M) mosquito-borne; (T) tick-borne; (N) no vector known or not vector-borne.

TABLE II. Flaviviruses Associated with Disease in Animals of Economic
Importance

Virus	Species affected	Disease	Distribution	Incidence
Japanese encephalitis	Horse	Encephalitis	Asia	Low-moderate, regionally high
	Pig	Stillbirth, decreased spermatogenesis	Asia	High
Louping ill	Principally sheep; also cows, pigs, horses, captive red grouse, and deer	Encephalitis	Scotland, northern Ireland	Moderate
Wesselsbron	Sheep, occasionally cows	Abortion, hepatitis/ hemorrhage, congenital malformation	Africa	Low-moderate
Israel turkey meningo- encephalitis	Turkeys	Encephalitis	Israel	Outbreaks in 1950s
Omsk hemorrhagic fever	Wild muskrats	Encephalitis	Western Siberia	Intermittent epizootics
West Nile	Horses	Encephalitis	Africa, Europe	Rare
Kunjin	Horses	Encephalitis[a]	Australia	Rare
Murray Valley encephalitis	Horses	?Encephalitis[b]	Australia	?Rare

[a] Badman et al. (1984).
[b] Association on serological grounds only; etiological relationship to MVE not established.

case fatality rate is approximately 20%, and a high proportion of survivors
are left with permanent neuropsychiatric impairment.

Three other flaviviral infections, West Nile (WN) fever, tick-borne
encephalitis (TBE), and Kyasanur Forest disease, are less prevalent but
assume considerable importance in certain regions. WN virus is widely
distributed in Africa, the Middle East, Europe, and Asia. In hyperendemic
areas, relatively mild infection occurs early in childhood, and adult populations are largely immune, but in regions of less intense virus activity,
epidemics occur, affecting all age groups (McIntosh et al., 1976). The
severity of disease increases with age; adolescents and young adults experience denguelike disease, whereas the elderly sometimes develop meningoencephalitis (Marburg et al., 1956).

TBE is a widespread endemic disease in eastern Europe, the U.S.S.R.,
and Scandinavia, causing hundreds to thousands of cases annually.
Human infection is acquired by tick bite or by consumption of unpas-
teurized goat or sheep milk or cheese. Case fatality rates vary from 1 to
30%; the disease in the far eastern regions of the U.S.S.R. is more severe
and residual neurological damage more frequent than in Europe.

FIGURE 1. Predominant disease patterns associated with flaviviral infections, with examples of individual viruses. (YF) Yellow fever; (WESS) Wesselsbron; (KFD) Kyasanur Forest disease; (OMSK) hemorrhagic fever; (JE) Japanese encephalitis; (SLE) St. Louis encephalitis; (TBE) tick-borne encephalitis; (BUSS) Bussuquara; (DEN) dengue; (WN) West Nile; (SPO) Spondweni; (BAN) Banzi.

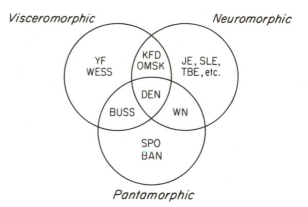

Kyasanur Forest disease, first discovered in 1957, affects human populations in several districts of western and southwestern India, where hundreds of cases occur annually. The largest epidemic occurred in South Kanara District in 1983–1984, with 1573 cases and 191 deaths. The disease is characterized by fever, headache, myalgia, lymphadenopathy, cough, and hemorrhage; a mild form of meningoencephalitis is a frequent complication. The case fatality rate is 1–10%.

Three other viruses listed in Table I, St. Louis encephalitis (SLE), Murray Valley encephalitis, and Rocio encephalitis, occur as intermittently epidemic diseases in North America, Australia, and Brazil, respectively. Omsk hemorrhagic fever, which caused an epidemic of human disease in the late 1940s, now occurs as an endemic affliction of hunters and trappers in western Siberia (Casals et al., 1970).

Eight flaviviruses have been reported to cause disease in domesticated or wild animals of economic importance (Table II). The most important of these is JE, which causes epizootic encephalitis in horses and stillbirth/abortion and decreased fecundity in swine in many areas of Asia. Wesselsbron virus is associated with abortion, congenital malformation, and severe neonatal disease in lambs in southern Africa. Louping ill virus produces an encephalitic illness in sheep in parts of the British Isles.

As shown in Table I, flavivirus infections are expressed in three main disease patterns: nondescript febrile illness (pantamorphic infections), encephalitis (neuromorphic infections), and hemorrhagic fever (visceromorphic infections). Although there is considerable overlap among these disease patterns (Fig. 1), they do reflect differing tropisms for critical target organs, virulence, and evolutionary direction.

III. PATHOBIOLOGICAL SIGNIFICANCE OF NATURAL VIRUS VARIATION AND HETEROGENEITY

Heterogeneity of virions contained in flavivirus strains isolated from natural hosts and vectors with respect to phenotypic markers, including neurovirulence, has been repeatedly demonstrated. Eckels *et al.* (1976) found a mixture of large and small plaques in dengue 2 virus isolated from a human and showed that the small-plaque marker correlated with low mouse virulence and temperature sensitivity, as well as attenuation for nonhuman primates (Harrison *et al.*, 1977). JE virus strains isolated from the brains of human patients in Thailand were found to differ significantly in their T_1-resistant RNA oligonucleotide maps from concurrently isolated pig and mosquito strains, suggesting selective replication of a neurovirulent subpopulation in the brain (Burke *et al.*, 1985a). This observation may be analogous to studies with Semliki Forest virus in which a virulent strain replicated to higher titer in brain organ cultures, whereas growth of both virulent and attenuated strains was similar in extraneural tissues (Fleming, 1977).

Wild dengue virus strains of different origin have not been compared with respect to their virulence characteristics, in part because of the lack of suitable or practical laboratory models. Biological variation can be inferred, however, from epidemiological observations. In Thailand, dengue hemorrhagic fever with shock syndrome (DHF/DSS) occurs almost exclusively in persons with secondary infection with a heterologous serotype; the risk of developing DHF/DSS following secondary infection with dengue type 2 is significantly higher than for other serotypes (Sangkawibha *et al.*, 1984). In contrast, in Indonesia, secondary infections with dengue type 3 are the most frequent cause of DHF/DSS (Gubler *et al.*, 1979), suggesting that virulence differences exist between dengue 2 and dengue 3 strains in the two countries. In addition, primary infection with dengue 2 virus was associated with DHF/DSS and deaths during an outbreak in Nieu island in 1972 (Barnes and Rosen, 1974), indicating that some strains of dengue possess unusual potential to cause hemorrhagic disease.

In a recent study, Halstead *et al.* (1984) compared antibody-dependent infection enhancement of seven dengue 2 strains in an Fc-receptor-bearing macrophage cell line. Dengue specific- and group-reactive monoclonal antibodies were used in the analysis, and enhancement of dengue replication was associated with epitopes of both specificities. The virus strains were heterogeneous with respect to the distribution of enchancement epitopes. Although variation in this group of viruses did not correlate with disease severity or geographic origin, the study provides a basis for testing and predicting immune enchancement and disease severity in sequentially infecting pairs of heterologous dengue viruses.

JE strains were found to vary in peripheral virulence for 3-week-old mice, with differences between intracerebral and subcutaneous LD_{50} ranging from 1.0 to 6.8 (Huang, 1957, 1982). Peripheral virulence in mice correlated with viremia level and duration and with thermostability of the virus. Virus strains isolated from humans and pigs (the principal amplifying host) were invariably neurovirulent, whereas 10% of strains from mosquitoes had low virulence. Huang (1982) speculated that the main-

TABLE III. Concordance between St. Louis Encephalitis Viremias in Nestling
House Sparrows and Neurovirulence for Mice[a]

Mouse virulence	Number of strains tested	Proportion viremic	Duration (days)	Peak titer[b]
Virulent	15	0.94	3.18	4.20
Intermediate	7	0.66	1.78	2.64
Avirulent	5	0.24	0.44	0.75

[a] After Bowen et al. (1980).
[b] Log_{10} PFU/ml.

tenance of neurovirulent strains in nature may require cycling through
biological hosts, such as young birds (ducklings), which sustain brain
infections, and that attenuated strains arise by persistent inherited (transovarial) infection of mosquitoes. The latter hypothesis is given weight
by laboratory (Stollar and Shenk, 1973; Stollar, 1980; Kuno, 1982) and
field studies (Reeves et al., 1958) with mosquito-borne togaviruses, which
indicate that viruses from persistently infected arthropods or arthropod
cells develop markers of attenuation (reduced neurovirulence, small
plaque size, temperature lability or sensitivity).

Oligonucleotide mapping of JE strains from Thailand (Burke et al.,
1985a) indicated a high level of genotypic conservation among isolates
from brains of fatal human cases compared to isolates from mosquito
vectors. A geographic difference was noted that appeared to correlate with
human pathogenicity: Virus strains recovered from pigs in southern Thailand, where intense virus transmission occurs in the virual absence of
human encephalitis, differed from strains in the north, where JE causes
annual epidemics.

Biological characteristics of SLE virus strains from different sources
and geographic localities have been described (Bowen et al., 1980; Monath
et al., 1980b; Mitchell et al., 1983). Virus strains weere classified as highly
virulent, intermediate, or attenuated for weanling mice on the basis of
their intraperitoneal/intracerebral LD_{50} ratios (Monath et al., 1980b). Virulence was associated with high viremia and replication in extraneural
tissues and earlier appearance of virus in the brains of mice. Mouse virulence correlated with clinical and histological markers of pathogenicity
for intracerebrally inoculated rhesus monkeys. The same virus strains
were used to inject a natural host of SLE, the house sparrow (Passer domesticus), as well as 3-week-old chickens (Bowen et al., 1980). A high
degree of concordance was found between level and duration of viremia
in birds and neurovirulence for mice and monkeys (Table III). Mitchell
et al. (1983) determined the infectivity of mouse-attenuated and virulent
SLE virus strains for Culex pipiens quinquefasciatus mosquitoes and
again found a correlation. All viruses replicated to high titer if the midgut
barrier was bypassed by intrathoracic inoculation, but only the mouse-

virulent strains were able to cause disseminated infections in mosquitoes fed virus by the oral route.

A number of points of epidemiological significance emerged from these studies. Virus strains isolated from birds (the usual viremic hosts for SLE) were highly virulent, whereas strains from unusual hosts, rodents and carnivores, were attenuated, possibly representing dead-end or persistent infections. All virus strains recovered during major epidemics in the eastern United States, where the virus is transmitted by *C. pipiens* or *C. nigripalpus*, were highly virulent, whereas strains from the western United States (transmitted by *C. tarsalis*) were relatively attenuated. This observation supports the hypothesis (Chamberlain, 1958) that the lower viremias of birds caused by western SLE strains favor transmission by the more highly susceptible vector species *C. tarsalis* over *C. pipiens*. Furthermore, the lower virulence for mice of these strains appears to correlate with the lower human case fatality rates of SLE in the western United States (Monath, 1980).

The biological characteristics of SLE virus strains were related to genetic analyses by Trent *et al.* (1980, 1981). On the basis of similarity among oligonucleotide fingerprints, 57 virus strains from North America could be classified into three "topotypes," representing different geographic areas (the Ohio–Mississippi Basin, Florida, and the western United States) and virus–vector relationships (*C. pipiens*, *C. nigripalpus*, and *C. tarsalis*, respectively). There was significant covariation among genetic, epidemiological, and biological markers (Table IV). Virus strains isolated within 5- to 10-year periods in a region or associated with an epidemic period spanning several years had similar RNA fingerprints, but over longer periods of time, genetic drift was evident. Monoclonal antibodies raised against a single strain of SLE virus were able to detect antigenic differences among virus strains from various geographic areas (Roehrig *et al.*, 1983).

Strains of WN virus from Nigeria were compared by Odelola and Fabiyi (1977). Two of seven strains showed similar patterns of virus growth in organs of baby mice, whereas the other strains varied. Kinetic hemagglutination-inhibition tests showed that the two biologically similar strains were also antigenically identical (Odelola and Fabiyi, 1978). Six strains of WN virus from India showed varying degrees of pathogenicity for adult mice inoculated intraperitoneally (Umrigar and Pavri, 1977).

Three strains of WN virus, all of which had undergone a large number of laboratory passages, were studied in human cancer patients (Parks *et al.*, 1958). In the case of these viruses, covariation was not observed between virulence for laboratory rodents and primates (Table V); the strain with the highest mouse neurovirulence (WN 1) was not associated with central nervous system (CNS) disease in humans and produced minimal symptomatology and histopathology in intracerebrally inoculated monkeys. These results stand in contrast to those with SLE (Monath *et al.*,

TABLE I. Flaviviruses Associated with Disease Syndromes in Humans, Estimated Total Number of Cases, and Vectors Involved in Transmission

Predominant syndrome	Human disease prevalence (total cases reported)[a]			
	Very high ($>10^5$)	High (10^4–10^5)	Moderate (10^2–10^4)	Low ($<10^2$)
Febrile illness	Dengue (M)	West Nile (M)	—	Banzi (M), Bouboui (M), Bussuquara (M), Ilheus (M), Sepik (M), Spondweni (M), Wesselsbron (M), Zika (M), Rio Bravo (N), Tyuleiny (N)
Encephalitis	Japanese encephalitis (M)	Tick-borne encephalitis (T)	St. Louis encephalitis (M),. Rocio (M), Murray Valley encephalitis (M), Kyasanur Forest disease (T)	West Nile, Powassan (T), Omsk hemorrhagic fever, Kunjin (M), louping ill (T), Ilheus, Apoi (N), Negishi (T), dengue, Rio Bravo, ?Modoc (N)
Hemorrhagic fever	Dengue Yellow fever (M)	Kyasanur Forest disese	Omsk hemorrhagic fever (T)	—

[a] Mode of transmission: (M) mosquito-borne; (T) tick-borne; (N) no vector known or not vector-borne.

TABLE V. Biological Properties of Three Laboratory Strains of West Nile Virus[a]

	Virulence/pathogenicity[b]				
Strain	Human	8- to 10-gram mice (ip/ic LD_{50})	Hamsters (ip LD_{50})	Chicken (viremia)	Monkeys (ic)
EG 101	High	0.08	5.2	High	High
WN 1	Intermediate	0.49	8.1	Intermediate	Low
Len 75	Low	0	3.0	Low	Intermediate

[a] Modified from Parks et al. (1958).
[b] High = meningoencephalitis in 9% of cases; intermediate = generalized febrile illness; low = low-grade without symptoms. (ip) Intraperitoneal; (ic) intracerebral.

1980b) and with wild strains of yellow fever virus (Fitzgeorge and Bradish, 1980), which showed correlations between virulence for mice and primates, and thus underscore the need for caution in interpretation of phenotypic differences.

Strains of TBE from the far eastern U.S.S.R. transmitted by *Ixodes persulcatus*, and from Europe, transmitted by *I. ricinus*, are distinguishable on the basis of serological tests with polyclonal (Clarke, 1960, 1964) and monoclonal antisera (Heinz *et al.*, 1982; Stephenson *et al.*, 1984), pathogenicity for sheep and monkeys (Zilber, 1960), and clinical expression in humans. A high degree of antigenic conservation and similarity in peptide maps of nonstructural proteins (Heinz and Kunz, 1982) was evident among strains representing the European subtype of TBE. The far eastern strains are characterized in humans by a predilection for infection of the brainstem and upper cervical cord, a high (up to 35%) case fatality rate, and residual damage in 30–60% of survivors. In contrast, the European subtype causes diffuse meningoencephalitis and is associated with a lower lethality (1–5%). These differences are mirrored in the response of sheep and monkeys to intracerebral inoculation (Zilber, 1960; Ilienko and Pokrovskaya, 1960). In addition, Chunikhin and Kurenkov (1979) showed that virus strains isolated from *I. persulcatus* (the vector in the Far East) produced high-titered viremias in voles (*Clethrionomys glareolus*), a common host species, whereas strains from *I. ricinus* did not; this ecological marker, which was stable on repeated vole–vole passages, may be important in the maintenance of vector associations and virus subtype identity in areas of sympatric distribution.

Biological heterogeneity of TBE strains from a single biotype has also been described. Strains isolated in the Khabarovsk region and Sakhalin Island from *I. persulcatus* could be divided on the basis of neurovirulence and viremia in mice, plaque size, and physiochemical markers into two allopatric groups representing different landscape zones (Vereta *et al.*, 1983). Similar findings were reported by Nawrocka (1975); strains from one natural focus in Poland showed reduced mouse neurovirulence compared to other strains. An unusual viral isolate (Skalica) recovered from *C. glareolus* in Czechoslovakia was found to be both nonpathogenic for mice by the peripheral route and temperature-sensitive (Gresikova and Sekeyova, 1980); these markers were retained after passage through *I. ricinus* (Gresikova and Nosek, 1983). The Skalica strain produced no detectable viremia in *C. glareolus* voles and negligible viremias in several other wild rodent species (Kozuch *et al.*, 1981). Isolations of temperature-sensitive (*ts*) arboviruses from natural sources have not been frequent, may represent adaptations to persistent infection of host or vector, and may in fact be more common than currently recognized, since virus isolation techniques employed in most laboratories select against *ts* properties.

The wide variation among flavivirus strains in virulence and antigenic markers reflects the widely accepted high mutation rates of RNA viruses (Holland *et al.*, 1982; Holland, 1984) and emphasizes the potential importance of virus evolution to biological and epidemiological phenomena. It is also important, however, to reiterate that flavivirus genomes can be classified geographically (Trent *et al.*, 1980, 1981; Repik *et al.*, 1983), that genotypes and virulence phenotypes covary (Monath *et al.*, 1980b; Trent *et al.*, 1980), and that successful or dominant virus variants appear to persist in a given area for 10–20 years or more. The requirement that flaviviruses be cyclically transmitted between vertebrate hosts mounting effective viremias and susceptible arthropod vector species appears to provide an element of evolutionary stability. Arthropods, which develop noncytopathic persistent infections, lack the antiviral defense mechanism of vertebrate hosts and may sustain active infections over prolonged periods (e.g., hibernation or transovarial passage to subsequent generations); these features appear to favor genetic drift or selection toward *ts* and avirulent variants. Such variants may occasionally be transmitted to or develop in vertebrate hosts as well, but these events will be relatively unsuccessful in sustaining virus transmission cycles. Changes in susceptibility of vertebrate hosts (usually acquired in the form of immunity) or arthropod populations (genetically determined changes in vector competence) may also exert strong pressure for selection of virus variants.

IV. PATHOGENESIS OF FLAVIVIRAL ENCEPHALITIS

Neurotropism and neuromorphic disease expression is a biological common denominator linking the flaviviruses. In clinical hosts, many viruses usually associated with generalized infections or hemorrhagic fevers (e.g., dengue, yellow fever, Kyasanur Forest disease, Omsk hemorrhagic fever) occasionally cause neurological disease. All flaviviruses are neurotropic in laboratory rodents, many biological hosts, and even arthropod vectors in which brain and ganglia are major sites of replication.

This remarkable property may reflect evolutionary conservation of viral envelope proteins involved in receptor interactions with host cells and a wide distribution in nature of cell membrane antigens or molecules, such as neutrotransmitters, which subserve virus–receptor interactions. Preliminary evidence suggests a role for neurotransmitter molecules as receptors in the case of some alphaviruses and rhabdoviruses (Tignor *et al.*, 1984), but no studies have been reported with flaviviruses. A wide natural distribution of nerve-cell membrane antigens, as shown by cross-reactivity with monoclonal antibodies, has been demonstrated not only among vertebrate species (McKenzie *et al.*, 1982), but also between mammalian and arthropod (*Drosophila*) brains (Miller and Benzer, 1983).

Invasion of the nervous system has generally been viewed as a "dead end" for the virus, inconsequential to transmission or perpetuation in nature (Mims, 1977). There is evidence, however, that neurological infections may modify mosquito behavior. *Aedes triseriatus* mosquitoes infected with LaCrosse virus (a bunyavirus) exhibit enhanced probing responses during feeding (Grimstad *et al.*, 1980), behavior that would favor horizontal transmission. Neurological infections of vertebrates are less likely to be adaptive to transmission, and most biological hosts involved in transmission cycles probably escape brain infections. The possibility that behavioral changes resulting from neurological infection of bat and rodent reservoir hosts of non-arthropod-borne flaviviruses may play a role in transmission patterns has, however, not been investigated.

A. Host Factors That Influence Pathogenesis

Flaviviruses produce a wide spectrum of infection and disease, depending on the specific virus–host pairing, virus dose, and multiple factors that influence host response (Fig. 2) (Albrecht, 1968; Nathanson, 1980). Three patterns of pathogenesis have been defined (Weiner *et al.*, 1970; Nathanson, 1980): (1) fatal encephalitis usually preceded by early viremia and extensive extraneural replication; (2) subclinical encephalitis, usually preceded by low viremia, late establishment of brain infection, and clearance with minimal destructive pathology; and (3) inapparent infection, with trace viremia, limited extraneural replication, and no neuroinvasion. The role of immune responses in limiting viral dissemination has been repeatedly demonstrated in experiments employing immunosuppressive agents, which can convert subclinical to lethal encephalitis. In the case of virus–host pairs that result in fatal encephalitis, immunosuppression may prolong survival time, but does not alter outcome, indicating a potential pathogenic role for immune responses. A fourth pattern of flaviviral CNS infection will be discussed below, namely, chronic persistent infections.

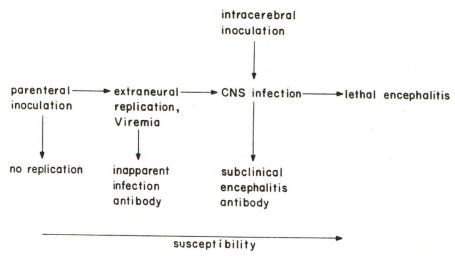

FIGURE 2. Spectrum of possible outcomes of neurotropic flavivirus infection.

1. Age

In general, neonatal or young animals are more susceptible to lethal encephalitis than older animals (O'Leary *et al.*, 1942; MacDonald, 1952; Grossberg and Scherer, 1966). This is best exemplified in the case of virus strains of low or moderate virulence in laboratory mice or hamsters. Neonatal animals inoculated by the peripheral route are highly susceptible until 3–4 weeks of age, when resistance develops (El Dadah *et al.*, 1967; Nathanson, 1980), but animals resistant by the peripheral route remain susceptible to lethal encephalitis when inoculated intracerebrally. Biological hosts also show age-related resistance to infection; nestling house sparrows infected with SLE (Bowen *et al.*, 1980) and Rocio virus (Monath *et al.*, 1978) develop viremias of longer duration and higher magnitude than adult birds.

Flaviviral infections in humans usually result in subclinical infection, the ratio of inapparent to apparent infection varying between 100 and 1000 to 1, depending on age of the host. In distinct contrast to experimentally infected mice, human beings exposed during infancy or childhood to SLE and WN viruses usually experience inapparent or mild infections, and susceptibility to encephalitis increases with advancing age, the elderly being most severely affected (Fig. 3). JE virus has a biomodal age distribution, affecting both children and old persons (Kono and Kim, 1969). The mechanisms that underlie the increasing susceptibility with age are not known. Underlying diseases (hypertension, arteriosclerotic cerebrovascular disease, diabetes mellitus, alcoholism, and chronic bronchopulmonary disease) appear to be strongly associated with severity of SLE infection (Brinker and Monath, 1980), possibly reflecting reduced effectiveness of blood–brain barrier or immunological functions.

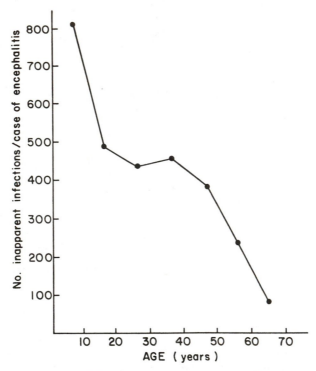

FIGURE 3. Increasing susceptibility of humans to neurological disease caused by St. Louis encephalitis virus with advancing age. Susceptibility is lowest in young children.

To the author's knowledge, the increased susceptibility to flaviviral encephalitis in the elderly has not been modeled in laboratory animals, although a single published observation suggests that it may occur (Zlotnik and Grant, 1976) (Fig. 4). Old hamsters showed increased susceptibility to Langat virus inoculated intranasally. This observation suggests the possibility of age-dependent changes at the level of the olfactory neuroepithelium.

2. Genetic Factors

Flavivirus-specific genetic resistance in mice, inherited as an autosomal dominant allele, is reviewed in Chapter 11. Although the molecular mechanisms are still unresolved, it appears from *in vitro* studies that resistance is associated with a greater production of defective interfering (DI) virus particles. Experiments with intact animals tend to support the concept that interfering virus production is involved in resistance. Early in infection of the congenic resistant mouse strain C3H/RV with Banzi virus, interfering virus appears in extraneural (lymphoreticular) tissues and is later amplified in the brain (A. Smith, 1981). Immunosuppression with cyclophosphamide, which converts an inapparent to a lethal Banzi

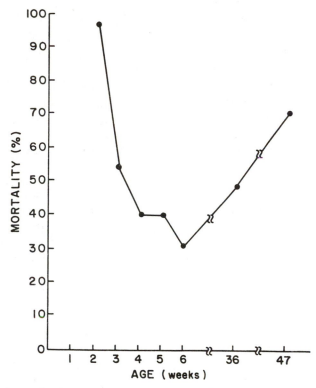

FIGURE 4. Relationship between age at inoculation and mortality in hamsters intranasally infected with Langat virus. The increased susceptibility of old hamsters is analogous to that seen in humans infected with SLE, WN, and JE viruses and may provide a model for study of the mechanisms involved. After Zlotnick and Grant (1976).

infection in RV mice (Bhatt and Jacoby, 1976), however, enhanced interfering virus production in the brain, indicating that immunological responses are essential to the expression of resistance. It is probable that in the intact RV animal, interfering virus production causes diminished extraneural viral replication and delayed neuroinvasion, allowing successful intervention of immunological responses.

As pointed out in Chapter 11, pathogenic viruses may be responsible for the selection of host resistance genes in nature and the evolution of stable host–virus relationships. The ecology of louping ill virus in Scotland exemplifies these phenomena. Woodland and forest species of birds (pheasants and capercaillie), which have had a long association with *I. ricinus* ticks and louping ill virus, have apparently developed resistance to infection, whereas moorland species (e.g., red grouse) are highly susceptible to lethal encepahlitis (Reid, 1975; Reid and Moss, 1980). The invasion of heather moorland by *I. ricinus*, and subsequent natural exposure of red grouse, is a recent development, coinciding with the development of sheep husbandry during the 19th century (Reid and Moss, 1980).

3. Other Host Factors

Andersen and Hanson (1974) showed that sexually mature female outbred mice demonstrated increased resistance to SLE virus compared to males; although a hormonal basis was suggested, neither pregnancy nor male castration altered resistance. Sex differences in susceptibility (as opposed to exposure to infected vectors) have not been reported in clinical hosts of flavivirus encephalitis.

Preexisting infections with unrelated agents have been shown to affect flavivirus pathogenesis. Several human cases of double infections with JE and herpes virus have been reported; doubly infected mice show colocalization of the two viruses, suggesting entry of JE viruses at sites of blood–brain barrier dysfunction caused by herpes virus infection (Hayashi and Arita, 1977). A similar mechanism may explain the increased susceptibility of mice to JE virus when dually infected with *Trichinella spiralis* (Cypress *et al.*, 1973; Lubiniecki *et al.*, 1974) or visceral larva migrans (Pavri *et al.*, 1975). In a human autopsy study of JE patients in India, Shakar *et al.* (1983) noted an association between JE and neurocysticercosis. On the basis of finding high immunoglobulin E (IgE) levels in acute-phase sera from patients, Pavri *et al.* (1980) and Shaikh *et al.* (1983) have postulated that chronic or intercurrent helminthic infections may predispose to clinically expressed JE infection. However, high IgE levels were present only during the acute phase of JE and declined in convalescence. This observation as well as the association of high IgE levels with several other viral diseases, including dengue, influenza, and herpes virus infections, suggests an immunological basis rather than a relationship to intercurrent parasitic infestation.

Elevated body temperature may restrict replication or increase inactivation of some flavivirus strains and convert a lethal to a sublethal infection (Cole and Wisseman, 1969). In general, virus strains that exhibit thermolability and temperature sensitivity also exhibit reduced virulence. Avian hosts of flavivirus encephalitis viruses have body temperatures in excess of 40°C and thus select against thermolabile or *ts* mutants.

Poisoning by heavy metals, including arsenicals (Gainer and Pry, 1972), lead (Thind and Singh, 1977), and cadmium (Suzuki *et al.*, 1981), potentiates flavivirus encephalitis in mice. The mechanisms involved are unknown; in the case of lead poisoning, interferon and antibody responses were suppressed and virus titers in extraneural tissues enhanced.

B. Major Aspects of Pathogenesis

1. Extraneural Infection

In natural infections with arthropod-borne flaviviruses, or following experimental intradermal or subcutaneous inoculation, virus first replicates in the inoculation site and in lymph nodes that drain the site (Albrecht, 1968). The specific cell types involved in replication of the inoculation site are unknown, but probably include smooth and striated muscle and connective tissue. The probing arthropod or needle deposits virus in both extravascular and intravascular spaces, and a brief viremia shower may result, followed within hours by release of virus from infected cells at the inoculation site and regional lymph nodes, seeding of other tissues, and appearance of a secondary viremia coincident with disseminated extraneural infection. In the case of TBE infection of sheep, virus is carried to the bloodstream by lymphatic channels (Malkova, 1960), but this pathway has not been examined in other models.

Knowledge of the main sites of flavivirus replication outside the CNS derives almost entirely from tissue titration and immunofluorescence studies on experimental animals (Albrecht, 1960, 1968; Huang and Wong, 1963; Kundin et al., 1963; Harrison et al., 1982). Principal sites are of meso-, endo-, and ectodermal origin, and include smooth and striated muscle, osteo- and chondroblasts, connective tissue, lymphoid and reticuloendothelial cells, epithelium, renal tubular epithelium, endocrine and exocrine glands, hair follicles and tooth pulp, adrenal medulla, and peripheral somatic and autonomic nerves. Viral antigen has been repeatedly demonstrated in the muscularis of arteries and arterioles and occasionally in capillary walls, but the contribution of vascular endothelium as a source of viremia (and invasion of the brain) remains uncertain.

In baby hamsters infected with SLE and Rocio viruses, pancreas and heart were the most severely affected organs. Cytopathological changes at the ultrastructural level were described by Harrison et al. (1980, 1982). In exocrine and endocrine areas of pancreas, virus particles, packed within secretory granules, are released by exocytosis. Virus production thus follows the same pathway as the cells' normal secretory products, a phenomenon also noted in mucus-secreting Bowman's gland cells in the olfactory epithelium (Monath et al., 1983 and Fig. 5). Infection of mammary glands and secretion of virus in milk of infected goats is an important mode of spread of TBE viruses (Gresikova, 1957; Woodall and Roz, 1977).

In the flavivirus–hamster model, myocardial degeneration and necrosis with productive viral infection of myocytes and endothelium of myocardial vessels were also prominent findings. Other target tissues included skeletal muscle and small intestinal lamina propria, muscularis mucosae, vascular endothelium, and parasympathetic ganglia.

Correlations are possible between these experimental infection models and the pathogenesis of infection in biological and clinical hosts. Interstitial myocarditis has been reported in a case of WN encephalitis (Albagai and Chaimoff, 1959), in pediatric human cases of JE (J. K. G. Webb and Pereira, 1956), and in horses and monkeys experimentally infected with JE virus (Miyake, 1964). A case of subacute thyroiditis was associated with SLE virus infection (Goldman et al., 1977), possibly in-

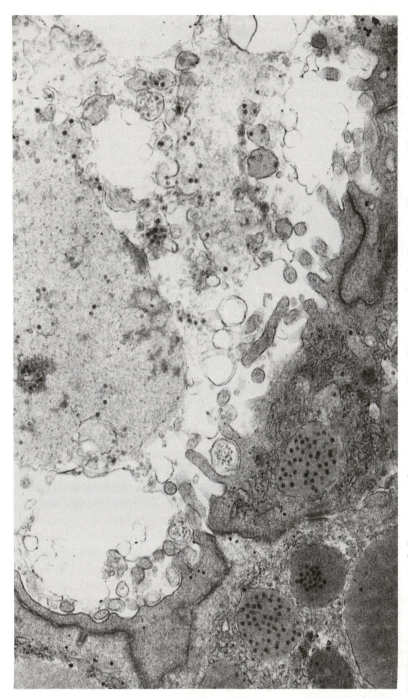

FIGURE 5. St. Louis encephalitis virus within secretory granules of Bowman's gland cell in the olfactory epithelium of a hamster (×31, 680). Virus particles are present in mucus within Bowman's gland duct. Shedding of flaviviruses in saliva and respiratory secretions is probably responsible for transmission of non-arthropod-borne flaviviruses of rodents and bats. From Monath et al. (1983).

dicating viral tropism for endocrine glands in humans as well as in experimental animals. WN virus infection has been associated with pancreatitis in humans (Perelman and Stern, 1974).

2. Viremia

In biological hosts involved in transmission cycles, viremia, the result of vascular uptake of virus released from infected cells in extraneural tissues, reaches and is sustained at levels sufficient to infect arthropod vectors. In humans infected with TBE or WN virus, viremia is frequently detected during the acute phase, but this is not the case for the other flavivirus encephalitides, in which viremia is probably confined mainly to the incubation period. Viremia levels in clinical hosts are rarely if ever of a magnitude sufficient to infect arthropod vectors.

The level of viremia represents a balance among opposing forces of virus uptake into the bloodstream, virus clearance, and thermal inactivation. The decilife of Langat virus in the vascular compartment of *Ateles* monkeys was found to be 60 min (Nathanson and Harrington, 1967), and extravascular clearance by the reticuloendothelial system (rather than intravascular inactivation) was primarily responsible for disappearance.

Studies of nonimmune clearance of alphaviruses by hepatic sinusoidal cells have shown that avirulent strains were cleared from the circulation more rapidly than virulent strains (Jahrling and Sherer, 1973; Jahrling and Gorelkin, 1975; Jahrling, 1976) and that plasma clearance rates *in vivo* correlated with both increased affinity for cultured cells and pH attachment optima for macrophages near physiological pH (Marker and Jahrling, 1979). Clearance by macrophages of neurotropic flaviviruses is also an important determinant of the course of infection (Zisman *et al.*, 1971; Monath and Borden, 1971; Olson *et al.*, 1975). In contrast to the findings with alphaviruses, however, no difference was demonstrable in plasma clearance rates between virulent and avirulent strains of SLE virus in the mouse (Fig. 6), indicating that other factors (e.g., virus-induced macrophage killing, differential release from extraneural sites, or complement-mediated virolysis) are responsible for strain differences in viremia levels.

3. Neuroinvasion

Studies of experimental flavivirus encephalitis in mice have shown a relationship among level of viremia, development of brain infection (Weiner *et al.*, 1970), and widespread or multisite simultaneous appearance of viral antigen in nervous tissue (Albrecht, 1960), supporting the concept of hematogenous spread to the CNS (Johnson, 1982). The process by which flavivirus particles cross the blood–brain barrier remains uncertain. The ability of these viruses to replicate in vascular endothelial cells suggests that they may infect and "grow across" capillaries in the

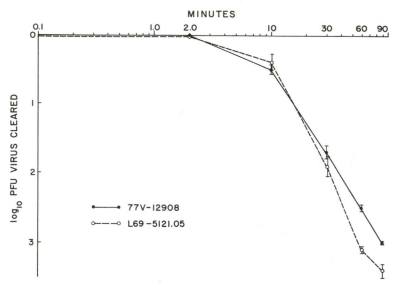

FIGURE 6. Clearance of virulent (77V–12908) and avirulent (L69–5121.05) St. Louis encephalitis virus strains from blood of weanling mice. Clearance rates do not differ significantly, indicating that other factors are responsible for the marked differences in height and duration of viremia between strains.

brain parenchyma. Viral antigen has been found only rarely in endothelial cells of brain capillaries by immunofluorescence or immunoperoxidase staining (Albrecht, 1968; Johnson *et al.*, 1985), but the perivascular recruitment of immunologically specific inflammatory cells suggests that antigen presentation occurs at this level.

The olfactory tract has long been recognized as an alternative pathway to the CNS (Peck and Sabin, 1947) and an important mode of spread following aerosol exposure, but has been dismissed in naturally acquired infections. In a recent study, however, adult hamsters and weanling mice infected with SLE virus by the peripheral route showed a clear progression of infection from extraneural tissues to sensory neurons in the olfactory epithelium (Monath *et al.*, 1983). Virus spread by axonal transport to the brain was documented by electron microscopy. There was no correlation between viremia and neuroinvasion. In this model, low (or undetectable) viremia, similar to levels that occur in human and equine hosts, results in infection of highly susceptible olfactory neurons, which are unprotected by a blood–brain barrier.

Dissemination of two SLE virus strains in weanling mice inoculated by the intraperitoneal route is compared in Fig. 7. Both strains produce indistinguishable virus replication and lethal encephalitis when inoculated directly into the brain. After intraperitoneal inoculation, however, replication of the avirulent virus is markedly restricted and is found only rarely in olfactory epithelium or brain. The virulent virus invades the brain on day 5–6 by way of the olfactory epithelium and olfactory bulb.

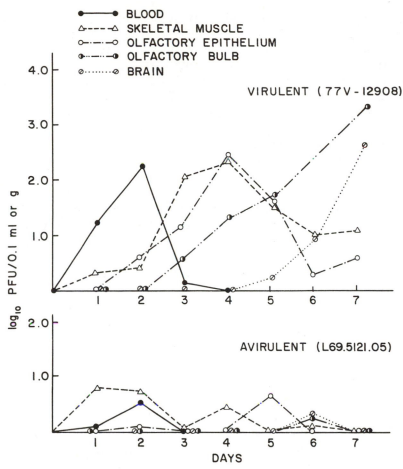

FIGURE 7. Sequence of infection of tissues in the mouse with two St. Louis encephalitis virus strains with different virulence. The olfactory neuroepithelium is infected early, in concert with extraneural tissues (e.g., skeletal muscle). Spread of virus is by axonal transport to the olfactory bulb.

The appearance of virus in olfactory epithelium parallels that in extraneural tissues (e.g., skeletal muscle), indicating absence of a barrier between blood and olfactory neural cells.

Few data are available on the routes of virus dissemination in humans. Embil *et al.* (1983) described an unusual case of Powassan virus infection that mimicked herpes encephalitis, having olfactory hallucinations and signs localizing to the temporal lobe. Sulkin *et al.* (1939) inoculated nasopharyngeal washings from SLE patients into mice and found evidence for subsequent immunity, but shedding of virus in the washings could have resulted from centrifugal (brain-to-olfactory-epithelium) spread of infection. Neuropathological lesions in the brains of humans are more extensive in thalamus, basal ganglia, and midbrain than in areas more proximal to the olfactory bulbs (Gardner and Reyes, 1980; Reyes *et al.*, 1981). Johnson *et al.* (1985) studied JE antigen distribution and histopathological lesions in the brains of JE patients and found no predominance in olfactory bulbs. In one early infection, the only viral antigen detected was in the brainstem, leading the authors to conclude that viral invasion occurred from the bloodstream. However, similar findings are seen in monkeys intranasally inoculated with SLE and other flaviviruses (Zlotnik *et al.*, 1970; Hambleton *et al.*, 1983), indicating that selective vulnerability of neuronal centers, rather than routes of viral spread, determine neuropathological patterns.

Interestingly, the olfactory route of neuroinvasion has been demonstrated not only in rodent models but also in a natural avian host species, the house sparrow (T. P. Monath and C. B. Cropp, unpublished data). Birds peripherally inoculated with an SLE virus strain with intermediate virulence were found to have inconsistent, minimal viremias; all birds developed olfactory neuroepithelial infections on days 3–5, but tissue titers were low. Brain infection followed on day 7–10 in only 5% of birds. Intracerebral inoculation of virus produced an asymptomatic, indolent infection with brain virus titers never exceeding 3–4 dex/g in 80% of birds; the remaining birds developed clinical encephalitis. Thus, in hosts with a low level of neuronal susceptibility, infection of olfactory epithelium may be terminated without establishment of brain infection.

4. Pathological Changes in the Central Nervous System

The extent, character, and distribution of lesions vary with host and virus strain. By light microscopy, inflammatory changes in meninges and along Virchow-Robin spaces of penetrating vessels are characterized by infiltration of lymphocytes, macrophages, and plasma cells. In the brain parenchyma, perivascular infiltrates surround small vessels and also contain mainly small lymphocytes as well as moderate numbers of histiocytes. Cellular aggregations or nodules in brain parenchyma consist of infiltrating mononuclear cells and possibly also activated resident microglial cells. Inflammatory changes may occur alone, or they may be accompanied by acute swelling, degeneration, and necrosis of neurons and neuronophagia. Endothelial-cell proliferation and necrosis, perineural and perivascular edema, spongy degeneration, and focal hemorrhages are evident in some models, e.g., mice infected with TBE virus (Vince and Grcevic, 1969).

In highly susceptible animals, such as the infant mouse, which succumbs rapidly to infection, inflammatory changes are often less striking than in more prolonged infections. In the infant mouse, changes at the light-microscopic level may be subtle at the time of death, but immunofluorescence and electron microscopy reveal extensive viral replication and ultrastructural cytopathology (Murphy *et al.*, 1968).

In monkeys experimentally infected with a number of flaviviruses

(Nathanson *et al.*, 1966; Zlotnik *et al.*, 1976; Monath *et al.*, 1980b; Hambleton *et al.*, 1983) and in fatal human cases (Haymaker and Sabin, 1947; Miyake, 1964; Rosemberg, 1977; Reyes *et al.*, 1981), lesions are global in distribution but most prominent in gray matter of midbrain, thalamus, pons, cerebellum, and medulla. The selective vulnerability of these regions correlates with clinical manifestations (Gardner and Reyes, 1980). In some human cases, the constellation of clinical neurological findings allows relatively precise localization to brainstem, midbrain, or cerebellum (Estrin, 1976; Kaplan and Koveleski, 1978). The predominant involvement of thalamus, brainstem, and cerebellum explains the frequent occurrence of tremors, motor incoordination, and dystonia in these infections. The Far Eastern form of TBE is characterized by a predilection for damage to the medulla and cervical spinal cord, resulting in upper motor neuron paralysis of the shoulder girdle.

The comparative pathology of flaviviruses in various clinical hosts reveals interesting differences. Louping ill virus causes neuronal degeneration and inflammatory changes in brainstem and cerebellum of sheep, producing a characteristic ataxic disease (Doherty and Reid, 1971). In the red grouse, however, lesions localize in the forebrain and resemble those associated with eastern equine encephalitis in pheasants, possibly indicating a common avian response to viral injury (Buxton and Reid, 1975).

Horses with clinical encephalitis due to JE virus have distribution and character of neuropathology similar to humans (Sugawa *et al.*, 1949; Miayake, 1964), whereas WN encephalitis in horses is characterized by a poliomyelitislike illness with prominent lesions restricted to the spinal cord (Guillon *et al.*, 1968). The molecular basis for the selective vulnerability of neural subsets to viruses remains largely unknown (Johnson, 1980). The possible involvement of differences in virus–receptor interactions, particularly those involving neurotransmitter molecules, has been suggested (Tignor *et al.*, 1984).

Residual neurological deficits, electroencephalographic changes (Lehtinen and Halonen, 1984), and psychiatric disturbances in humans frequently persist after recovery from acute encephalitis. Pathological descriptions of such cases have been few. Ishii *et al.* (1977) studied four cases with neurological residua 12–67 years after recovery from acute JE. Lesions were characterized by small areas of rarification, representing areas of neuronal loss, surrounded by dense microglial scarring, and distributed in areas typically affected during the acute phase.

Residual damage has been more difficult to assess in animal models. In mice, hindlimb paralysis occurs frequently during the acute phase, and such animals survive with this defect if saved from inanition. Changes in behavior and orientation, learning disabilities, and memory disorders have been documented in mice and rats infected with WN, Langat, yellow fever, and Murray Valley encephalitis viruses (Duffy *et al.*, 1958; Duffy and Murphree, 1959; Seamer and Peto, 1969; Museteanu *et al.*, 1979). Changes in catecholamine metabolism, implicated in behavioral abnormalities with other viruses, have not been investigated in the case of the flaviviruses.

Demyelination, a feature of several alphavirus infections in rodent models (del Canto and Rabinowitz, 1982; Suckling *et al.*, 1978; Sheahan *et al.*, 1983), has not been described in experimental flaviviral infections. One reported human SLE case had clinical features resembling cranial nerve root demyelination as seen in Guillain–Barré syndrome (Sanders *et al.*, 1953), but this has not been a typical feature of human flaviviral infections. Either oligodendrocytes are not targets for flaviviral replication or infection of neural cells with these viruses fails to initiate autoimmune injury. Subacute and chronic forms of encephalitis have been described in animals and humans, especially in association with tick-borne flaviviruses. Louping ill in sheep and Far Eastern TBE in humans (Ogawa *et al.*, 1973) may produce a subacute or chronic, progressive encephalitis. Hamsters given Langat virus by various routes (Zlotnik *et al.*, 1973; Zlotnik and Grant, 1976) develop a subacute sclerosing disease characterized by typical inflammatory changes early on, followed by a progressive degeneration over 3 months with astrocytic proliferation, perivascular granulomatous infiltrates, and neuronal vacuolation. Chronic encephalitis with similar pathological lesions was found in monkeys surviving acute infection with tick-borne viruses (Ilienko *et al.*, 1974; Zlotnik *et al.*, 1976; Asher, 1979).

5. Persistent and Congenital Infections

TBE strains vary in their capacity to produce chronic encephalitis in monkeys, but apparently not in their ability to induce persistent infections (Ilienko *et al.*, 1974; Pogodina *et al.*, 1981a). In monkeys surviving encephalitis, virus was recovered for as long as 2 years (Pogodina *et al.*, 1981a). Persistent infection was demonstrated after resolution of motor deficits in this study, but not in another (Asher, 1979). Some chronically infected animals had titers of 3–5 dex/g in brain homogenates; however, reactivation techniques (cocultivation with cells or explantation) were often required for recovery of virus, especially from visceral organs (Fokina *et al.*, 1982). Among the latter, lymphoreticular tissues appeared to be especially important sites of latent infection, a strategy adopted by many other viruses to avoid immune elimination.

Relatively little is known, however, about the mechanisms involved in persistence in the intact host. Mice with chronic brain infections and paralysis due to Kyasanur Forest disease virus appear not to develop serum neutralizing antibodies (Price, 1966), again suggesting, in some model systems, a relationship between persistent virus infection and suppression of immune elimination. In support of this view is the demonstration that JE virus persists for prolonged periods in lymph nodes of athymic nude mice in the absence of clinical signs (Hotta *et al.*, 1981a). Changes in phenotypic markers of strains recovered from chronically infected monkeys have been reported (Pogodina *et al.*, 1981b). Some but not all strains exhibited reduced mouse neurovirulence or absence of hemagglutinating antigen or both. No information is available regarding temperature markers or DI properties of these strains.

Similar features have been described in intracerebrally and peripherally inoculated monkeys that developed asymptomatic or postencephalitic persistent infections with a variety of WN virus strains (Pogodina *et al.*, 1983). Pathological changes in the CNS of persistently infected animals took the form of a subacute, progressive process with lesions at various stages of development indicating fresh, ongoing neuronal destruction as well as foci of reparation and scarring. A number of the persistently infected animals never developed neutralizing antibodies or reverted to seronegative. After $5\frac{1}{2}$ months of persistence, virus recovered from tissues was avirulent for mice and noncytopathic in cell culture, but synthesized antigen detectable by immunofluorescence. The evidence suggests that multiple mechanisms may be involved *in vivo*, but generation of mutant viruses having characteristics associated with persistence in cell culture may occur (reviewed by Brinton, 1983). These observations raise concerns about the development of live attenuated vaccines against neurotropic flaviviruses and underscore the need to elucidate host- and virus-specified mechanisms that underlie persistence.

Viral persistence in vertebrate hosts has long attracted attention as one possible explanation for survival of viruses over periods adverse to transmission by arthropod vectors. Persistent infections with prolonged viremia or recrudescent viremia following arousal from hibernation have been demonstrated in bats infected with SLE and JE viruses (Sulkin and Allen, 1974), hedgehogs, dormice, and bats infected with TBE (Kozuch *et al.*, 1963; Nosek *et al.*, 1960), and snakes and lizards infected with JE virus (Lee, 1968; Doi *et al.*, 1983).

Congenital infections with neurotropic flaviviruses are medically and ecologically important. In Asia, JE virus is an important cause of epizootic abortion and stillbirth in swine, which serve as major amplifying viremic hosts in nature. JE virus has also been isolated from brain, liver, and placental tissues of aborted human fetuses (Chaturvedi *et al.*, 1980a). Transplacental infection of several neurotropic flaviviruses has been studied in mice, hamsters (Tamura *et al.*, 1977), and bats (Sulkin *et al.*, 1966). Congenitally infected progeny of SLE-virus-infected mice generally show minimal neuropathological lesions, but exhibit significant behavioral changes (Andersen and Hanson, 1970, 1975). Mice infected with JE virus show the highest frequency of resorption, stillbirth, and congenital malformations when inoculated 9–16 days before parturition (Sagamata and Miura, 1982). Mice have been shown to transmit JE virus to offspring of consecutive pregnancies (Mathur *et al.*, 1982), and immune suppression during pregnancy has been postulated to play a role in establishment of persistent maternal infection in this model (Mathur *et al.*, 1983a).

Transplacental transmission in swine is not the only way in which reduced fertility is produced by JE virus. This agent has been suspected on epidemiological grounds to cause reduced libido and hypospermia in boars (Habu *et al.*, 1977). Experimentally infected boars develop fever, viremia, and shedding of virus in sperm for up to 17 days. After recovery, they show defective spermatogenesis and testicular inflammatory histopathological changes (Ogasa *et al.*, 1977). Artificial insemination with infected sperm resulted in viremia and lack of fertilization of recipient sows.

C. Immune Responses

Three general approaches have been used to characterize humoral and cell-mediated responses in neurotropic flavivirus infections: (1) direct measurement of the immune response or its *in vitro* correlates, using specimens taken at various times after infection; (2) additive studies, in which antibodies or cells from an immunized donor are passively transferred to a recipient host; and (3) subtractive studies, in which the course of viral infection in the host is altered by immunosuppression. Considerable attention has been focused on the role of immune responses in age-related and genetically controlled resistance in mice, on the pathological role of inflammation in the CNS, and on the role of suppressor T cells in the induction of persistent infections. Other specialized aspects of flaviviral immunity that have been studied include antibody-mediated enhancement of viral replication and the appearance of autoantibodies.

A schematic representation of the sequence of viral replication and immune responses that have been shown to occur in flaviviral encephalitis is shown in Fig. 8.

1. Humoral Immunity in Animal Models

In murine models of flaviviral encephalitis, hemagglutination-inhibiting (HI) and neutralizing (N) antibodies appear 4–6 days after peripheral inoculation of virus (H. E. Webb *et al.*, 1968; Bhatt and Jacoby, 1976; Monath and Borden, 1971; A. L. Smith and Jacoby, 1986). Detection of antibodies is closely associated temporally with termination of viremia and appearance of virus in the brain. Serum antibodies with biological (HI, N) functions in both the IgM and IgG classes appear nearly simultaneously, but IgM antibodies predominate in the early phase of infection (H. E. Webb *et al.*, 1968; Monath and Borden, 1971; Ishii *et al.*, 1968). During a brief period of immune clearance, both antibodies and virus may cocirculate in blood. Measurements of antibody plaque-forming cells (PFC) in mice infected with JE virus indicate initiation of a specific immune response in spleen as early as 24 hr postinfection, reaching a peak

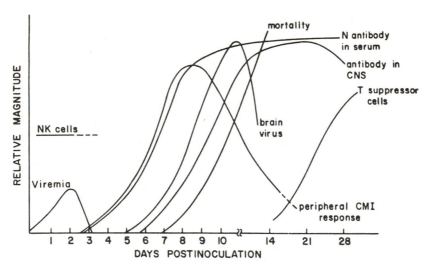

FIGURE 8. Schematic representation of course of viral replication and immune responses during infection with a neurotropic flavivirus. (NK cells) Natural killer cells. Peripheral cell-mediated immune (CMI) response includes delayed-type hypersensitivity, leukocyte-migration inhibition, cell-mediated cytotoxicity, and adoptively transferable protection. CMI responses in the CNS probably occur later.

on day 5–6 coincident with detection of serum antibodies (Mathur *et al.,* 1983b).

The consistency and timing of humoral responses to primary infection suggest that they play a major role in recovery. Survival of sheep infected with louping ill virus was shown to correlate with rapidity of the serum N antibody response (Reid and Doherty, (1971). Studies of systemic and local CNS antibody responses in humans with JE have shown strong correlations between survival and both early and quantitatively strong antibody responses (Burke *et al.,* 1985b,c). Patients who survived JE were found to have serum IgG antibodies during the first 5 days of illness, whereas fatal cases became seropositive later.

Administration of cyclophosphamide, which has a number of immunomodulating effects but primarily impairs the humoral antibody response, has been shown to potentiate flaviviral encephalitis (Bhatt and Jacoby, 1976; Camenga *et al.,* 1974; Cole and Nathanson, 1968). Treated animals have higher and more persistent viremia, a progressive increase in brain virus titers and neuronal necrosis, and higher mortality ratios. Genetic resistance of C3H/RV mice to Banzi virus is abrogated by cyclophosphamide treatment (Bhatt and Jacoby, 1976).

The ability of passively administered antibody to confer protection has also been established (Hammon and Sather, 1973). Antibody transferred to mice as late as 6 days after peripheral inoculation with WN virus prevented lethal infection, even though brain infection had already been established (Camenga *et al.,* 1974). Serum from donor mice taken 1 and

2 weeks after infection protected recipients against lethal JE virus challenge given 24 hr after antibody (Mathur *et al.*, 1983b); the protective capacity of the serum was abolished by removal of IgM with 2-mercapthoethanol.

Is antibody that appears during the critical *early* phase of infection effective in the recovery process? A. L. Smith and Jacoby (1986) compared the early immune responses of congenic flavivirus-susceptible (C3H/He) and -resistant (C3H/RV) mice. In this model, the mice succumb to parenteral Banzi virus infection on day 7–8, whereas RV mice develop on day 6 subclinical brain infection, which peaks on day 9 and then clears (Jacoby and Bhatt, 1976). N antibodies in serum are first detectable on day 5 or 6 in both RV and He mice, but antibody with capacity to protect adoptively immunized mice did not appear until day 10 postinfection. Protective capacity was associated with complement-dependent cytotoxicity for infected mouse neuroblastoma cells. These observations suggested that the appearance of protective and cytolytic antibody coincided with onset of viral clearance from the brain, but was not present early enough to be critically involved in the primary control of infection in resistant mice.

The biological role of N antibody and antibodies involved in cytotoxicity and antibody-dependent cell-mediated cytotoxicity (ADCC) reactions, as well as the viral proteins to which these antibodies are directed, require further investigation. Cytolysis of dengue-virus-infected Vero cells by complement and antibody was found to be relatively inefficient (Cantanzaro *et al.*, 1974). Brandriss and Schlesinger (1984), however, have recently described protection of mice against encephalitis by complement-dependent cytotoxic antibodies directed against a nonstructural polypeptide (NV3) of yellow fever virus. Antibodies directed against nonstructural virus-specified proteins on the surface of infected cells appear relatively late, but probably play an important role in viral clearance. ADCC was not detected in a study of Banzi-virus-infected mice (A. L. Smith and Jacoby, 1986), but Kurane *et al.*, (1984) demonstrated antibody-dependent lysis of dengue-infected Raji cells by human peripheral-blood mononuclear cells. Although demonstrable *in vitro*, the biological significance of ADCC *in vivo* remains unknown.

2. Cell-Mediated Immunity in Animal Models

A role for cell-mediated immunity (CMI) in primary flaviviral infection has been postulated on the basis of measurement of delayed-type hypersensitivity (DTH) responses, T-cell cytotoxicity, *in vitro* correlates of CMI (leukocyte-migration inhibition), and immunosuppression–reconstitution experiments. Most studies have shown onset of CMI coincident with or shortly before the appearance of serum antibodies. In Banzi-virus-infected C3H/RN and He mice, T-cell immunity demonstrated by cytotoxicity assays with splenic (Sheets *et al.*, 1979) or peritoneal exudate

(A. L. Smith and Jacoby, 1986) effector cells appeared on day 6 after infection. Generation of cytotoxic T cells has also been demonstrated in mice infected with a member of the TBE complex (Gajdosova *et al.*, 1981). In JE-virus-infected mice, transient antigen-specific leukocyte-migration inhibition was found beginning on day 3 and peaking on day 9 (Mathur *et al.*, 1983b). CMI responses to dengue viruses were also investigated using this assay by P. S. Nagarkatti *et al.* (1978) and Chaturvedi *et al.* (1978). DTH responses determined by measuring inflammatory infiltration after footpad challenge have been demonstrated in mice infected with SLE (Hudson *et al.*, 1979), JE (Mathur *et al.*, 1983a), and dengue viruses (Pang *et al.*, 1982). The maximum DTH response in these models occurred on day 6 and then rapidly declined; responses were enhanced by cyclophosphamide and splenectomy, probably by eliminating T suppressor cells or their precursors. It should be emphasized that the CMI responses measured in these studies represent peripheral responses and that CMI in the CNS probably occurs later.

Selective depletion of T lymphoctyes with antiserum has been shown to potentiate flaviviral encephalitis, while prolonging survival time (see Section IV.C.4). Treatment with antithymocyte serum (Jacoby *et al.*, 1980) or thymectomy (Bhatt and Jacoby, 1976) converted RV mice infected with Banzi virus to the susceptible phenotype. Susceptible He mice were protected after adoptive immunization with spleen cells (Jacoby *et al.*, 1980); however, transfer of donor cells was effective only beginning 5–7 days after priming and only in recipient mice adoptively immunized within 24 hr after virus challenge. Taken together, studies of the Banzi–C3H mouse model indicate that T-cell immunity (like humoral antibody) is required for recovery, playing a role in late viral clearance, but that other factors, such as production of DI virus (A. Smith, 1981)(see also Chapter 11), are critical early determinants of genetic resistance.

The contribution of CMI to recovery and protection from encephalitis induced by yellow fever and dengue viruses has been assessed using athymic nude mice (Bradish *et al.*, 1980; Hotta *et al.*, 1981a,b). The participation of T cells in determining the course and outcome of infection is influenced by the viral strain. Depending on the viral strain used, susceptibility to lethal infection may be enhanced or unaffected. Presumably, host defense mechanisms that are not T-cell-dependent (e.g., macrophages, IgM antibody) are sufficient to abort infection with viral strains of marginal neuroinvasiveness, whereas CMI, together with other responses, is required for clearance of more virulent viral strains.

Other cellular effector mechanisms that may play a role in flavivirus encephalitis include natural killer (NK) cells and macrophages (Mϕ). Cells with NK functions against virus-infected cells or NK-sensitive target-cell lines have been described in the case of Kunjin (McFarland and White, 1980), yellow fever (Fagraeus *et al.*, 1982), and dengue (Kurane *et al.*, 1984). The functions of Mϕ in flaviviral encephalitis are essentially unknown. Resistance of mice to encephalitis caused by yellow fever was abrogated

TABLE VI. Permissiveness of Stimulated Mouse Peritoneal Macrophages for Two Strains of St. Louis Encephalitis Virus

Virus	Yield[a] (PFU/ml) in the presence of:			
	TCF	NS	77V-12908 Aby	L69-5121.05 Aby
77V-12908 (virulent)	2.6 ± 0.2	2.6 ± 0.2	3.7 ± 0.2	2.4 ± 0.2
L69-5121.05 (avirulent)	1.8 ± 0.2	1.0 ± 0.4	3.6 ± 0.2	3.4 ± 0.1

[a] Yield is expressed in plaque-forming units (PFU)/ml at 72 hr postinfection. (TCF) Tissue-culture fluid; (NS) normal serum.

by Mφ blockade with silica (Zisman et al., 1971). Similarly, viral multiplication of SLE virus in mice was enhanced by treatment with thorotrast (Monath and Borden, 1971). These measures may have broad effects (e.g., on circulating monocytes and NK cells), and thus the role of nonspecific resistance of tissue Mφ remains uncertain.

The intrinsic and extrinsic interactions of Mφ with encephalitis viruses have received considerably less attention than those with dengue and yellow fever viruses. Infection of Mφ-like cell lines with TBE (Phillpotts et al., 1985) and with WN virus (Peiris and Porterfield, 1979; Cardosa et al., 1983; Gollins and Porterfield, 1984) has been studied in relation to antibody and complement-receptor-mediated enhancement of replication, but less is known about the role of Mφ in vivo. Macrophages in the host are functionally diverse, as are primary Mφ obtained from animals by various techniques (bacille Calmette Guérin activation or thioglycollate-stimulated inflammation). The complexity of WN virus–Mφ interactions has been demonstrated recently by Cardosa et al., (1986), who found that permissiveness to virus growth varied with physiological state of Mφ, age and strain of mouse, and receptor pathway of virus entry. Other observations suggest a role for Mφ in determining outcome of infection with flaviviruses having different virulence characteristics. Thioglycollate-stimulated peritoneal Mφ from 4-week-old mice were found to be relatively less permissive for an avirulent strain of SLE virus than for a virulent virus strain in the absence of antibody (Monath, unpublished data) (Table 6). Interestingly, homologous antibody-mediated enhancement of virus growth was demonstrated, but reciprocal heterologous enhancement occurred in one direction only, indicating heterogeneity of enhancing epitopes on SLE, as was found for dengue viruses (Halstead et al., 1984). The mechanisms involved in intrinsic resistance are not known, but might include interferon production or DI particles.

3. Immune Response in Humans

Relatively little is known about humoral and cellular immunity in human flaviviral encephalitis. As in experimentally infected mice, antibodies in humans are often detectable at the time of onset of neurological symptoms, and it has been rarely possible to study the chronology of immune responses that begin during the incubation period. Antibodies have been measured by standard assays (HI, N, complement-fixation, immunoassays) and have not been characterized by viral protein specificity or employed in cytotoxicity or ADCC assays. Several investigators have described the distribution of antibodies in the IgM and IgG classes in sera from cases of flavivirus encephalitis (Ishii *et al.*, 1968; Burke *et al.*, 1985c; Monath *et al.*, 1984). Serum N antibody titers in convalescent sera from cases with overt CNS disease are usually higher than titers in persons who have sustained inapparent infections (Mayer *et al.*, 1976; Chatuverdi *et al.*, 1979), probably reflecting greater viral replication and antigenic stimulation in the former group. Antibodies are also locally produced in the CNS (see below). Recent studies of antibody responses in human cases of JE (Burke *et al.*, 1985b,c) led to several important conclusions: (1) Persons sustaining inapparent infections do not develop local CNS antibody and thus have not experienced brain infection; (2) during the acute phase of infection, persons with fatal encephalitis have absent, low, or delayed levels of virus-specific IgM and IgG in serum and cerebrospinal fluid; whereas (3) patients with encephalitis who survive have vigorous and brisk systemic and local CNS antibody responses. These observations illustrate some of the dynamics of the "race" between humoral immune response, extraneural replication, and neuroinvasion. T-cell responses in humans with TBE (Mayer *et al.*, 1976; Sipos *et al.*, 1981) and JE (Chaturvedi *et al.*, 1979) have been reported.

4. Immune Responses of the Central Nervous System: Role in Recovery and Pathogenesis

Cellular and humoral immune responses within the CNS have been extensively studied in mice infected with an alphavirus (Sindbis) (McFarland *et al.*, 1972; Griffin *et al.*, 1983; Griffin, 1981; Moench and Griffin, 1984) (see also Chapter 8).

Although not well characterized, similar events are probably involved in immune clearance of flaviviruses from the brain. Local production of antibody had been demonstrated after louping ill virus infection in sheep (Reid *et al.*, 1971), SLE in mouse (R. Bowen and T. Monath, unpublished data), and SLE (Ehrenkrantz *et al.*, 1974), JE (Burke *et al.*, 1985a), and TBE (Hofmann *et al.*, 1979) virus infection in humans. During the acute phase of infection, an early and vigorous antibody response in the CNS is an important component of virus clearance and recovery (Burke *et al.*, 1985b). High or persisting IgM antibody levels appear to correlate with the severity of the encephalitis and thus, presumably, with the amount of antigenic presentation (Ehrenkrantz *et al.*, 1974; Edelman *et al.*, 1976). However, Burke *et al.* (1985c) found local synthesis of IgM in the CNS of patients long after recovery from JE, raising the possibility of virus persistence.

The demonstration of high IgE levels in acute JE infections (Shaikh *et al.*, 1983) raises the possibility that IgE-mediated release of histamine from sensitized mast cells may play a role in the CNS inflammatory response. Although not described for flaviviruses, other viruses are known to induce IgE (Perelmutter *et al.*, 1978).

The cellular immune response in flavivirus encephalitis coincides with peak peripheral cytotoxic T-cell and DTH responses (Hudson *et al.*, 1979). Kitamura *et al.* (1972) and Kitamura (1975) studied the fate of ^{3}H-labeled circulating mononuclear cells in mice with JE. Both perivascular infiltrates and brain parenchymal infiltrates, including the rod cells in glial nodules (usually believed to be activated resident microglial cells), were found to be of hematogenous origin. Adoptive transfer of immune spleen cells taken from mice at the peak of the cytotoxic T-cell response was shown to protect against challenge with Banzi (Jacoby *et al.*, 1980) and JE viruses (Mathur *et al.*, 1983b). Concanavalin A, which has a wide variety of potential effects on T-cell function, has been reported to protect against JE in mice (Kelkar, 1982).

Clearance of flaviviuses depends in part on recognition and destruction of infected cells by T effector cells in concert with activated macrophages. T helper cells, B cells, and antibodies also represent an important component of flavivirus clearance and recovery from primary infection. Passively transferred antibodies can abort flavivirus encephalitis even when given after neuroinvasion has occurred, whereas transfer of immune cells is not effective at this stage. Among other factors, diffusion of antibodies into the brain substance must be relatively more efficient than movement of cytotoxic T cells and macrophages.

Johnson *et al.* (1985) studied the cellular components of CNS inflammation in fatal human cases of JE. Their findings were similar in many respects in those in the Sindbis–mouse model (Moench and Griffin, 1984). At the time of death (3–9 days after onset), approximately 30–40% of cells in perivascular cuffs had T-cell markers, 5–10% were suppressor/cytotoxic cells, 15–20% were macrophages, and 10% were B cells.

The role of interferon in flaviviral clearance has not been clearly defined. A protective effect of exogenously administered interferon has been demonstrated in mice challenged with JE virus (J.-L. Liu, 1972). However, no difference in interferon production or sensitivity has been found between genetically resistant and susceptible mice (Darnell and Koprowski, 1974). Interferon levels in serum and brain are highest in cases of severe infection or infection with virulent virus strains (Monath *et al.*, 1980b) and thus reflect the extent of virus replication, rather than a component of the recovery process.

5. Immunopathology

Inflammatory responses engaged in viral clearance may also play a pathogenic role when they involve a target organ with critical tolerances such as the brain. A number of studies have shown prolongation of the mean time to death from flaviviral encephalitis in immunosuppressed animals (Hirsch and Murphy, 1967; H. E. Webb et al., 1968; Camenga et al., 1974; Camenga and Nathanson, 1975; Semenov et al., 1975; Jacoby et al., 1980). The pathological role of the cellular immune response varies with virus–host pairing and also, probably, with the degree of expression of viral antigens on critical neuronal targets for antibody- or T-cell-mediated cytotoxicity or Mφ-mediated ADCC. Thus, immunosuppression reduced inflammation in both WN and Langat encephalitis of mice, but prolonged survival of Langat-virus-infected mice only (Camenga and Nathanson, 1975). Moreover, high grades of inflammation may occur without producing symptomatology (Zlotnik et al., 1970; Reid et al., 1982).

Another potential immunopathological mechanism is antibody- (or complement-) mediated enhancement of viral replication in Mφ. Although demonstrated in vitro with a number of flaviviruses (Peiris and Porterfield, 1979) and an important component of host responses to dengue viruses (see Section V.6), there is as yet no evidence for a role in the pathobiology of encephalitis. Antibody-mediated enhancement requires monocytes/macrophages bearing Fc receptors and noncytophilic, nonneutralizing antiviral type-specific or cross-reactive IgG antibodies [or CR3 receptors and complement (Cardosa et al., 1983)]. In the case of flaviviral encephalitis, nonneutralizing antibody from a prior heterologous virus infection could enhance extraneural replication and increase the risk of neuroinvasion. Against this possibility is the allopatric distribution of most members of the TBE group and the SLE–JE–WN–Murray Valley encephalitis complex, making sequential heterologous infections unusual events. The one common sympatric heterologous infection (dengue) appears to cross-protect against SLE and JE, rather than to enhance these infections (Bond, 1969; Tarr and Hammon, 1974; Edelman et al., 1975a). This observation is supported by a study showing lack of immune enhancement of dengue infection by JE virus antibodies (Putvatana et al., 1984).

Whether nonneutralizing antibodies produced early in the immune response to primary infection could enhance virus replication in vivo is not known; this does not appear to play a major role, however, in dengue, an infection that utilizes Mφ as cellular targets for replication to a much greater extent than the encephalitis viruses.

6. Suppressor-T-Cell Functions and Viral Persistence

As with all other balanced biological reactions in vivo, mechanisms exist to regulate or suppress the immune response. Generation of suppressor cells was demonstrated in JE-virus-infected mice, both in vitro and by adoptive transfer of spleen cells (Mathur et al., 1983c, 1984). The ability of antigen-specific T lymphocytes from JE-virus-primed mice to suppress DTH responses and IgM antibody plaque formation against JE virus in spleen appeared 18 days after priming and persisted for 6 weeks. The suppressor T cells are cyclophosphamide-sensitive and mediate suppression through soluble products (Mathur *et al.*, 1984). These findings correlated with previous reports (Mathur *et al.*, 1983a,b) showing disappearance of CMI and protective antibodies after the 3rd week.

Pregnant mice develop persistent JE virus infections, which are reactivated during a consecutive pregnancy with transplacental infection of the fetus (Mathur *et al.*, 1982). Pregnant mice had markedly impaired DTH and delayed but preserved leukocyte-migration inhibition responses to JE virus (Mathur *et al.*, 1983a). Congenitally infected offspring had reduced T-cell populations in their spleens and impaired CMI responses to JE and to unrelated antigens (sheep erythrocytes). These observations suggest that impaired immune responses early in pregnancy followed by the appearance of T suppressor cells may be responsible for establishment and maintenance of persistent infection in this model. Impaired viral clearance by CMI mechanisms is similarly responsible for the persistence of JE viral infections in athymic nude mice (Hotta *et al.*, 1981a).

7. Protection from Subsequent Infection and Heterologous Cross-
   Immunity

Long-lasting immunity follows infection with flaviviruses; even in long-lived species, protection from disease is assumed to be lifelong, although this may not always be the case (Ishii *et al.*, 1968). Reexposure results in a rapid anamnestic humoral antibody response and quenching of viremia. A number of studies have demonstrated the protective capacity of antibodies passively administered before challenge (Lubiniecki *et al.*, 1973).

Animals immunized by the peripheral route and then challenged intracerebrally may or may not be protected. In one study, passive administration of IgG antibodies was effective in protecting against lethal intracerebral challenge, whereas adoptive transfer of immune spleen cells was not (Chaturvedi *et al.*, 1978). In contrast, protection from intracerebral challenge is complete in animals that have survived a prior brain infection, indicating local immunological memory in the CNS (Gerhard and Koprowski, 1977). This observation raises the potential problem of immunizing arbovirus laboratory workers against aerosol (olfactory route) exposure, especially with inactivated vaccines.

Cellular mechanisms are also involved in defense against subsequent challenge. It is not established whether intrinsic resistance of Mϕ to flaviviruses increases after immunization. The growth of WN and yellow fever viruses was reduced in organ cultures of meninges from immunized mice (Rubenstein *et al.*, 1972), an effect possibly mediated by Mϕ and immune interferon. Cell-mediated responses are probably important in containing and eliminating infection after reexposure, since virus replication occurs at the site of inoculation and in regional lymph nodes (Mal-

TABLE VII. Comparative Features of the Pathobiology of Dengue, Yellow Fever, and Wesselsbron Disease[a]

	DHF/DSS	Yellow fever	Wesselsbron
Hepatitis	+ + +	+ + +	+ + +
Lymphoid, hyperplasia, necrosis	+ + +	+ +	+ + +
Myocarditis	+	+ +	+
Hemorrhagic diathesis	+ + +	+ + +	+ +
Increased capillary permeability, shock	+ + +	+ + +	+ +
Complement consumption	+ + +	?	?
Renal pathology	+	+ +	0
Encephalitis	+	+	+ +
Congenital malformation, abortion	0[b]	0	+ +

[a] (?) Not studied; (0) absent; (+) reported; (+ +) frequent; (+ + +) constant feature.
[b] Reports of abortion and birth defects associated with classic dengue in the older literature have not been substantiated (Mirovsky et al., 1962).

kova and Kolman, 1964). As measured by DTH, leukocyte-migration inhibition, and adoptive transfer of cells, CMI responses are relatively short-lived after primary infection (Mathur et al., 1983b), but memory functions and secondary CMI responses have not been studied.

Heterologous cross-protection among flaviviruses has been extensively investigated (Vorndam, 1980), and its molecular basis has been explored using monoclonal antibodies (Roehrig et al., 1983). While antibodies appear to play the principal role in cross-protection, cellular immunity and interference phenomena may also be involved. Cross-reactivity between Langat and yellow fever viruses was found in a T-cell cytotoxicity assay (Gajdosova et al., 1980).

Significant protection against intracerebral challenge with Langat and WN viruses has even been reported to follow intracerebral immunization with alphaviruses (Oaten et al., 1976, 1980). The interference mechanism is unclear, but cross-immunity to host brain-cell glycolipid antigens incorporated into viral envelope (H. E. Webb et al., 1984) may be involved.

V. PATHOGENESIS OF FLAVIVIRAL HEMORRHAGIC FEVERS

Three mosquito-borne viruses, dengue, yellow fever (YF), and Wesselsbron (WSL), share certain pathogenetic features and are compared in this section and in Table VII. The emphasis will be on dengue hemorrhagic fever (DHF) and its most severe form, dengue shock syndrome (DSS), which have been shown, largely by Halstead and his collaborators, to have a unique immunopathological basis: antibody enhancement of viral infection (Halstead, 1980).

DHF/DSS, an epidemic disease that principally affects children, is distinguished from classic dengue fever by the presence of (1) increased vascular permeability with hemoconcentration, hypovolemia, hypotension, and serous effusions; (2) a bleeding diathesis of complex etiology, but always accompanied by thrombocytopenia, usually mild, but in about 5–10% of cases causing significant hemorrhage into skin, gastrointestinal tract, or urine; (3) hepatic enlargement and dysfunction of mild degree; (4) activation of the complement system; and (5) depletion or necrosis of lymphoid tissues. The case fatality ratio in untreated cases with shock is approximately 10%, but early recognition and correction of hypovolemia significantly reduce the lethality. Myocarditis has been reported as a complication of dengue infection.

YF is an endemic and epidemic disease of humans and some species of monkeys characterized in its severe form by: (1) marked hepatic necrosis, dysfunction, and jaundice; (2) a bleeding diathesis, often with severe gastrointestinal hemorrhage; (3) vascular collapse and shock; (4) albuminuria and renal impairment; (5) depletion or necrosis of lymphoid tissues; and (6) myocarditis. The case fatality ratio is approximately 20%.

WSL virus causes disease in sheep (principally newborn lambs) characterized by fever, severe hepatic necrosis and jaundice, hemorrhagic manifestations, myocarditis, signs of increased vascular permeability, and lymphoid necrosis. Mortality in experimentally infected lambs is as high as 30%. Maternal deaths occur in pregnant ewes and abortion and congenital neurological defects in fetuses (Coetzer and Barnard, 1977). WSL virus generally causes a mild, febrile illness in calves, but occasionally a more severe infection resembling the disease in lambs (Blackburn and Swanepoel, 1980).

A. Comparative Pathology

The pathological manifestations in fatal cases of DHF/DSS are relatively unimpressive and generally insufficient to explain death. Whereas YF cases manifest more extensive pathology in at least one vital organ (liver), the immediate cause of death is also not clear and probably reflects a metabolic disturbance rather than "organ failure." Gross autopsy findings in DHF/DSS, YF, and WSL disease include hepatic enlargement or altered appearance, serous effusions, edema, and hemorrhages. The latter are usually of minor degree in DHF/DSS, but prominent gastrointestinal hemorrhage, similar to that seen frequently in human YF and ovine WSL disease (LeRoux, 1959), has been increasingly emphasized in some reports (Sumarmo et al., 1983; Guzman et al., 1984). Morphological evidence of disseminated intravascular coagulation (intravascular thrombi) has been rarely recognized in DHF/DSS (Fresh et al., 1969) and is not reported in YF or WSL. Skin biopsies from DHF patients with rashes show evidence of microvascular damage, with swelling and occasional necrosis of endothelial cells and perivascular edema and infiltration of mononuclear leukocytes (Boonpucknavig *et al.*, 1979a,b). Similar lesions are inconsistently seen in other tissues from fatal cases.

At the microscopic level, the principal changes in DHF/DSS affect the liver, bone marrow, lymphoid organs, heart, lungs, and kidneys (Bhamarapravati *et al.*, 1967). Lesions in the liver resemble the early phase of experimental YF in the rhesus monkey (Bearcroft, 1957; Tigertt *et al.*, 1960; Monath *et al.*, 1981), in which hepatocellular necrosis is minimal, and include hypertrophy and necrosis of Kupffer cells, central or paracentral focal coagulative necrosis of hepatocytes, Councilman bodies, and mild fatty metamorphosis. The pathology of fatal YF is qualitatively similar to DHF/DSS, but with much more severe and extensive hepatocellular necrosis (Ishak *et al.*, 1982). Typically, a ring of preserved hepatocytes, one to two cells thick, surrounds the central veins and portal areas, and the distribution of necrosis can be described as "midzonal." In fatal human YF, the proportion of the liver lobule actually necrosed varies from 5 to 100%, with a mean of 80% (Klotz and Belt, 1930). In dengue and YF, inflammatory changes are minimal. Cholestasis of mild to moderate degree may be present in YF, especially after the 8th day of illness (Camain and Lambert, 1966), but bile-duct proliferation is not observed. Recovery is associated with granulomatous changes and rapid regeneration of hepatocytes. Prolonged hyperbilirubinemia has been noted in some cases (Elton *et al.*, 1955); the contribution of underlying chronic liver disease in such cases must be considered (Francis *et al.*, 1972).

The liver pathology of WSL disease in lambs (LeRoux, 1959; Coetzer *et al.*, 1978; Coetzer and Theodoridis, 1982) is characterized by diffusely scattered necrosis of individual hepatocytes, Councilman bodies, Kupffer cell proliferation, fatty infiltration, and cholestasis. In contrast to YF and dengue, infiltration of leukocytes is more prominent and there is proliferation of bile ducts.

In cases of DHF/DSS, the bone marrow shows increased cellularity and megakaryocytic arrest (Bierman and Nelson, 1965). Marked changes are evident in thymus and T-cell-dependent areas of nodes and spleen, including lymphocytolysis in the cortex of the thymus, germinal centers of lymph nodes, and splenic white pulp, phagocytosis of lymphoctes, and hyperplasia of reticulum cells and plasmacytoid cells (Bhamarapravati *et al.*, 1967; Aung-Khin *et al.*, 1975). Similar changes have been described in YF-infected monkeys (Monath *et al.*, 1981) (Fig. 9) and in sheep with WSL virus disease (Coetzer and Theodoridis, 1982).

A transient proliferative glomerulonephritis due to deposition of immune complexes has been found in renal biopsies from DHF patients (Boonpucknavig *et al.*, 1976a). Glomerular lesions in YF, consisting of Schiff-positive alteration in the basement membrane, have been associated with the marked proteinuria in this disease (Barbareschi, 1957). Deposition of immune complexes in kidneys has not been studied in YF, but in comparison with dengue would not be expected to play an im-

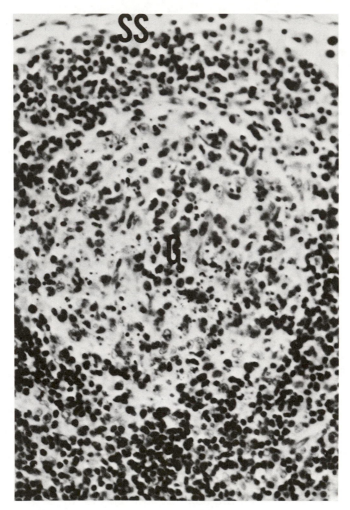

FIGURE 9. Necrosis of germinal center (G) of mesenteric lymph node of a rhesus monkey 6 days after inoculation of yellow fever virus. (SS), subcapsular sinus. From Monath *et al.* (1981).

portant role. Tubular epithelial changes in YF are more severe than in dengue and appear to represent a progression of shock and prerenal failure to acute tubular necrosis (Monath *et al.*, 1981). Why this complication occurs in YF but not in DHF/DSS is unclear, but it may relate to the severity of hepatic necrosis in YF and the attendant hemodynamic changes affecting renal function. No renal pathology has been reported in WSL disease.

Hyaline degeneration, vacuolation, and necrosis of myocardial fibers, only rarely associated with mononuclear-cell infiltration, have been described in DHF/DSS (Bhamarapravati *et al.*, 1967), YF (Lloyd, 1931), and WSL (Coetzer and Theodoridis, 1982). Cardiac muscle is probably a site of viral replication and pathological lesions the result of direct viral injury. Myocardial injury, bradyarrhythmia, and diminished cardiac output may contribute to the shock state in these infections.

B. Sites of Viral Replication

In dengue, the role of mononuclear phagocytes as target cells and principal sites of replication seems established (Halstead, 1980). Antigen has been detected in the cytoplasm of mononuclear leukocytes in skin (Boonpucknavig et al., 1979b), in spleen (Bhamarapravati and Boonpucknavig, 1966), and in lymph nodes, hepatic sinusoids, alveolae, and thymic cortex (Bhamarapravati, 1981). Denguelike viral particles have been demonstrated in renal glomerular Mφ (Boonpucknavig et al., 1976a). Dengue viral antigen was demonstrated on the surface of unfixed circulating B lymphocytes of DHF patients (Boonpucknavig et al., 1976b), but attachment of virus vs. replication in these cells was not differentiated. Scott et al. reported isolation of dengue viruses from peripheral-blood leukocytes (mainly glass-adherent monocytes) from DHF patients.

The tropism of dengue viruses for mononuclear leukocytes is supported by experimental infections of monkeys (Marchette et al., 1973) and further clarified by a number of in vitro studies (Halstead et al., 1973a, 1976, 1977; Theofilopoulos et al., 1976). In monkeys, virus replication increased progressively in lymphoid tissues and leukocytes and peaked at the time viremia was terminated, corresponding to the onset of shock in DHF/DSS patients. There was also evidence for viral replication in sites favored by other flaviviruses (skeletal and smooth muscle, salivary gland, adrenal gland). Increased levels of serum acid phosphatase of osteoclast origin in human patients (Lam et al., 1982) suggest that these cells may also be targets of viral replication or injury and are consistent with the bone pain associated with the disease.

Sequential histopathological and immunofluorescence studies of monkeys infected with YF virus have shown the earliest lesions and antigen in Kupffer cells in the liver and in scattered reticuloendothelial cells in lymph nodes and spleen (Tigertt et al., 1960; Monath et al., 1981). Infection of hepatocytes follows, but widespread necrosis is a late event in the rhesus monkey model, occurring during the last 1–2 days before death. Although far from complete, the available data suggest certain parallel features with DHF/DSS: (1) early infection of mononuclear phagocytes (of which Kupffer cells appear most important as a gateway of spread to the hepatic parencyhyma); (2) lymphoid organs, including lymph nodes and spleen, are sites of viral growth and in late stages of infection show necrosis of germinal centers and reticulum-cell hyperplasia; (3) the onset of clinical symptoms (jaundice, oliguria, hypotension) coincides with the beginning of viral clearance from blood and tissues; (4) virus persists in tissues after clearance of viremia. Both antibody and virus (and thus presumably immune complexes) have been demonstrated in the sera of human YF patients in the course of the disease (World Health Organization, 1971). These observations taken together suggest that immune clearance of intracellular virus and, possibly, the formation of immune complexes may play a pathogenic role in human YF as in DHF/DSS. However, the very rapid course of infection in the rhesus monkey model culminating in death only 5 or 6 days after inoculation, as well as the virtual absence of inflammatory lesions in YF, indicate that direct viral injury and secondary physiological effects of injury rather than immunopathological mechanisms are predominant events in this disease.

The experimental pathology of WSL virus has been described in sheep, cattle, and goats (Coetzer et al., 1978; Coetzer and Theodoridis, 1982; Blackburn and Swanepoel, 1980), but virological studies to define sites of virus replication have not been reported. On the basis of the pathological lesions, liver, heart, and possibly lymphoid tissues may be implicated as sites of viral replication. WSL virus is also clearly neurotropic in the ovine and bovine fetus. Both wild virus and live attenuated vaccine strains cross the placenta and produce neurological infection and malformations, including hydranencephaly and arthrogryposis (Coetzer and Barnard, 1977).

In mice, dengue, YF, and WSL viruses cause primarily encephalitic infections and have pathogeneses similar to those of other neurotropic flaviviruses. A spectrum of neurotropism is evident among the three agents: Wild strains of dengue virus often require sequential passage in mouse brain before full expression of neurovirulence (reviewed by R. W. Schlesinger, 1977, 1980), whereas unadapted YF strains do not. Age-dependent resistance to encephalitic infection develops in mice to parenteral inoculation of both dengue and YF viruses, whereas adult mice remain susceptible to WSL virus. Although not extensively studied, the extraneural sites of replication of these viruses in the mouse appear similar to other flaviviruses and include lymph nodes, spleen, and cardiac and skeletal muscle (Hotta et al., 1981a,b; David-West, 1975). Dengue-infected nu/nu and nu/+ mice develop infection (demonstrated by immunofluorescence) and acidophilic hyaline necrosis of Kupffer cells without evidence for progression to hepatocellular infection (Hotta et al., 1981b), and suckling mice infected with YF virus may show focal degeneration of liver cells (David-West and Smith, 1971); in these models, however, the brain is the eventual principal target organ for viral growth and pathogenesis. The only lower mammal that manifests a viscerotropic response to YF virus similar to monkeys and humans is the European hedgehog (Erinaceus europeaus) (Findlay and Clarke, 1934). Boonpucknavig et al. (1981) studied the course of clinically inapparent dengue viral infection of adult outbred mice inoculated by the intraperitoneal route. Virus could not be isolated from tissues, but viral antigen was found by immunofluorescence in mononuclear leukocytes as late as 3 weeks after inoculation. Evidence was obtained for immune-complex deposition and proliferative glomerulonephritis, analagous to that described in humans with DHF (Boonpucknavig *et al.*, 1976a).

The neurotropism of dengue and YF viruses must also be considered with respect to primate hosts. Dengue virus strains require adaptation by mouse-brain passage to become neurovirulent for monkeys, whereas unadapted YF viruses exhibit a higher degree of neurovirulence. Rhesus monkeys inoculated by the intracerebral route succumb to typical viscerotropic YF unless protected by immune serum, in which case they die of encephalitis (Theiler, 1951). Nevertheless, human cases of encephalitis due to YF virus are very unusual, and only a single case of naturally acquired YF encephalitis has been reported (Stefanopoulo and Mollaret, 1934). Post vaccinal encephalitis was a relatively frequent complication of children with the French neurotropic strain (LeMercier *et al.*, 1966). Neuroinvasion by 17D vaccine, which exhibits reduced neurovirulence for monkeys is, however, very unusual in humans, and cases have been limited to infants and young children. This suggests that in this age group, encephalitis may be a heretofore unrecognized complication of infection with wild YF virus strains. Signs of CNS dysfunction, including stupor, coma, hyperexcitability, and convulsive seizures, are a frequent component of severe YF, but pathological examination of brain, showing small perivascular hemorrhages and cerebral edema, without significant inflammation or neuronal changes, has not indicated viral replication and encephalitis (Stevenson, 1939). A wide variety of neurological syndromes, including encephalitis during acute dengue or DHF, and postinfectious complications (encephalitis, polyneuritis, Reye's syndrome), have been associated with dengue (Gubler *et al.*, 1983), but neither incontrovertible evidence for an etiological association nor documentation of brain infection (as opposed to secondary effects of hemorrhage and edema) has been obtained.

It thus appears that in primate hosts, dengue and YF viruses rarely if ever cross the blood–brain barrier and establish CNS infection, despite the development of exceedingly high levels of viremia and extraneural infection and (in the case of YF) the encephalitogenic potential of some unadapted virus strains. The susceptibility of cells at the gateway of the brain (capillary endothelial cells and olfactory neurons) to dengue and YF viruses has not been clearly determined in primates, although neurotropic strains of YF virus are pathogenic following intranasal inoculation (Findlay and Clarke, 1935). However, a more likely explanation is provided by the genetic heterogeneity of parental virus populations and the selective replication advantage of viscerotropic virions. Studies in which mixtures of neurotropic and viscerotropic YF virus strains were inoculated by the intraperitoneal route into mice, hedgehogs, and monkeys showed that the viscerotropic virus subpopulation replicated more rapidly than the neurotopic strain (Theiler, 1951).

C. Pathogenesis of Dengue Hemorrhagic Fever/Dengue Shock Syndrome

The pathogenesis of DHF/DSS has been the subject of many comprehensive reviews (Halstead, 1980, 1982a,b; Bhamarapravati, 1981; Pang, 1983), and a brief synopsis is thus appropriate here. DHF/DSS occurs almost exclusively in persons who experience a secondary infection with a heterologous dengue serotype; the seroepidemiological evidence for this is reviewed by Halstead (1980) and is strongly supported by recent studies in Thailand (Sangkawibha *et al.*, 1984). The overall risk of acquiring DHF/DSS in cases of secondary infection is 2–6%, more than 100 times higher than in cases of primary infection. Young age, female sex, good nutritional status, an interval of less than 5 years between infections, and sequence of infection (e.g., dengue 1 followed by dengue 2 in Thailand), and strain of virus are identified risk factors (Halstead 1980, 1982a). That genetic factors may be involved was suggested by the significantly higher incidence of DHF in whites than in blacks in Cuba in 1981 (Guzman *et al.*, 1985). Human leukocyte antigen (HLA)-associated factors in DHF have not been defined.

The basis for the DHF/DSS syndrome is an abnormal immunological response involving virus–antibody complexes, leukocytes, and complement. Central to the pathogenesis of the syndrome is the concept of IgG antibody-dependent enhancement (ADE) of dengue viral replication in Fc-receptor-bearing monocytes (Halstead *et al.*, 1973a, 1976; Halstead and O'Rourke, 1977; Halstead, 1980). Preformed nonneutralizing heterotypic IgG antibodies complexed with virus promote both attachment to and infection of monocytes (Gollins and Porterfield, 1984), leading to progressive amplification of replication in these target cells in the host. The mechanisms involved, derived largely from *in vitro* studies employing peripheral-blood leukocytes and Mϕlike cell lines, are reviewed in Chapter 11. A second type of enhancement, dependent on the CR3 complement receptor, has been described (Cardosa *et al.*, 1983). Evidence for an *in vivo* role for ADE comes from experimental sequential infections of monkeys (Halstead *et al.*, 1973b; Marchette *et al.*, 1973) and challenge of monkeys after passive administration of antibodies (Halstead, 1979), in which viremia and tissue virus titers were higher than in nonimmune controls. Peripheral-blood leukocytes from immune humans and monkeys have also been shown to replicate dengue virus to higher titer than leukocytes from nonimmunes (Halstead *et al.*, 1973a). These observations have culminated in the hypothesis that ADE plays a role in DHF/DSS in persons with past heterotypic immunity or in young infants with either passively acquired heterotypic antibody or homotypic antibody that has waned in titer below the end point of neutralization.

The reasons for expression of severe disease in a small subset (2–6%) of persons with secondary infections are problematic. Several observations, including the finding of enhanced disease in monkeys with lymphoproliferative disorders and enhanced viremias after pertussis immunization (Halstead, 1980), as well as enhanced immune phagocytosis and replication in macrophages following activation with peptidoglycans and bacterial cell wall products (Hotta *et al.*, 1983), indicate that a variety of host-related factors could influence pathogenesis by increasing the number of permissive cells.

DHF/DSS usually occurs in individuals who have previously sustained only one heterotypic dengue infection. Both the interval between prior infection and the sequence of infecting serotypes are important determinants of disease expression (Halstead, 1980). Anamnestic antibody responses early in the course of secondary infection are typically broadly cross-reactive, but examination of early sera for neutralizing antibody titers may reveal the original antigenic sin phenomenon (Halstead *et al.*, 1983), thereby revealing the serotype responsible for primary infection. Original antigenic sin appears to be a general phenomenon for closely related flaviviruses (Inouye *et al.*, 1984). Aside from its practical usefulness in unraveling infection sequences in DHF/DSS, the phenomenon may exaggerate immunological enhancement.

In patients with DHF/DSS, the clinical syndrome, including signs of increased vascular permeability, appear coincident with activation of complement, fibrinogen consumption, and evidence for viral clearance. Halstead (1980, 1982a) has proposed that viral clearance mechanisms involving activation of infected monocytes/macrophages by cytotoxic T cells results in the release of complement-activating enzymes, thromboplastin, and vascular permeability factors. This cascade of events, illustrated in Fig. 10, amplifies itself and mediates pathophysiological events including plasma leakage, hypotension, and abnormal coagulation.

Evidence that CMI responses are involved in viral clearance has been recently reviewed (Pang, 1983) and may be summarized as follows: (1) Circulating killer T cells have been demonstrated in monkeys 5–7 days after dengue infection (Halstead, 1980); (2) human peripheral-blood mononuclear cells participate in cell-mediated cytotoxicity and ADCC reactions (Kurane *et al.*, 1984); (3) transformed lymphocytes are found in large numbers in buffy coat cells from DHF/DSS patients, and blast transformation in response to dengue antigen occurs in peripheral-blood leukocytes from sensitized monkeys (Bhamarapravati, 1981); (4) there is pathological evidence for perivascular inflammatory responses in skin and other tissues resembling a DTH reaction (Bhamarapravati *et al.*, 1967; Bhamarapravati, 1981); (5) the striking lymphocytolysis involving T-cell-dependent areas of nodes and spleen in autopsy material may reflect an active CMI response and production of cytotoxic factors; and (6) *in vivo* DTH responses demonstrated in experimental dengue infection of mice (Pang *et al.*, 1982) are mediated by two types of T cells (Ly $+ - 1.1^+$ and Ly $+ - 2.1^+$) requiring H-2 histocompatibility (Pang *et al.*, 1984).

The identity of the vascular permeabiity factor presumably released

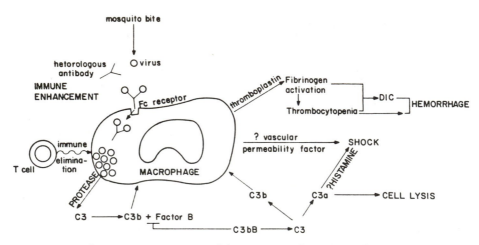

FIGURE 10. Schematic representation of the immune enhancement phenomenon and amplification and effector mechanisms in the pathogenesis of DHF/DSS. Heterologous dengue antibody mediates infection of macrophages. Infected macrophages express dengue viral antigens on plasma membrane and are activated by sensitized T cells. During the course of immune elimination, infected macrophages release proteases and thromboplastin and initiate activation cascades involving the complement and coagulation systems. Release of a vascular permeability factor may play a role in the development of shock. From Halstead (1982a).

by activated dengue-infected macrophages remains unknown; Pang (1983) proposed that it may be a leukotriene. The kinin system does not appear to be involved (Edelman *et al.*, 1975b). Russell and Brandt (1973) emphasized the role of histamine released from basophils/mast cells in response to complement activation and generation of C3A anaphylatoxin. In support of this concept, Bhamarapravati *et al.* (1967) noted that some cells in perivascular infiltrates of skin from fatal cases resembled degranulated mast cells, and Tuchinda *et al.* (1977) have reported increased urinary excretion of histamine by DHF patients. Increased levels of IgE have been found in sera from DHF/DSS patients compared to controls (Pavri *et al.*, 1979), leading Pavri and Prasad (1980) to speculate that IgE-mediated histamine release plays a role in the pathogenesis of shock and that underlying parasitic disease may be responsible for elevated IgE. Against the role of histamine-mediated type I hypersensitivity in this disease are observations that DHF/DSS patients do not respond clinically to corticosteroids or antihistamines, nor do they manifest cutaneous signs such as urticaria (Halstead, 1982a).

Hemodynamic and metabolic disturbances associated with increased vascular permeability and shock have been investigated. Patients with DHF/DSS have increased peripheral vascular resistance and diminished cardiac output, presumably due to decreased venous return (Pongpanich and Kumponpant, 1973; Futrakul *et al.*, 1977). In light of other evidence, however, direct viral injury and myocardial dysfunction would appear to deserve further study. Metabolic alterations including hyperkalemia, metabolic acidosis, and hypoxemia (Cohen and Halstead, 1966) are probable terminal events in patients with shock (see also Section V.D).

The etiology of the hemorrhagic diathesis in DHF (reviewed by Halstead, 1982b) includes vascular injury, thrombocytopenia, and coagulopathy. Increased capillary fragility (positive tourniquet test) and petechial hemorrhages may be mediated by vasocative mediators, since light- and electron-microscopic studies have not implicated direct viral injury or replication (Sahaphong et al., 1980). Thrombocytopenia may be due to decreased platelet production, but megakaryocytic arrest observed in bone marrow does not appear to coincide temporally with the fall in platelet count (Nelson et al., 1964). Increased turnover of platelets and sequestration in the liver were demonstrated in kinetic studies by Mitrakul et al. (1977) and increased platelet adhesiveness by Doury et al. (1976). Funahara et al. (1982) have presented evidence for direct attachment of dengue viruses to platelets with subsequent immune elimination, thus explaining the increased severity of thromobocytopenia in cases of secondary infection. Conflicting reports have appeared for an autoimmune basis for thrombocytopenia due to antiplatelet antibodies (Basanta Otero et al., 1983). Defects in platelet function have also been described (Mitrakul et al., 1977; Almagro Vasquez et al., 1983).

A degree of consumption coagulopathy has been demonstrated in a number of clinical studies (World Health Organization, 1973; Srichaikul et al., 1977; Doury et al., 1980); generally, this has been mild, and the coagulation defect has not correlated with severity of illness. In unusual cases involving older patients with severe gastrointestinal hemorrhage, however, disseminated intravascular coagulation may be an important event. Release of thromboplastin by activated dengue-infected macrophages has been suggested as a mechanism for the coagulopathy in DHF/DSS (Halstead 1982a,b).

Autoantibodies against smooth muscle or mitochrondria have been reported in association with both DHF (Boonpucknavig and Udomsangpetch, 1983) and YF (J. A. Smith et al., 1973). No clinical significance has been attached to these findings, and a number of possible mechanisms may be involved. A recent report showing a monoclonal antibody cross-reaction between flaviviral-specified and host-cell proteins (Gould et al., 1983) is of interest in this regard.

D. Yellow Fever: Pathophysiological Correlations

The availability of suitable nonhuman primate models of YF (which do not exist for DHF/DSS) provides the opportunity to define pathogenesis of this disease, but has not been exhaustively utilized. The evidence suggests similiarity between certain aspects of YF and dengue pathogenesis, although two elements stand apart: (1) lack of evidence for a pathological role for antibody-mediated enhancement and (2) the greater hepatotropism and more severe hepatocellular injury in YF.

YF virus has been shown to replicate in monocytes, phytohemagglutinin (PHA)-stimulated lymphocytes, and Mϕlike cell lines (Wheelock and Edelman, 1969; Yamamoto and Hotta, 1981; J. J. Schlesinger and Brandriss, 1981; Liprandi and Walder, 1983), as well as in lymph nodes *in vivo* (Theiler, 1951). Kupffer cells are permissive to viral growth and are important sites of early replication and spread to surrounding liver parenchymal cells. Although YF occurs in geographic areas that are hyperendemic for a number of related flaviviruses, cross-protection rather than immunological enhancement occurs (Monath *et al.*, 1980a). Experimental studies in monkeys support this conclusion (Henderson *et al.*, 1970; Theiler and Anderson, 1975) and show that cross-protection (and absence of enhancement by heterologous viruses) occurs in the absence of YF neutralizing antibodies.

In experimental YF, as in human DHF/DSS, there is marked cytolysis of thymus and T-cell-dependent areas of lymph nodes, spleen, and Payer's patches, but these changes are less marked in human YF. The degree of liver-cell injury, reflected by hyperbilirubinemia and elevation of serum transminases, correlates temporally and quantitatively with disease severity (Oudart and Rey, 1970; Monath *et al.*, 1981). Kupffer cell and hepatocellular necrosis may be central events leading to a number of secondary physiological disturbances, due to impaired biosynthetic and detoxification mechanisms.

Multiple coagulation defects occur in experimental and human YF, including thrombocytopenia and reduction in clotting factors. Although consumption coagulopathy has been documented in monkeys (Dennis *et al.*, 1969) and in humans (Santos *et al.*, 1973), decreased synthesis of vitamin-K-dependent clotting factors by the diseased liver is probably the more important mechanism.

The basis for the renal injury in YF is not entirely clear, but evidence from the monkey model suggests that it may be secondary and physiological in nature (Monath *et al.*, 1981). Studies during the 24 hr before death indicate that renal failure is due to a functional decrease in glomerular filtration rate and renal plasma flow, possibly reflecting increased renovascular resistance, and that renal tubular necrosis occurs as a terminal event associated with shock.

As in DHF/DSS, profound hemodynamic and metabolic alterations occur in severe YF and are probable terminal events. Experimentally infected monkeys become progressively hypotensive, hypoxic, and hyperkalemic during the final hours before death (Monath *et al.*, 1981). Significant changes in cell metabolism reflected in the distribution of tissue water, electrolytes, and trace metals in vital organs (cardiac muscle, medulla oblongata) have been observed (C. T. Liu and Griffin, 1982) and may contribute to cardiorespiratory failure. In contrast to DHF/DSS, in which hemoconcentration, leakage of plasma proteins into the extravascular space, and hypovolemia occur, YF-infected monkeys were found to have reduced hematocrits and expanded blood and plasma volumes (C. T. Liu and Griffin, 1982).

There are no studies of circulating mediators of shock in YF. Because of the lesions in reticuloendothelial tissues, Monath *et al.* (1981) speculated that impaired detoxification of bacterial endotoxin may be involved. In an alphavirus (Venezuelan encephalitis) infection of hamsters, lymphoid necrosis and depletion of reticuloendothelial function have been shown to cause bacterial endotoxemia and death (Gorelkin and Jahrling, 1975). Endotoxin has varied effects on many systems, including mediation of release of effector substances (procoagulants, prostaglandins, and proteolytic enzymes) from Mφ (Nolan, 1981). This potential mechanism for shock in YF and other hemorrhagic fevers should be investigated because of the implication for therapy.

The role of immune responses in recovery from YF has not been characterized. Clearance of viremia coincides with the appearance of neutralizing antibodies and approximately with the onset of severe organ dysfunction (period of intoxication); both antibody and virus (and thus immune complexes) have been found in plasma. Virus replication continues in tissues after disappearance of viremia, but partial or complete clearance from tissues may be evident by the time of death. Both antibody- and cell-mediated immune responses may be involved in viral clearance. Activated T cells have been described in humans given 17D vaccine (Ehrnst *et al.*, 1978), including cells that exhibit cytotoxicity for NK-sensitive target-cell lines (Fagraeus *et al.*, 1982). PHA-stimulated peripheral-blood leukocytes taken between 7 and 11 days after YF vaccination are refractory to virus infection (Wheelock *et al.*, 1970). Unlike DHF/DSS, inflammatory-cell infiltration is not a significant component of the histopathological picture in YF, and direct viral injury rather than immunopathology would appear to explain the pathogenesis of this acute infection. It is possible, however, that deposition of virus–antibody complexes in renal glomeruli could be responsible for morphological changes (Barbareschi, 1957) and albuminuria. Moreover, the contribution of immune responses to the lymphoid necrosis in YF remains uncertain (see below).

E. Immune Modulation, Suppression, and Lymphocytolysis

A large body of information has resulted from studies showing immune suppression to homologous and heterologous antigens in mice infected with dengue virus (M. Nagarkatti and P. S. Nagarkatti, 1979; M. Nagarkatti *et al.*, 1980; Chaturvedi *et al.*, 1978; Tandon *et al.*, 1979a,b). Two types of immune suppression have been shown, one dengue-virus-specific and a second nonspecific. A T suppressor (TS_1)-cell response occurs in spleens of dengue-primed mice. TS_1 lymphocytes produce a soluble antigen-specific suppressor factor (Chaturvedi and Shukla, 1981) that is presented by direct contact of Mφlike cells to a second subset of T suppressor cells (TS$_2$); these cells in turn release prostaglandin, which mediates suppression of dengue-virus-specific IgM antibody PFC *in vitro* and *in vivo* (Chaturvedi and Shukla, 1981; Chaturvedi *et al.*, 1981a, 1982a; Shukla and Chaturvedi, 1982, 1983). A third generation of T cells is apparently required to mediate suppression in this system (Shukla and Chaturvedi, 1984).

In contrast, nonspecific immunosuppression is mediated by release of a lymphokine cytotoxic factor (CF) from a different T-cell Ly phenotype (Shukla *et al.*, 1982). CF kills T lymphoctes and Mφ in spleen, lymph nodes, and thymus (Chaturvedi *et al.*, 1980b,c; 1981a,c) and suppresses the immune response to sheep erythrocytes (Chaturvedi *et al.*, 1981b). CF does not kill circulating leukocytes of mice or humans, but adversely affects various functions including phagocytosis by Mφ and E-rosette formation (Chaturvedi *et al.*, 1982b; Gulati *et al.*, 1982). CF, produced by splenic T lymphocytes in mice, peaks between 9 and 11 days after inoculation of dengue virus. Contact with CF induces macrophages to produce an effector cytotoxic factor (CF2), thus providing an amplification step (Gulati *et al.*, 1983a). CF has been partially characterized and found to be heat-labile, trypsin-sensitive, and dialyzable (Chaturvedi *et al.*, 1983a) and has been shown to act on target cells through damage to plasma membranes (Gulati *et al.*, 1983b). In other investigations with mice, dengue virus infection reduced macrophage Fc-mediated attachment and ingestion of opsonized sheep erythrocytes (Chaturvedi *et al.*, 1983b).

In similar studies by Wong *et al.* (1984), infection of mice with dengue virus resulted in transient immunosuppression of the DTH response to sheep erythrocytes, beginning on day 3 after priming. In contrast to the studies described above, there was no evidence for a soluble CF, since suppression was transferable by spleen cells, but not by serum. The cells responsible for suppression had both Ly-1+ and Ly-2+ markers. P. S. Nagarkatti and M. Nagarkatti (1983) have also demonstrated enhanced T-suppressor-cell activity in dengue-virus-infected mice, regulating both T- and B-cell responses and adoptively transferable to normal mice.

Little is known about the relevance of these observations from experiments with mice to human dengue, in which heightened immune responsiveness rather than immunosuppression appears to occur. Nevertheless, regulation of immune responses in human dengue undoubtedly incorporates T-suppressor-cell activity, and disturbances of regulation could be involved in pathological reactions. A preliminary study of human DHF cases in Cuba (Santos Lagresa *et al.*, 1983) showed reduced T-lymphocyte responsiveness to mitogens and reversal by indomethacin, an inhibitor of prostaglandins [the presumed mediator of dengue-specific T-suppressor activity (Chaturvedi *et al.*, 1981a)]. Another form of auto-regulation of the immune response may be provided by autocytotoxic antilymphocyte antibodies found in human DHF cases (Boonpucknavig and Udomsangpetch, 1983; Gilbreath *et al.*, 1983).

Little information of a similar nature is available for YF viral infection in mice or primates. Vaccination with YF 17D virus did not suppress tuberculin skin test reactivity in one study (Marvin *et al.*, 1968).

Several possible mechanisms may be put forward to explain the lymphocytolysis and reduction in circulating T cells observed in dengue and YF, including: (1) virus-induced release of cytotoxic factors by sensitized T cells (Chaturvedi *et al.*, 1980b, 1981c), (2) complement-mediated cytolysis by antilymphocyte antibodies (Gilbreath *et al.*, 1983), (3) local production of lymphotoxins (e.g., C3a) by activated macrophages (Halstead, 1982b), and (4) cytolysis mediated by high levels of endogenous or iatrogenically administered corticosteroids.

F. Persistent and Congenital Infections

Persistent infections with dengue and YF viruses are readily established in a number of cell cultures, but there is only meager evidence for their occurrence *in vivo*. Prolonged synthesis of IgM antibodies has been reported following YF 17D vaccination of humans (Monath, 1971), suggesting that the remarkable longevity of immunity to the live vaccine may reflect repeated or continuous antigen stimulation. Persistence of 17D virus in brains of intracerebrally inoculated monkeys for up to 159 days, without change in mouse virulence markers, has been reported (Penna and Bittencourt, 1943); this may reflect the importance of neural tissue in flaviviral persistence and is not an element of natural YF infections.

Congenital infections of lambs and calves with WSL virus have been described in an earlier section. There is no clinical or experimental evidence for congenital infection with dengue (Mirovsky *et al.*, 1962) or YF viruses.

G. Genetic Host Resistance

Coevolution of YF virus with its nonhuman primate hosts in Africa has apparently led to the development of host resistance and a balanced host–virus relationship. Although most species develop high titers of viremia, African primates are not susceptible to clinical illness or death. In contrast, many species of neotropical monkeys succumb to lethal YF infection, consistent with the probable recent introduction of YF virus from Africa during the 16th century (Warren, 1951).

Associations between HLA genes and susceptibility to YF or dengue have not been investigated. However, an analysis of gene frequencies among descendants of Dutch settlers in Surinam, who had survived outbreaks of YF, showed variations that were unlikely to be due to drift (De Vries *et al.*, 1979). The authors suggested that these variations may be due to selection of resistance genes.

Another interesting observation indicates that genetic factors, while they do not affect host resistance/susceptibility *per se*, may play a role in viral transmission. *Aedes aegypti*, the urban vector of dengue and YF viruses, preferentially takes blood meals from humans with blood group O rather than group A or B (Wood, 1976).

As mentioned previously, there was a significant excess of DHF cases among whites compared to blacks during the Cuban epidemic in 1981; this was not explained by racial differences in immunological background.

REFERENCES

Albagai, C., and Chaimoff, R., 1959, A case of West Nile myocarditis, *Harefuah* **57**:274–275.

Albrecht, P., 1960, Pathogenesis of experimental infection with tick-borne encephalitis virus, in: *Biology of Viruses of the Tick-Borne Encephalitis Complex* (H. Libikova, ed.), pp. 247–257, Academic Press, New York.

Albrecht, P., 1968, Pathogenesis of neurotropic arbovirus infections, *Curr. Top. Microbiol. Immunol.* **43**:44–91.

Almagro Vasquez, D., Gonzalez Cabrera, I., Cruz Gomez, Y., and Castaneda Morales, M., 1983, Platelet function in dengue hemorrhagic fever, *Acta Haematol.* **70**:276–277.

Andersen, A. A., and Hanson, R. P., 1970, Experimental transplacental transmission of St. Louis encephalitis virus in mice, *Infect. Immun.* **2**:230–325.

Andersen, A. A., and Hanson, R. P., 1974, Influence of sex and age on natural resistance to St. Louis encephalitis virus infection in mice, *Infect. Immun.* **9**:1123–1125.

Andersen, A. A., and Hanson, R. P., 1975, Intrauterine infection of mice with St. Louis encephalitis virus: Immunological, physiological, neurological, and behavioral effects in progeny, *Infect. Immun.* **12**:1173–1183.

Asher, D. M., 1979, Persistent tick-borne encephalitis infection in man and monkeys: Relation to chronic neurologic disease, in: *Arctic and Tropical Arboviruses*, Proceedings of the 2nd International Symposium on Arctic Arboviruses (E. Kurstak, ed.), pp. 179–195, Academic Press, New York.

Aung-Khin, M., Khin, Ma-Ma, Thant-Zin, 1975, Changes in the tissues of the immune system in dengue hemorrhagic fever, *J. Trop. Med. Hyg.* **78**:256–261.

Badman, R. T., Campbell, J., and Alfred, J., 1984, Arbovirus infections in horses—Victoria, 1984, Communicable Disease Intell No. 84/17, Australia.

Barbareschi, G., 1957, Glomenulosi tossica in febbre gialla, *Rev. Biol. Trop.* **5**:201–209.

Barnes, W. J. S., and Rosen, L., 1974, Fatal hemorrhagic disease and shock associated with primary dengue infection on a Pacific Island, *Am. J. Trop. Med. Hyg.* **23**:495–506.

Basanta Otero, P., Gonzalez Villalonga, C., and Orbeal Aldama, L., 1983, Platelet autoantibodies in dengue hemorrhagic fever, *Acta Haematol.* **70**:141–142.

Bearcroft, W. G. C., 1957, The histopathology of the liver of yellow fever infected rhesus monkey, *J. Pathol. Bacteriol.* **74**:295–303.

Bhamarapravati, N., 1981, Pathology and pathogenesis of DHF, in: *Dengue Hemorrhagic Fever, 1981* (S. Hotta, ed.), pp. 207–214, International Center for Medical Research Kobe University School of Medicine Kobe, Japan.

Bhamarapravati, N., and Boonpucknavig, V., 1966, Immunofluorescent study of dengue virus in human cases, *Bull. W.H.O.* **35**:50.

Bhamarapravati, N., Tuchinda, P., and Boonyapaknavik, V., 1967, Pathology of Thailand haemorrhagic fever: A study in 100 autopsy cases, *Ann. Trop. Med. Parasitol.* **61:**500–510.

Bhatt, P. N., and Jacoby, R. O., 1976, Genetic resistance to lethal flavivirus encephalitis. II. Effect of immunosuppression, *J. Infect. Dis.* **134:**166–173.

Bierman, H. R., and Nelson, E. R., 1965, Hematodepressive virus diseases of Thailand, *Ann. Intern. Med.* **62:**867–884.

Blackburn, N. K., and Swanepoel, R., 1980, An investigation of flavivirus infections of cattle in Zimbabwe, Rhodesia with particular reference to Wesselsbron virus, *J. Hyg.* **85:**1–33.

Bond, J. O., 1969, St. Louis encephalitis and dengue fever in the Caribbean area: Evidence of possible cross-protection, *Bull. W.H.O.* **40:**160–163.

Boonpucknavig, S., and Udomsangpetch, R., 1983, Autoantibodies in viral infection, *J. Clin. Lab. Immunol.* **10:**171–172.

Boonpucknavig, V., Bhamarapravati, N., Boonpucknavig, S., Futradul, P., and Tanpaichitr, P., 1976a, Glomerular changes in dengue hemorrhagic fever, *Arch. Pathol. Lab. Med.* **100:**206–212.

Boonpucknavig, S., Bhamarapravati, N., Nimmannitya, S., Phalavadhtana, A., and Siripont, J., 1976b, Immunofluorescent staining of the surfaces of lymphocytes in suspension from patients with dengue hemorrhagic fever, *Am. J. Pathol.* **85:**37–47.

Boonpucknavig, S., Lohachitranond, C., and Nimmannitya, S., 1979a, The pattern and nature of the lymphocyte population response in dengue hemorrhagic fever, *Am. J. Trop. Med. Hyg.* **28:**885–889.

Boonpucknavig, S., Boonpucknavig, V., Bhamarapravati, N., and Nimmaranitya, S., 1979b, Immunofluorescent study of skin rash in patients with dengue hemorrhagic fever, *Arch. Pathol. Lab. Med.* **103:**463–466.

Boonpucknavig, S., Vettiviroj, O., and Boonpucknavig, V., 1981, Infection of young adult mice with dengue virus type 2, *Trans. R. Soc. Trop. Med. Hyg.* **75:**647–653.

Bowen, G. S., Monath, T. P., Kemp, G. E., Kerschner, J. H., and Kirk, L. J., 1980, Geographic variation among St. Louis encephalitis virus strains in the viremic responses of avian hosts, *Am. J. Trop. Med. Hyg.* **29:**1411–1419.

Bradish, C. J., Fitzgeorge, R., and Titmuss, D., 1980, The responses of normal and athymic mice to infections by togaviruses: Strain differentiation in active and adoptive immunization, *J. Gen. Virol.* **46:**255–265.

Brandriss, M. W., and Schlesinger, J. J., 1984, Passive protection of mice to intracerebral challenge with 17D yellow fever viruses by monoclonal antibodies to 17D yellow fever and to dengue 2, *Abstr. XI Int. Cong. Trop. Med. Malaria*, Calgary, Alb., p. 13.

Brinker, K. R., and Monath, T. P., 1980, The acute disease, in: *St. Louis Encephalitis* (T. P. Monath, ed.), pp. 503–534, *American Public Health Association*, Washington, D.C.

Burke, D. S., Schmaljohn, C. S., and Dalrymple, J. M., 1985a, Strains of Japanese encephalitis virus isolated from human brains have a highly conserved genotype compared to strains isolated from other natural hosts, *Abstr. Ann. Meet. Am. Soc. Virol.*, Univ. New Mex., Albuquerque.

Burke, D. S., Lorsomondee, W., Leake, C., Hoke, C. H., Nisalak, A., Chongswasdi, V., and Laorakpongse, T., 1985b, Fatal outcome in acute Japanese encephalitis, *Am. J. Trop. Med. Hyg.* **34:**1203–1210.

Burke, D. S., Nisalak, A., Ussery, M., Laorakpongse, T., and Chantaribul, S., 1985c, Kinetics of Japanese encephalitis virus immunoglobulin M and G antibodies in human serum and cerebrospinal fluid, *J. Infect. Dis.* (in press).

Buxton, D., and Reid, H. W., 1975, Experimental infection of red grouse with louping ill virus (flavivirus group). II. Neuropathology, *J. Comp. Pathol.* **85:**231–235.

Camain, R., and Lambert, D., 1966, Histopathologie des foies amarils preleves postmortem et par ponction-biopsie hepatique au cours de l'epidemie de Diourbel (Senegal), Novembre–Decembre, 1965, *Bull. Soc. Med. Afr. Noire Lang. Fr.* **11:**522–540.

Camenga, D. L., and Nathanson, N., 1975, An immunopathologic component in experimental togavirus encephalitis, *J. Neuropathol. Exp. Neurol.* **34**:492–500.

Camenga, D. L., Nathanson, N., and Cole, G. A., 1974, The relative influence of cellular and humoral factors in the modification of cyclophosphamide-potentiated West Nile virus encephalitis, *J. Infect. Dis.* **130**;634–641.

Cantanzaro, P. J., Brandt, W. E., Hogrefe, W. R., and Russell, P. K., 1974, Detection of dengue cell surface antigens by peroxidase labelled antibodies and immune cytolysis, *Infect. Immun.* **10**:381–388.

Cardosa, M. J., Porterfield, J. S., and Gordon, S., 1983, Complement receptor mediates enhanced flavivirus replication in macrophage, *J. Exp. Med.* **158**:258–263.

Cardosa, M. J., Gordon, S., Hirsch, S., Springer, T. A., and Porterfield, J. S., 1986, Interaction of West Nile virus with primary murine macrophages: Role of cell activation and receptors for antibody and complement, *J. Virol.* (in press).

Casals, J., Henderson, B. E., Hoogstraal, H., Johnson, K., and Shelokov, A., 1970, A review of Soviet hemorrhagic fevers, 1969, *J. Infect. Dis.* **122**:437–453.

Chamberlain, R. W., 1958, Vector relationships of the arthropod-borne encephalitides in North America, *Ann. N. Y. Acad. Sci.* **70**:312–319.

Chaturvedi, U. C., and Shukla, M. I., 1981, Characterization of the suppressor factor produced in the spleen of dengue virus infected mice, *Ann. Immunol. (Paris)* **132C**:245–255.

Chaturvedi, U. C., Tandon, P., Mathur, A., and Kumar, A., 1978, Host defense mechanisms against dengue virus infection in mice, *J. Gen. Virol.* **39**:293–302.

Chaturvedi, U. C., Mathur, A., Tandon, P., Natu, S. M., Rajvanshi, S., and Tandon, H. O., 1979, Variable effect on peripheral blood leukocytes during JE virus infection of man, *Clin. Exp. Immunol.* **38**:492–498.

Chaturvedi, U. C., Mathur, A., Chandra, A., Das, S. K., Tandon, H. O., and Singh, U. K., 1980a, Transplacental infection with Japanese encephalitis virus, *J. Infect. Dis.* **141**:712–715.

Chaturvedi, U. C., Bhargava, A., and Mathur, A., 1980b, Production of cytotoxic factor in the spleen of dengue virus-infected mice, *Immunology* **40**:653–658.

Chaturvedi, U. C., Dalakoti, H., and Mathur, A., 1980c, Characterization of the cytotoxic factor produced in the spleen of dengue virus-infected mice, *Immunology* **41**:387–392.

Chaturvedi, U. C., Shukla, M. I., and Mathur, A., 1981a, Thymus dependent lymphocytes of the dengue virus infected mouse spleen mediate suppression through prostaglandin, *Immunology* **42**:1–6.

Chaturvedi, U. C., Shukla, M. I., Mathur, K. R., and Mathur, A., 1981b, Dengue virus-induced cytotoxic factor suppresses immune response of mice to sheep RBC, *Immunology* **43**:311–316.

Chaturvedi, U. C., Mathur, K. R., Gulati, L., and Mathur, A., 1981c, Target lymphoid cells for the cytotoxic factor produced in the spleen of dengue virus-infected mice, *Immunol. Lett.* **3**:13–16.

Chaturvedi, U. C., Shukla, M. I., and Mathur, A., 1982a, Role of macrophages in transmission of dengue virus-induced suppressor signal to a subpopulation of T lymphocytes, *Ann. Immunol. (Paris)* **133C**:83–92.

Chaturvedi, U. C., Gulati, L., and Mathur, A., 1982b, Inhibition of E-rosette formation and phagocytosis by human blood leukocytes after treatment with dengue virus-induced cytotoxic factor, *Immunology* **45**:679–685.

Chaturvedi, U. C., Gulati, L., and Mathur, A., 1983a, Further studies on the production of dengue virus-induced macrophage cytotoxin, *Ind. J. Exp. Biol.* **21**:235–279.

Chaturvedi, U. C., Nagar, R., and Mathur, A., 1983b, Effect of dengue virus infection on Fc receptor functions of mouse macrophages, *J. Gen. Virol.* **64**:2399–2407.

Chunikhin, S. P., and Kurenkov, V. B., 1979, Viraemia in *Clethrionomys glareolus*—a new ecological marker of tick-borne encephalitis virus, *Acta Virol.* **23**:257–260.

Clarke, D. H., 1960, Antigenic analysis of certain group B arthropod-borne viruses by antibody absorption, *J. Exp. Med.* **111**:21–32.

Clarke, D. H., 1969, Further studies on antigenic relationships among the viruses of the group B tick-borne complex, *Bull. W.H.O.* **31**:45–56.

Coetzer, J. A. W., and Barnard, J. H., 1977, *Hydrops amnii* in sheep with hydranencephaly and arthrogryposis with Wesselsbron and Rift Valley fever as aetiological agents, *Onderstepoort J. Vet. Res.* **44**:119–126.

Coetzer, J. A. W., and Theodoridis, A., 1982, Clinical and pathological studies in adult sheep and goats experimentally injected with Wesselsbron disease virus, *Onderstepoort J. Vet. Res.* **49**:19–22.

Coetzer, J. A. W., Theodoridis, A., and Van Heerden, A., 1978, Wesselsbron disease: Pathological, haematological and clinical studies in natural lcases and experimentally infected newborn lambs, *Onderstepoort J. Vet. Res.* **45**:93–106.

Cohen, S. N., and Halstead, S. B., 1966, Shock associated with dengue infection. 1. Clinical and physiologic manifestations of dengue hemorrhagic fever in Thailand, 1964, *J. Pediatr.* **68**:448–456.

Cole, G. A., and Nathanson, N., 1968, Potentiation of experimental arbovirus encephalitis by immunosuppressive doses of cyclophosphamide, *Nature (London)* **220**:399–401.

Cole, G. A., and Wisseman, C. L., Jr., 1969, The effect of hyperthermia on dengue virus infection of mice, *Proc. Soc. Exp. Biol. Med.* **130**:359–363.

Cypress, R. H., Lubiniecki, A. S., and Hammon, W. McD., 1973, Immunosuppression and increased susceptibility to Japanese B encephalitis virus in *Trichinella sprialis*-infected mice, *Proc. Soc. Exp. Biol. Med.* **143**:469–473.

Darnell, M. B., and Koprowski, H., 1974, Genetically determined resistance to infection with group B arboviruses. II. Increased production of interfering particles in cell culture from resistant mice, *J. Infect. Dis.* **129**:208–256.

David-West, T. S., 1975, Concurrent and consecutive infection and immunization with yellow fever and UGMP-359 virus, *Arch. Virol.* **48**:21–28.

David-West, T. S., and Smith, J. A., 1971, Yellow fever virus infection: A correlation of complement fixing antigen with histopathology, *Br. J. Exp. Pathol.* **52**:114–121.

Del Canto, M. C., and Rabinowitz, S. G., 1982, Experimental models of virus-induced demyelination of the central nervous system, *Ann. Neurol.* **11**:109–127.

Dennis, L. H., Reisberg, B. E., Crosbie, J., Crozier, D., and Conrad, M. E., 1969, The original haemorrhagic fever: Yellow fever, *Br. J. Haematol.* **17**:455–462.

DeVries, R. R. P., Meera Khan, P., Bernini, L. F., van Loghem, E., and van Rood, J. J., 1979, Genetic control of survival in epidemics, *J. Immunogenet.* **6**:271–287.

Doherty, P. C., and Reid, H. W., 1971, Louping ill encephalomyelitis in the sheep. II. Distribution of virus and lesions in nervous tissue, *J. Comp. Pathol.* **81**:531–536.

Doi, R., Oya, A., Shirasaka, A., Yabe, S., and Sasa, M., 1983, Studies on Japanese encephalitis virus infection of reptiles. II. Role of lizards on hibernation of Japanese encephalitis virus, *Jpn. J. Exp. Med.* **53**:125–134.

Doury, J. C., Teyssier, J., Forcain, A., and Doury, F., 1976, Modifications de l'adhésivité plaquettaire au cours de la dengue à forme hémorrhagiques, *Bull. Soc. Pathol. Exot.* **69**:493–495.

Doury, J. C., Teyssier, J., Doury, F., Gentile, B., and Forçain, M., 1980, Dengue à forme hémorrhagique: Mise en evidence d'un syndrome de coagulopathie de consommation, *Med. Trop. (Marseille)* **40**:127–135.

Duffy, C. E., and Murphree, O. D., 1959, Maze performance of mature rats recovered from early postnatal infection with Murray Valley encephalitis, *J. Comp. Physiol. Psychol.* **52**:175–180.

Duffy, O. D., Murphree, O. D., and Morgan, P. N., 1958, Learning deficit in mature rats recovered from early post-natal infection with West Nile virus, *Proc. Soc. Exp. Biol. Med.* **98**:242–244.

Eckels, K. H., Brandt, W. E., Harrison, V. R., McConn, J. M., and Russell, P. K., 1976, Isolation of a temperature-sensitive dengue 2 virus under conditions suitable for vaccine development, *Infect. Immun.* **14**:1221–1227.

Edelman, R., Schneider, R. J., and Chieowanich, P., 1975a, The effect of dengue virus infection on the clinical sequelae of Japanese encephalitis: A one year followup study in Thailand, *Southeast Asian J. Trop. Med. Public Health* **6:**308–315.

Edelman, R., Nimmannitya, S., Colman, R. W., Talamo, R. C., and Top, F. H., Jr., 1975b, Evaluation of the plasma kinin system in dengue hemorrhagic fever, *J. Lab. Clin. Med.* **86:**410–421.

Edelman, R., Schneider, R. J., Vejjajiva, A., Pornpibul, R., and Voodhikul, P., 1976, Persistence of virus specific IgM and clinical recovery after Japanese encephalitis, *Am. J. Trop. Med. Hyg.* **23:**733–738.

Ehrenkrantz, N. J., Zemel, E. S., Bernstein, C., and Slater, K., 1974, Immunoglobulin M in the cerebrospinal fluid of patients with arbovirus encephalitis and other infections of the central nervous system, *Neurology* **24:**976–980.

Ehrnst, A., Lambert, B., and Fagraeus, A., 1978, DNA synthesis in subpopulations of blood mononuclear leucocytes in human subjects after vaccination against yellow fever, *Scand. J. Immunol.* **8:**339–346.

El Dadah, A. H., Nathanson, N., and Sarsitis, R., 1967, Pathogenesis of West Nile encephalitis in mice and rats. I. Influence of age and species on mortality and infections, *Am. J. Epidemiol.* **86:**765–775.

Elton, N. W., Romero, A., and Trejos, A., 1955, Clinical pathology of yellow fever, *Am. J. Clin. Pathol.* **25:**135–146.

Embil, J. A., Camfield, P., Artsob, H., and Chase, D. P., 1983, Powassan virus encephalitis resembling herpes simplex encephalitis, *Arch Intern. Med.* **143:**341–343.

Estrin, W. J., 1976, The serological diagnosis of St. Louis encephalitis in a patient with the syndrome of opsoclonia, body tremulousness and benign encephalitis, *Ann. Neurol.* **1:**596–598.

Fagraeus, A., Ehrnst, A., Klein, E., Patarroyo, M., and Goldstein, G., 1982, Characterization of blood mononuclear cells reacting with K562 cells after yellow fever vaccination, *Cell. Immunol.* **67:**37–48.

Findlay, G. M., and Clarke, L. P., 1934, Susceptibility of hedgehog to yellow fever: Viscerotropic virus, *Trans. R. Soc. Trop. Med. Hyg.* **28:**193–200.

Findlay, G. M., and Clarke, L. P., 1935, Infection with neurotropic yellow fever virus following instillation into nares and conjunctival sac, *J. Pathol. Bacteriol.* **40:**55–64.

Fitzgeorge, R., and Bradish, C. J., 1980, The *in vivo* differentiation of strains of yellow fever virus in mice, *J. Gen. Virol.* **46:**1–13.

Fleming, P., 1977, Age-dependent and strain-related differences of virulence of Semliki Forest virus in mice, *J. Gen. Virol.* **37:**93–105.

Fokina, G. I., Malenko, G. V., Levina, L. S., Koreshkova, G. V., Rzhakhova, O. E., Mamonenko, L. L., Pogodina, V. V., and Frolova, M. P., 1982, Persistence of tick-borne encephalitis virus. V. Virus localization after subcutaneous inoculation, *Acta Virol.* **26:**369–375.

Francis, T., Moore, D. L., Edington, G. M., and Smith, J. A., 1972, A clinicopathological study of human yellow fever, *Bull. W.H.O.* **46:**659–667.

Fresh, J. W., Reyes, V., Clarke, E. J., and Uylangco, C. V., 1969, Philippine hemorrhagic fever: A clinical, laboratory and necropsy study, *J. Lab. Clin. Med.* **73:**451–458.

Funahara, Y., Fujita, N., Okuno, Y., Ogawa, K., Hirata, M., Miki, M., and Kitaguchi, H., 1982, Virus-induced thrombocytopenia: *In vitro* studies on the dengue virus–platelet interaction, *Blood Vessel* **13:**341–344.

Futrakul, P., Mitrakul, C., Chumderpadetsuk, S., and Sitprija, V., 1977, Studies on the pathogenesis of dengue hemorrhagic fever: Hemodynamic alteration and effect of alpha blocking agent, *J. Med. Assoc. Thailand* **60:**610–614.

Gainer, S. H., and Pry, T. W., 1972, Effect of arsenicals on viral infections in mice, *Am. J. Vet. Res.* **33:**2299–2307.

Gajdosova, E., Mayer, V., and Oravec, C., 1980, Cross-reactive killer T lymphocytes in a flavivirus infection, *Acta Virol.* **24:**291–293.

Gajdosova, E., Oravec, C., and Mayer, V., 1981, Cell mediated immunity in flavivirus infections. I. Induction of cytotoxic T lymphocytes in mice by an attenuated virus from the tick-borne encephalitis complex and its group-reactive character, *Acta Virol.* **25:**10–18.

Gardner, J. J., and Reyes, M. G., 1980, Pathology, in: *St. Louis Encephalitis* (T. P. Monath, ed.), pp. 551–569, American Public Health Association, Washington, D.C.

Gerhard, W., and Koprowski, H., 1977, Persistence of virus-specific memory B cells in mice CNS, *Nature (London)* **266:**360–361.

Gilbreath, M. J., Pavarand, K., MacDermott, R. P., Ussery, M., Burke, D. S., Nimmannitya, S., and Tulyayon, S., 1983, Cold-reactive immunoglobulin M antilymphocyte antibodies directed against B cells in Thai children with dengue hemorrhagic fever, *J. Clin. Microbiol.* **17:**672–676.

Goldman, J., Bochna, A., and Becker, F. O., 1977, St. Louis encephalitis and subacute thyroiditis, *Ann. Intern. Med.* **87:**250.

Gollins, S. W., and Porterfield, J. S., 1984, Flavivirus infection enhancement in macrophages: Radioactive and biological studies on the effect of antibody on viral fate, *J. Gen. Virol.* **65:**1261–1272.

Gorelkin, L., and Jahrling, P. B., 1975, Virus-initiated septic shock: Acute death of Venezuelan encephalitis virus-infected hamsters, *Lab. Invest.* **32:**78–85.

Gould, E. A., Chanas, A. C., Buckley, A., and Clegg, C. S., 1983, Monoclonal immunoglobulin M antibody to Japanese encephalitis virus that can react with a nuclear antigen in mammalian cells, *Infect. Immun.* **41:**774–779.

Gresikova, M., 1957, Elimination of tick-borne encephalitis virus by goat's milk, *Vet. Cas.* **6:**177–182 (in Slovak).

Gresikova, M., and Nosek, J., 1983, Marker stability of the Skalica strain (from the tick-borne encephalitis complex) propagated in *Ixodes ricinus* ticks, *Acta Virol.* **27:**180–182.

Gresikova, M., and Sekeyova, M., 1980, Characteristics of some tick-borne encephalitis virus strains isolated in Slovakia, *Acta Virol.* **24:**72–75.

Griffin, D. E., 1981, Immunoglobulins in the cerebrospinal fluid: Changes during acute viral encephalitis in mice, *J. Immunol.* **126:**27–31.

Griffin, D. E., Mokhtarian, F., Park, M. M., and Hirsch, R. L., 1983, Immune responses to acute alphavirus infection of the central nervous system: Sindbis virus encephalitis in mice, *prog. Brain Res.* **59:**11–21.

Grimstad, P. R., Ross, Q. E., and Craig, G. B., Jr., 1980, *Aedes triseriatus* (Diptera: Culicidae) and La Crosse virus. II. Modification of mosquito feeding behavior by virus infection, *J. Med. Entomol.* **17:**1–7.

Grossberg, S. E., and Scherer, W. F., 1966, The effect of host age, virus dose and route of inoculation on inapparent infection in mice with Japanese encephalitis virus, *Proc. Soc. Exp. Biol. Med.* **123:**118–124.

Gubler, D. J., Suharyono, W., Sumarmo, Wulur, H., Jahja, E., and Sulianti Saroso, J., 1979, Virological surveillance for dengue haemorrhagic fever in Indonesia using the mosquito inoculation technique, *Bull. W.H.O.* **57:**931–936.

Gubler, D. J., Kuno, G., and Waterman, S., 1983, Neurologic disorders associated with dengue infection, in: *Procedings of the International Conference on Dengue/Dengue Haemorrhagic Fever* (T. Pang and R. Pathmanathan, eds.), pp. 290–306. University of Malaya, Kuala Lampur.

Guillon, J. C., Oudar, J., Joubert, L., and Hannoun, C., 1968, Lesions histologiques du systeme nerveux dans l'infection a virus West Nile chez le cheval, *Ann. Inst. Past.* **114:**539–550.

Gulati, L., Chaturvedi, U. C., and Mathur, A., 1982, Depressed macrophage functions in dengue virus-infected mice: Role of the cytotoxic factor, *Br. J. Exp. Pathol.* **63:**194–202.

Gulati, L., Chartuvedi, U. C., and Mathur, A., 1983a, Dengue virus-induced cytotoxic factor induces macrophages to produce a cytotoxin, *Immunology* **49:**121–130.

Gulati, L., Chaturvedi, U. C., and Mathur, A., 1983b, Plasma membrane-acting drugs inhibit the effect of dengue virus-induced cytotoxic factor, *Ann. Immunol. (Paris)* **134C:**227–235.

Guzman, M. G., Kouri, G., Morier, L., Soler, M., and Fernandez, A., 1984, A study of fatal hemorrhagic dengue cases in Cuba, 1981, *Bull P.A.H.O.* **18**:213–220.

Guzman, M. G., Kouri, G., Bravo, J., Soler, M., and Vasquez, S., 1985, Dengue hemorrhagic fever in Cuba, 1981, II. Study of patients clinically diagnosed with dengue hemorrhagic fever and dengue shock syndrome, *Trans. R. Soc. Trop. Med. Hyg.* (in press).

Habu, A., Murakanu, Y., Ogasa, A., and Fujisaki, Y., 1977, Disorder of spermatogenesis and viral discharge into semen in boars infected with Japanese encephalitis, *Virus* **27**:21–26.

Halstead, S. B., 1979, *In vivo* enchancement of dengue virus infection in rhesus monkeys by passively transferred antibody, *J. Infect. Dis.* **140**:527–533.

Halstead, S. B., 1980, Immunopathological parameters of togavirus disease syndromes, in: *The Togaviruses: Biology, Structure, Replication* (R. W. Schleslinger, ed.), pp. 107–174, Academic Press, New York.

Halstead, S. B., 1982a, Immunopathology in viral disease: Immune enhancement of dengue virus infection, in: *Virus Infections: Modern Concepts and Status* (L. C. Olson, ed.), pp. 41–85, Marcel Dekker, New York.

Halstead, S. B., 1982b, Dengue: Hematological aspects, *Semin. Hematol.* **19**:116–131.

Halstead, S. B., and O'Rourke, E. J., 1977, Dengue viruses and mononuclear phagocytes. I. Infection enhancement by non-neutralizing antibody, *J. Exp. Med.* **146**:201–217.

Halstead, S. B., Chow, J. S., and Marchette, M. J., 1973a, Immunological enhancement of dengue virus replication, *Nature (London)* **243**:24–25.

Halstead, S. B., Shotwell, H., and Casals, J., 1973b, Studies on the pathogenesis of dengue infection in monkeys. II. Clinical laboratory responses to heterologous infection, *J. Infect. Dis.* **128**:15–22.

Halstead, S. B., Marchette, N. J., Chow, J. S., and Lolekha, S., 1976, Dengue virus replication enhancement in peripheral blood leukocytes from immune human beings, *Proc. Soc. Exp. Biol. Med.* **151**:136–139.

Halstead, S. B., O'Rourke, E. J., and Allison, A. C., 1977, Dengue viruses and mononuclear phagocytes. II. Identity of blood and tissue leukocytes supporting *in vitro* infection, *J. Exp. Med.* **146**:218–229.

Halstead, S. B., Rojanasuphot, S., and Sangkawibha, N., 1983, Original antigenic sin in dengue, *Am. J. Trop. Med. Hyg.* **32**:154–156.

Halstead, S. B., Venkateschan, C. N., Gentry, M. K., and Larsen, L. K., 1984, Heterogeneity of infection enhancement of dengue 2 strains by monoclonal antibodies, *J. Immunol.* **132**:1529–1532.

Hambleton, P., Stephenson, J. R., Baskerville, A., and Wiblin, C., 1983, Pathogenesis and immune response of vaccinated and unvaccinated rhesus monkeys to tick-borne encephalitis virus, *Infect. Immun.* **40**:995–1003.

Hammon, W. McD., and Sather, G. E., 1973, Passive immunity for arbovirus infection. I. Artificially induced prophylaxis in man and mouse for Japanese encephalitis, *Am. J. Trop. Med. Hyg.* **22**:524–534.

Harrison, V. R., Eckels, K. H., Sagartz, J. W., and Russell, P. K., 1977, Virulence and immunogenicity of a temperature-sensitive dengue 2 virus in lower primates, *Infect. Immun.* **18**:151–156.

Harrison, A., Murphy, F. A., Gardner, J. J., and Bauer, S. P., 1980, Myocardial and pancreatic necrosis induced by Rocio virus, a new flavivirus, *Exp. Mol. Pathol.* **32**:102–113.

Harrison, A., Murphy, F. A., and Gardner, J. J., 1982, Visceral target organs in systemic St. Louis encephalitis virus infection of hamsters, *Exp. Mol. Pathol.* **37**:292–304.

Hayashi, K., and Arita, T., 1977, Experimental double infection of Japanese encephalitis and herpes simplex virus in mouse brain, *Jpn. J. Exp. Med.* **47**:9–13.

Haymaker, W., and Sabin, A. B., 1947, Topographic distribution of lesions in the central nervous system in Japanese B encephalitis, *Arch. Neurol. Psychol.* **57**:673–692.

Heinz, F. X., and Kunz, C., 1982, Molecular epidemiology of tick-borne encephalitis virus: Peptide mapping of large non-structural proteins of European isolates and comparison with other flaviviruses, *J. Gen. Virol.* **62**:271–285.

Heinz, F. X., Berger, R., Majdic, O., Knapp, W., and Kunz, C., 1982, Monoclonal antibodies to the structural glycoprotein of tick-borne encephalitis virus, *Infect. Immun.* **37:**869–874.

Henderson, B. E., Cheshire, P. P., Kirya, A. B., and Lule, M., 1970, Immunologic studies with yellow fever and selected African group B arboviruses in rhesus and vervet monkeys, *Am. J. Trop. Med. Hyg.* **19:**110–118.

Hirsch, M. S., and Murphy, F. A., 1967, Effects of anti-thymocyte serum on 17D yellow fever infection in adult mice, *Nature (London)* **216:**179–180.

Hofmann, H., Frisch-Niggemeyer, W., Heinz, F., and Kunz, C., 1979, Immunoglobulins to tick-borne encephalitis in the cerebrospinal fluid of man, *J. Med. Virol.* **4:**241–245.

Holland, J. J., 1984, Continuum of change in RNA virus genomes, in: *Concepts in Viral Pathogenesis* (A. L. Notkins and M. B. A. Oldstone, eds.), pp. 137–143, Springer-Verlag, New York.

Holland, J., Spindler, V., Horodyski, F., Grabau, E., Nichol, S., and VandePol, S., 1982, Rapid evolution of RNA genomes, *Science* **215:**1577–1585.

Hotta, H., Murakami, I., Miyasaki, K., Takeda, Y., Shirane, H., and Hotta, S., 1981a, Inoculation of dengue virus into nude mice, *J. Gen. Virol.* **52:**71–76.

Hotta, H., Murakami, I., Miyasaki, K., Takeda, Y., Shirane, H., and Hotta, S., 1981b, Localization of dengue virus in nude mice, *Microbiol. Immunol.* **25:**89–93.

Hotta, H., Hotta, S., Matsumura, T., Wiharta, A. S., Sujudi, Kotani, S., Takada, H., and Tsuji, M., 1983, Increased production of dengue virus in mouse peritoneal macrophage cultures: A possible mechanism underlying the pathogenesis of severe dengue infection, *Proceedings of the International Conference on Dengue and Dengue Hemorrhagic Fever* (T. Pang and R. Pathanathan, eds.), pp. 320–324, University of Malaysia, Kuala Lampur.

Huang, C. H., 1957, Studies of virus factors as causes of inapparent infections in Japanese B encephalitis: Virus strains, viraemia, stability to heat and infective dosage, *Acta Virol.* **1:**36–45.

Huang, C. H., 1982, Studies of Japanese encephalitis in China, *Adv. Virus Res.* **27:**71–101.

Huang, C. H., and Wong, C., 1963, Relation of the peripheral multiplication of Japanese B encephalitis virus to the pathogenesis of the infection in mice, *Acta Virol.* **7:**322–330.

Hudson, B. W., Wolff, K., and DeMartini, J. C., 1979, Delayed-type hypersensitivity responses in mice infected with St. Louis encephalitis virus; Kinetics of the response and effects of immunoregulatory agents, *Infect. Immunol.* **24:**71–76.

Ilienko, V. I., and Pokrovskaya, O. A., 1960, Clinical picture in *Macaccus rhesus* monkeys infected with various strains of tick-borne encephalitis virus, in: *Biology of the Viruses of the Tick-Borne Encephalitis Complex* (H. Libikova, ed.), pp. 266–269, Academic Press, New York.

Ilienko, V. I., Komandenko, N. I., Platonov, V. G., Prozorova, I. N., and Panov, A. G., 1974, Pathogenetic study on chronic forms of tick-borne encephalitis, *Acta Virol.* **18:**341–346.

Inouye, S., Matsuno, S., and Yashito, T., 1984, "Original antigenic sin" phenomenon in experimental flavivirus infections of guinea pigs: Studies by enzyme-linked immunosorbent assay, *Microbiol. Immunol.* **28:**569–574.

Ishak, K. G., Walker, D. H., Coetzer, J. A. W., Gardner, J. J., and Gorelkin, L., 1982, Viral hemorrhagic fevers with hepatic involvement: Pathologic aspects with clinical correlations, *Prog. Liver Dis.* **7:**495–515.

Ishii, K., Matsunaga, Y., and Kono, R., 1968, Immunoglobulins produced in response to Japanese encephalitis virus infections of man, *J. Immunol.* **101:**770–775.

Ishii, K., Matsushita, M., and Hamada, S., 1977, Characteristic residual neuropathological features of Japanese B encephalitis, *Acta Neuropathol.* **38:**181–186.

Jacoby, R. O., Bhatt, P. N., and Schwartz, A., 1980, Protection of mice from lethal flavivirus encephalitis by adoptive transfer of splenic cells from donors infected with live virus, *J. Infect. Dis.* **141:**617–624.

Jahrling, P. B., 1976, Virulence heterogeneity of a predominantly avirulent western equine encephalitis virus population, *J. Gen. Virol.* **32:**121–127.

Jacoby, R. O., and Bhatt, P. N., 1976, Genetic resistance to lethal flavivirus encephalitis. I. Infection of congenic mice with Banzi virus, *J. Infect. Dis.* **134:**158–169.

Jahrling, P. B., and Gorelkin, L., 1975, Selective clearance of a benign clone of Venezuelan equine encephalitis virus from hamster plasma by hepatic reticuloendothelial cells, *J. Infect. Dis.* **132:**667–676.

Jahrling, P. B., and Scherer, W. F., 1973, Growth curves and clearance rates of virulent and benign Venezuelan encephalitis viruses in hamsters, *Infect. Immun.* **8:**456–462.

Johnson, R. T., 1980, Selective vulnerability of neural cells to viral infections, *Brain* **103:**447–472.

Johnson, R. T., 1982, *Viral Infections of the Nervous System*, Raven Press, New York, 433 pp.

Johnson, R. T., Burke, D. S., Elwell, M., Leake, C. J., Nisalak, A., Hoke, C. H., and Lorsomrudee, W., 1985, Japanese encephalitis: Immunocytochemical studies of viral antigen and inflammatory cells in fatal cases, *Ann. Neurol.* **18:**567–573.

Kaplan, A. M., and Koveleski, J. T., 1978, St. Louis encephalitis with particular involvement of the brain stem, *Arch. Neurol.* **35:**45–46.

Kelkar, S., 1982, Protection against Japanese encephalitis virus in infant mice by concanavalin A., *Indian. J. Med. Res.* **76:**47–52.

Kitamura, T., 1975, Hematogenous cells in experimental Japanese encephalitis, *Acta Neuropathol.* **32:**341–346.

Kitamura, T., Hattori, H., Fujita, S., 1972, EM-autoradiographic studies on the inflammatory cells in experimental Japanese encephalitis, *J. Electron Microsc.* **21:**315–322 (in Japanese).

Klotz, O., and Belt, T. H., 1930, Pathology of the liver in yellow fever, *Am. J. Pathol.* **6:**663–687.

Kono, R., and Kim, K. H., 1969, Comparative epidemiological features of Japanese encephalitis in the Republic of Korea, China (Taiwan) and Japan, *Bull. W.H.O.* **40:**263–277.

Kozuch, O., Nosek, J., Ernek, E., Lichard, M., and Albrecht, P., 1963, Persistence of tick-borne encephalitis virus in hibernating hedgehogs and dormice, *Acta Virol.* **7:**430–433.

Kozuch, O., Chunikhin, S. P., Gresikova, M., Nosek, J., Kuenkov, V. B., and Lysy, J., 1981, Experimental characteristics of viremia caused by two strains of tick-borne encephalitis virus in small rodents, *Acta Virol.* **25:**219–224.

Kundin, W. D., Liu, C., Hysell, P., and Hamachige, S., 1963, Studies on West Nile virus infection by means of fluorescent antibodies, *Arch. Gesamte Virusforsch.* **12:**514–528.

Kuno, G., 1982, Persistent infection of a nonvector mosquito cell line (TRA-171) with dengue viruses, *Intervirology* **18:**45–55.

Kurane, I., Hebblewhite, D., Brandt, W. E., and Ennis, F. A., 1984, Lysis of dengue virus-infected cells by natural cell-mediated cytotoxicity and antibody-dependent cell-mediated cytotoxicity, *J. Virol.* **52:**223–230.

Lam, K.-W., Burke, D. S., Siemans, M., Cipperly, V., Li, C.-Y., and Lam, L. T., 1982, Characterization of serum acid phosphatase associated with dengue hemorrhagic fever, *Clin. Chem.* **28:**2296–2299.

Lee, H. W., 1968, Multiplication and antibody formation of Japanese encephalitis virus in snakes, II. Proliferation of the virus, *Seoul J. Med.* **9:**147–161.

Lehtinen, I., and Halonen, J.-P., 1984, EEG findings in tick-borne encephalitis, *J. Neurol, Neurosurg. Psychiatry* **47:**500–504.

LeMercier, G., Guerin, M., and Collomto, H., 1966, Etude histopathologique de l'encéphalite consécutive à l'inoculation du vaccin antiamaril de l'Institut Pasteur de Dakar, *Bull. Soc. Med. Afr. Noire Lang. Fr.* **11:**601–609.

LeRoux, J. M. W., 1959, The histopathology of Wesselsbron disease in sheep, *Onderstepoort J. Vet. Res.* **28:**237–243.

Liprandi, F., and Walder, R., 1983, Replication of virulent and attenuated strains of yellow fever virus in human monocytes and macrophage-like cell lines (4937), *Arch. Virol.* **76:**51–61.

Liu, C. T., and Griffin, M. J., 1982, Changes in body fluid compartments, tissue water and electrolyte distribution, and lipid concentrations in rhesus macaques with yellow fever, *Am J. Vet. Res.* **43:**2013–2018.

Liu, J.-L., 1972, Protective effect of interferon on mice experimentally infected with Japanese encephalitis virus, *Chinese J. Microbiol.* **5:**1–9.

Lloyd, W., 1931, The myocardium in yellow fever, *Am. Heart J.* **6:**504–516.

Lubiniecki, A. S., Cypress, R. H., and Hammon, W. McD., 1973, Passive immunity for arbovirus infection. I. Artificially acquired protection in mice for Japanese (B) encephalitis virus, *Am. J. Trop. Med. Hyg.* **22:**535–542.

Lubiniecki, A. S., Cypress, R. H., and Lucas, J. P., 1974, Synergistic interaction of two agents in mice: Japanese B encephalitis virus and *Trichinella spiralis, Am. J. Trop. Med. Hyg.* **23:**235–241.

MacDonald, F., 1952, Murray Valley encephalitis infection in the laboratory mouse. I. Influence of age on susceptibility to infection, *Aust. J. Exp. Biol. Med.* **30:**319–326.

Malkova, D., 1960, The role of the lymphatic system in experimental infections with tick-borne encephalitis. I. The tick-borne encephalitis virus in the lymph and blood of experimentally infected sheep, *Acta Virol.* **4:**233–240.

Malkova, D., and Kolman, J. M., 1964, Role of the regional lymphocytic system of the immunized mouse in penetration of the tick-borne encephalitis virus into one blood strain, *Acta Virol.* **8:**10–13.

Marburg, K., Goldblum, N., Sterk, V. V., Jasinka-Klingberg, W., and Klingberg, M. A., 1956, The natural history of West Nile fever. I. Clinical observations during an epidemic in Israel, *Am. J. Hyg.* **64:**259–269.

Marchette, N. J., Halstead, S. B., Falkler, W. A., Jr., Stenhouse, A., and Nash, D., 1973, Studies on the pathogenesis of dengue infection in monkeys. III. Sequential distribution of virus in primary and heterologous infections, *J. Infect. Dis.* **128:**23–30.

Marker, S. C., and Jahrling, P. B., 1979, Correlation between virus–cell receptor properties of alphaviruses *in vitro* and virulence *in vivo, Arch. Virol.* **62:**53–62.

Marvin, J. A., Zvolanek, E. E., Nowosiwsky, T., and Greenberg, J. H., 1968, Tuberculin sensitivity (TINE) in apparently healthy subjects after yellow fever vaccination, *Am. Rev. Respir. Dis.* **98:**703–706.

Mathur, A., Arora, K. L., and Chaturvedi, U. C., 1982, Transplacental Japanese encephalitis virus (JEV) infection in mice during consecutive pregnancies, *J. Gen. Virol.* **59:**213–217.

Mathur, A., Arora, K. L., and Chaturvedi, U. C., 1983a, Immune response to Japanese encephalitis virus in mother mice and their congenitally infected offspring, *J. Gen. Virol.* **64:**2027–2031.

Mathur, A., Arora, K. L., and Chaturvedi, U. C., 1983b, Host defense mechanisms against Japanese encephalitis virus infection in mice, *J. Gen. Virol.* **64:**805–811.

Mathur, A., Rawat, S., and Chaturvedi, U. C., 1983c, Induction of suppressor cells in Japanese encephalitis virus infected mice, *Br. J. Exp. Pathol.* **69:**336–343.

Mathur, A., Rawat, S., and Chaturvedi, U. C., 1984, Suppressor T cells for delayed-type hypersensitivity to Japanese encephalitis virus, *Immunology* **52:**395–402.

Mayer, V., Gajdosova, E., and Slavik, I., 1976, *In vitro* studies on cell-mediated immune response to tick-borne encephalitis virus: Findings in convalescents and human subclinical infections, *Acta Virol.* **20:**395–401.

McFarland, R. I., and White, D. O., 1980, Further characterization of natural killer cells induced by Kunjin virus, *Aust. J. Exp. Biol. Med. Sci.* **58:**77–89.

McFarland, H. F., Griffin, D. E., and Johnson, R. T., 1972, Specificity of the inflammatory response in viral encephalitis. I. Adoptive immunization of immunosuppressed mice infected with Sindbis virus, *J. Exp. Med.* **136:**216–226.

McIntosh, B. M., Jupp, P. G., Dos Santos, I., and Meenehan, G. M., 1976, Epidemics of West Nile and Sindbis viruses in South Africa with *Culex (Culex) univittatus* Theobold as vector, *S. Afr. J. Sci.* **72:**295–300.

McKenzie, J. L., Dalchau, R., and Fabre, J. W., 1982, Biochemical characterization and localization in brain of a human brain–leucocyte membrane glycoprotein recognized by a monoclonal antibody, *J. Neurochem.* **39:**1461–1466.

Miller, C. A., and Benzer, S., 1983, Monoclonal antibody cross-reactions between *Drosophila* and human brain, *Proc. Natl. Acad. Sci. U.S.A.* **80:**7641–7645.

Mims, C. A., 1977, *The Pathogenesis of Infectious Disease,* Academic Press, New York.

Mirovsky, J., Holub, J., and Nguyen-Ba-Can, 1962, The influence of dengue on pregnancy and fetus, *Cesk. Pediatr.* **17:**985–988 (in Czech).

Mitchell, C. J., Gubler, D. J., and Monath, T. P., 1983, Variation in infectivity of St. Louis encephalitis viral strains for *Culex pipiens quinquefasciatus* (Diptera: Culicidae), *J. Med. Entomol.* **20:**526–533.

Mitrakul, C., Poshyachinda, M., Futrakul, P., Sangkawibha, N., and Ahandrik, S., 1977, Hemostatic and platelet kinetic studies in dengue hemorrhagic fever, *Am. J. Trop. Med. Hyg.* **26:**975–984.

Miyake, M., 1964, The pathology of Japanese encephalitis: A review, *Bull. W.H.O.* **30:**153–160.

Moench, T. R., and Griffin, D. E., 1984, Immunocytochemical identification and quantification of the mononuclear cells in the cerebrospinal fluid, meninges, and brain during acute viral meningo-encephalitis, *J. Exp. Med.* **159:**77–88.

Monath, T. P., 1971, Neutralizing antibody response in the major immunoglobulin classes to yellow fever 17D vaccination of humans, *Am. J. Epidemiol.* **93:**122–129.

Monath, T. P., 1980, Epidemiology, in: *St. Louis Encephalitis* (T. P. Monath, ed.), pp. 239–312, American Public Health Association, Washington, D.C.

Monath, T. P. C., and Borden, E. C., 1971, Effects of thorotrast on humoral antibody, viral multiplication, and interferon during infection with St. Louis encephalitis virus in mice, *J. Infect. Dis.* **123:**297–300.

Monath, T. P., Kemp, G. E., Cropp, C. B., and Bowen, G. S., 1978, Experimental infection of house sparrows (*Passer domesticus*) with Rocio virus, *Am. J. Trop. Med. Hyg.* **27:**1251–1254.

Monath, T. P., Craven, R. B., Adjukiewicz, A., Germain, M., Francy, D. B., Ferrara, L., Samba, D. M., N'Jie, H., Cham, K., Fitzgerald, S. A., Crippen, P. H., Simpson, D. I. H., Bowen, E. T. W., Fabiyi, A., and Salaun, J.-J., 1980a, Yellow fever in the Gambia, 1978–1979: Epidemiologic aspects with observations on the occurrence of Orungo virus infections, *Am. J. Trop. Med. Hyg.* **29:**912–928.

Monath, T. P., Cropp, C. B., Bowen, G. S., Kemp, G. E., Mitchell, C. J., and Gardner, J. J., 1980b, Variation in virulence for mice and rhesus monkeys among St. Louis encephalitis virus strains of different origin, *Am. J. Trop. Med. Hyg.* **29:**948–962.

Monath, T. P., Brinker, K. R., Chandler, F. W., Kemp, G. E., and Cropp, C. B., 1981, Pathophysiologic correlations in a rhesus monkey model of yellow fever with special observations on the acute necrosis of B cell areas of lymphoid tissues, *Am. J. Trop. Med. Hyg.* **30:**431–443.

Monath, T. P., Cropp, C. B., and Harrison, A. K., 1983, Mode of entry of a neurotropic arbovirus into the central nervous system: Reinvestigation of an old controversy, *Lab. Invest.* **48:**399–410.

Monath, T. P., Nystrom, R. R., Bailey, R. E., Muth, D. J., and Calisher, C. H., 1984, IgM antibody capture ELISA for diagnosis of St. Louis encephalitis, *J. Clin. Microbiol.* **20:** 784–790.

Murphy, F. A., 1979, Viral pathogenetic mechanisms, in: *Mechanisms of Viral Pathogenesis and Virulence* (P. A. Bachmann, ed.), pp. 7–19, WHO Collaborating Center for Collection and Evaluation of Data on Comparative Virology, Munich.

Murphy, F. A., Harrison, A. K., Gary, F. W., Jr., Whitfield, S. G., and Forrester, F. T., 1968, St. Louis encephalitis virus infection of mice: Electron microscopic studies of the central nervous system, *Lab. Invest.* **19:**652–662.

Museteanu, C., Welte, M., Henneberg, G., and Haase, J., 1979, Relation between decreased mental efficiency in mice and the presence of cerebral lesions after experimental encephalitis caused by yellow fever virus, *J. Infect. Dis.* **139:**320–323.

Nagarkatti, M., and Nagarkatti, P. S., 1979, Suppression of intrinsic B cell function in dengue infected mice, *Experientia* **35:**1518–1519.

Nagarkatti, M., Nagarkatti, P. S., and Rao, K. M., 1980, Effects of experimental dengue virus infection on humoral and cell-mediated immune response to thymus-dependent antigen, *Int. Arch. Allergy Appl. Immunol.* **62:**361–369.

Nagarkatti, P. S., and Nagarkatti, M., 1983, Effect of experimental dengue virus infection on immune response of the host. I. Nature of changes in T suppressor cell activity regulating the B and T cell responses to heterologous antigens, *J. Gen. Virol.* **64:**1441–1447.

Nagarkatti, P. S., D'Souza, M. B., and Rao, K. M., 1978, Use of sensitized spleen cells in capillary tube migration inhibition tests to demonstrate cellular sensitization to dengue virus in mice, *J. Immunol. Methods* **23:**341–348.

Nathanson, N., 1980, Pathogenesis, in: *St. Louis Encephalitis* (T. P. Monath, ed.), pp. 201–236, American Public Health Association, Washington, D.C.

Nathanson, N., and Harrington, B., 1967, Experimental infection of monkeys with Langat virus. II. Turnover of circulating virus, *Am. J. Epidemiol.* **85:**494–503.

Nathanson, N., Davis, M., Thind, I. S., and Price, W. H., 1966, Histological studies of the neurovirulence of group B arboviruses. II. Selection of indicator centers, *Am. J. Epidemiol.* **84:**524–540.

Nawrocka, E., 1975, Characteristics of tick-borne encephalitis virus circulating in Poland, *Acta Microbiol. Pol. Ser. A* **7:**237–245.

Nelson, E. R., Bierman, H. R., and Chulajata, R., 1964, Hematologic findings in the 1960 hemorrhagic fever epidemic (dengue) in Thailand, *Am. J. Trop. Med. Hyg.* **13:**642–649.

Nolan, J. P., 1981, Endotoxin, reticuloendothelial function, and liver injury, *Hepatology* **1:**458–465.

Nosek, J., Gresikova, M., and Rehacek, J., 1960, Persistence of tick-borne encephalitis virus in hibernating bats, in: *Biology of Viruses of the Tick-Borne Encephalitis Complex* (H. Libikova, ed.), pp. 394–396, Academic Press, New York.

Oaten, S. W., Webb, H. E., and Bowen, E. T. W., 1976, Enhanced resistance of mice to infection with Langat (TP21) virus following pre-treatment with Sindbis or Semliki Forest virus, *J. Gen. Virol.* **23:**381–388.

Oaten, S. W., Webb, H. E., and Jagelman, S., 1980, Resistance of mice to infection with West Nile virus following pre-treatment with Sindbis, Semliki Forest and chikungunya viruses, *Microbios Lett.* **13:**85–90.

Odelola, H. A., and Fabiyi, A., 1978, Kinetic haemagglutination-inhibition technique as a means of detecting antigenic variations among strains of Nigerian flaviviruses, *Arch. Virol.* **56:**291–295.

Ogasa, A., Yokoki, Y., Fujisaki, Y., and Habu, A., 1977, Reproductive disorders in boars infected experimentally with Japanese encephalitis virus, *Jpn. J. Anim. Reprod.* **23:**171–175.

Ogawa, M., Okubo, H., Tsuji, Y., Yasui, N., and Someda, K., 1973, Chronic progressive encephalitis occurring 13 years after Russian spring summer encephalitis, *J. Neurol. Sci.* **19:**363–373.

O'Leary, J. L., Smith, M. G., and Reames, H. R., 1942, Influence of age on susceptibility of mice to St. Louis encephalitis virus and on the distribution of lesions, *J. Exp. Med.* **75:**233–247.

Olson, L. C., Sithisarn, P., and Djinawi, N. K., 1975, Role of macrophages in Wesselsbron and Germiston virus infections of mice, *J. Infect. Dis.* **131:**119–127.

Oudart, J.-L., and Rey, M., 1970, Proteinurie, proteinémie, et transaminasemies dans 23 cas de fièvre jaune confirmées, *Bull. W.H.O.* **42:**95–102.

Pang, T., 1983, Delayed type hypersensitivity: Probable role in the pathogenesis of dengue hemorrhagic fever/dengue shock syndrome, *Rev. Infect. Dis.* **5:**346–352.

Pang, T., Wong, P. Y., and Pathmanathan, R., 1982, Induction and characterization of delayed type hypersensitivity to dengue virus in mice, *J. Infect. Dis.* **146:**235–242.

Pang, T., Devi, S., Yeen, W. P., McKenzie, I. F. C., and Leong, Y. K., 1984, Lyt phenotype and H-2 compatability requirements of effector cells in the delayed-type hypersensitivity response to dengue virus infection, *Infect. Immun.* **43:**429–431.

Parks, J. J., Ganaway, J. R., and Price, W. H., 1958, Studies on immunologic overlap among certain arthropod-borne viruses. III. A laboratory analysis of three strains of West Nile virus which have been studied in human cancer patients, *Am. J. Hyg.* **68:**106–119.

Pavri, K. M., and Prasad, S. R., 1980, T suppressor cells: Role in dengue hemorrhagic fever and dengue shock syndrome, *Rev. Infect. Dis.* **2:**142–146.

Pavri, K. M., Ghalsasi, G. R., Dastur, D. K., Goverdhan, M. K., and Lalitha, V. S., 1975, Dual infections of mice: Visceral larva migrans and sublethal infections with Japanese encephalitis virus, *Trans. R. Soc. Trop. Med. Hyg.* **69:**99–110.

Pavri, K. M., Swe Than, Ramamoorthy, C. L., and Chodankar, V. P., 1979, Immunoglobulin E in dengue hemorrhagic fever (DHF) cases, *Trans. R. Soc. Trop. Med. Hyg.* **73:**451–452.

Pavri, K. M., Ramamoorthy, C. L., and Dhorje, S., 1980, Immunoglobulin E in patients with Japanese encephalitis, *Infect. Immun.* **28:**290–291.

Peck, J. L., Jr., and Sabin, A. B., 1947, Multiplication and spread of the virus of St. Louis encephalitis in mice with special emphasis on its fate in the alimentary tract, *J. Exp. Med.* **85:**647–662.

Peiris, J. S. M., and Porterfield, J. S., 1979, Antibody mediated enhancement of flavivirus replication in macrophage-like cell lines, *Nature (London)* **282:**509–511.

Penna, H. A., and Bittencourt, A., 1943, Persistence of yellow fever virus in brains of monkeys immunized by cerebral inoculation, *Science* **97:**448–449.

Perelman, A., and Stern, J., 1974, Acute pancreatitis is West Nile fever, *Am. J. Trop. Med. Hyg.* **23:**1150–1152.

Perelmutter, L., Phipps, P., and Potvin, L., 1978, Viral infections and IgE levels, *Ann. Allergy* **41:**158–159.

Phillpotts, R. J., Stephenson, J. R., and Porterfield, J. S., 1985, Antibody dependent enhancement of tick-borne encephalitis infectivity, *J. Gen. Virol.* **66:**1831–1837.

Pogodina, V. V., Frolova, M. P., Malenko, G. V., Fokina, G. I., Levina, C. S., Mamonenko, L. L., Koreshkova, G. V., and Ralf, N. M., 1981a, Persistence of tick-borne encephalitis virus in monkeys. I. Features of experimental infection, *Acta Virol.* **25:**337–343.

Pogodina, V. V., Levina, L. S., Fokina, G. I., Koreshkova, G. V., Malenko, G. V., Bochkova, N. A., and Rzhakhova, O., 1981b, Persistence of tick-borne encephalitis virus in monkeys. III. Phenotypes of the persisting virus, *Acta Virol.* **25:**352–360.

Pogodina, V. V., Frolova, M. P., Malenko, G. V., Fokina, G. I., Koreshkova, L. L., Kiseleva, N. G., Bochkova, N. G., and Ralph, N. M., 1983, Study of West Nile virus persistence in monkeys, *Arch. Virol.* **75:**71–86.

Pongpanich, B., and Kumponpant, S., 1973, Studies of dengue hemorrhagic fever. V. Hemodynamic studies of clinical shock associated with dengue hemorrhagic fever, *J. Pediatr.* **83:**1073–1077.

Price, W. H., 1966, Chronic disease and virus persistence in mice inoculated with Kyasanur Forest disease virus, *Virology* **29:**679–681.

Putvatana, R., Yoksan, S., Chayaydohin, T., Bhamarapravati, N., and Halstead, S., 1984, Absence of dengue 2 infection enhancement in human sera containing Japanese encephalitis, *Am. J. Trop. Med. Hyg.* **33:**288–294.

Reeves, W. C., Bellamy, R. E., and Scrivani, R. P., 1958, Relationships of mosquito vectors to winter survival of encephalitis viruses. I. Under natural conditions, *Am. J. Hyg.* **67:**78–89.

Reid, H. W., 1975, Experimental infection of red grouse with louping ill virus (flavivirus group). I. The viraemia and antibody response, *J. Comp. Pathol.* **85:**223–229.

Reid, H. W., and Doherty, P. C., 1971, Louping ill encephalomyelitis in the sheep. I. The relationship of viremia and the antibody response to susceptibility, *J. Comp. Pathol.* **81:**521–527.

Reid, H. W., and Moss, R., 1980, The response of four species of birds to louping ill, in: *Arbovirus in the Mediterranean Countries* (J. Vesenjak-Hirjan, J. S. Porterfield, and E. Arslanagic, eds.), *Abh. Bakteriol. Suppl.* **9:**219–223, Gustav Fischer Verlag, Stuttgart.

Reid, H. W., Doherty, P. C., and Dawson, A. M., 1971, Louping ill encephalomyelitis in the sheep. III. Immunoglobulins in cerebrospinal fluid, *J. Comp. Pathol.* **81:**537–543.

Reid, H. W., Buxton, D., Pow, I., and Finlayson, J., 1982, Experimental louping ill virus infection in two species of British deer, *Vet. Rec.* **111:**61.

Repik, P. M., Dalrymple, J. M., Brandt, W. E., McCown, J. M., and Russell, P. K., 1983, RNA fingerprinting as a method for distinguishing dengue 1 virus strains, *Am. J. Trop. Med. Hyg.* **32:**577–589.

Reyes, M. G., Gardner, J. J., Poland, J. D., and Monath, T. P., 1981, St. Louis encephalitis: Quantitative histologic and immunofluorescent studies, *Arch. Neurol.* **38:**329–334.

Roehrig, J. T., Mathews, J. H., and Trent, D. W., 1983, Identification epitopes on the E glycoprotein of St. Louis encephalitis virus using monoclonal antibodies, *Virology* **128:**118–126.

Rosemberg, S., 1977, Neuropathological study of a new viral encephalitis: The encephalitis of Sao Paulo South Coast, *Rev. Inst. Med. Trop. Sao Paulo* **19:**280–282.

Rubenstein, D., Wheelock, E. F., and Tyrrell, D. A. J., 1972, The growth of arboviruses in organ culture of mouse meninges and the influence on *in vitro* virus growth of previous vaccination, *Proc. Soc. Exp. Biol. Med.* **140:**1123–1126.

Russell, P. K., and Brandt, W. E., 1973, Immunopathologic processes and viral antigens associated with sequential dengue virus infection, *Perspect. Virol.* **7:**263–277.

Sagamata, M., and Miura, T., 1982, Japanese encephalitis virus infection in fetal mice at different stages of pregnancy. I. Stillbirth, *Acta Virol.* **26:**279–282.

Sahaphong, S., Riengrojpitak, S., Bhamarapravati, N., and Chirachariyavej, T., 1980, Electron microscopic study of the vascular endothelial cell in dengue hemorrhagic fever, *Southeast Asian J. Trop. Med. Public Health* **11:**194–211.

Sanders, M., Blumberg, A., and Haymaker, W., 1953, Polyradiculopathy in man produced by St. Louis encephalitis virus (SLE), *South. Med. J.* **46:**606–611.

Sangkawibha, N., Rojanasuphot, S., Ahandrik, S., Viriyapongse, S., Jatanasen, S., Salitul, V., Phanthumachinda, B., and Halstead, S. B., 1984, Risk factors in dengue shock syndrome: A prospective epidemiologic study in Rayong, Thailand. I. The 1980 outbreak, *Am. J. Epidemiol.* **120:**653–669.

Santos, F., Lima, C. P., Paiva, M., Costa e Silva, M., and Nery de Castro, C., 1973, Coagulacao intravascular disseminada aguda na febre amarela: Dosagem dos fatores da coagulacao, *Bras.-Med.* **9:**9–16.

Santos Lagresa, M. N., Villaescusa, R., Ballester, J. M., and Hernandez, P., 1983, Indomethacin-mediated enhancement of lymphocyte response to phytohemagglutinin in dengue haemorrhagic fever patients, *Br. J. Haematol.* **55:**379–380.

Schlesinger, J. J., and Brandriss, M. W., 1981, Growth of 17D yellow fever virus in a macrophage-like cell line U937: Role of Fc and viral receptors in antibody mediated infection, *J. Immunol.* **127:**659–665.

Schlesinger, R. W., 1977, *The Dengue Viruses. Virology Monographs*, Vol. 16, Springer-Verlag, Vienna and New York.

Schlesinger, R. W., 1980, Virus–host interactions in natural and experimental infections with alphaviruses and flaviviruses, in: *The Togaviruses* (R. W. Schlesinger, ed.), pp. 83–106, Academic Press, New York.

Scott, R. McN., Nisalak, A., Cheamudon, U., Seridhoranakul, S., and Nimmannitya, S., 1980, Isolation of dengue viruses from peripheral blood leukocytes of patients with hemorrhagic fever, *J. Infect. Dis.* **141:**1–6.

Seamer, J., and Peto, S., 1969, A method of assessment of central nervous function in mice with viral encephalomyelitis, *Lab. Anim.* **3:**129–140.

Semenov, B. F., Khozinsky, V. V., and Vargin, V. V., 1975, The damaging action of cellular immunity in flavivirus infections of mice, *Med. Biol.* **53**:331–336.

Shaikh, B. H., Pavri, K. M., Ramamoorthy, C. L., Verma, S. P., and Deuskar, N. J., 1983, Total serum immunoglobulins in Japanese encephalitis patients with high IgE levels in acute phase, *Indian J. Med. Res.* **77**:765–769.

Shankar, S. K., Rao, T. V., Mruthyun-Jayanna, B. P., Devi, M. G., and Deshpande, D. H., 1983, Autopsy study of brains during an epidemic of Japanese encephalitis in Karnataka, India, *Indian J. Med. Res.* **78**:431–440.

Sheahan, B. J., Gates, M. C., Caffrey, J. F., and Atkins, G. J., 1983, Oligodendrocyte infection and demyelination produced in mice by the M9 mutant of Semliki Forest virus, *Acta Neuropathol.* **60**:257–265.

Sheets, P., Schwartz, A., Jacoby, R. O., and Bhatt, P. N., 1979, T cell-mediated cytotoxicity for L929 fibroblasts infected with Banzi virus (*flavivirus*), *J. Infect. Dis.* **140**:384–391.

Shukla, M. I., and Chaturvedi, U. C., 1982, *In vivo* role of macrophages in transmission of dengue virus-induced suppressor signal to T lymphocytes, *Br. J. Exp. Pathol.* **63**:522–530.

Shukla, M. I., and Chaturvedi, U. C., 1983, Transmission of dengue-virus induced suppressor signal from macrophage to lymphocyte occurs by cell contact, *Br. J. Exp. Pathol.* **64**:87–92.

Shukla, M. I., and Chaturvedi, U. C., 1984, Study of the target cell of the dengue virus-induced suppressor signal, *Br. J. Exp. Pathol.* **65**:267–273.

Shukla, M. I., Dalakoti, H., and Chaturvedi, U. C., 1982, Ly phenotype of T lymphocytes producing dengue virus-induced immunosuppressive factors, *Indian J. Exp. Biol.* **20**:525–528.

Sipos, J., Ribiczey, P., Gabor, V., Toth, Z., and Bartok, K., 1981, Investigations on blood and cerebrospinal fluid lymphocytes in patients suffering from tick-borne encephalitis, *Infection* **9**:258–263.

Smith, A., 1981, Genetic resistance to lethal flavivirus encephalitis: Effect of host age and immune status and route of inoculation on production of interfering Banzi virus *in vivo*, *Am. J. Trop. Med. Hyg.* **30**:1319–1323.

Smith, A. L., and Jacoby, R. O., 1986, Immune responses of mice genetically resistant or susceptible to lethal Banzi virus infection (in prep.).

Smith, J. A., Francis, T. I., and David-West, T. S., 1973, Auto-antibodies in acute viral hepatitis, yellow fever, and hepatocellular carcinoma: Clinical and experimental findings, *J. Pathol.* **109**:83–91.

Srichaikul, T., Nimmannitya, S., Artchararit, N., Siriasawakul, T., and Sungpeuk, P., 1977, Fibrinogen metabolism and disseminated intravascular coagulation in dengue hemorrhagic fever, *Am. J. Trop. Med. Hyg.* **26**:525–532.

Stefanopoulo, G. J., and Mollaret, P., 1934, Hemiplegie d'origine cerebrale et névrite optique en cours d'un cas de fièvre jaune, *Bull. Mem. Soc. Med. Hop.* (Paris) **50**:1463–1466.

Stephenson, J. R., Lee, J. M., and Wilton-Smith, P. D., 1984, Antigenic variation among members of the tick-borne encephalitis complex, *J. Gen. Virol.* **63**:81–89.

Stephenson, L. D., 1939, Pathologic changes in nervous system in yellow fever, *Arch. Pathol.* **27**:249–266.

Stollar, V., 1980, Togaviruses in cultured arthropod cells, in: *The Togaviruses: Biology, Structure, Replication* (R. W. Schlesinger, ed.), pp. 584–622, Academic Press, New York.

Stollar, V., and Shenk, T. E., 1973, Homologous viral interference in *Aedes albopictus* cultures chronically infected with Sindbis virus, *J. Virol.* **11**:592–595.

Suckling, A. J., Pathak, S., Jagelman, S., and Webb, H. E., 1978, Virus-associated demyelination: A model using avirulent Semliki Forest virus infection of mice, *J. Neurol. Sci.* **39**:147–154.

Sugawa, Y., Hiroshi, M., and Yamamoto, S., 1949, Histopathological studies on naturally affected horses with Japanese encephalitis, *Bull. Natl. Inst. Anim. Health* (Tokyo) **22**:9–25.

Sulkin, S. E., and Allen, R., 1974, Virus infections in bats, *Monogr. Virol.* **8**:1–103.

Sulkin, S. H., Harford, C. G., and Bronfenbrenner, J. J., 1939, Immunization of mice by intranasal instillation of nasopharyngeal washings from cases of St. Louis encephalitis, *Proc. Soc. Exp. Biol. Med.* **41:**427–429.

Sulkin, S. E., Sims, R., and Allen, R., 1966, Studies of arthropod-borne virus infections in Chiroptera. II. Experiments with Japanese B and St. Louis encephalitis in the gravid bat and evidence for transplacental transmission, *Am. J. Trop. Med. Hyg.* **13:**475–481.

Sumarmo, Wilur, H., Jahja, E., Gubler, D. J., Suharyono, W., and Sorenson, K., 1983, Clinical observations on virologically confirmed fatal dengue infections in Jakarta, Indonesia, *Bull. W.H.O.* **61:**693–701.

Suzuki, M., Simizu, B., Yabe, S., Oya, A., and Seto, H., 1981, Effect of cadmium on Japanese encephalitis virus infection in mice. I. Acute and single-dose exposure experiment, *Toxicol. Lett.* **9:**231–235.

Tamura, H., Koyama, T., Kuwanizu, I., Nakamura, I., Ema, M., and Miura, T., 1977, Effect of methylmercury chloride and Japanese encephalitis virus infection on fetus of hamsters, *Med. Biol.* **9:**161–164.

Tandon, P., Chaturvedi, U. C., and Mathur, A., 1979a, Differential depletion of T lymphcytes in the speen of dengue virus-infected mice, *Immunology* **37:**1–6.

Tandon, P., Chaturvedi, U. C., and Mathur, A., 1979b, Dengue virus induced thymus-derived suppressor cells in the spleens of mice, *Immunology* **38:**653–658.

Tarr, G. C., and Hammon, W. McD., 1974, Cross-protection between group B arboviruses: Resistance in mice to Japanese B encephalitis and St. Louis encephalitis viruses induced by dengue virus immunizations, *Infect. Immun.* **9:**909–915.

Theiler, M., 1951, The virus, in *Yellow Fever* (G. K. Strode, ed.), pp. 39–136, McGraw-Hill, New York.

Theiler, M., and Anderson, C. R., 1975, The relative resistance of dengue-immune monkeys to yellow fever virus, *Am. J. Trop. Med. Hyg.* **24:**115–117.

Theofilopoulos, A. N., Brandt, W. E., Russell, P. K., and Dixon, F. J., 1976, Replication of dengue 2 virus in cultured human lymphoblastoid cells and subpopulations of human peripheral leukocytes, *J. Immunol.* **117:**953–961.

Thind, I. S., and Singh, N. P., 1977, Potentiation of Langat virus infection by lead intoxication—influence on host defenses, *Acta Virol.* **21:**317–325.

Tigertt, W. D., Berge, T. O., Gochenour, W. S., Gleiser, C. A., Eveland, W. D., Vorder Bruegge, C., and Smetana, H. F., 1960, Experimental yellow fever, *Trans. N.Y. Acad. Sci.* **22:**323–333.

Tignor, G. H., Smith, A. L., and Shope, R. E., 1984, Utilization of host proteins as virus receptors, in: *Concepts in Viral Pathogenesis* (A. L. Notkins and M. B. A. Oldstone, eds.), pp. 109–116, Springer-Verlag, New York.

Trent, D. W., Monath, T. P., Bowen, G. S., Vorndam, A. V., Cropp, C. B., and Kemp, G. E., 1980, Variation among strains of St. Louis encephalitis virus: Basis for a genetic, pathogenetic and epidemiologic classification, *Ann. N. Y. Acad. Sci.* **354:**219–237.

Trent, D. W., Grant, J. A., Vorndam, A. V., and Monath, T. P., 1981, Genetic heterogeneity among St. Louis encephalitis virus isolates of different geographic origin, *Virology* **114:**319–332.

Tuchinda, M., Dhorranintra, B., and Tuchinda, P., 1977, Histamine content in 24-hour urine in patients with dengue hemorrhagic fever, *Southeast Asian J. Trop. Med. Public Health* **8:**80–83.

Umrigar, M. D., and Pavri, K. M., 1977, Comparative biological studies on Indian strains of West Nile virus isolated from different sources, *Indian J. Med. Res.* **65:**596–602.

Vereta, L. A., Ostrorskaya, O. V., Kikolaeva, S. P., and Pukhovskaya, N. M., 1983, Detection of natural heterogeneity of natural tick-borne encephalitis virus populations and grouping of strains, *Vopr. Virusol.* **28:**706–710 (in Russian).

Vince, V., and Grcevic, N., 1969, Development of morphological changes in experimental tick-borne meningoencephalitis induces in white mice by different virus doses, *J. Neurol. Sci.* **9:**109–130.

Vorndam, A. V., 1980, Immunization, in: *St. Louis Encephalitis* (T. P. Monath, ed.), pp. 623–635, American Public Health Association, Washington, D.C.

Warren, A. J., 1951, Landmarks in the conquest of yellow fever, in: *Yellow Fever* (G. K. Strode, ed.), pp. 6–37, McGraw-Hill, New York.

Webb, H. E., Wight, D. G. D., and Wiernik, G., 1968, Langat virus encephalitis in mice. II. The effect of irradiation, *J. Hyg.* **66**:355–364.

Webb, H. E., Mehta, S., Gregson, N. A., and Leibowitz, S., 1984, Immunological reaction of the demyelinating Semliki Forest virus with immune serum to glycolipids and its possible importance to central nervous system viral autoimmune disease, *Neuropathol. Appl. Neurobiol.* **10**:77–84.

Webb, J. K. G., and Pereira, S., 1956, Clinical diagnosis of an arthropod-borne type of encephalitis in children of North Arcot District, Madras State, India, *Indian J. Med. Sci.* **10**:573–581.

Weiner, L. P., Cole, G. A., and Nathanson, N., 1970, Experimental encephalitis following peripheral inoculation of West Nile virus in mice of different ages, *J. Hyg.* **68**:435–446.

Wheelock, E. F., and Edelman, R., 1969, Specific role of each human leukocyte type in viral infections. III. 17D yellow fever virus replication and interferon production in homogenous leukocyte cultures treated with phytohemagglutinin, *J. Immunol.* **103**:429–436.

Wheelock, E. F., Toy, S. T., and Stjernholm, R. L., 1970, Lymphocytes and yellow fever. I. Transient virus refractory state following vaccination of man with the 17D strain, *J. Immunol.* **105**:1304–1306.

Wong, P. Y., Devi, S., McKenzie, F. C., Yap, Y. L., and Pang, T., 1984, Induction and Ly phenotype of suppressor T cells in mice during primary infection with dengue virus, *Immunology* **51**:51–56.

Wood, C. S., 1976, ABO blood groups related to selection of human hosts by yellow fever vectors, *Hum. Biol.* **48**:337–341.

Woodall, J. P., and Roz, A., 1977, Experimental milk-borne transmission of Powassan virus in the goat, *Am. J. Trop. Med. Hyg.* **26**:190–192.

World Health Organization, 1971, *Third Report, Expert Committee on Yellow Fever, WHO Tech Rep. Ser.*, No. 479, Geneva.

World Health Organization, 1973, Pathogenesis mechanisms in dengue hemorrhagic fever: Report of an international collaborative study, *Bull. W.H.O.* **58**:117–123.

Yamamoto, M., and Hotta, S., 1981, Response of human leucocytes to yellow fever virus infection *in vitro*, *Kobe J. Med. Sci.* **27**:165–172.

Zilber, L. A., 1960, Pathogenicity of Far East and Western (European) tick-borne encephalitis viruses in sheep and monkeys, in: *Biology of Viruses of the Tick-Borne Encephalitis Complex* (H. Libikova, ed.), pp. 260–264, Academic Press, New York.

Zisman, B., Wheelock, E. F., and Allison, A. C., 1971, Role of macrophages and antibody in resistance of mice against yellow fever virus, *J. Immunol.* **107**:236–243.

Zlotnik, I., and Grant, D. P., 1976, Further observations on subacute sclerosing encephalitis in adult hamsters: The effects of intranasal infections with Langat virus, measles virus and SSPE-measles virus, *Br. J. Exp. Pathol.* **57**:49–66.

Zlotnik, I., Smith, C. E. G., Grant, D. P., and Peacock, S., 1970, The effect of immunosuppression on viral encephalitis with special reference to cyclophosphamide, *Br. J. Exp. Pathol.* **51**:434–439.

Zlotnik, I., Grant, D. P., Carter, G. B., and Batter-Hatton, D., 1973, Subacute sclerosing encephalitis in adult hamsters infected with Langat virus, *Br. J. Exp. Pathol.* **54**:29–39.

Zlotnik, I., Grant, D. P., and Carter, G. B., 1976, Experimental infection of monkeys with viruses of the tick-borne encephalitis complex: Degenerative cerebellar lesions following inapparent forms of the disease or recovery from clinical encephalitis, *Br. J. Exp. Pathol.* **57**:200–210.

Index